Organic Nanophotonics

NATO Science Series

A Series presenting the results of scientific meetings supported under the NATO Science Programme.

The Series is published by IOS Press, Amsterdam, and Kluwer Academic Publishers in conjunction with the NATO Scientific Affairs Division

Sub-Series

I. Life and Behavioural Sciences	IOS Press
II. Mathematics, Physics and Chemistry	Kluwer Academic Publishers
III. Computer and Systems Science	IOS Press
IV. Earth and Environmental Sciences	Kluwer Academic Publishers
V. Science and Technology Policy	IOS Press

The NATO Science Series continues the series of books published formerly as the NATO ASI Series.

The NATO Science Programme offers support for collaboration in civil science between scientists of countries of the Euro-Atlantic Partnership Council. The types of scientific meeting generally supported are "Advanced Study Institutes" and "Advanced Research Workshops", although other types of meeting are supported from time to time. The NATO Science Series collects together the results of these meetings. The meetings are co-organized bij scientists from NATO countries and scientists from NATO's Partner countries – countries of the CIS and Central and Eastern Europe.

Advanced Study Institutes are high-level tutorial courses offering in-depth study of latest advances in a field.
Advanced Research Workshops are expert meetings aimed at critical assessment of a field, and identification of directions for future action.

As a consequence of the restructuring of the NATO Science Programme in 1999, the NATO Science Series has been re-organised and there are currently Five Sub-series as noted above. Please consult the following web sites for information on previous volumes published in the Series, as well as details of earlier Sub-series.

http://www.nato.int/science
http://www.wkap.nl
http://www.iospress.nl
http://www.wtv-books.de/nato-pco.htm

Series II: Mathematics, Physics and Chemistry – Vol. 100

Organic Nanophotonics

edited by

Fabrice Charra

Direction for Matter Sciences (DSM),
Atomic Energy Commission,
Saclay, France

Vladimir M. Agranovich

Institute of Spectroscopy,
Russian Academy of Sciences,
Troisk, Russia

and

François Kajzar

Direction for Technological Research (DRT),
Atomic Energy Commission (CEA),
Saclay, France

Kluwer Academic Publishers

Dordrecht / Boston / London

Published in cooperation with NATO Scientific Affairs Division

Proceedings of the NATO Advanced Research Workshop on
Organic Nanophotonics
Aix-en-Provence, Frande
25–29 August 2002

A C.I.P. Catalogue record for this book is available from the Library of Congress.

ISBN 1-4020-1279-9

Published by Kluwer Academic Publishers,
P.O. Box 17, 3300 AA Dordrecht, The Netherlands.

Sold and distributed in North, Central and South America
by Kluwer Academic Publishers,
101 Philip Drive, Norwell, MA 02061, U.S.A.

In all other countries, sold and distributed
by Kluwer Academic Publishers,
P.O. Box 322, 3300 AH Dordrecht, The Netherlands.

Printed on acid-free paper

Printed in the Netherlands.

TABLE OF CONTENTS

Introduction

Single-molecule devices

Photonics at surfaces

Macromolecules

Subwavelength photonics

Single-molecule photophysics

Photochromic and photoreactive molecules

Photonic crystals

Photonics, the science and technology of generation, transport, processing and detection of light, drives a large amount of industrial activity, mainly devoted to information technology, telecommunications, environmental monitoring and biomedical science and instrumentation.

This field benefits from the nanotechnology revolution and has been strongly renewed recently with the introduction of nanoscale concepts. The new prospects include nanostructured materials with original or improved opto-electronic properties, single quantum devices or light sources, nanolasers, and near-field optical techniques with developments towards applications in material analysis as well as in high-density data storage.

Organic materials recently appeared as key components in photonic devices such as light-emitting diodes, integrated lasers, photovoltaic cells or photodetectors. Organic molecular systems offer unique opportunities in nanophotonics since both top-down and bottom-up strategies can be pursued towards the nanoscale. Physical approaches consist in down-scaling the patterning of materials in order to build either single nanoobjects (e.g. nanolasers) or nanostructured materials (e.g. organic photonic band gap materials, donor-acceptor blends for photovoltaics). Conversely chemical approaches consist in building increasingly large molecules up to the nanoscale domain (e.g. self-assembling molecules, carbon nanotubes, dendrimers, engineered DNA), thus permitting an atomic-scale control of material architecture.

This book gathers the proceedings of the first scientific meeting explicitly and exclusively devoted to "Organic Nanophotonics", *i.e.* nanophotonics based on organic molecules and materials. It was held in Aix-en-Provence, France, August 25th - 29th , 2002. One main goal of this workshop was to define routes towards organic nanophotonics taking advantage of both top-down and bottom-up techniques in order to bridge the gap between these two converging approaches. More generally, this workshop constituted a forum aimed at evaluating the potentials and outlooks of molecular-based nanotechnologies for photonics. This book thus constitutes a snapshot of the state of the art in this novel emerging research area of organic nanophotonics.

The organizers of this workshop are highly indebted to its main financial support by the NATO Scientific Affairs Division. Thanks are also due to the European Office of Aerospace Research and Development of the Air-Force Office of Scientific Research, USA, the Direction for Matter Sciences Research (DSM) and the Direction for Technological Research (DRT) from "Commissariat à l'Énergie Atomique", France, and the "Délégation Générale pour l'Armement" (DGA), France. We would like to thank also the members of the scientific committee, Prof. N. V. Kamanina, Prof. P. N. Prasad and Prof. C. Taliani. We address special thanks to Dr. P. A. Chollet who undertook many aspects of the organization of this meeting as organization coordinator and to K. Perronet for her essential assistance in editing these proceedings. Thanks are also due to the staff of the Aquabella hotel for assistance and a friendly receiving of the workshop participants.

Saclay, December 2002

Fabrice Charra
Vladimir M. Agranovich
François Kajzar

NANOPHOTONICS: AN EXCITING EMERGING FIELD

PARAS N. PRASAD
Institute for Lasers, Photonics and Biophotonics, Departments of Chemistry, Physics, Electrical Engineering and Medicine, University at Buffalo, Buffalo, New York 14260

Abstract

Nanophotonics is an exciting new frontier which is defined by optical science and technology at nanoscale. This field has received considerable attention worldwide from both academic and industrial communities. The Institute for Lasers, Photonics and Biophotonics at Buffalo has a comprehensive multidisciplinary program in Nanophotonics funded by the United State Department of Defense. This program focuses on three major areas of Nanophotonics: (i) interactions involving nanoscale confined radiation, (ii) use of nanoscale photoexcitation for nanofabrication and (iii) design and control of excitation dynamics in nanostructured optical materials.

Selected examples of accomplishments under this program are presented here to illustrate the potential of Nanophotonics.

1. Introduction

Today, major scientific breakthrough frequently occurs at the interface between various disciplines. Nanophotonics defined as nanoscale optical science and technology is the interface between Photonics and NanoScience and Technology.[1] It is a new frontier offering challenging opportunities for studying fundamental processes of interaction between the radiation field and matter on a scale much smaller than the wavelength of radiation as well as for design, synthesis and processing of novel nanostructured optical materials. Furthermore, the use of such confined interactions to spatially localized photochemical processes also offers exciting technological opportunities for nanofabrication. These potentials for Nanophotonics are being recognized worldwide. As a result, there has been a considerable growth of new programs both at academia and industries. A search on the internet lists a number of industries engaged in product development.

Our Institute for Lasers, Photonics and Biophotonics is leading a multiinstitutional, multidisciplinary program in organic and polymeric Nanophotonics, which is very comprehensive in nature. The focus of this program is to provide new capabilities for information technology covering communication, signal processing, storage and display. Each of these aspects involve nanotechnology. The comprehensive approach involved is

F. Charra et al. (eds.), Organic Nanophotonics, 1–9.

summarized in Table I, which integrates a number of technologies that have natural ties by either change of length scale or frequency.

TABLE I.

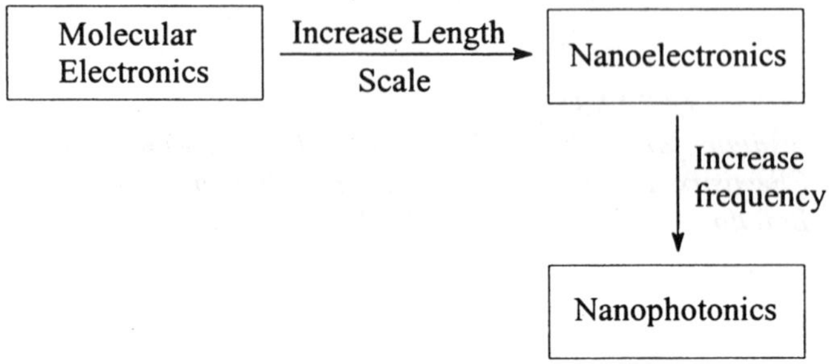

The three major areas of nanophotonics and the features achieved are described in Table II.[1]

TABLE II. The three major areas of Nanophotonics

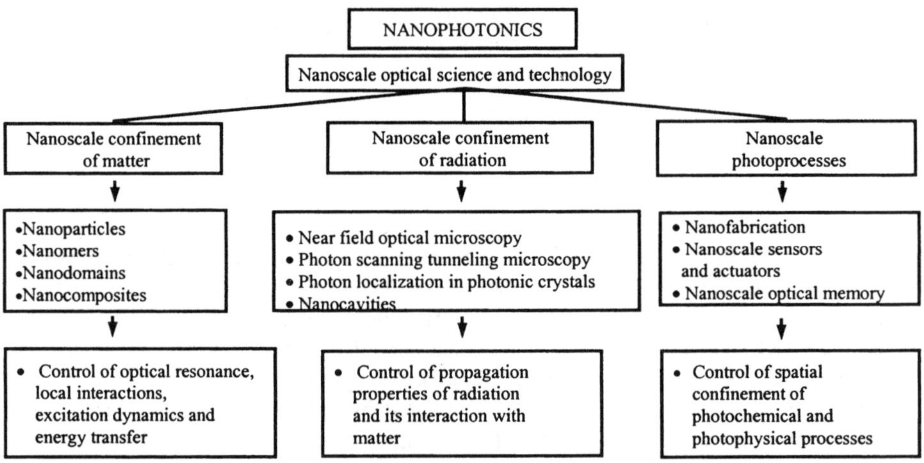

Some selected results from our program in these areas are presented here. Our program in nanoscale confinement of radiation focuses on the use of a near field propagation of short pulses to study nonlinear optical interactions. In the area of nanoscale photoprocesses, we have used near field linear and two-photon induced photopolymerization to produce nanoscale structures for optical memory and for nanolithography.

The work in the nanoscale confinement of matter for photonics has focused on nanoparticles, nanocomposites, nanoscale block co-polymers and self-assemblying or externally directed ordering of nanostructures. In the last case, the emphasis is on producing a periodic or predetermined aperiodic nanoassembly for co-operative amplification and synergism between different functions.

2. Nanoscale Nonlinear Optical Interactions

Theoretical modeling of nonlinear optical processes induced by a near field excitation through a tapered optical tip has been modeled by using multiple-multipole interactions.[2] The light transmitted through the tip has two characteristics. A central portion of the radiation has the wave vector with a real value. This is called the propagating (allowed) light. Light at the wider angle has an imaginary wave vector, and thus is exponentially decaying as an evanescent wave. It is also referred to as the forbidden light. The nonlinear coupling, such as second harmonic generation in a non-centrosymmetric medium can be introduced both by the allowed and the forbidden light. But in the case of the forbidden light, the second-harmonic generation has a much stronger dependence on the separation between the fiber tip and the sample, decaying rapidly as this distance is increased.

For experimental studies, the arrangement used is shown in Figure 1 which shows a combination of near-field scanning optical microscopy (NSOM) and photon scanning tunneling microscopy (PSTM).

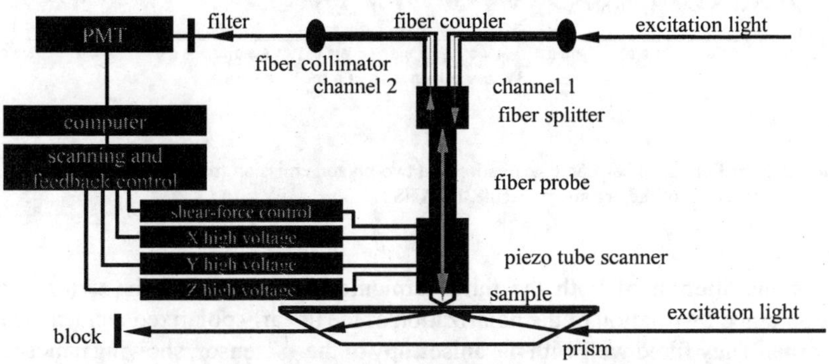

Figure 1. Schematics of experimental arrangement used for combined NSOM and PSTM.

In NSOM, either the sample is excited in the near field by the light emanating from the fiber tip (near field excitation), or the emitted light from the sample, now excited in the far-field, is collected by the fiber tip (near field collection). In the PSTM geometry, the sample is excited by an evanescent wave (tunneling photon) emanating from a prism surface at which light is incident at a total internal reflection angle. A combination of NSOM and PSTM has been utilized in our laboratory in which a sample interacts with the incident light in the PSTM geometry, and the nonlinear optical response (e.g. second harmonic generation) is collected in the near field by the fiber tip (near field collection). The spatial resolution is determined by the fiber tip size. This method provides the advantage that high intensity pulses can be conveniently used for nonlinear optical interaction studies.

An example of the nonlinear optical interaction on nanoscale is provided by the study of third order nonlinear optical interactions in organic nanocrystals of DEANST having dimensions ~200 nm. Excitation in this case was provided at 1064 nm, from a nanosecond Nd:Yag laser.[3]

Figure 2 shows the spectra of the nonlinear optical response from a single nanocrystal. It generates two types of nonlinear optical response: (i) third-harmonic generation, seen as a sharp peak at precisely $\lambda/3 = 355$ nm, derived from the absolute value of $\chi^{(3)}$ of the sample and (ii) two-photon emission, seen as a broad peak at ~585 nm, derived from the imaginary part of $\chi^{(3)}$.

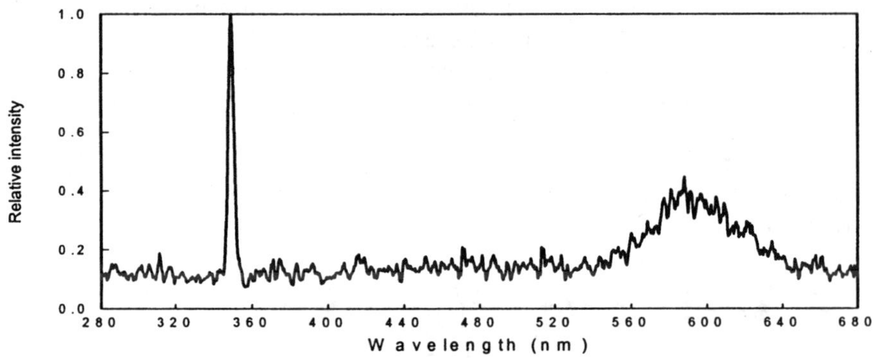

Figure 2. Near field third harmonic generation and two-photon emission from a ~100 nm size nanocrystal of the organic molecule, DEANST.

The angular distribution of both the third-harmonic generation and the two-photon emissions obtained by rotation of the polarization of the linearly polarized incident beam were the same. They fitted well with the anisotropy of the $\chi^{(3)}$ tensor, showing nanoscale ordering in the nanocrystal with the same $\chi^{(3)}$ tensor property as in the bulk sample.

Our group has also observed the effect of surface plasmon enhancement on nanodomain two-photon emission.[4] For this purpose, the PSTM geometry was modified by the deposition of a 500Å silver film on the top of a film substrate which served as a substrate for the organic nanodomains. The sample was excited by the surface plasmon wave produced by the incident light at the surface plasmon coupling angle. Many orders of magnitude enhancement of the two-photon emission were observed when the incident light was introduced in the prism at the surface plasmon coupling angle.

More recently, our study has focused on the study of dynamical processes in nanodomains using time-resolved pump-probe experiments. In a recent study, we used femtosecond pulses at 800 nm to saturate, by two-photon excitation, a transition in a highly efficient two-photon absorbing material.[5] The recovering of saturation was monitored by using linear absorption of a 400 nm probe pulse, time-delayed with respect to the pump pulse. The transmitted probe was collected by the fiber tip (near field collection). The advantage of this approach is that the pump and the probe, derived from the same laser source (~400 nm is obtained by frequency doubling of 800 nm), are widely separated in frequency to be easily discriminated. The recovery time of the saturation provides information on the excited state lifetime. It was found that the excited state lifetime is strongly dependent on the distance between the fiber tip and the sample, the shorter the distance the faster the observed decay. The result was quantitatively fitted into a non-radiative decay model involving interaction between the sample dipole and the metal coating of the outer surface of the fiber tip.

3. Nanoscale-Confinement of Optical Materials

Nanomaterials hold a significant promise for photonics. Some applications of polymeric nanomaterials are listed in Table III.

TABLE III. Applications of polymeric nanomaterials

• Nanoparticle technology for lasing and optically guided drug delivery
• Nanocomposites and nanoscale block co-polymers for optimization of electronic, photonic, mechanical, and thermal properties
• Nanoscale surface modification for multi-level heterogeneous integration
• Self-assembling or externally directed ordering of nanostructures for co-operative amplification and synergism between functions

Our program in the development of nanoscale optical materials have focussed on the following:

- Quantum confined structures
- Nanoparticles
- Nanocomposites
- Photonics crystals

Nanoparticle technology allows one to achieve the following: (i) produce quantum confined structures with the prospect to judiciously control the band gap, (ii) control the excitation dynamics and photo-induced carrier dynamics, (iii) disperse in a polymer matrix or a sol-gel glass to form nanocomposites and (iv) control the dynamics at the hybrid interface between an inorganic and an organic components.

In our earlier work, we utilized a fluorescent dye in the form of a ZnS-dye core-shell structure encapsulated in a silica bubble.[6] When dispersed in water, the emission lifetime of the dye was 400 ps compared to 84 ps found for the dye itself in water, indicating a significant reduction of the non-radiative effect by encapsulation. Consequently, the emission efficiency was also enhanced by more than an order of magnitude for the encapsulated dye. This result is shown in Figure 3.

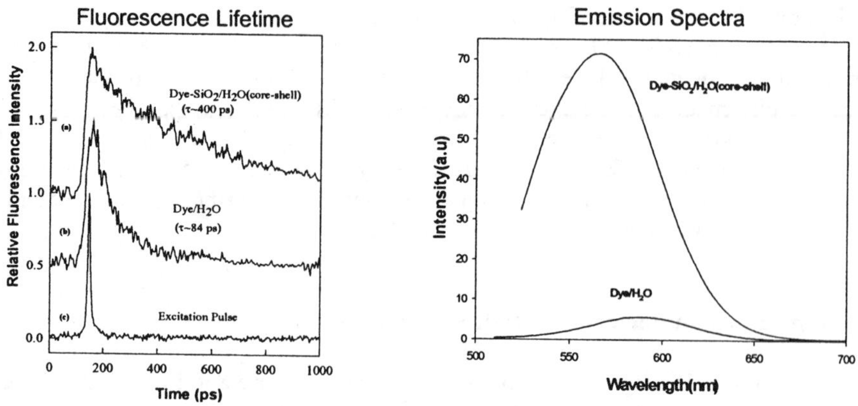

Figure 3. Fluorescence lifetime and emission efficiency of dye encapsulated in a silica bubble and those of the dye itself.

Another major thrust of our program in development of nanomaterials is on inorganic:organic nanocomposites. We are focusing on nanocomposites containing inorganic quantum dots dispersed in a polymer matrix. These are novel systems providing opportunities for fundamental studies of photogeneration of charge-carriers, charge-transfer across the inorganic:organic interface and the carrier mobilities.[7,8] We are also exploring the applications of these hybrid nanocomposites for photorefractivity and for electroluminescence. From the point view of charge carrier mobility, a quantum dot:polymer nanocomposite is a novel system with the inorganic nanoparticles providing photosensitization. In these systems, we are investigating if the nanoparticles enhance charge carrier transport or they contribute to trapping. Another aspect being investigated is if one can improve the speed of response of a hybrid nanocomposite photorefractive material by enhancing the mobility of the polymer host in the presence of the quantum dots. The carrier mobility was measured in a CdS quantum dot:polymer nanocomposite using a time-of-flight measurement.[8] Our results of charge-carrier mobility provides the following results:

- A dispersive transport is observed for holes.
- The mobility data, activation energies, field dependences are similar to those of PVK.
- There is a discernible enhancement of carrier mobility (a factor of less than an order of magnitude at the best) due to the presence of nanoparticles.
- The concentration of nanoparticles too low for percolation. The enhancement more likely arises from local field effects.

We have used a photorefractive nanocomposite containing HgS or PbS nanocrystals dispersed in the PVK matrix containing electro-optic chromophore NPP and the TCP platicizer to obtain photorefractivity at the communication wavelength of 1.3 μm.[9] A net two-beam coupling gain was observed.

More recently, we have used a polymer dispersed liquid crystal nanocomposite for photorefractivity, which consists of the following components: (i) a quantum dot such as CdS for photosensitization, (ii) liquid crystal nanodroplets for electro-optic effect and (iii) a hole transporting matrix consisting of hole transporting molecules, ethyl carbazole (ECZ) dispersed in the PMMA polymer.[10] This nanocomposite exhibited a very photorefractive figure of merit with a net diffraction efficiency reaching a very high value of over 70% as the applied bias field was increased. This result is shown in Figure 4.

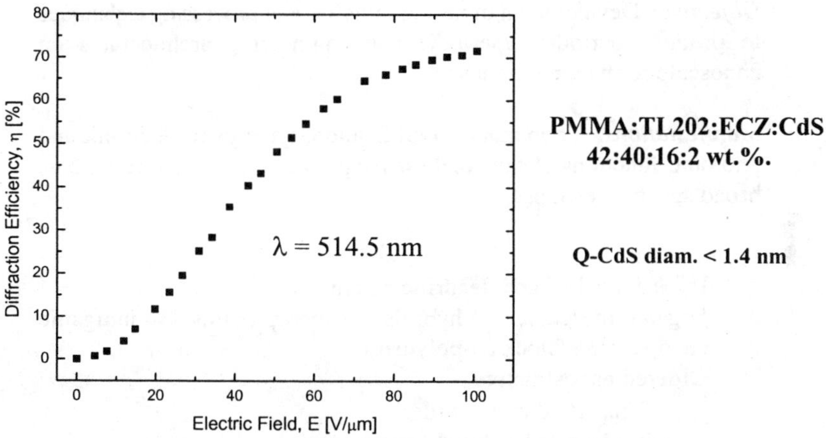

Figure 4. Net diffraction efficiency of the polymer dispersed liquid crystal nanocomposite containing CdS quantum dots for photosensitizations. The wavelength used is 514.5 nm from an argon-ion laser.

Another area of nanoscale material development is ordering of nanoparticles to produce amplification or co-operative interactions leading to new manifestations. An example is provided by our work on investigating nonlinear optical effects in photonics crystals. We have observed enhancement of two-photon emission at the edge of the pseudogap in a photonic crystal.[11] These types of effects utilizing local field enhancement near the edge of the band gap are being actively investigated.

4. Nanomaterials for Biophotonics

A major application of nanoparticles being investigated at our Institute is in the area of biophotonics. Biophotonics is an emerging hot field dealing with interaction of light with biological matters. The applications range from optical bioimaging to light activation of drugs using a nanoparticle platform.

A major development at our Institute is the use of nanochemistry to form Nanoclinics.[12] Nanoclinics are nanoscale silica bubbles containing optical probes, magnetic probes and drug delivery systems. The surface of these nanoparticles are functionalized with appropriate carrier groups to carry them to cancer (or diseased) cells. Another development targeting application to biophotonics is that of up-converting Y_2O_3 nanoparticles containing Er^{3+}. Er^{3+} ions in these nanoparticles can be excited at 970 nm in the IR to generate efficient emission in the visible (green and red).[13] We are using these nanoparticles for bioimaging and for photodynamic therapy.

Some future directions of development of optical nanomaterials are listed in Table IV.

TABLE IV. Future Directions for Optical Nanomaterials

Objective: Development of new assemblies and processing techniques to produce periodic, aperiodic and engineering architectures on nanoscale for nanophotonics

Manifestations: Cooperative amplification, synergism, electronic and photonic functions at the smallest length scale, new manifestations, broad spectral response

Approaches:

- Hyperbranched and dendritic structures
- Organic:Inorganic hybrids incorporating inorganic nanoparticles/block co-polymers
- Ordered nanostructure
    - Template directed order
    - Bio-inspired ordered nanostructures
    - Surface directed assembly
    - Externally directed assembly

Acknowledgements

The work reported in this paper was supported by a DURINT grant from the Directorate of Chemistry and Life Sciences of the Air Force Office of Scientific Research.

5. References

1. Shen, Y., Friend, C.S., Jiang, Y., Jakubczyk, D., Swiatkiewicz, J., and Prasad, P.N. (2000) Nanophotonics: Interactions, Materials, and Applications. *J. Phys. Chem. B* **140**, 7577.
2. Jiang, Y., Jakubczyk, D., Shen, Y., and Prasad, P.N. (2000) Nanoscale Nonlinear Optical Processes: Theoretical Modeling of Second-Harmonic Generation for Both Forbidden and Allowed Light. *Opt. Lett.* **25** 640.
3. Shen, Y., Swiatkiewicz, J., Markowicz, P., and Prasad, P.N. (2001) Near-Field Microscopy and Spectroscopy of Third-Harmonic Generation and Two-Photon Excitation in Nonlinear Organic Crystals. *Appl. Phys. Lett.* **79**, 2681.

4. Shen, Y., Swiatkiewicz, J., Lin, T.-C., Markowicz, P., and Prasad, P.N. (2002) Near-Field Probing Surface Plasmon Enhancement Effect on Two-Photon Emission. *J. Phys. Chem. B* **106**, 4040.
5. Shen, Y., Dai, J., Lin, T.C., Markowicz, P., and Prasad, P.N. Near-field optical imaging and spectroscopy of transient absorption dynamics in organic nanostructures (submitted to *Appl. Phys. Lett.*)
6. Lal, M., Levy, L., Kim, K.S., He, G.S., Wang, X., Min, Y.H., Pakatchi, S., and Prasad, P.N. (2000) Silica Nanobubbles Containing an Organic Dye in a Multilayered Organic/Inorganic Heterostructure with Enhanced Luminescence. *Chem. Mater.* **12**, 2632.
7. Winiarz, J.G., Zhang, L., Lal, M., Friend, C.S., and Prasad, P.N. (1999) *Chem. Phys.* **245**, 417.
8. Roy Choudhary, K., Winiarz, J.G., Samoc, M., and Prasad, P.N. (in press), *Appl. Phys. Lett.*
9. Winiarz, J.G., Zhang, L., Park, J., and Prasad, P.N. (2002) Inorganic: Organic Hybrid Nanocomposites for Photorefractivity at Communication Wavelengths. *J. Phys. Chem. B* **106**, 967.
10. Winiarz, J. and Prasad, P.N. (2002) Photorefractive inorganic-organic polymer-dispersed liquid-crystal nanocomposite photosensitized with cadmium sulfide quantum dots. *Opt. Lett.* **27**, 1330.
11. Markowicz, P., Friend, C., Shen, Y., Swiatkiewicz, J., Prasad, P.N., Toader, O., John, S., and Boyd, R.W. (2002) Enhancement of two-photon emission in photonic crystals. *Opt. Lett.* **27**, 351.
12. Levy, L., Sahoo, Y., Kim, K.S., Bergey, E.J., and Prasad, P.N. (2002) Synthesis and Characterization of Multifunctional Nanoclinics for Biological Applications. *Chem. Mater.* **14**, 3715.
13. Kapoor, R., Friend, C., Biswas, A., and Prasad, P.N. (2000) Highly Efficient Infrared-to-Visible Energy Upconversion in Er^{3+}:Y_2O_3. *Opt. Lett.* **25**, 338.

AFOSR INTERESTS IN ORGANIC NANOPHOTONICS

DR. CHARLES LEE
AFOSR/NL
801 N. Randolph St., Ste 732
Arlington, VA 22203-1977
USA

Acknowledgements

The author wishes to thank Professor Paras N Prasad, Dr. Haridas Pudavar, and Ms. Patricia Randall of the New York State University at Buffalo, Professors David Carroll of Clemson University and Yang Yang of University of California, Los Angeles, for their assistance in the preparation of this manuscript.

1. Introduction

In 2001, the US Government initiated a National Nanotechnology Initiative to encourage research activities related to Nanotechnology within the United States. This initiative involves many government agencies that include National Science Foundation, Department of Energy, Department of Commerce, Department of Defense, National Aeronautics and Space Administration and many others. US government support in this area was at \$422M in Fiscal Year (FY) 2001 and \$604M in FY 2002. The projected level of support in FY03 will be \$710M.

Department of Defense has supported research in the Nanotechnology area for a number of years. In 2001, as part of the National Nanotechnology Initiative, DoD initiated a multidisciplinary research program, The Defense University Research Initiative on Nanotechnology (DURINT), to encourage focused research in topics relating to Nanotechnology. A total of 15 topics were selected. One of them is "Organic Nanophotonics and Nanoelectronics", managed by the Air Force Office of Scientific Research (AFOSR).

Many programs at AFOSR have interests in Nanotechnology including the Polymer Chemistry Program. The Polymer Chemistry Program at AFOSR has supported research relating to Nanotechnology in Organics and Polymers in the last decade. The supported research included block copolymers and polymer blends with nanoscale phase separation, novel dendrimer molecules, and nanocomposites. Properties of interest include mechanical properties, photonics and electronic properties. This area of interest was further augmented since 1966 under a Multidisciplinary University Research Initiative (MURI) on Polymeric Nanostructures. The team members of this MURI include University of Southern California, University of California, Berkeley,

F. Charra et al. (eds.), Organic Nanophotonics, 11–15.

California Institute of Technology, State University of New York at Buffalo, and North Carolina State University. This effort ended in 2001. Under this program, exciting advances in photonic nanocomposites (ref 1), novel dendrimers (ref 2) and electro-optical polymers (ref 3) have been reported.

2. Multidisciplinary Efforts

The new effort on Organic Nanophotonics and Nanoelectronics was started in 2001 under the Defense University Research Initiative on NanoTechnology (DURINT). This effort is headed by University of Buffalo, with team members of University of Washington, University of California, Berkeley, Yale University and Massachusetts Institute of Technology. The objective of this effort is to develop device-design methodologies at the molecular and nanometer scale for producing new materials/structures to yield future generation technology for applications such as display, communication, power generation and data processing and storage.

A related MURI project at University of Washington was also initiated in 2001 in the area of Smart Skin. Participants in this project include University of California, Berkeley, University of Southern California, University of California, Santa Barbara, and Rice University. While this program is not specifically targeting nanotechnology, several projects under this program are nanotechnology related, including dendrimers, single wall carbon nanotubes and photonic bandgap structures.

3. Organic Nanophotonics and Nanoelectronics DURINT

There are many projects under the Organic Nanophotonics and Nanoelectronics DURINT, many of them are collaborations between different groups to foster interdisciplinary interaction. Examples of projects include collaboration between Professor Paras Prasad of University of Buffalo and Professor Jean Frechet of University of California, Berkeley to use novel dendrimer chemistry to enhance two-photon absorption induced fluorescence by coupling chromophores with large two photon absorption cross-section as the shell of a dendrimer molecule. The chromophores on the shell will absorb longer wavelength light through a two-photon process and re-emit through another chromophore at the core of the dendrimer. This can enhance the upconversion fluorescence efficiency in a material (ref 4).

Through the interaction with Professor Yuonon Xia of University of Washington, Professor Prasad was able to fabricate a photonic crystal structure immersed in a solution of two-photon dye. The resulting two-photon fluorescence spectrum of the structure shows a shift in gain from the photonic bandgap wavelength to a wavelength near the band edge. This shift and enhancement in gain can be used to form a high Q cavity for low threshold two-photon lasing (ref 5).

Inorganic and organic hybrid nanocomposites are also part of this project. A glass-polymer nanocomposites have been used to demonstrate high density data storage. Glass structures with nanopores was fabricated and then back-filled with monomers doped with two-photon dyes. After polymerization, the glass-polymer nanocomposites can be used to store multiple layers (20□m apart) of digital and analog information using a two-photon recording process and a one-photon read out (ref 6).

Similarly, a Polymer Dispersed Liquid Crystal (PDLC) material was modified with semiconductor nanoparticles to form a nanocomposite. Liquid Crystal droplets of diameter of 200 nm were dispersed in a polymer (PMMA) matrix. The polymer matrix was modified to include N-ethylcarbazole (ECZ) for charge transport and 10nm CdS nanoparticles for optical sensitivity. The resulting PDLC composites showed photorefractive behavior with very high diffraction efficiency (ref 7).

4. Carbon Nanotubes

There are other projects within the Polymer Chemistry Program that are nanophotonics related. Two of them will be highlighted here. At Clemson University, Professor David Carroll studied the interaction of carbon nanotubes with polymers relevant to photonic applications. Many nanotube polymer blends (ref 8) and their properties in various applications have been investigated, including photovoltaic and light emitting devices. In the case of incorporating carbon nanotubes in polymer LED structures (ref 9-10), it was found that the carbon nanotubes enhance the charge mobility in the device, resulting in a more efficient structure with lower turn-on voltage. The carbon nanotubes can also modulate the charge mobility in the nanocomposites, making it feasible to tune the color of a single pixel that has multiple emissive layers by adjusting the field. Another discovery is that nanotubes can prolong the operational lifetime of the device by countering the effects of oxidation through the establishment of charge balance conditions beneficial to the OLED and through quenching of the triplet excitations associated with photo-induced oxidation processes.

His research also showed carbon nanotubes to exhibit optical limiting behavior in suspension (ref 11). Carbon nanotubes were also found to readily form blends with fluoropolymers. Dispersing carbon nanotubes in a fluoropolymer matrix will yield an optically clear solid that shows optical limiting behavior as in a suspension. These studies are the first to demonstrate feasibility of solid state limiters based on nanotubes. It was also found that N- and B-doped carbon nanotubes show optical limiting behavior at an even lower laser power level (ref 12).

5. Organic Bistable Device

Another interesting project in nanophotonics is at University of California, Los Angeles. Professor Yang Yang was able to control the layer-thicknesses in a multilayer structure in the nanometer range to show bistable conductivity states. The three-layer structure have two layers of organic molecules of 10's nm thickness sandwiching a metal layer that is also in the same range. The three-layer structure is then sandwiched between two electrodes. With two different biased voltages, the structure shows two distinct conductive states with a difference in conductivity in 6-7 orders of magnitude (ref 13). Both states persist even when the biased voltage is turned off. The response time of the switching was determined to be in the nanosecond regime. This makes the device ideal as a nonvolatile memory. This device can also be sandwiched between a LED structure and one of its electrodes, making optical readout of this nonvolatile memory possible (ref 14). These features make the organic bistable light-emitting device a promising candidate for several applications, such as digital memories, opto-electronic books, and recordable papers.

6. Summary

In summary, AFOSR is supporting a wide range of research activities related to organic nanophotonics. The areas of research include nanocomposites, block copolymers, organic-inorganic hybrids, carbon nanotubes and fullerenes and photonic bandgap structures. Some of the research activities are highlighted here.

7. References

1. Prasad P. N., He, G. S., Joshi, M. P., Swiatkiewicz, J., Manchala, G., Lal, M., Biswas, A., and Kim K-S. (1999), Nanostructured Materials and Composites for Optical Power Limiting, *Nonlinear Opt.* **21**, 39-47.

2. Adronov, A., Malenfant, P.R.L., Robello, D. R., Fréchet, J.M.J. (2001) Light-Harvesting and Energy-Transfer within Coumarin-Labeled Polymers. *J. Polym. Sci. A* **39**, 1366-1373.

3. B. H. Robinson, L. R. Dalton, A. W. Harper, A. Ren, F. Wang, C. Zhang, G. Todorova, M. Lee, R. Aniszfeld, S. M. Garner, A. Chen, W. H. Steier, S, Houbrecht, A. Persoons, I. Ledoux, J Zyss, and A. K. Y. Jen (1999) The Molecular and Supramolecular Engineering of Polymeric Electro-Optic Materials, *Chem. Phys.* **245**, 35-50.

4. G.S. He, S-J. Chung, T-C Lin, D. W. Brusmiche, J. M. Serin, J.M.J. Frechet and P.N. Prasad, unpublished work.

5. P. Markowicz, C. Friend, Y. Shen, J. Swiatkiewicz, P. N. Prasad, O. Toader, S. John, and R. W. Boyd (2002) Enhancement of two-photon emission in photonic crystals, *Opt. Lett.* **27**, 351-353

6. H. Pudavar and P.N. Prasad, unpublished work

7. J. Winiarz and P. N. Prasad (2002) Photorefractive inorganic-organic polymer-dispersed liquid-crystal nanocomposite photosensitized with cadmium sulfide quantum dots, *Opt.Lett,* **27** 1330-1332

8. B. McCarthy, J. Coleman, R. Czerw, A Dalton, D.L. Carroll, and W. Blau (2001) Microscopy studies of nanotube – conjugated polymer interactions, *Synthetic Metals* **121**, 1225.

9. H. S. Woo, R. Czerw, S. Webster, D. L. Carroll, J. Ballato, A. E. Strevens, and W. J. Blau (2000) Hole blocking in carbon nanotube-polymer composite organic light

emitting diodes based on poly (m-phenylene vinylene-co-2, 5-dioctoxy-p-phenylene vinylene), *Applied Physics Letters* 77 **9**, 1393,

10. H.S. Woo, R. Czerw, S. Webster, D.L. Carroll, J.W. Park, J.H. Lee (2001) Organic Light Emitting Diodes Fabricated with Single Walled Carbon Nanaotubes Dispersed in a Hole Conducting Buffer: The Role of Carbon Nanotubes in a Hole Conducting Polymer, *Synthetic Metals* **116** , 369–372.

11. Riggs, J.E., Walker, D.B., Carroll, D.L., Sun, Y.P. (2000) Optical limiting properties of suspended and solubilized carbon nanotubes, *Journal of Physical Chemistry B* **104 (30)** 7071-7076.

12. Blase, X., Charlier, J.-C., De Vita, A., Car, R., Redlich, Ph., Terrones, M., Hsu, W. K., Terrones, H., Carroll, D. L., and Ajayan, P. M. (1999) Boron-mediated growth of long helicity-selected carbon nanotubes, *Physical Review Letters* **83, 24,** 2348–2352.

13. Ma, L., Liu, J., and Yang, Y. (2002) Organic electrical bistable devices and rewritable memory cells, *Appl. Phys. Lett.* **80**, 2997.

14. Ma, L., Liu, J., Pyo, S., and Yang, Y. (2002) Organic bistable light-emitting devices, *Appl. Phys. Lett.* **80**, 362.

OPTICAL PROPERTIES AND ENERGY TRANSPORT IN DENDRIMERS

P. REINEKER, A. ENGELMANN
Abteilung für Theoretische Physik, Universität Ulm
Albert-Einstein-Allee 11, 89069 Ulm, Germany

AND

V. I. YUDSON
Center for Frontier Science, Chiba University
1-33 Yayoi-cho, Inage-ku, Chiba 263-8522, Japan
Institute of Spectroscopy, Russian Academy of Sciences
Troitzk/Moscow 142090, Russia

1. Introduction

In recent years, dendrimers – a new class of engineered macromolecules – have attracted the interest of many research groups in chemistry and physics. The molecules are imitating on a nanometer scale a branching pattern present in nature on a macroscopic scale. They consist of repeat units, e.g. phenylacetylene molecules, arranged in a self-similar manner around a core. Two approaches for synthesizing dendrimers have been reported. One is the so called divergent approach, developed independently by Vögtle and Tomalia [1,2], where the dendrimer is built up divergently – generation by generation – from the core to the periphery. The other route due to Hawker and Fréchet is called the convergent approach [3]. In the first step of this process the different branches are synthesized and in the last step they are connected with the core molecule.

The growing interest in these structures is due to the progress in chemical engineering; influences on shape, size and solubility can be controlled and suitably modified [4–9]. By providing this controllable structure and the possibility of functionalization of core and periphery, dendrimers are considered to allow for a broad range of applications such as molecular recognition, catalysis, electron transfer, organic light emitting diodes or photochemical reactions (see ref. [10–13]). Applications of dendrimers in bio-organic chemistry are dicussed in [14].

For a first classification, dendrimers are divided into two families: *compact* and *extended* dendrimers. The former family has the same bond length between successive generations whereas for the extended dendrimers, the bond length increases from the periphery to the core [4,15].

F. Charra et al. (eds.), Organic Nanophotonics, 17–30.

In the focus of the present paper we will have optical absorption and energy transfer in compact dendrimers. We take into account the topological structure, the local electronic excitation energies, the electronic interaction between the local subunits and the influence of vibrational degrees of freedom. We restrict ourselves to the three lowest dendrimers D4, D10 and D22. The linear and nonlinear optical susceptibilities of compact dendrimers were dicussed in [16, 17], however, only for coherent excitons, i.e. the influence of vibrational degrees of freedom was neglected. On the other hand, energy transport in compact dendrimers for the completely incoherent case was examined in [18, 19] on the basis of a set of master equations with phenomenological transition rates between the local subunits. In this approach the dependence of the mean-first-passage time on the generation number was calculated.

In our approach we take the influence of vibrational degrees of freedom explicitly into account. This allows to describe the coupled coherent and incoherent motion of excitons in the dendrimer. As a first approach to real dendrimers not only electronic interactions within their branches are discussed, but for D22 also peripheric interaction are considered. The paper is organized as follows: in section 2 we present our model, based on the Frenkel exciton picture. Section 3 discusses optical properties. A short overview of our method is given in section 3.1 and the results are discussed in section 3.2. In section 4 some results of our energy transfer investigations are reported and the paper ends in section 5 with concluding remarks.

2. Model

We describe the electronic excitations by Frenkel exciton [20] and model the dephasing according to Haken, Strobl and Reineker in a stochastic manner [21–24]. The Hamiltonian for an arbitrary dendrimer with 3 branches in the core, branching ratio 2 in higher generations and $n_{\max}$ generations reads

$$\begin{aligned} H(t) \;=\; & [\varepsilon_0 + \varepsilon_0(t)]\, b_0^\dagger b_0 + \sum_{n=1}^{n_{\max}} \sum_{\alpha=1}^{3\cdot 2^{n-1}} [\varepsilon_{n_\alpha} + \varepsilon_{n_\alpha}(t)]\, b_{n_\alpha}^\dagger b_{n_\alpha} \\ & + \sum_{\alpha=1}^{3} [J_1 + J_{1_\alpha}(t)] \left(b_{1_\alpha}^\dagger b_0 + b_0^\dagger b_{1_\alpha} \right) \\ & + \sum_{n=2}^{n_{\max}} \sum_{\alpha=1}^{3\cdot 2^{n-2}} \Big\{ \Big[\left(J_n + J_{n_{2\alpha-1}}(t) \right) b_{n_{2\alpha-1}}^\dagger \\ & \quad + \left(J_n + J_{n_{2\alpha}}(t) \right) b_{n_{2\alpha}}^\dagger \Big] b_{n-1_\alpha} + \text{h.c.} \Big\} \end{aligned} \qquad (1)$$

The molecules are labelled according to their generation and the number in that generation, i.e. n_α corresponds to the molecule α in the generation n. The core molecule 0 has no subscript α. The numbering is shown explicitly in figure 1 for the compact dendrimer D22 consisting of phenylacetylene units. $b_{n_\alpha}^\dagger$ and b_{n_α} are creation and annihilation operators, respectively, for an exciton at molecule n_α.

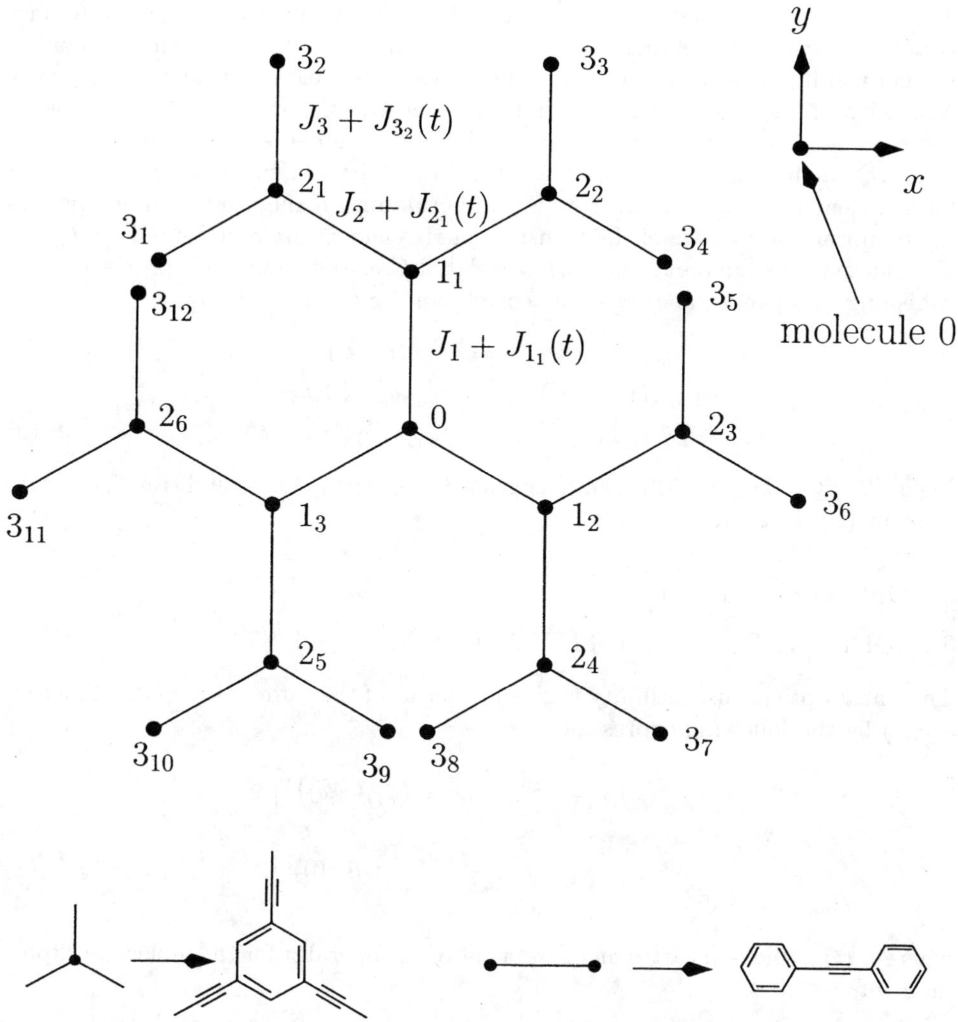

Figure 1. Schematic picture of the compact dendrimer D22 made of phenylacetylene units together with the real chemical compounds. Transfer matrix elements are given on one branch only for brevity. Molecules are labelled according to the generation number n and the number α in that generation, i.e. n_α stands for molecule α in generation n.

These operators are used as Bose-type operators for convenience, because the exact (Pauli) commutation relations play no role as long as only single excitations of the system and its linear response to the electromagnetic field is considered. ε_{n_α} is the excitation energy of molecule n_α and J_n is the transfer matrix element between generations $n-1$ and n. The influence of vibrational degrees of freedom is taken into account by allowing for fluctuations of the exitation energy and the transfer matrix elements. $\varepsilon_{n_\alpha}(t)$ represents local fluctuations of the excitation energy and the nonlocal fluctuations of the transfer matrix element are represented by $J_{n_\alpha}(t)$. The fluctuations are described by δ-correlated Gaussian stochastic processes with vanishing meanvalues and two-time correlation functions given by

$$\begin{aligned} \langle \varepsilon_0(t)\, \varepsilon_0(t') \rangle &= \gamma_0^0\, \delta(t-t') \\ \left\langle \varepsilon_{n_\alpha}(t)\, \varepsilon_{m_\beta}(t') \right\rangle &= \delta_{n_\alpha, m_\beta}\, \gamma_n^0\, \delta(t-t') \\ \left\langle J_{n_\alpha}(t)\, J_{m_\beta}(t') \right\rangle &= \delta_{n_\alpha, m_\beta}\, \gamma_n^1\, \delta(t-t') \end{aligned} \tag{2}$$

In eq. (2) γ_n^0 and γ_n^1 describe the strengths of the energy (local) and transfer matrix element (nonlocal) fluctuations.

3. Optical absorption

3.1. OPTICAL SUSCEPTIBILITY AND DENSITY OF STATES

The linear optical susceptibility is given in terms of the retarded response function $\phi_{ij}^{>}(\omega)$ by the following expression

$$\begin{aligned} \chi_{ij}(\omega) &= i\left[\phi_{ij}^{>}(\omega) - \left(\phi_{ij}^{>}(-\omega)\right)^*\right] \\ \phi_{ij}^{>}(\omega) &= \int_0^\infty dt\, e^{i\omega t}\, \langle \mu_i(t)\mu_j(0) \rangle \end{aligned} \tag{3}$$

where $\mu_i(t)$ is the i-th cartesian component of the operator for the molecular dipole moment

$$\mu_i(t) = \sum_{n_\alpha} \mu_i^{n_\alpha} \left(b_{n_\alpha}^\dagger(t) + b_{n_\alpha}(t)\right) \tag{4}$$

We derive a system of equations for the Green's functions

$$\begin{aligned} \widetilde{G}^{>}_{n_\alpha\, m_\beta}(t) &= \left\langle b_{n_\alpha}(t)\, b_{m_\beta}^\dagger(0) \right\rangle \\ \widetilde{G}^{>}_{n_\alpha\, m_\beta}(0) &= \delta_{n_\alpha, m_\beta} \end{aligned} \tag{5}$$

and, after averaging this system over the stochastic process, solve this system analytically or numerically. Details can be found in [25]. The real space interpretation of the absorption spectra is performed by defining three different densities of states (DOS) according to

$$\mathcal{N}_{n_\alpha}(\omega) = -\frac{1}{\pi}\, \mathrm{Im}\, G^R_{n_\alpha n_\alpha}(\omega) \tag{6}$$

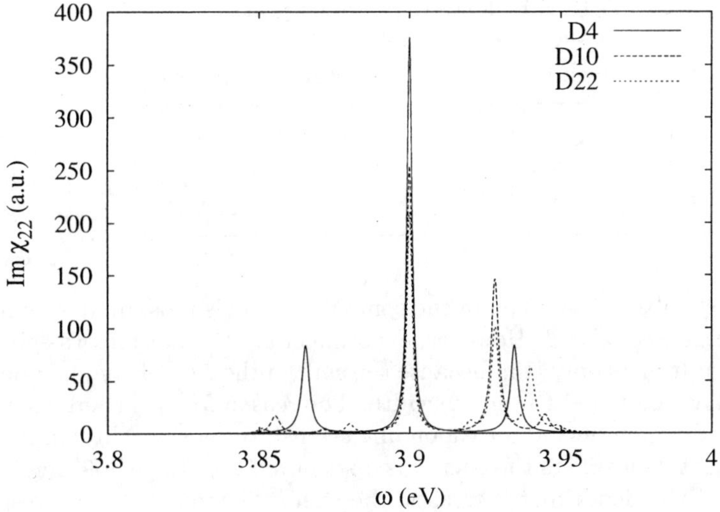

Figure 2. $\text{Im}\chi_{22}(\omega)$ for compact dendrimers scaled to the total number of molecules. $\varepsilon_{0,1,2,3} = 3.9, J_{1,2,3} = 0.02, \gamma^0_{0,1,2,3} = 0.01, \gamma^1_{1,2,3} = 0.001$ in eV. Dipole moments in units of μ^0: $\mu^{n_\alpha} = 1$.

$$\mathcal{N}_n(\omega) = -\frac{1}{\pi}\sum_\alpha \text{Im}\, G^R_{n_\alpha n_\alpha}(\omega) \tag{7}$$

$$\mathcal{N}(\omega) = \frac{1}{N}\sum_n \mathcal{N}_n(\omega) \tag{8}$$

with the Fourier transformed retarded Green's function

$$G^R_{n_\alpha m_\beta}(\omega) = \int_{-\infty}^{\infty} dt\, e^{i\omega t}\, G^R_{n_\alpha m_\beta}(t) = -i\, G^>_{n_\alpha m_\beta}(\omega) \tag{9}$$

$\mathcal{N}_{n_\alpha}(\omega)$ is the contribution to the DOS at the molecule n_α, $\mathcal{N}_n(\omega)$ the contribution to the DOS in the n-th generation and $\mathcal{N}(\omega)$ is the total DOS, normalized to the total number of molecules N. The key point for a real space analysis is the DOS in the n-th generation, eq. (7). $\mathcal{N}_n(\omega)$ enables us to infer the distribution of the excitation energy over the various generations of the dendrimer.

3.2. DISCUSSION OF THE LINE SHAPE

The results for the linear optical susceptibility are shown in figure 2 for compact dendrimers. For the parameters we used numerical values as estimated from optical absorption spectra [15] and assumed that the external electric field is directed along the y-axis, compare figure 1. The induced dipole moments $\vec{\mu}^{n_\alpha}$ are orientated along the direction of the chemical bonds.

TABLE 1. Lowest absorption peaks for compact dendrimers.

Dendrimer	Energy (compact)
D4	3.865 eV
D10	3.855 eV
D22	3.851 eV

The lowest absorption peak in the compact family is redshifted with increasing dendrimer size, see table 1. However, experimentally the redshift was observed for extended dendrimers only [15], because the shift in the case of compact dendrimers is too small as compared to the linewidth. The reason for this energy shift is excitation delocalization: in the corresponding eigenstates of the Schrödinger equation of the compact dendrimers the energy is distributed over the whole dendrimer and as the size of the dendrimer increases, the energy is redshifted as a result of the delocalization. The real space interpretation of the absorption spectra is given in figures 3(a) to 3(d) for compact dendrimers. The DOS is normalized according to

$$\int d\omega\, \mathcal{N}(\omega) \quad = \quad \sum_{n=0}^{n=n_{\max}} \int d\omega\, \mathcal{N}_n(\omega) = 1$$

with $n_{\max} = 1, 2, 3$ respectively for D4, D10 and D22. The most intensive peak in the absorption spectrum figure 3(a) shows up at 3.9 eV for D4, D10 and D22. Figure 3(b) shows that for D4 the density of states in the 1st generation $\mathcal{N}_1(\omega)$ has a maximum at the same energy. For D10 the highest value of the density of states shows up in the 2nd generation $\mathcal{N}_2(\omega)$ represented in figure 3(c). Figure 3(d) finally shows that the density of states in the 3rd generation, $\mathcal{N}_3(\omega)$, again peaks up at an energy of 3.9 eV. Obviously, the most intensive absorption peak stems from the molecules at the surface of the dendrimer. Figures 3(a) and 3(b) demonstrate that for D4 all molecules contribute equally to the lowest and the highest energy peaks. Similarly, all molecules contribute to the lowest energy absorption peak for D10 and D22, demonstrating once more the effect of delocalization.

Finally, we consider the influence of the interaction between closely neighbored peripheric molecules on the absorption spectra. With increasing size of the dendrimer, peripheric molecules approach each other and – due to the resonant dipole-dipole interaction between these molecules – an energy exchange becomes possible. This effect should be present for large dendrimers and therefore we included such an interaction for D22 only. Quantitatively, we added the following term to the Hamiltonian for D22 in eq. (1) (compare figure 1)

$$J_R \left(b^{\dagger}_{3_1} b_{3_{12}} + b^{\dagger}_{3_5} b_{3_4} + b^{\dagger}_{3_9} b_{3_8} + \text{h.c.} \right) \tag{10}$$

This term preserves the symmetry of the dendrimer. For dipole-dipole interaction $J_R \sim 1/R^3$. The imaginary part of $\chi_{22}(\omega)$ for three different magnitudes of J_R

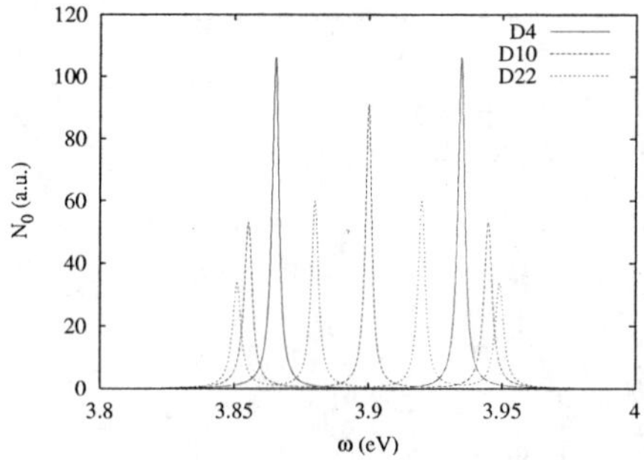

(a) $\mathcal{N}_0(\omega)$ for compact dendrimers.

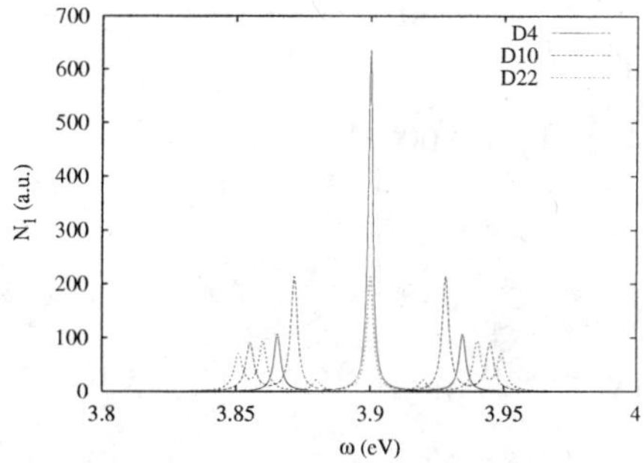

(b) $\mathcal{N}_1(\omega)$ for compact dendrimers.

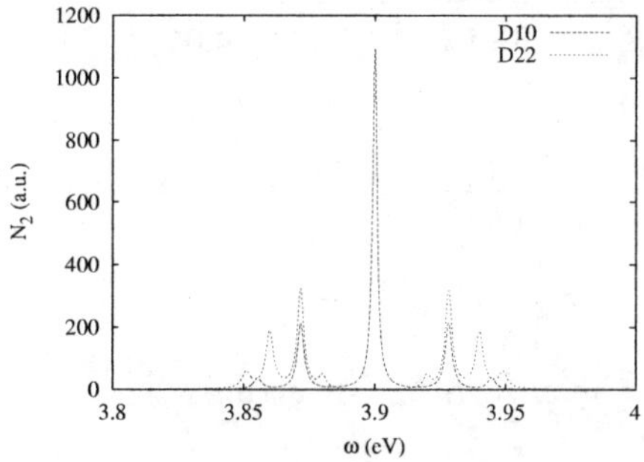

(c) $\mathcal{N}_2(\omega)$ for compact dendrimers.

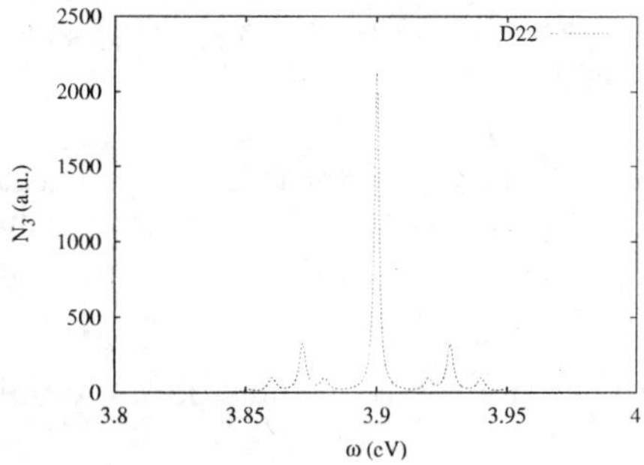

(d) $\mathcal{N}_3(\omega)$ for compact dendrimers.

Figure 3. $\mathcal{N}_0(\omega) \ldots \mathcal{N}_3(\omega)$ for compact dendrimers. Numerical values as in figure 2.

and for D22 compact is shown in figure 4, the following figures 5(a) to 5(d) display $\mathcal{N}_{0,1,2,3}(\omega)$. The absorption spectrum for D22 compact does not change significantly for small J_R, only a very slight shift of ≈ 0.002 eV can be seen between the spectra for $J_R = 0$ eV and $J_R = 0.002$ eV. The DOS $\mathcal{N}_{0,1,2,3}(\omega)$ shows no change at all for $J_R = 0.002$ eV as compared to $J_R = 0$ eV. The absorption spectrum changes if J_R becomes comparable to the other transfer matrix elements $J_{1,2,3}$. The two absorption peaks just below 3.94 eV merge into a single peak at 3.94 eV originating from all molecules except the core molecule. These molecules and the core molecule contribute to another peak at 3.88 eV. Qualitatively, the peripheric interaction for compact D22 results in a slight broadening of the DOS in higher generations, especially in the 2nd and 3rd generation, and in a weak asymmetry around the central peak at 3.9eV which is shown by a detailed inspection of $\mathcal{N}_n(\omega)$. Due to J_R there are more states below than above 3.9 eV and the DOS in the 3rd generation widens. It is only in the case of J_R being comparable to $J_{1,2,3}$ that the energy is *not* mainly concentrated at the peripheric molecules, compare figure 5(d),

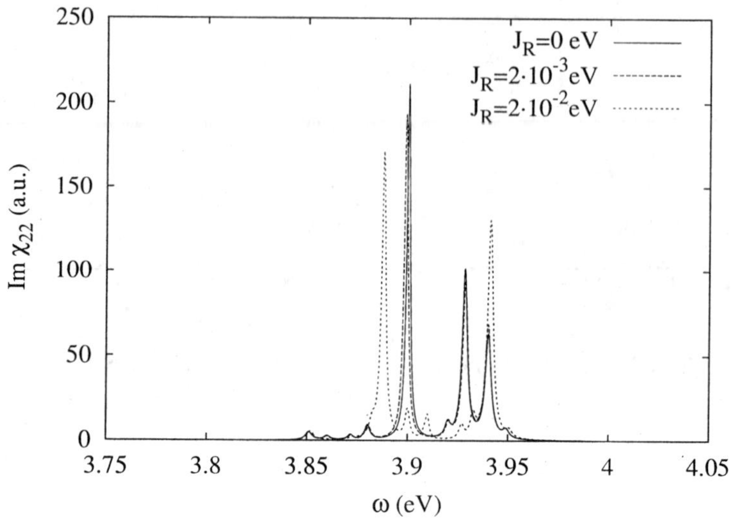

Figure 4. $\mathrm{Im}\chi_{22}(\omega)$ for D22 compact scaled to the total number of molecules for three different magnitudes of J_R. Numerical values as in figure 2.

and that the rim molecules do *not* dominate the absorption spectrum.

4. Energy transfer

The transport dynamics of the electronic excitation in the dendrimer is described by its density operator. Within the stochastic model of the Haken, Strobl and Reineker the equation of motion reads

$$\partial_t|\rho(t)) = L|\rho(t)) \tag{11}$$

The Liouville operator is non-hermitean. Its (in general complex) eigenvalues λ_i and eigenvectors $|\sigma_i)$ satisfy

$$L|\sigma_i) = \lambda_i|\sigma_i) \tag{12}$$

They are obtained numerically using program packages from Ref. [26]. The general solution for the density operator in terms of its eigensolutions is given by

$$|\rho(t)) = \sum_i c_i e^{\lambda_i t}|\sigma_i) \tag{13}$$

The coefficients c_i are determined by the initial conditions. As an extension of the previous model, we have allowed the exciton to decay (e.g. via flourescence) at the center of the dendrimer with a decay constant γ_F. We have calculated the occupation probabilites for excitons, when the exciton is initially created at the

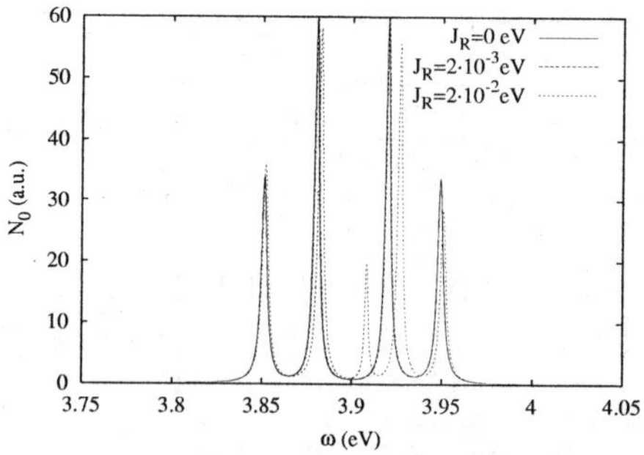

(a) $\mathcal{N}_0(\omega)$ for D22 compact.

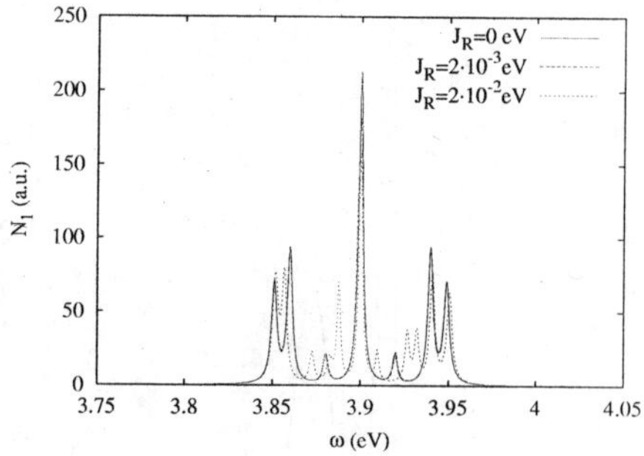

(b) $\mathcal{N}_1(\omega)$ for D22 compact.

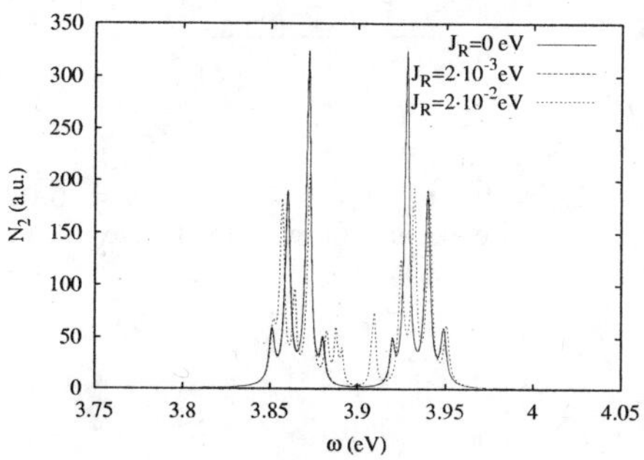

(c) $\mathcal{N}_2(\omega)$ for D22 compact.

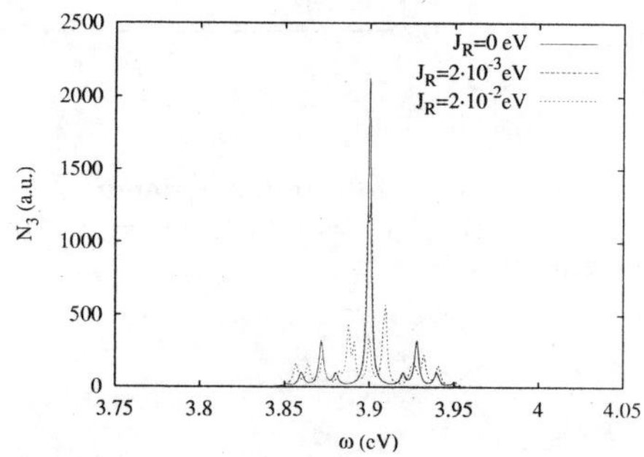

(d) $\mathcal{N}_3(\omega)$ for D22 compact.

Figure 5. $\mathcal{N}_0(\omega)$ ... $\mathcal{N}_3(\omega)$ for D22 compact. Numerical values as in figure 4.

center of the dendrimer and when it resides initially at the periphery. In this contribution we present only the first situation. Figure 6 shows the time dependence of the occupation probability at the center for D4, D10, and D22. For the parameter values chosen and given in the caption the occupation probability shows a an oscillatory behavior in all cases expressing the coherence of the exciton motion. The oscillations level off and result in the occupation probabilities of the stationary states of the various dendrimers. Figure 7 and 8 show the time dependence of the occupation probabilities of the various generations of D10 and D22. The exchange of the excitation energy between the various generations is obvious. After several periods again the stationary state is obtained with the occupation probabilities summing up to 1.

Finally in figure 9 we see the occupation probabilities at the spatially neighbored but in the bond topology distant sites 3_1 and 3_{12} (see figure 1) for various values of the dipole-dipole interaction between peripheric molecules. The figure shows that for small values of the interaction the occupation probabilities are

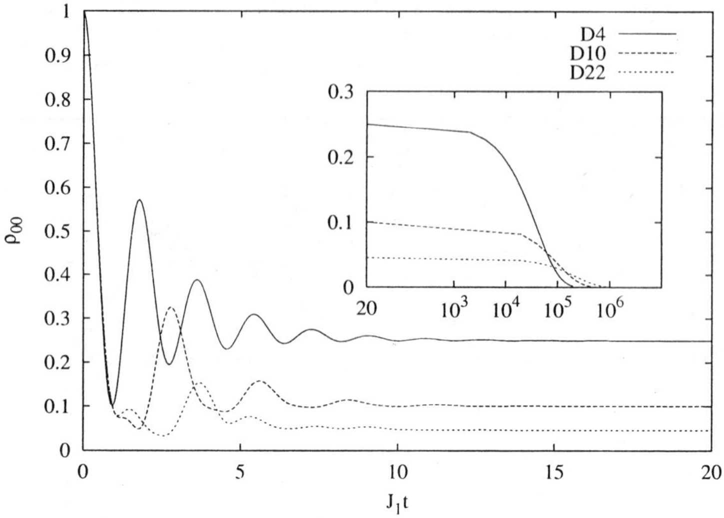

Figure 6. $\varrho_{00}(t)$ for compact dendrimers if $|\psi(0)\rangle = |0\rangle$. $\varepsilon_{0,1,2,3} = 3.9$, $J_{1,2,3} = 0.02$, $\gamma^0_{0,1,2,3} = 0.01, \gamma^1_{1,2,3} = 0.001$, $\gamma_F = 2 \cdot 10^{-6}$, in eV. The insert shows the behavior for large times.

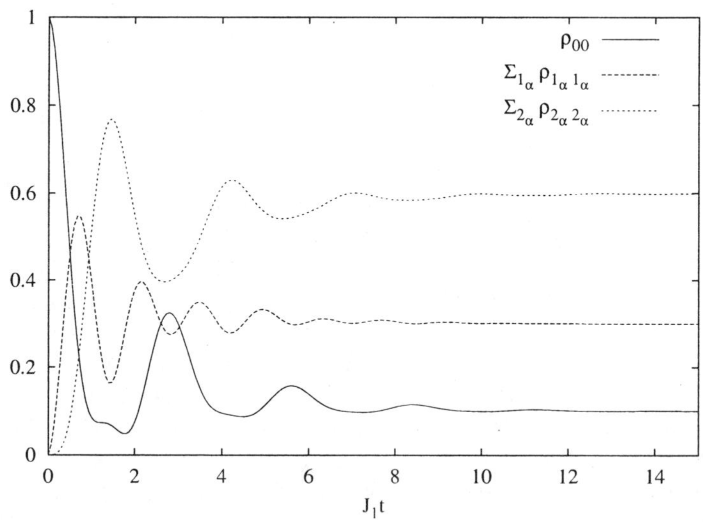

Figure 7. Occupation probabilities for different generations of D10 compact and small times. Parameters as in figure 6.

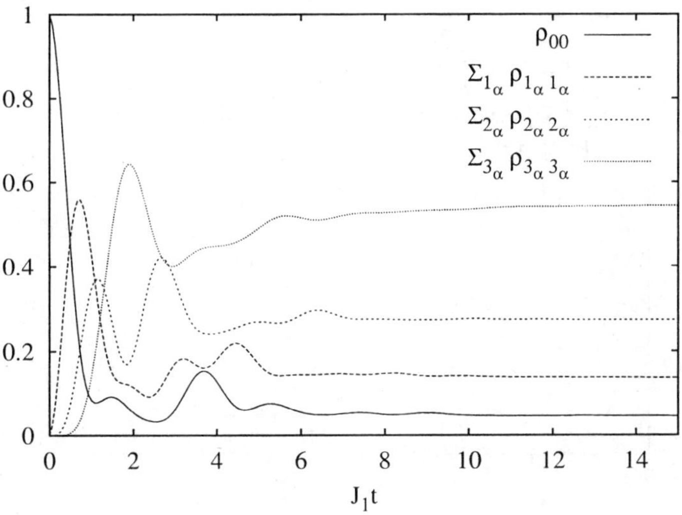

Figure 8. Occupation probabilities for different generations of D22 compact and small times. Parameters as in figure 6.

almost unchanged as compared to the case with no interaction at all. For values comparable to those of the other bonds, we have a direct energy exchange between the two sites.

5. Concluding remarks

In this paper we present an investigation of the optical absorption spectra and of the energy transport of dendrimers. Our aim was to enlighten the role of the geometry and to describe the distribution of energy after the absorption process. We showed that for compact dendrimers the optical absorption is dominated by peripheric molecules and that the energetically lowest absorption peak is redshifted with increasing dendrimer size. We identified the states responsible for this lowest absorption peak and explained the redshift in terms of the delocalization of energy over the whole dendrimer. This was confirmed by inspection of the densities of states $\mathcal{N}_n(\omega)$. Since with increasing size of the dendrimer rim molecules approach each other, we took an interaction between neighboring peripheric molecules into account. If this interaction is comparable to the other transfer matrix elements, energy is distributed over all molecules and there is no dominant absorption peak due to rim molecules any more. The DOS widens and smoothes out, especially in the 3rd generation.

Finally, we investigated the time dependence of the energy transfer in the dendrimer. For the parameter values chosen, we found an oscillatory behavior of the occupation number, indicating a coherent energy transport. We observed also a strong dependence of the energy transfer on the magnitude of the interaction

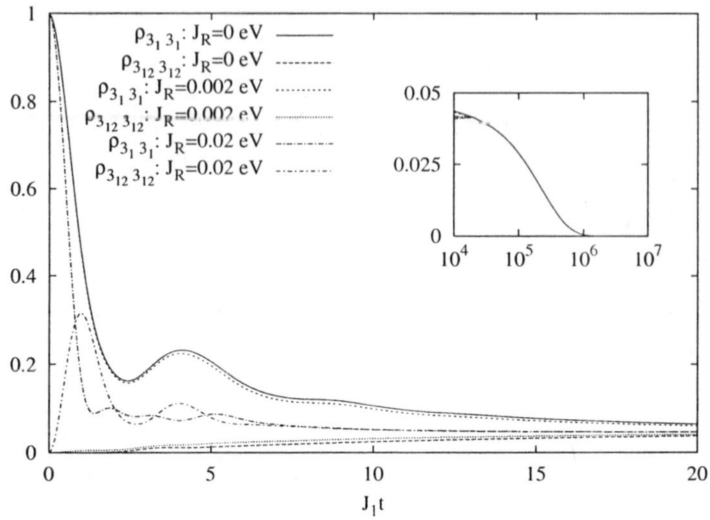

Figure 9. Occupation probabilities of the molecules 3_1 and 3_{12} for various values of J_R. D22 (compact) with $\varrho(0) = |3_1\rangle\langle 3_1|$, $\varepsilon_{0,1,2,3} = 3.9$, $J_{1,2,3} = 0.02$, $\gamma^0_{0,1,2,3} = 0.01$, $\gamma^1_{1,2,3} = 0.001$, all values in eV. Three different values of the transfer matrixelements are shown $J_R = 0, 0.002, 0.02$ eV.

between peripheric molecules.

In this paper we have investigated optical properties and the energy transport of well ordered compact dendrimers with low density in the center and high density at the periphery under the influence of dynamic disorder. The results of the investigation of extended dendrimers will be published elsewhere. Furthermore in recent publications [27,28] it was shown that dendrimers also exist in configurations with a high density in the center due to conformational disorder. The influence of such a spatial disorder as well as of the functionalization of groups of molecules are under consideration.

Acknowledgment
The authors are grateful to the Deutsche Forschungsgemeinschaft for financial support. Discussions with R. Steib are gratefully acknowledged.

References

1. Buhleier, E., Wehner, W. and Vögtle, F.(1978) Cascade and nonskid-chain-like syntheses of molecular cavity topologies, *Synthesis*, 155-158.
2. Tomalia, D.A., Naylor, A.M. and Goddard III, W.A. (1990) Starburst-Dendrimere: Kontrolle von Größe, Gestalt, Oberflächenchemie, Topologie und Flexibilität beim Übergang von Atomen zu makroskopischer Materie, *Angew. Chemie* **102**, 119-157.

3. Hawker, C.J. and Fréchet, M.J.J. (1990) Preparation of polymers with controlled molecular architecture. A new convergent approach to dentritic macromolecules, *J. Am. Chem. Soc.* **112**, 7638-7647.
4. Xu, Z. and Moore, J.S. (1993) Synthese und Charakterisierung eines steifen Dendrimers mit hohem Molekulargewicht, *Angew. Chem.* **105**, 261-264.
5. Börzsönyi, T., Tóth-Katona, T., Buka, A. and Gránásy, L. (1999) Dentrites regularized by spatially homogeneous time-periodic forcing, *Phys. Rev. Lett.* **83**, 2853-2856.
6. Percec, V., Ahn, C.H., Ungar, G., Yeardley, D.J., Möller, M. and Sheiko, S.S. (1998) Controlling polymer shape through the self-assembly of dentritic side-groups, *Nature* **391**, 161-164.
7. Karim, A., Liu, D.W., Bauer, B.J., Douglas, J.F. and Amis, E.J. (1997) Influence of generation number on the formation of dendrimer monolayers, *Polymeric Materials Science and Engineering, Washington* **77**, 181.
8. Zimmerman, S.C., Zeng, F., Reichert, D.E.C. and Kolotuchin, S.V. (1996) Self-assembling dendrimers, *Science* **271**, 1095-1098.
9. Bell, T.W. (1996) Molecular trees: a new branch of chemistry, *Science* **271**, 1077-1078.
10. Devadoss, C., Bharathi, P. and Moore, J.S. (1996) Energy transfer in dentritic macromolecules: molecular size effects and the role of an energy gradient, *J. Am. Chem. Soc.* **118**, 9635.
11. Stewart, G.M. and Fox, M.A. (1996) Chromophore labeled dendrons as light harvesting antennae, *J. Am. Chem. Soc.* **118**, 4354.
12. Mukamel, S. (1997) Trees to trap photons, *Nature* **388**, 425-427.
13. Jiang, D.L. and Aida, (1997) Photoisomerization in dendrimers by harvesting of low-energy photons *Nature* **388**, 454-456.
14. Kim, Y. and Zimmerman, S.C. (1998) Applicationof dendrimers in bio-organic chemistry, *Current Opinion in Chemical Biology* **2**, 733-742.
15. Kopelman, R., Shortreed, M., Shi, Z.Y., Tan, W., Xu, Z., Moore, J.S., Bar-Haim, A. and Klafter, J. (1997) Spectroscopic evidence for excitonic localization in fractal antenna supermolecules, *Phys. Rev. Lett.* **78**, 1239-1242.
16. Tretiak, S., Chernyak, V. and Mukamel, S. (1998) Localized electronic excitations in Phenylacetylene dendrimers, *J. Phys. Chem. B* **102**, 3310-3315.
17. Chernyak, V., Poliakov, E.Y., Tretiak, S. and Mukamel, S. (1999) Two-exciton states and spectroscopy of phenylacetylene dendrimers, *J. Chem. Phys.* **111**, 4158-4168.
18. Bar-Haim, A., Klafter, J. and Kopelman, R. (1997) Dendrimers as controlled artificial energy antennae *J. Am. Chem. Soc.* **119**, 6197-6198.
19. Bar-Haim, A. and Klafter, J. (1998) On mean residence and first passage times in finite one-dimensional systems, *J. Chem. Phys.* **109**, 5187.
20. Davydov, A.S. (1971) *Theory of molecular excitons*, Plenum Press, New York.
21. Haken, H. and Strobl, G. (1967) Exact treatment of coherent and incoherent triplet exciton migration, in Zahlan, A.B. (ed.), *The Triplet State*, Cambridge University Press, Cambridge, pp. 311-314.
22. Haken, H. and Reineker, P. (1972) The coupled coherent and incoherent motion of excitons and its influence on the line shape of optical absorption, *Z. Phys.* **249**, 253-268.
23. Haken, H. and Strobl, G. (1973) An exactly solvable model for coherent and incoherent exciton motion, *Z. Phys.* **262**, 135-148.
24. Reineker, P. (1982) Stochastic Liouville equation approach: coupled coherent and incoherent motion, optical line shapes, magnetic resonance phenomena, in Höhler, G. (ed.), *Exciton Dynamics in Molecular Crystals and Aggregates*, Springer Verlag, Berlin, Vol. 94, pp.111-226.
25. Engelmann, A. (2000) *Energietransport und optische Absorption von Dendrimeren*, Ph.D. Thesis, University of Ulm, Germany.
26. Press, W.H., Teukolsky, S.A., Vetterling, W.T. and Flannery, B.P. (1992) *Numerical Recipes in C*, Cambridge University Press, Cambridge.

27. Pötschke, D., Ballauf, M., Lindner, P., Fischer, M. and Vögtle, F. (2000) Analysis of the structure of dendrimers in solution by small-angle neutron scattering using contrast variation, *J. Appl. Cryst.* **33**, 605-608.
28. Sheiko, S.S. and Möller, M. (2001) Hyperbranched macromolecules: soft particles with adjustable shape and persistent motion capability, *Topics in Current Chemistry* **212**, 137-176.

PHOTOSWITCHABLE MOLECULAR RECEPTORS AND NANOSTRUCTURES ON THEIR BASIS

M.V. ALFIMOV
Photochemistry Center of the Russian Academy of Sciences
Ul. Novatorov, 7a
117421 Moscow, Russia

Nanostructures / photoswitchable molecular receptors / cation-induced molecule self-organization / crown-containing dyes / spectroscopy / fluorescence / photoizomerization / photocycloaddition

In the article crown-containing photoswitchable styryl, butadienyl, srironapnthoxazine dyes and chromenes are described. The addition of metal cations to dye solutions initiates self-organizing molecules of dyes, formation of intramolecular and intermolecular complexes. The structure of formed nanostructures is characterized by NMR methods, UV and fluorescence spectroscopy. It is shown, that the stereostructure of nanostructures depends on chemical composition and structure of participating molecules and metal cations. The spectral properties and fluorescence nanostructures, direction and efficiency of photochemical reactions in them are determined by their stereostructure. It is shown, that nanostructures have a number of photoswitchable steady states .It is planned to use photocontrolled nanostructures as cells for optical memory.

1. Introduction

One of the promising directions of photonics is related to use of photocontrolled molecules. Such molecules work in photoswitchable optical gates and light modulators as photoswitchable enzyme analogeus.

If photoswitchable molecules have additional receptor properties and an ability to bond selectively cations or molecules, it can be used for creation of photocontrolled sensors, photocontrolled extraction, photocontrolled transport of charges through membranes, etc.

Photocontrolled receptors can be also bricks for an assembly of supramolecular systems – photocontrolled organic nanostructures. Such nanostructures can be of interest for nanophotonics, including elements of optical computers, elements of photocontrolled nanomechanic devices, logic elements, and in nanocomposite materials.

F. Charra et al. (eds.), Organic Nanophotonics, 31–45.

Photocontrolled nanostructures consisting from n molecules can have more than 2^n steady states. The transition between those states is carried out by light.

The most effective way of creation of nanostructures consists in self-organizing molecules in a supramolecular structure (nanostructure). The self-organization of molecules in a supramolecular system is possible under condition of intermolecular interaction. In biology and chemistry there are many examples of such molecular self-organization. A basis of this phenomenon are various non-covalent interactions such as hydrogen bonding, hydrophobic interaction, donor-acceptor interaction, ion-ion, dipole-dipole, π-π stacking, and other interactions.

Last years ionophore groups including crown ethers are widely used as functional groups, ensuring self-organization of molecules. The structure of the molecule is infused by one or two crown ether substituents capable to ensure selective bonding of the molecule with metal cations or the other molecules. An ability of crown ethers to bond selectively only certain metal cations makes it possible to control process of self-assembly of organic nanostructures.

In our researches construction of supramolecular systems is realized using photoswitchable molecules, containing in the structure various ionophore groups. These molecules have two functions – they are capable to change their structure due to photoexcitation and to connect selectively cations of metals and the other molecules. It is possible to name them photoswitchable molecular receptors.

One can use as photoswitchable structures photochromic molecules of various kinds – dihydroindolizines, spiropyranes, spirooxazines, chromenes, fulgides, and diarylethylenes (Figure 1).

Such photoswitchable molecules containing ionophore groups in their structure represent a new class of molecular receptors – photoswitchable receptors. Their properties are described in a number of reviews [1-3].

In the present report we consider research directed on creation and properties of nanostructures (supramolecular structures) on the basis of molecular receptors.

The task of creation of photoswitchable organic nanostructures includes:

1. Development of methods for chemical synthesis of photocontrolled molecular receptors,
2. Study of conditions for self-assembly of such receptors in organic nanostructures,
3. Study of a structure and photonics of such organic nanostructures.

The work is devoted to the analysis of those problems.

2. Complexes of molecular receptors with metal cations as elementary organic nanostructures (supramolecular systems)

The elementary example of supramolecular system is the complex of a crown-containing photochromic molecule with metal cations. This complex includes a photoswitchable receptor molecule, metal cation and one or more anions. How the properties of such system differ from the properties of molecular receptor? How the properties of such supramolecular systems can be controlled by light?

Investigations of complexes of various composition and structures have shown, that if ionophore group is not included in a coupling chain of a photochromic molecule, the

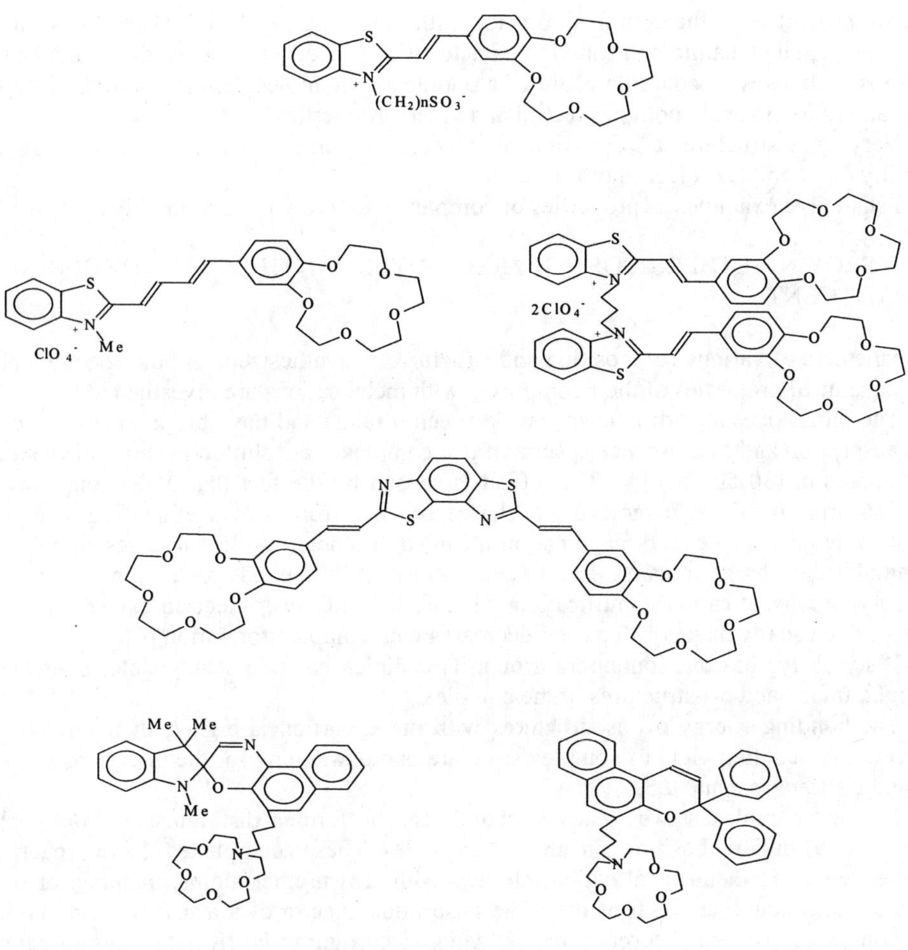

Figure 1. Chromophores

spectral properties of the complex coincide with the properties of initial photochromic molecule. Their dynamic and kinetic characteristics can differ. In case that ionophore group is included in a coupling chain of a photochromic molecule, the complex differs from an initial molecule both in spectral and kinetic properties.

Varying a structure it is possible to receive systems, which increase or reduce stability of a complex under photoirradiation.

I shall give examples of properties of complexes for crown-containing dyes.

2.1. CROWN ETHERS OF STYRYL DYES WITH ONE IONOPHORE SUBSTITUENT

Crown ethers of various composition and structure are synthesized, and the spectral and photochemical properties of their complexes with metal cations are investigated.

The initial dyes absorb light in a wide spectral range and have bright fluorescence. The absorption and fluorescence spectra of dye complexes are shifted in short-wave side of a spectrum (30-50 nm) [4]. This effect is caused by the fact that under long-wave light absorption in a dye molecule a charge transfer from crown ether fragment on heterocycle takes place, and metal cation included into heterocycle, increases energy of intramolecular charge transfer. The size of spectral shift depends on a cation type and can serve a way of cation identification [5]. Life time of lower electron-excited singlet state of dye usually makes 100 ps and decreases with complex formation [6].

If styryl dye has one ionophore group, its complex has two steady states switched by light, trans– and cis-structures of the complex.

The bonding energy of cis-structures with metal cartion is higher, than for trans-structure. A quantum yield of complex structure photoswitching for the directions trans-cis and cis-trans is equal 0.5.

In some complexes two steady structures can be formed distinguished from each other by a cation position inside macrocycle. These complexes have spectral differences, as in them metal cation interacts with crown ether atoms, included or not included in a coupling chain of dye. The absorption spectra of a azacrown-containing dye complex have two absorption bands, which according to NMR data can be related to complex structures with different location of cation in crown ether cavity. With photoexcitation as a result of electron transfer from crown-substituent on heterocycle there is a cation repulsion from nitrogen atom of crown ether, and cation departs to oxygen atoms, forming new structure of a complex. This photoreaction effectively proceeds in a solid matrix. Thus, the complex of azacrown-containing dye has four steady states switched by light, which spectrally differ from each other: trans– and cis-structures with two various cation positions in macrocycle.

Other examples of crown-containing dyes with one ionophore group are azacrown-containing spironaphthoxazines (Figure 2) and chromenes. In these compounds crown ether is coupled with a dye molecule by polymethylene chain containing one or more methylene groups. In photochromic transformation the initial neutral molecule turns to a zwitter-ion (open form). Thus, an oxygen anion is formed capable to interact with metal cations. If distance between this atom and metal cation localized in a crown ether cavity is small, there is an additional interaction, which increases stability constant K of the open form.

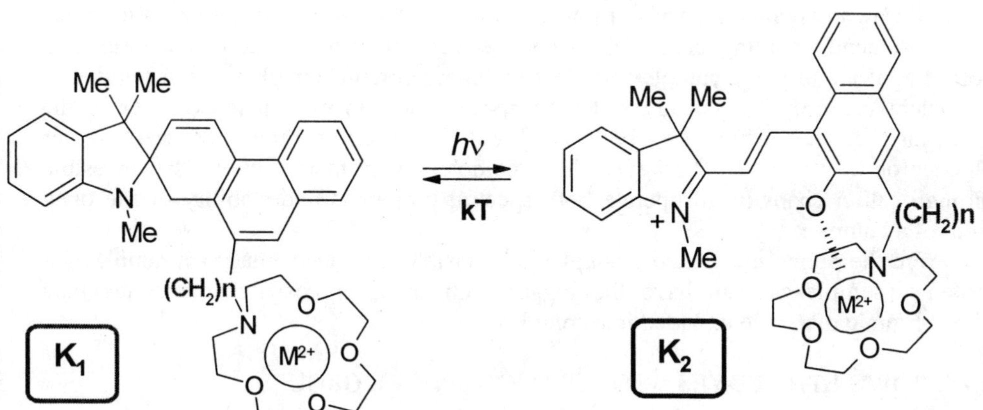

$$n = 1 \qquad \frac{K_2}{K_1} \sim 1$$

$$n = 5 \qquad \frac{K_2}{K_1} > 10^2$$

Figure 2. Anion-capped complex formation of crown-containing spironaphthoxazines

For crown-containing chromenes, where a nitrogen atom of crown ether is included in a chromene coupling chain, electronic density on nitrogen atom decreases and accordingly stability of a complex decreases in photochromic complex transformation.

There are only two steady states of spiropirane and chromene complexes, the transition between which can be carried out with help of light, as well as for photochromic molecules. However, in a complex with metal cation it is possible through cation changing to operate both spectral properties and stability of the open form of a complex.

From the above mentioned examples it is visible, that elementary supramolecular systems (complexes) can have the bigger number of photoswitchable states than photochromic molecule included in a complex.

2.2. CROWN ETHER DYES WITH TWO IONOPHORE GROUPS

For crown-containing dyes having two ionophore groups (sulphoxyl and crown ether), the structure of possible complexes with structure 1:1 can differ from the structure of a complex with one ionophore group. The trans-structure complex of dye with one ionophore group turns to a cis-structure complex under photoexcitation. The absorption spectrum of a cis-structure complex is close to a trans-structure complex, however it has considerably smaller oscillator strength of transition [7].

For dye with two ionophore groups photoexcitation of a trans-dye complex will form two cis-structure dyes with various stability and structure. The absorption spectra of these two forms are different, and their structures correspond to the open cis-form, when cation is coupled only with crown ether group or only with sulphoxyl group, and when both ionophore groups are coupled with one metal cation (anion-"capped" cis-isomer) (Figure 3). In the latter case the ring structure will be formed, thus the angles of rotation around of single bonds are so deformed, that intramolecular conjugation is destroyed, and as follows from experiment and calculations, the intensity of long-wave transition decreases more than in 100 times.

The stability constant of such type of complex exceeds stability constants of open cis-complex and trans-structure complex in 10-1000 times. The difference in stability constants depends on a structure of photochromic compound (structure and composition of crown ether group and length of arm on which sulphoxyl group is immobilized).

Taking into account an opportunity of the fast (in a nanosecond range) photoswitching of molecular structure from trans– to cis-form and large difference of spectra, considered supramolecular systems can find application in processes of photocontrolled cation extraction from solutions and for record of optical information.

Some of given examples demonstrate that our approach is promising for creation of systems with many photoswitchable states. Even on the basis of one photochromic molecule having two photoswitchable states it is possible to create supramolecular system having more than two steady states, photoswitchable by light.

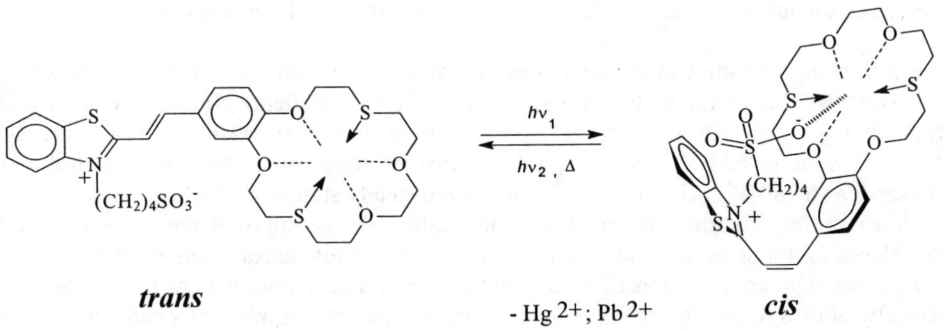

$$\frac{K_{cis}}{K_{trans}} = 11 \text{ (for Hg } ^{2+}\text{)}$$

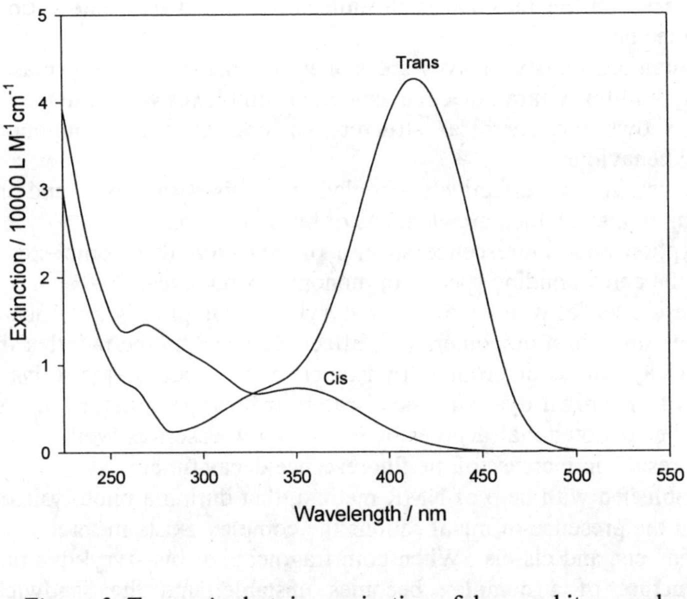

Figure 3. Trans-cis photoisomerization of dyes and its complexes
Anion-"capped" *cis*-isomers

3. Supramolecular systems containing two photoswitchable molecules

The following supramolecular system in complexity hierarchy is a system, containing two photoswitchable molecule – dimer structure. From two identical molecules of styryl dyes having two steady structures it is possible to design many various stereostructures. If it is possible to carry out photoswitching between these stereostructures, such dimer can serve as a cell of optical memory with eleven steady states.

Two untied identical photoswitchable molecules having two steady states and considered as single system have only three various steady states: trans-trans, trans-cis, and cis-cis. The construction of nanostructures from many photoswitchable molecules basically allows to create organic nanostructures with large number of steady states and correspondingly with large capacity of operational memory, and also logic elements for optical computers.

3.1. COMPLEXES OF CROWN-CONTAINING BIS-STYRYL DYES WITH METAL CATIONS [8]

The idea of creation of supramolecular systems with use of chemically bonded molecules is based on the fact that such molecule coupled with one cation can form stable supramolecule.

Crown-containing bis-styryl dyes are sometimes named as molecular pincers, as they effectively will form intramolecular sandwich complexes with cations.

Figure 4 shows the chemical structure of one of such compounds and its photochemical behaviour.

The large cations are effectively coupled with bis-styryl dyes and inefficiently coupled with monomers, which are elements of bis-styryl dye.

The absorption and fluorescence spectra of bis-styryl dye complexes practically coincide with the corresponding spectra of monomer complexes.

The fluorescence decay time of a bis-styryl dye complex is significantly bigger, than appropriate time for a monomer. This effect is caused by the fact that the complex fluorescence decay time is determined by isomerization speed of styryl dye. Formation of a complex of bis-crown dye with metal cation makes the structure of bis-dye more rigid, the number of rotational degrees of freedom decreases, as well as isomerization constant. This results in increase of the fluorescence decay time.

It was established with help of NMR method, that during a photolysis of bis-styryl dye solution in the presence of metal cations the complex exists in three various states: trans-trans, trans-cis, and cis-cis.. When both fragments of bis-styryl dye pass into cis-form, the structure of a complex becomes unstable, and the sandwich complex collapses, turning to a complex, in which a cation is coupled only with one ionophore (crown ether) group.

It is necessary to note, that the isomerization process in a complex, when freedom of fragment rotation is probably considerably limited, proceeds according to hula-twist mechanism. In this mechanism isomerization is caused by change of double bond conformation and related unitary bonds, and moving the large substituents is not required. Such isomerization mechanism is usually realized in a solid state.

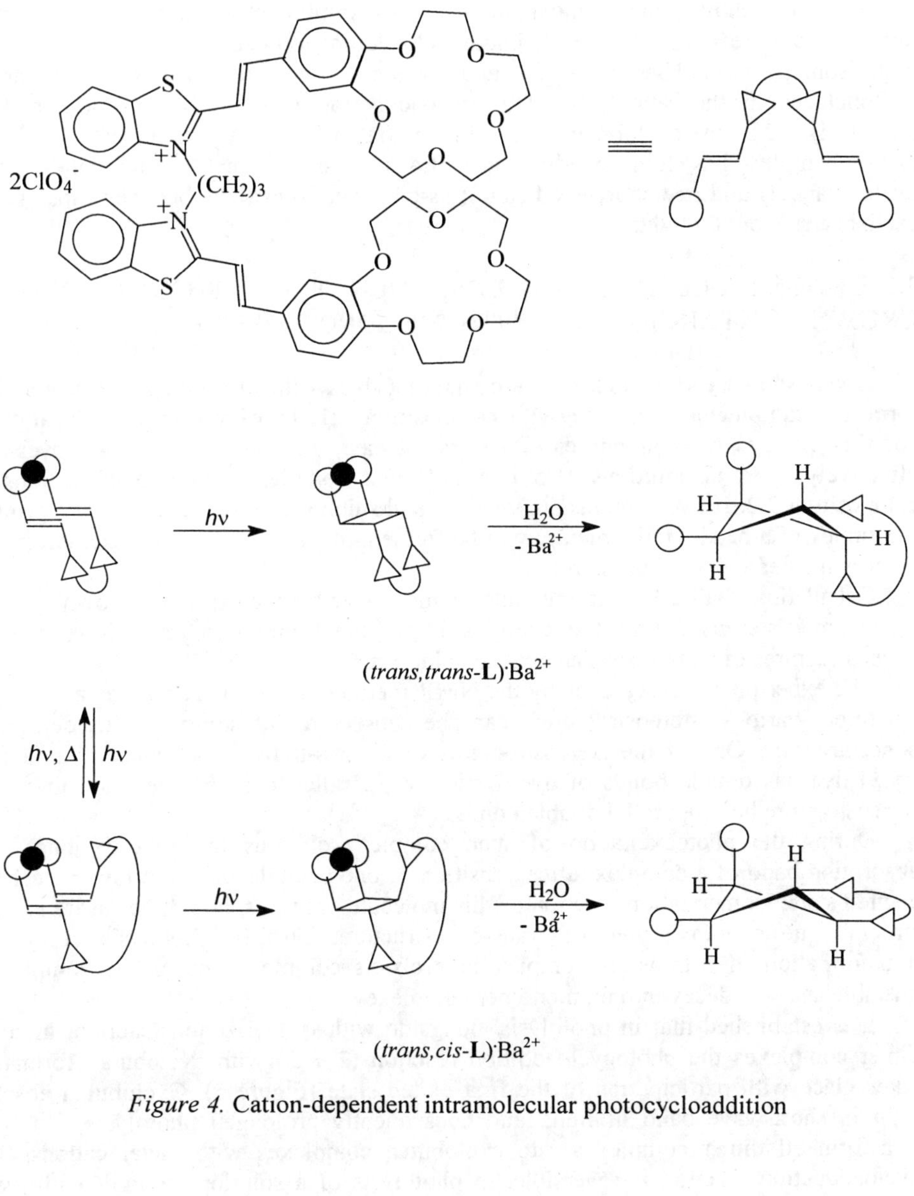

Figure 4. Cation-dependent intramolecular photocycloaddition

It was also shown, that during a photolysis in a solution of bis-styryl dye complexes intramolecular reaction of (2 + 2) photocycloaddition proceeds effectively alongside with isomerization. The trans-trans and trans-cis complexes turn into appropriate cyclobutanes in the ratio 1: 4. The cycloadduction reaction is reversible and in photolysis (331 nm) cyclobutane turns into initial bis-styryl dye structure. Thus, the dimer complex has four steady states (trans-trans, trans-cis, cyclobutane-1, and cyclobutane-2) and one short-lived state (cis-cis). The transitions between those states can be carried out by light.

3.2. CATION-INDUCED SELF-ORGANIZATION OF DIMER COMPLEXES OF CROWN-CONTAINING DYES WITH TWO IONOPHORE GROUPS

It was shown earlier, that crown-containing dyes with two ionophore groups can form a intramolecular complex with composition 1:1, in which one metal cation is coordinated by two ionophores. It is established, that such ionophore dyes are effectively form in solutions also intermolecular complexes with metal cations of composition 2:2. If dye contains crown ether and sulphoxyl group as ionophore groups, a complex of a head – tail structure will be formed, in which each cation is coordinated by crown ether and sulphoxyl groups.

Calculations based on the methods of molecular mechanics have shown that for each dimer stereostructure the energies in a global minimum are close, and all stereostructures of complexes can exist in solutions.

The experimental researches by the NMR method have shown, that, as a rule, two or three various stereostructures can be observed in solutions in detectable concentrations. One of the stereostructures of a trans-trans crown ether complex of styryl dye has double bonds of dyes which are parallel to each other, and the other stereostructure has unparallel double bonds.

During the photoexcitation of such complex solutions by light in long-wave absorption band of a complex, after transition of one of molecules in a lower electron-excited state, isomerization reaction of this molecule proceeds, and dimer of trans-trans structure turns into dimer of trans-cis structure. Further photolysis results in transformation of a trans-cis complex in a cis-cis complex. The cis-cis complex is unstable and will decay into on monomer complexes.

It is established that in photolysis alongside with isomerization reaction, in some dimer complexes the photocycloaddition reaction (2 + 2) with cyclobutan formation takes place with participation of the first exited state (Figure 5). Cyclobutan absorbs light in short-wave band of light, and consequently prolonged photolysis allows to transform all dimer complexes into cyclobutan complexes with metal cations. The cycloadduction reaction is reversible, in photolysis of a solution by light with wave length 313 nm a cyclobutan complex turns into a dimer complex. The photochemical cycloadduction reaction proceeds only in electron-excited state. An activation barrier of reaction in dimer with participation of one electron-excited and one not exited molecule determined basically by styric factors, i.e. by possibility of approach of double bonds. The photocycloaddition reaction can proceed only in those dimers where the stereostructure provides approach of double bonds necessary for reaction.

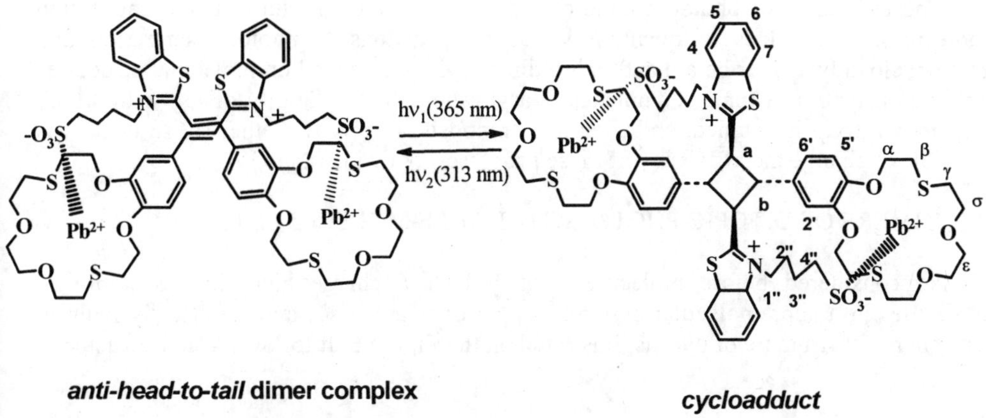

Quantum yields $\varphi_{PCA} \times 10^4$ of the photocycloaddition for complexes of CSDs (C = 4×10^{-5} M/l) in acetonitrile

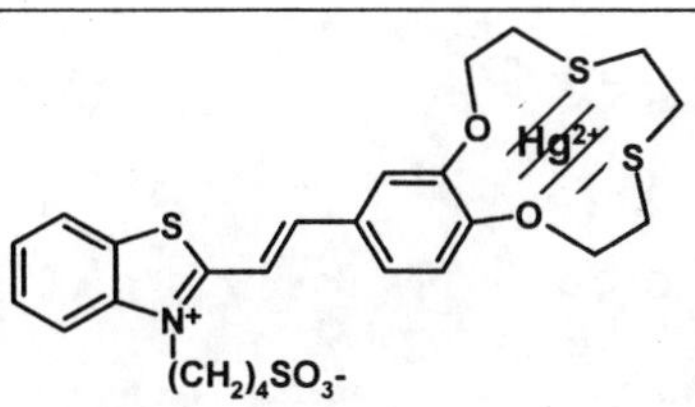	**1.3**	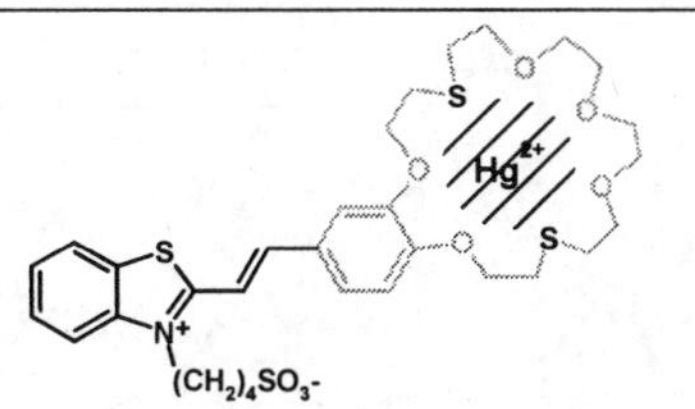	**0.79**
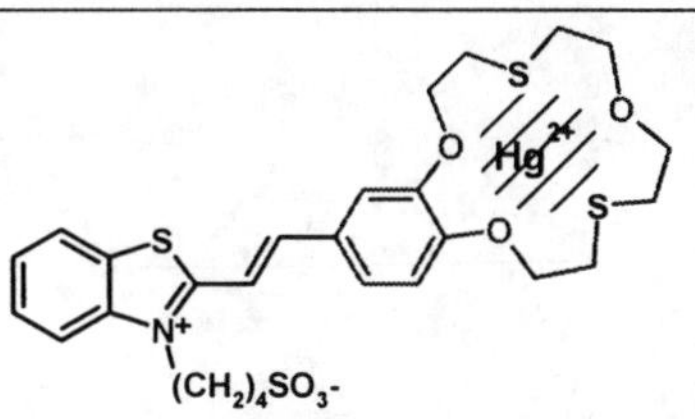	**28**	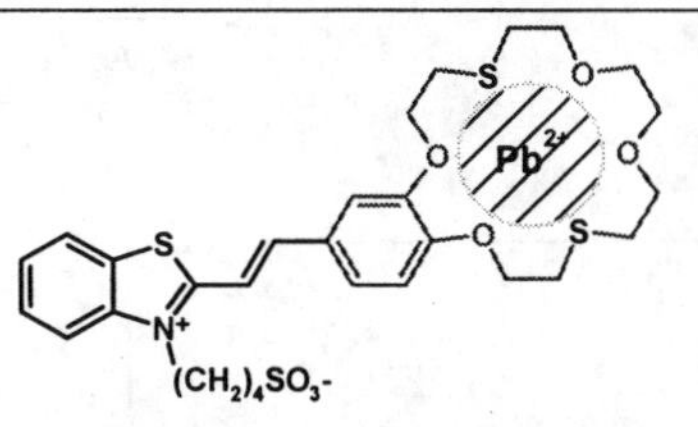	**10**
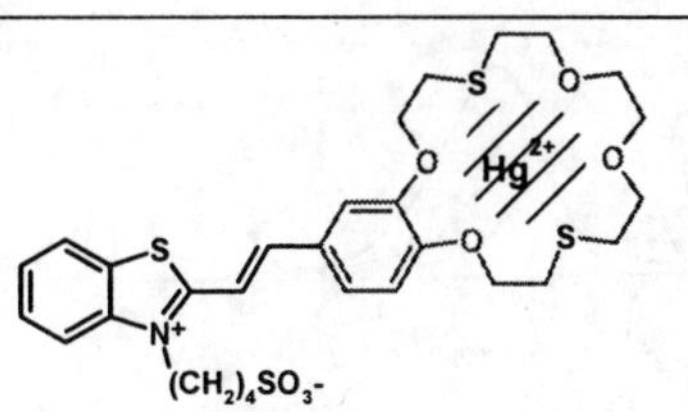	**1.3**	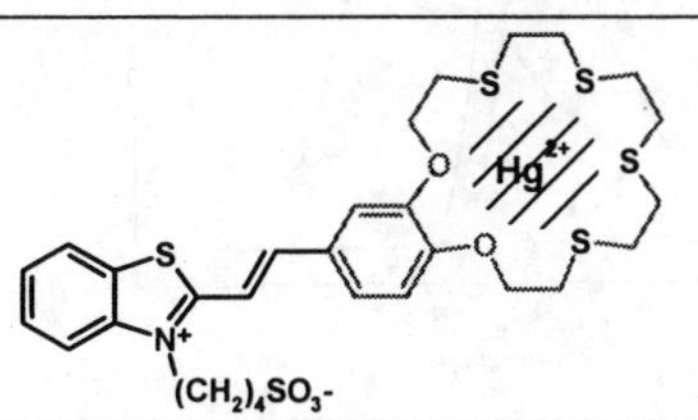	**0**

Figure 5. Photochemistry: photocycloadduct formation

The efficiency of photocycloaddition depends on the structure of dye and cation molecules. The values of quantum yields for reactions of photoisomerization and photocycloaddition, for betain and quinoldin styryl dyes depend on molecular structures and cation typess in dimer complexes. Varying length and rigidity of leg with which sulphoxyl group is coupled, and chemical composition of crown ether substituent, it is possible to change the efficiency of these processes in a wide range [9].

3.3. DIMERS OF DITOPIC PHOTOSWITCHABLE MOLECULES [10]

Unlike considered before molecules, which bear a charge hindering assembly of molecules in a supramolecular system, ditopic dyes are bases and electrically neutral. The chemical structure of dyes is illustrated on the Figure 6. It follows from a structure,

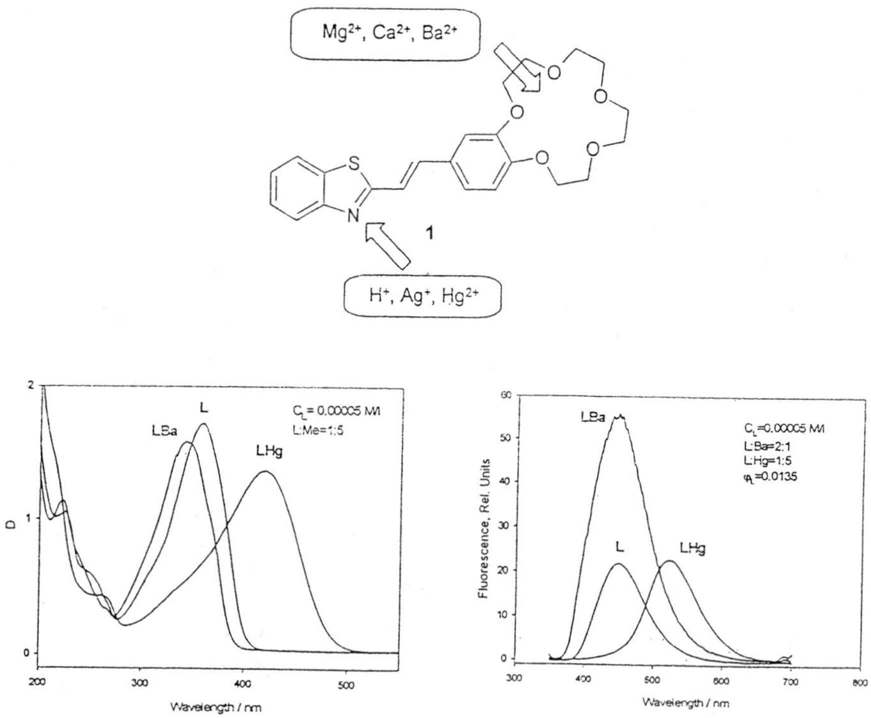

Figure 6. Ditopic receptor

that there are two places in a molecule for cation binding – nitrogen atom of a heterocycle and crown ether substituent. It is shown, that in solutions these compounds form two types of sandwich complexes at the presence of cations – due to binding two nitrogen atoms of heterocycle by a cation or two crown ether substituents.

The NMR research have shown that the complexes exist in solutions as sandwich trans-trans and trans-cis structures. During photolysis in complexes an isomerization and cycloadduction reactions flow effectively, causing formation of two cyclobutans with different stereostructures.

A stereogeometry of a sandwich complex created by complex formation of a cation with two crown ether groups can be controlled by adding dicarboxilic acids, forming hydrogen bonds with heterocycle nitrogen, in a solution. Varying a length of bridge, connecting hydroxylic groups, it is possible to change the ratio of photocycloproducts considerably.

Advantage of this class of compounds is their high solubility in organic solvents and polymers.

3.4. DIMER COMPLEXES OF BUTADIENYL DYES

Crown-containing butadien dyes are a new class of photoswitchable molecular receptors. In solutions, when strontium or barium salts are added, these dyes form complexes of 1:1 and 2:1 composition. Unlike styryl dyes of a similar structure, butadien complexes of 2:1 composition have sandwich head-to-head structure, probably due to heterocycle stacking interaction. Butadien dyes, containing sulphoxyl group instead of ethylene substituent in heterocycle, form also dimer complexes head-to-tail of 2:2 composition at the presence of magnesium cations. During photolysis isomerization and (2 + 2) cycloadduction (trans,trans-trans,trans and trans,cis-trans,cis) take place in complexes with a quantum yield 0.3.

3.5. DONOR-ACCEPTOR COMPLEXES OF CROWN-CONTAINING DYES

A derivative of dipyridylethylene and bis-crown stilbene in acetonitrile solution can form a supramolecular complex by coordination crown ether groups and amino-groupes. Emission quenching and appearance of a new absorption band in visible spectral region indicate the formation of a charge-transfer complex.

The calculations executed by semiempirical methods and by a method of density functional have shown, that during complex formation the charge redistribution goes in such a manner that the crown ether fragments appear positively charged. It specifies that during complex formation the charge transfer to alkylammonium group dipyridinethylene takes place. These compounds alongside with a complex of 1:1 structure can form a trimolecular complex of a sandwich structure, in which the acceptor is connected to two molecules of the bis-crown stilbene. The measurements by femtosecond spectroscopy methods have shown, that the life time corresponding to full electron transfer is equal 1-2 ps. The adding of barium cations in a solution results in a complex destruction, and a charge transfer band vanishes.

4. Conclusion

It is shown that photoswitchable molecules containing ionophore groups in solutions are capable to form nanostructures if metal catios (molecules) are added. The directed modification of a photoswitchable molecule chemical structure and selection of metal cation (molecule) allows to create nanosystems of the given stereostructure. The created nanostructures have several steady states corresponding to various structures, and have various spectral and photochemical properties. The nanostructure transition from one state to another can be carried out by light. It is offered to use received photoswitchable nanostructures as memory cells for optical computers.

5. References

1. Ctano I.C., Guglielmetti R., eds. (1999) Organic Photochromic and Thermochromic Compounds, vol. 2, Plenum Press, New York.
2. Alfimov M.V., Gromov S.P. (1992) Photochemistry of crown ether styryl dyes. *Supramolecular Chemistry (Edited by V.Balzani and L.de Cola), NATO ASI Series.* 343-358.
3. Gromov S.P., Alfimov M.V. (1997) Supramolecular organic photochemistry of crown-ether-containing styryl dyes. *RUSS CHEM B+***46,** (4), 611-636.
4. Baskin I.I., Burshtein K.Ya., Bagatur`yants A.A., Gromov S.P. and Alfimov M.V. (1992) Molecular simulation of the complexation effects on conformations and electronic absorption spectra of crown ether styryl dyes. *J.Molecular Structure,* **274**, 93-104.
5. Gromov S.P., M Fomina.V., and M Alfimov.V. (1997) Crown ether-containing styryl dyes. 9. Synthesis of chromogenic indolenine crown ethers from nitropyridinium salts. *RUSS CHEM B+***42** (9) 1561-156.
6. Alfimov M.V., Kamalov V.F., Struganova I.A., Yoshihara K. (1992) Excited state relaxation of crown ether styryl dyes: photoisomerization. *Chem. Phys. Lett.* **195**, N.2,3, 262-266.
7. Alfimov M.V., Vedernikov A.I., Gromov S.P., Fedorov Yu.V., and Fedorova O.A. (1997) Crown-containing styryl dyes. 23. Synthesis and complexation of cis-isomers of photochromic dithia-15(18)-crown-5(6) ethers. *RUSS CHEM B+***46**: (12), 2099-2106.
8. Gromov S.P., Fedorova O.A., Ushakov E.N., Buevich A.V., Baskin I.I., Pershina Y.V., Eliasson B., Edlund U., Alfimov M.V. (1999) Photoswitchable molecular pincers: synthesis, self-assembly into sandwich complexes and ion-selective intramolecular [2+2]-photocycloaddition of an unsaturated bis-15-crown-5 ether. *J. Chem. Soc., Perkin Trans.* 2, 7,1323-1331.

9. Fedorov Yu.V., Fedorova O.A., Andryukhina E.N., Gromov S.P., Alfimov M.V. Kuzmina L.G., Churakov A.V., Howard I.A.K., Aaron I.-I. (2002) Ditopic complex formation of the crown-containing 2-styrylbenzo-thiazole. *RUSS CHEM B* (in press)

10. Gromov S.P., Vedernikov A.I., Ushakov E.N., Kuz'mina L.G., Feofanov A.V., Avakyan V.G., Churakov A.V., Alaverdian Y.S., Malysheva E.V., Alfimov M.V., Howard J.A.K., Eliasson B., and Edlund U.G. (2002) Synthesis, Structure, Spectroscopic Studies, and Complexation of Novel Crown Ether Butadienyl Dyes, *Helv. Chim. Acta*, **85**, 60-81.

11. Gromov S.P., Ushakov E.N., Vedernikov A.I., Lobova N.A., Alfimov M.V., Strelenko Y.A., Whitesell J.K. and Fox M.A. (1999) A Novel Optical Sensor for Metal Ions Based on Ground-State Intermolecular Charge-Transfer Complexation. *Org. Lett.,* **1**, N 11, 1697-1699.

MOLECULES IN MOTION : TOWARDS HYDROGEN BOND-ASSEMBLED MOLECULAR MACHINES

DAVID A LEIGH
School of Chemistry,
The University of Edinburgh, West Mains Road, Edinburgh EH9 3JJ, United Kingdom.
Email: David.Leigh@ed.ac.uk

1. Introduction

Many phenomena of biological interest originate directly from mechanical motions at the molecular level. Celebrated examples include the *trans-cis* isomerisation of double bonds that triggers the visual signal and the rotary motion of the enzyme F_1-ATPase, the cornerstone of ADP to ATP conversion. This dependence on molecular level motion in key natural processes is inspiring scientists to try and bridge the gap between synthetic chemistry, which by and large relies upon electronic and chemical effects and does not exploit molecular motions, and the macroscopic world, where our everyday machines rely upon the synchronized motions of their components to perform their designated tasks. Accordingly, there is great current interest in trying to make molecular analogues of some of the fundamental components of machinery from the macroscopic world (cogs, wheels, shuttles, pistons etc) [1]. The idea is that such structures could form the basis of synthetic devices or materials that, like biological systems, could function through molecular level mechanical motion.

In the last few years many significant advances have been made in the aim to control, and then subsequently exploit, such molecular level mechanical motion. Some key contributions include (i) the demonstration of unidirectional rotary motion in synthetic molecules using light [2] and chemical [3] stimuli, (ii) the control of large amplitude linear "shuttling" of components up and down molecular "threads" using light, electrical stimuli or the nature of the environment [4-7], (iii) the use of temperature, polarity and alternating electric fields to control the speed of rotating components in molecules [8], and (iv) the controlled release of substances by mechanical expansion of molecular "capsules" [9].

F. Charra et al. (eds.), Organic Nanophotonics, 47–56.

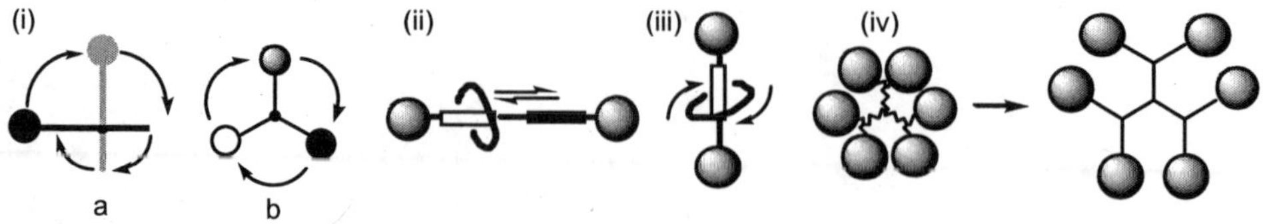

Figure 1. Molecules in motion: (i) unidirectional rotation in molecular systems powered by (a) light or (b) chemical stimuli; (ii) molecular 'shuttles', controlled by heat, light, electrons, pH or polarity of the environment; (iii) rotaxanes where the rate of macrocycle spinning is influenced by an alternating electric field; (iv) dendrimers which expand to release their contents in response to $h\nu$.

2. Catenanes and Rotaxanes

Molecules with mechanically interlocked components could potentially be very useful for molecular machine-type applications because, in principle, they permit the controlled, large amplitude, movement and positioning of one mechanically interlocked component with respect to another. They fall into two classes: "catenanes" which consist of two or more interlocked rings and "rotaxanes" where a macrocycle is locked onto a linear "thread" by bulky "stoppers" (Figure 2).

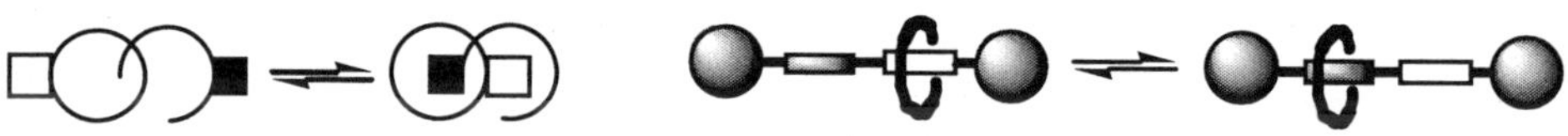

Rotational motion in a catenane

Linear motion in a rotaxane (a "molecular shuttle")

Figure 2. Possibly controllable motions in mechanically-interlocked architectures.

In the 1980's and early 90's, ingenious routes to both types of architectures were developed by the groups of Sauvage and Stoddart using the assembly of π-electron deficient cyclophanes and π-electron rich aromatic polyether units [4] (Stoddart) and transition metal-coordinated ligands [6] (Sauvage). More recently, Sauvage has shown [10] that the metal-based system can undergo 'positional isomerism' by varying the oxidation level (and thereby the coordination requirements) of a bound transition metal by electronic or photochemical redox reactions. Similarly, Stoddart has prepared 'shuttles' that can be switched *via* electrical, photochemical or chemical (H^+ transfer) stimuli [11] and which form the basis of remarkable macroscopic electronic devices which appears to work through mechanochemical switching of a rotaxane or catenane assembled as a thin film on silicon [12].

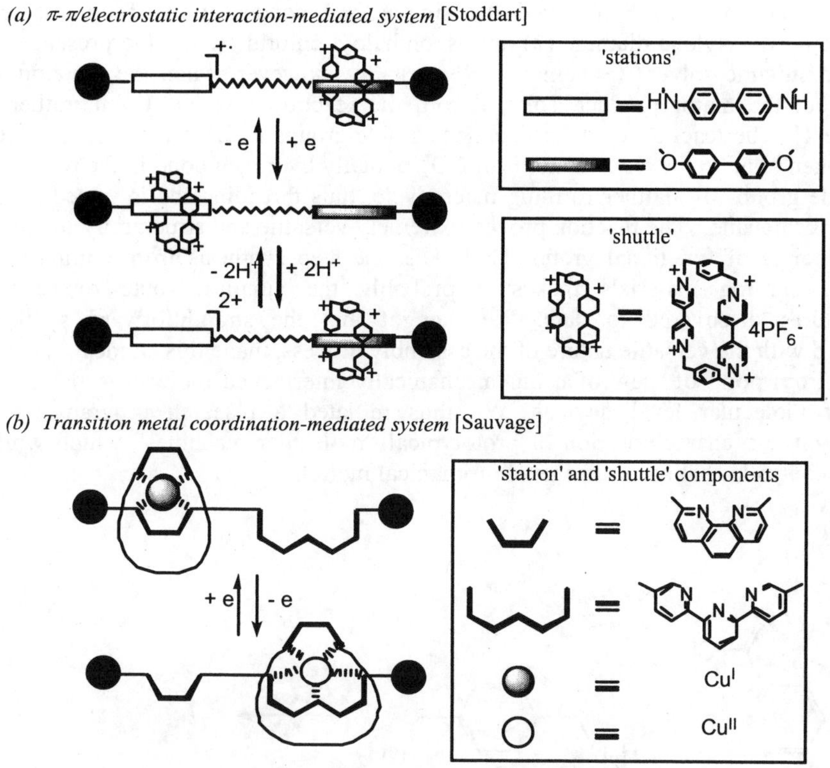

Figure 3. Interlocked systems which switch the relative positions of their components using external stimuli.

3. A Serendipitous Beginning – The Discovery of a New Catenane Synthesis

In 1995, as part of a programme to prepare chemical sensors for small neutral molecules, our group attempting to prepare the macrocycle **1**, which molecular modelling had identified as a potential receptor for CO_2 [13].

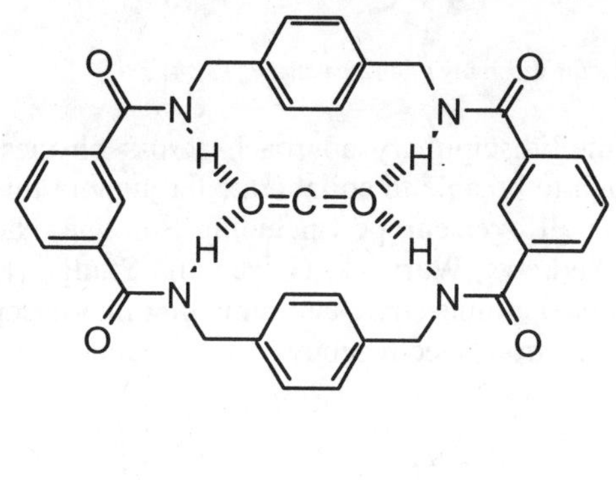

1 + CO_2

The proposed route to **1** involved the simple 2+2 condensation of commercially available *para*-xylylene diamine (**2**) and isophthaloyl chloride (**3**) in the presence of a base and suitable solvent (Scheme 1). Remarkably, however, when the reaction was carried out the major product isolated from the reaction was not **1**, but rather the catenane **4**! The reaction works because the amide groups which were designed into **1** to hydrogen bond to the oxygen atoms of CO_2, actually hydrogen bond in the reaction to the amide groups of another forming macrocycle, thus threading the intermediate and forming a catenane. The reaction proved extremely versatile and structurally tolerant to a broad range of functional groups [14]. The one step synthesis from commercially available starting materials makes it probably the simplest route to catenane architectures discovered to date. We thought that the straightforward synthesis, combined with the versatile nature of the assembly process, made this an ideal system to explore from point of view of using mechanically interlocked molecules to form the basis of molecular level devices. We thus initiated a research program on the preparation and characterisation of prototypical "molecular machines" which worked through controlled submolecular level mechanical motion.

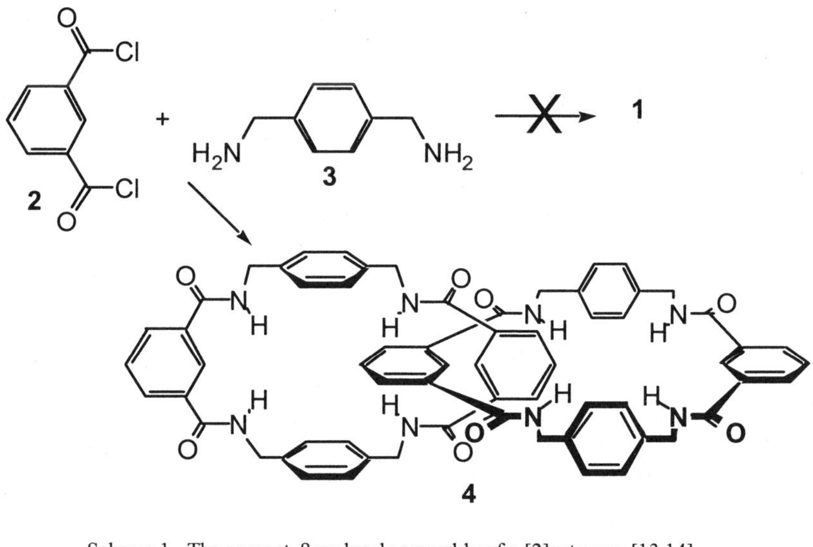

Scheme 1. The one pot, 8 molecule assembly of a [2]catenane [13,14]

To carry out such a study requires a multidisciplinary approach across physics and chemistry, and we have been lucky enough to be able to collaborate throughout this on-going project with excellent groups from all over europe (including Bologna, Namur, Paris, Amsterdam, Trieste, Louvain, St Andrews, Warwick, Geleen and Stuttgart) who are experts in surface and solid state science, atomic force and tunneling microscopy, a broad range of optical spectroscopies, electron spectroscopy, nuclear and electronic magnetic resonance, electrochemistry, and computation and modeling studies to clarify and explain the experimental results.

4. Characterising a [2]Catenane - What are the Effects of Mechanical Interlocking at the Molecular Level ?

The first phase of our work entailed the synthesis of a range of novel benzylic amide catenanes and their comprehensive characterisation by a range of techniques including (i) infrared, Raman, UV-visible electronic spectroscopy and photophysical studies [15-18], (ii) inelastic neutron scattering [19,20], (iii) x-ray crystallography [21], (iv) electrochemistry [22], (v) solution and solid state nuclear magnetic resonance [23], (vi) surface studies [24-27], (vii) doping processes [28], (viii) linear and non-linear optics [29-32], all supported by detailed simulations of the catenanes dynamics properties [33] and the effects these had on bulk properties using computational quantum chemistry [34,35].

The studies revealed that mechanically interlocked architectures can impart unexpected and highly unusual effects on physical properties : for example, (i) the IR and Raman spectra are unusually matrix-dependent (*e.g.* large frequency shifts are observed between KBr and CsI) but matrix-insensitive in the case of their non-interlocked macrocyclic components [18]. This is a very rare sort of result that could prove useful in the design of sensors for specialist applications such as a molecular 'pressure gauge' for inorganic salts. (ii) The electrochemical behaviour of macrocycles and catenanes are very different, with electrochemically-induced braking of circumrotation in the catenane providing an important means of being able to modulate the dynamic properties in response to a convenient external stimuli [22]. (iii) Some catenanes could conveniently be sublimed to form high optical quality films of unusually high refractive index (1.6) and of thicknesses (0.8 μm) sufficient for guided wave applications [29-31], indicating considerable promise for these materials as versatile photonic materials. (iv) IR and NMR results showed that amphiphilic catenanes could behave as 'smart' molecules, changing their surface properties to suit the nature (the polarity) of their environment [36,37]. (v) NMR and dynamic modelling studies revealed that the frequency of ring rotation in benzylic amide catenanes could be varied over *nine* orders of magnitude by judicious choice of chemical structure and operating environment [23,33]. This unprecedented control augured well for tuning specific dynamics for various applications.

As a whole this program gave us a good - probably unprecedented - understanding of the key structure-function relationships in a series mechanically-interlocked materials - *i.e.* the determination of the key physical, electronic and dynamic properties influenced (and controllable) by the architecture and functionality of catenanes.

In addition, the information provided by the characterisation program, particularly X-ray crystallography but also NMR data and the molecular modelling work, led to another breakthrough in synthesis.

5. From Catenanes to Rotaxanes

The hydrogen bonding patterns in the solid state structures of the catenanes suggested that modification of the synthetic procedure might allow rotaxanes to be synthesized. Sure enough, we discovered that if the catenane reaction is carried out in the presence of a suitable stoppered benzylic 1,3-diamide thread (*e.g.* **5**) rotaxanes such as **6** are also formed (Scheme 2) [38].

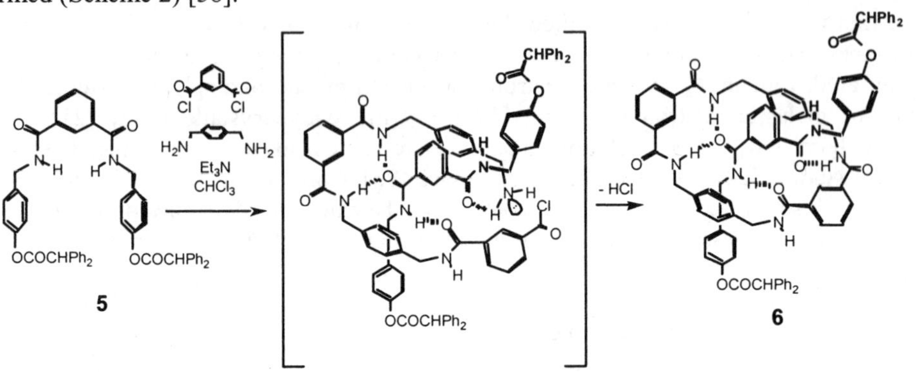

Scheme 2. Hydrogen bond-directed rotaxane formation

Like catenane formation, a rotaxane architecture also brings about remarkable changes in chemical and physical properties; for example, the rotaxane **6** is 100000x more soluble in nonpolar organic solvents as a rotaxane than its individual components [38] !

The comprehensive spectroscopic data showed that the mechanism for the formation of the interlocked products is primarily the templated assembly of the macrocycle around two amide bonds. Divergent hydrogen bonding sites in a similar spatial arrangement occur in adjacent amino acid residues in peptide chains and, indeed, we found that simple peptides are also able to template the cyclisation of benzylic amide macrocycles to give *peptide* rotaxanes (Scheme 3) [39,40]. There are possible significant implications for this discovery in biological and medical applications (*e.g.* rotaxanes as potential drug delivery systems).

Following a similar line of thought, the degree of flexibility in the peptide backbone cannot be optimal for rotaxane formation. Replacing the peptide motif with a fumaramide unit would maintain the required hydrogen bonding groups but with a rigidified backbone. When this was carried out experimentally the results were extraordinary; a world record yield of 97% for a rotaxane was obtained, for a five component assembly process in just two steps from commercial starting materials (figure 4) [41]. The X-ray crystal structure (inset) shows the four amide-amide hydrogen bonds from macrocycle to thread.

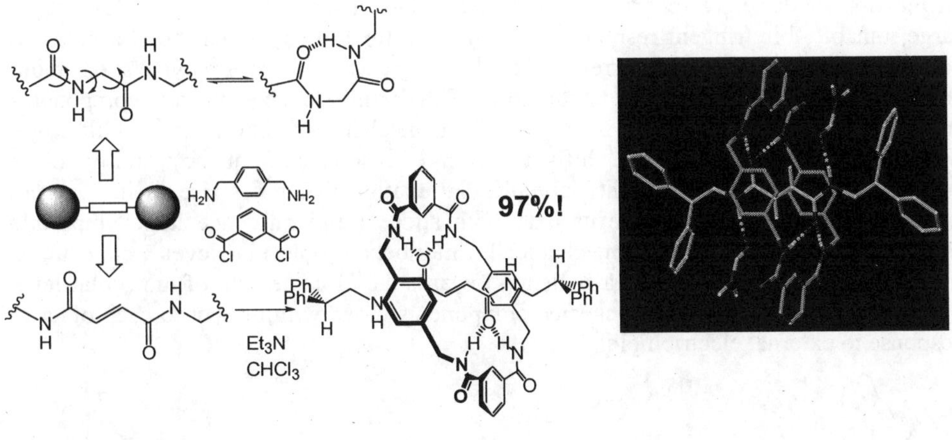

Figure 4. Stiff, and sticky in the right places! The dramatic effect of preorganizing binding sites on the efficiency of rotaxane synthesis [41].

By using other rigid threads, rotaxanes were obtained containing amide-ester hydrogen bonds. By measuring the rate of macrocycle spinning in these systems, a direct measure of amide-amide versus amide-ester hydrogen bonding strengths could be measured [41]. This information could be valuable in deciding which groups to substitute in designing artificial replacements for natural agonists (i.e. pharmaceuticals).

6. From Rotaxanes to Stimuli-Responsive Shuttles

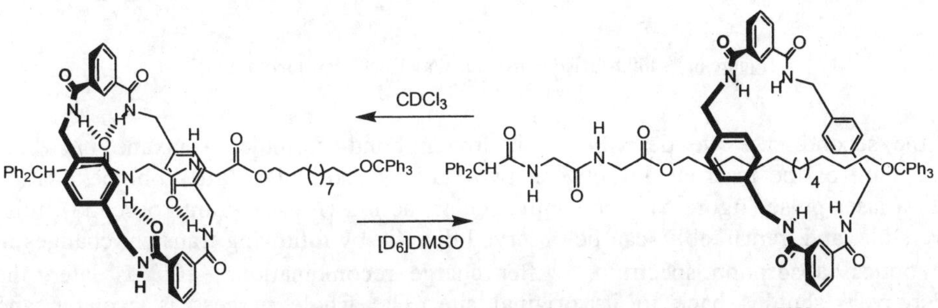

Figure 5. Translocation of a macrocycle along a thread in a amphiphilic peptide rotaxane [42].

These advances in rotaxane synthesis enabled the development of stimuli-responsive molecular shuttles where the macrocycle can be translocated along the backbone of the thread using an external stimulus. The first of these were amphiphilic peptide rotaxanes where the shuttling was carried out by varying the polarity of the environment (figure 5) [42]. However, we later developed methods for controlling the motion of the components using stimuli more suited for device applications, including oscillating electric fields [7] and light [43,44].

In the first of these we reported the unexpected result that AC electric fields trigger large, tunable, birefringent responses in dilute solutions of hydrogen bond-assembled rotaxanes at frequencies that correspond to the rate at which the macrocycle spins about the thread. The effect is absent in solutions of the uninterlocked rotaxane components indicating that the phenomenon is intrinsic to the mechanically interlocked architecture. Variation of the applied voltage shifts the observed resonance frequency, and therefore by inference also the rate of the submolecular motion that gives rise to it. Thus an oscillating electric field can be used to both address and influence large amplitude intercomponent movements in mechanically interlocked molecular level architectures. These molecules thus represent a first step towards the ultimate goal of molecular level systems where mechanically connected components move with respect to each other in response to external electric fields.

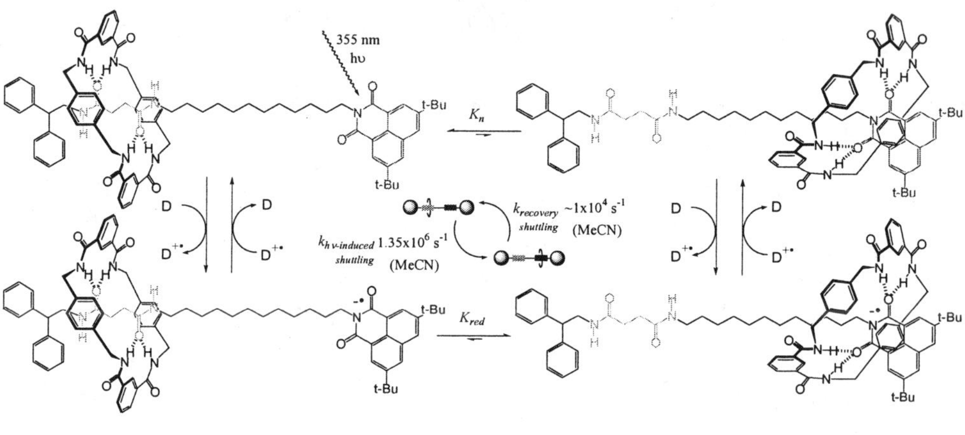

Figure 6. A light-fueled translational molecular level motor [43].

In the second case we described a hydrogen bond-assembled rotaxane in which translation of the macrocycle between two stations takes place after photoexcitation with a laser pulse (figure 6). The submolecular motion is fast (~1 microsecond), fully reversible, and, remarkably, can be observed directly by following transient changes in the optical absorption spectrum. After charge recombination, ~100 μs later, the macrocycle shuttles back to its original site. The whole process is cyclable and reminiscent of the workings of a piston, where a powered stroke (during which mechanical work can be performed) is followed by a recovery stroke that returns the system to its starting position. Practical applications of such light-induced mechanical motion at the molecular level might involve the rearrangement of the structure of surfaces or the "fetching-and-carrying" of molecules or clusters of atoms between specific locations (e.g. across membranes).

7. Conclusions and Perspectives

Hydrogen bonding is an extraordinarily effective means of both directing the synthesis of mechanically interlocked architectures and controlling the relative motions of their components. Transmission of the effects of mechanical motion from the molecular level through to the macroscopic world could lead to breakthroughs in areas of technology that today sound like science fiction. For example, imagine (i) being able to move objects along a surface by simply tracing the path you want them to take with a laser pointer! (by the molecules reacting to the light and changing shape/property so that they pull or push the object along the surface) or, (ii) clothing made of polymers which expand or contract to fit the wearer in response to an electrical stimulus, or, (iii) surfaces that respond in real-time to incoming information and change their properties accordingly. These are technologies that sound like they are far in the future, but they are certainly closer because of the studies currently going on all around the world in the growing area of 'molecular machines'.

References

[1] See, for example, *Acc. Chem. Res.* Special Issue on "*Molecular Machines*" **2001**, *34*(6), 409-522; V. Balzani, A. Credi, F.M. Raymo, J.F. Stoddart, *Angew. Chem. Int. Ed.* **2000**, *39*, 3348-3391.

[2] N. Koumura, R.W.J. Zijlstra, R.A. van Delden, N. Harada, B.L. Feringa, *Nature*, **1999**, *401*, 152-155.

[3] T.R. Kelly, H. De Silva, R.A. Silva, *Nature*, **1999**, *401*, 150-152.

[4] V. Balzani, M. Gómez-López, J.F. Stoddart, *Acc. Chem. Res.* **1998**, *31*, 405-414.

[5] C.P. Collier, G. Mattersteig, E.W. Wong, Y. Luo, K. Beverley, J. Sampaio, F.M. Raymo, J.F. Stoddart, J.R. Heath, *Science*, **2000**, *289*, 1172-1175.

[6] J.P. Sauvage, *Acc. Chem. Res.* **1998**, *31*, 611-619.

[7] A.S. Lane, D.A. Leigh, A. Murphy, *J. Am. Chem. Soc.* **1997**, *119*, 11092-11093.

[8] V. Bermudez, N. Capron, T. Gase, F.G. Gatti, F. Kajzar, D.A. Leigh, F. Zerbetto, S. Zhang, *Nature*, **2000**, *406*, 608-611.

[9] F. Vogtle, S Gestermann, C Kauffmann, P Ceroni, V Vicinelli, L De Cola, V Balzani, *J Am Chem Soc* **121**, 12161-12166 (1999)

[10] M.C. Jiménez, C. Dietrich-Buchecker, J.P. Sauvage, *Angew. Chem. Int. Ed.* **2000**, *39*, 3284-3287

[11] R.A. Bissell, E. Córdova, A.E. Kaifer, J.F. Stoddart, M.S. Tolley, *Nature*, **1994**, *369*, 133-137.

[12] C. P. Collier, G. Mattersteig, E. W. Wong, Y. Luo, K. Beverly, J. Sampaio, F. M. Raymo, J. F. Stoddart, J. R. Heath, *Science* **2000**, *289*, 1172-1175.

[13] A.G. Johnston, D.A. Leigh, R.J. Pritchard, M.D. Deegan, *Angew. Chem., Int. Edn. Engl.* **1995**, *34*, 1209-1212; For a remarkable related catenane system where the rings cannot rotate see F. Vögtle, T. Dünnwald, T, Schmidt, *Acc. Chem. Res.* **1996**, *29*, 451-460.

[14] A G Johnston, D A Leigh, L Nezhat, J P Smart and M D Deegan, *Angew Chem, Int Ed Engl*, **34**, 1212-1216 (1995).

[15] M Fanti, C-A Fustin, D A Leigh, A Murphy, P Rudolf, R Caudano, R Zamboni and F Zerbetto, *J Phys Chem A*, **102**, 5782-5788 (1998).

[16] A M Brouwer, W J Buma, R Caudano, M Fanti, C A Fustin, D A Leigh, A Murphy, P Rudolf, F Zerbetto and J M Zwier, *Chem Phys* **238**, 421-428 (1998).

[17] F Biscarini, W Gebauer, D DiDomenico, R Zamboni, J I Pascual, D A Leigh, A Murphy and D Tetard *Synthetic Metals* **102**, 1466-1467 (1999).

[18] C-A Fustin, D A Leigh, P Rudolf, D Timpel, F Zerbetto *ChemPhysChem*, 97-100 (2000).

[19] R Caciuffo, A Degli Esposti, M S Deleuze, D A Leigh, A Murphy, B Paci, S F Parker and F Zerbetto, *J Chem Phys*, **109**, 11094-11100 (1998).

[20] D A Leigh, S F Parker, D Timpel and F Zerbetto, *J. Chem. Phys.*, **114** , 5006-5011 (2001)

[21] F Biscarini, M Cavallini, D A Leigh, S León, S J Teat, J K Y Wong, F Zerbetto, *J. Am. Chem. Soc.*, **124**, 225-233 (2002).
[22] P Ceroni, D A Leigh, L Mottier, F Paolucci, S Roffia, D Tetard and F Zerbetto, *J Phys Chem B*, **103**, 10171-10179 (1999).
[23] D A Leigh, A Murphy, J P Smart, M S Deleuze and F Zerbetto, *J Am Chem Soc*, **120**, 6458-6467 (1998).
[24] C A Fustin, P Rudolf, A F Taminiaux, F Zerbetto, D A Leigh and R Caudano, *Thin Solid Films*, **329**, 321-325 (1998).
[25] F Biscarini, W Gebauer, D DiDomenico, R Zamboni, J I Pascual, D A Leigh, A Murphy and D Tetard *Synthetic Metals* **102**, 1466-1467 (1999).
[26] C De Nadai, C M Whelan, C Perollier, G Clarkson, D A Leigh, R Caudano, P Rudolf, *Surf Sci*, **454-456**, 112-117 (2000).
[27] C M Whelan, F Cecchet, G J Clarkson, D A Leigh, R Caudano and P Rudolf, *Surf Sci*, **474**, 71-80 (2001)
[28] C-A Fustin, R Gouttebaron, C De Nadaï, R Caudano, P Rudolf, F Zerbetto and D A Leigh, *Surf Sci*, **474**, 37-40 (2001)
[29] T Gase, D Grando, P A Chollet, F Kajzar, A Lorin, D Tetard and D A Leigh, *Photonics Sci News*, **3**, 16-21 (1998).
[30] T Gase, D Grando, P-A Chollet, F Kajzar, A Murphy and D A Leigh, *Adv Mater* **11**, 1303-1306 (1999).
[31] D Grando, T Gase, F Kajzar, M Fanti, F Zerbetto, A Murphy, D A Leigh *Mol. Crys. Liq. Crys.* **353**, 545-559 (2000).
[32] M Asakawa, G Brancato, M Fanti, D A Leigh, T Shimizu, A M Z Slawin, J K Y Wong, F Zerbetto, and S Zhang, *J. Am. Chem. Soc.*, **124**, 2939-2950 (2002).
[33] M S Deleuze, D A Leigh and F Zerbetto, *J Am Chem Soc*, **121**, 2364-2379 (1999).
[34] D A Leigh, A Troisi and F Zerbetto, *Angew Chem Int Ed*, **39**, 350-353 (2000).
[35] D A Leigh, A Troisi and F Zerbetto, *Chem. Eur. J.*, **7**, 1450-1454 (2001)
[36] D A Leigh, K Moody, J P Smart, K J Watson and A M Z Slawin, *Angew Chem, Int Ed Engl*, **35**, 306-310 (1996)
[37] T J Kidd, D A Leigh and A J Wilson, *J Am Chem Soc*, **121**, 1599-1600 (1999)
[38] A G Johnston, D A Leigh, A Murphy, J P Smart and M D Deegan *J Am Chem Soc* **118**, 10662-10663 (1996).
[39] D A Leigh, A Murphy, J P Smart and A M Z Slawin, *Angew Chem, Int Ed Engl,* **36**, 728-732 (1997).
[40] W Clegg, C Gimenez-Saiz, D A Leigh, A Murphy, A M Z Slawin and S J Teat, *J Am Chem Soc*, **121**, 4124-4129 (1999).
[41] F G Gatti, D A Leigh, S A Nepogodiev, A M Z Slawin, S J Teat and J K Y Wong, *J Am Chem Soc*, **123**, 5983-5989 (2001).
[42] A.S. Lane, D.A. Leigh, A. Murphy, *J. Am. Chem. Soc.* **1997**, *119*, 11092-11093.
[43] A.M. Brouwer, C. Frochot, F.G. Gatti, D.A. Leigh, L. Mottier, F. Paolucci, S. Roffia and G.W.H. Wurpel, *Science* **291**, 2124-2128 (2001).
[44] G W H Wurpel, A M Brouwer, I H M van Stokkum, A Farran and D A Leigh *J. Am. Chem. Soc.*, **123**, 11327-11328 (2001).

POLYMERIC β-DICARBONYL COMPOUNDS AS STRUCTURAL BASIS FOR INTELLIGENT SYSTEMS WITH MULTIFUNCTIONAL NATURE

Dr. I. PETKOV
University "St.Kliment Ohridski", Faculty of Chemistry, Department of Organic Chemistry, Laboratory of Organic Photochemistry
1 James Bourchier Blvd., 1126 Sofia, Bulgaria

The development of new materials for application in ***high-tech*** technologies first of all will involve the design and synthesis of molecules which possess the aforementioned properties, followed by incorporating these compounds into supramolecular assemblies, in order to obtain ordered molecular systems. The role of organic materials in this field has become increasingly important with main emphasis on dyes and polymers as photoactive components. The potential of organic molecular devices in the field of nanotechnology can hardly be exaggerated. This development is further stimulated by the notion that a large increase in information storage, theoretically to the molecular level, might be obtained. The multifunctional polymer systems, based on recent advances in new organic sensitive materials will beyond doubt play a key role in future (nano)-technology developments. These systems, with independent and different sensing, processing, and actuating functions, enhance performance capabilities. Such material systems are said to be **'Intelligent'**. One of the most important challenges in this field is the development of materials and techniques to place as much data as possible on the least amount of material. The ultimate goal would be to achieve information storage at the molecular or even at the atomic level, whereas processing of data should occur at close to the speed of light by the use of all optical switching devices. Future developments will aim at further increasing the rates of data storage and storage densities, for example by applying frequency doubled lasers, all optical writing/reading systems, and other materials. During recent years the design and synthesis of organic compounds for applications as materials has been a field of enormous growth. Organic molecules are now widely accepted as useful synthetic building blocks in non-linear optics, liquid crystals in modern displays, organic (super)conductors and ferromagnets, optical sensors and supramolecular structures, which are ordered molecular assemblies based on molecular recognition and self-organization, although inorganic solids still provide most of the materials for optical, electrical, magnetic and the desired properties of inorganic solids, for example, has led to a large increase in research activities towards the development of future organic materials. Some typical features of organic materials are: the ease of fabrication; the possibility to shape organic compounds into the desired structures by molecular engineering; the introduction and fine tuning of a large variety of physical properties by small changes in the

F. Charra et al. (eds.), Organic Nanophotonics, 57–71.

molecular structure; and the construction and characterization of simple isolated systems, which can provide solutions for fundamental problems. Disadvantages associated with stability and reliability of organic materials might be solved by structural changes.

On the other side, the typical properties of polymers like ease of processability, mechanical strength, long-term stability etc. combined with the above mentioned advantages of organic molecules will be the basis for a whole new range of molecular devices.

Light has many superior properties such as wavelength, polarization and coherency (ultra short pulse, focusing). In order to further increase the memory density in optical recording systems, so-called photon-mode optical memory employing photochemical reactions instead of thermal reactions should be developed. Photochromic systems employing polymers or Longmuir-Blodgett(LB) films have recently attracted much interest in view of their promising applicability to such purposes. The photochromism reported so far involves changes of chemical bonds such as ring opening and closure in spiropyrans or cis-trans isomerization in azobenzene[1-3].

New, rich possibilities give combination ***polymeric β-dicarbonyls/sensitive organic compounds***. The polymeric ***β*** -dicarbonyl part possesses keto-enol tautomerism. The equilibrium can be shifted under the influence of different actions. The process is very simple - transfer of X atom into the molecule(**Figure 1**).

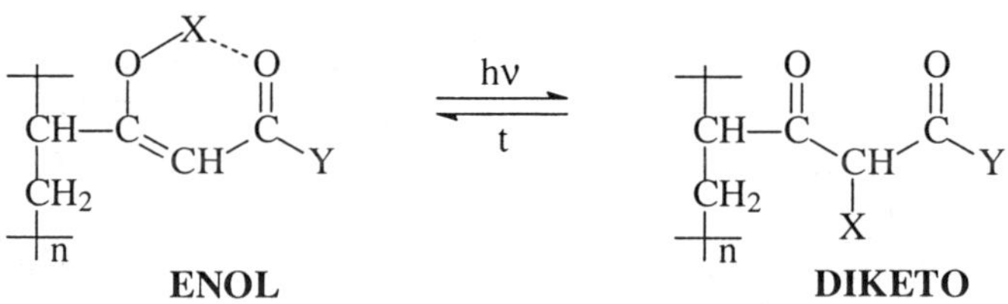

Figure 1.

The polymer β-dicarbonyl part can be modified in different structures with different functions. The combination of these structural characteristics and the multifunctional properties of the functional parts, connected with the β-dicarbonyl part can be base of the development of the sensitive multifunctional systems. In such systems can be realized several depending and independent processes:

1. Processes connected with the basis-polymeric chain: a) photoinduced proton or groups, cations, anion*s* transfer(keto-enol, enamin-ketimin, thiol-thione, enhydrazine-hydrazone, acylotropic-, cationotropic-, anionotropic tautomerism). b) photoinduced metallotropic tautomerism (photoinduced isomerization, pyrazolone derivatives) (**Figures 2-11**)[4-17].

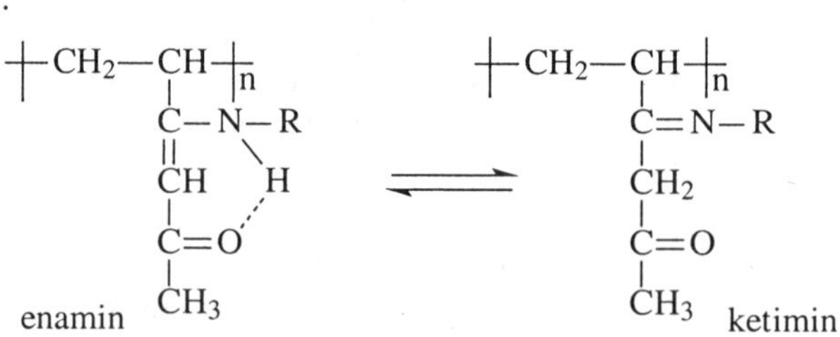

Figure 2. Enamin - ketimin modification

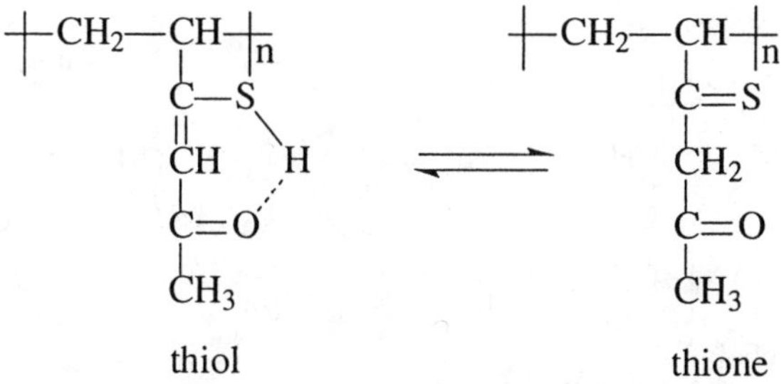

Figure 3. Thiol-thione modification

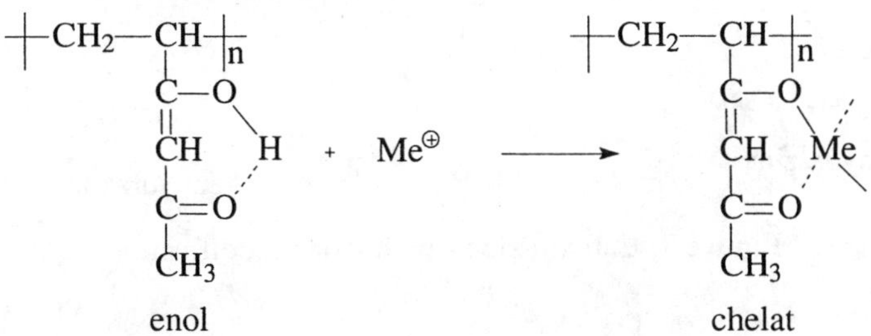

Figure 4. Metallotropic modification

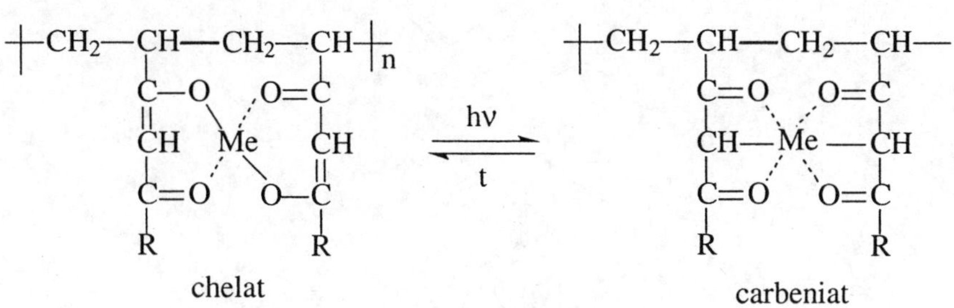

Figure 5. Metallotropic modification. Between two fragments in one polymer chain

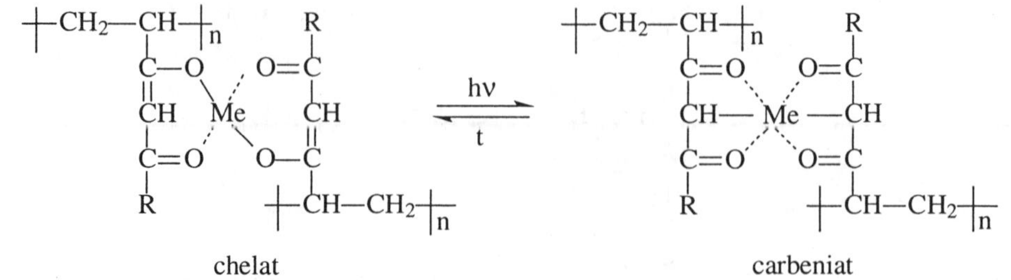

Figure 6. Metallotropic modification. Between fragments in different polymer chains

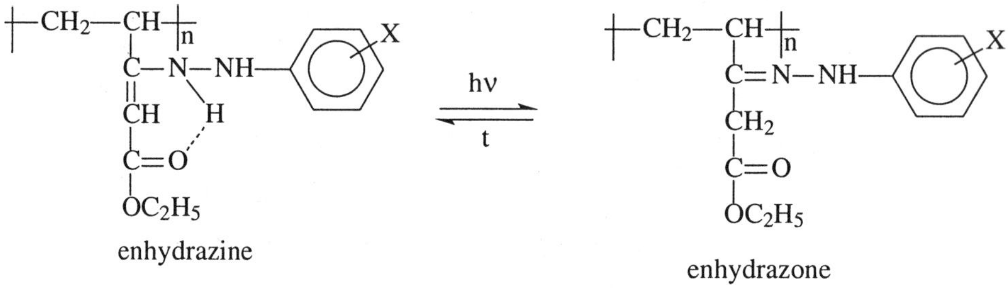

Figure 7. Enhydrazine - hydrazone modification

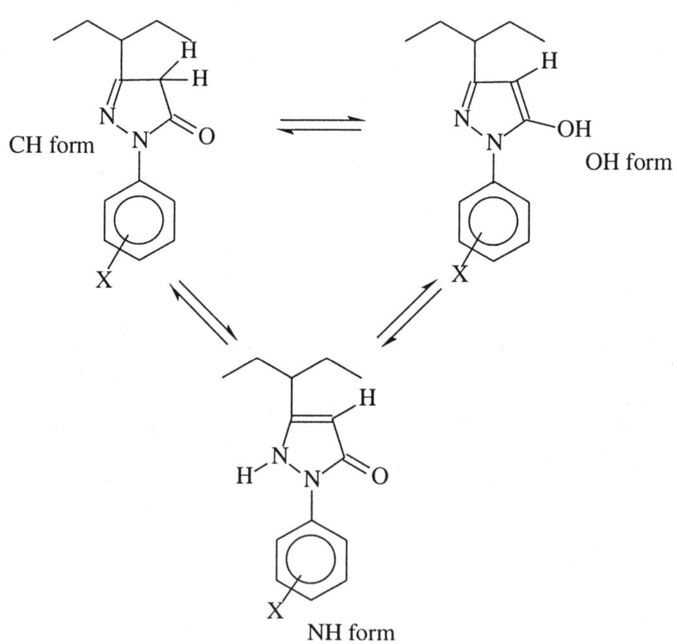

Figure 8. Polymeric pyrazolones

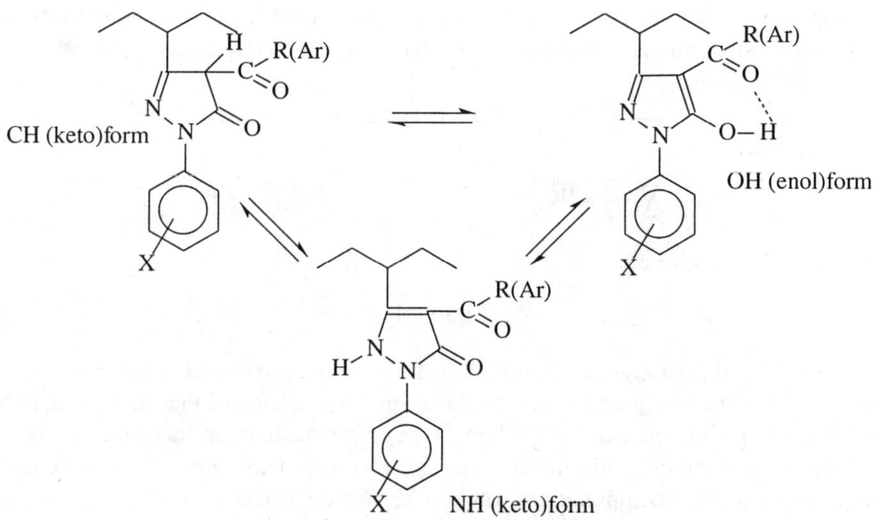

Figure 9. Polymeric acylpyrazolones

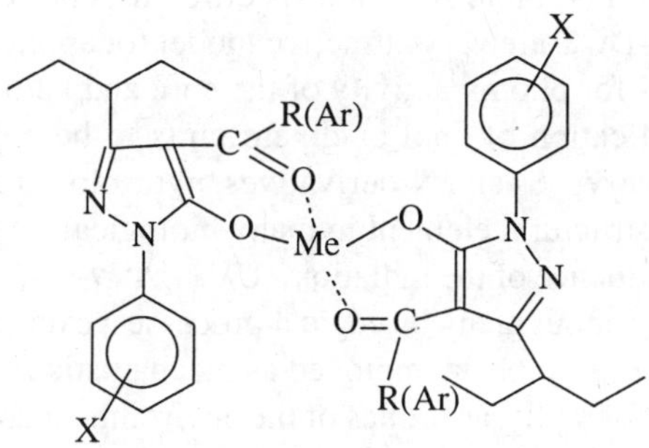

Figure 10. Polymeric acylpyrazolones. Chelate complexes

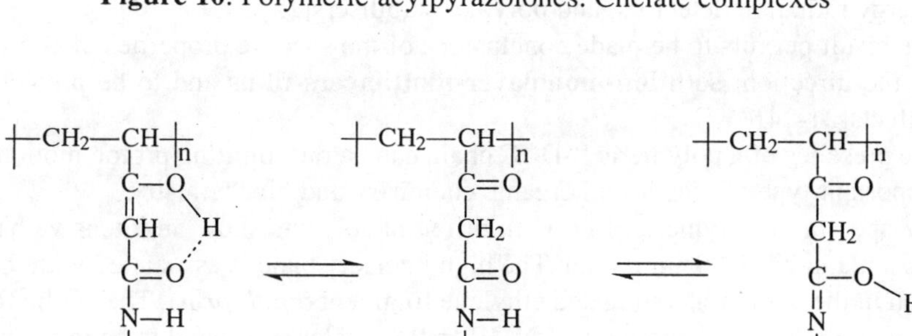

Figure 11. Polymeric anilides

The transfer between the tautomeric forms is reversible and can be realized many times: between two fragments(**Figure 12**) or between several fragments as cascade process.

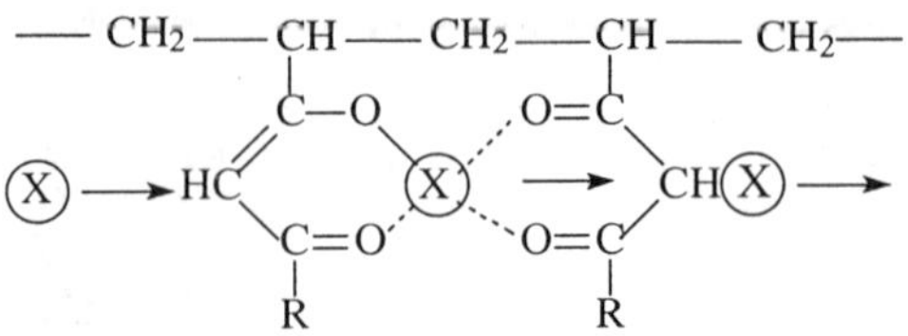

Figure 12.

Polymer β-dicarbonyl compounds can be examined as novel multifunctional systems for application in the different areas of the chemistry and the physics. The basic process under the influence of the UV light is the reversible photoinitiated tautomerization. The comparative study of the mechanism and kinetics of the photochemical enol-keto tautomerization of the compounds in solution, organized in a monolayer of closely packed monomer units and polymer films is made on the basis of the changes in the spectral behaviour of the system. The effects are results of the effective consecutive ***enol to keto*** photoconversion, increase of the area per unit during and slow interfacial reorganization of these forms to a more favorable state and specific influence of the media on the process[18-24]. Polymer β-DCC are very attractive model for application and investigation, because of: possibilities for photoreactivity of the enol and keto units; possibilities during the chemical modification of enol or diketo units to be received many functional polymers – metal complexes, S- and N-derivatives, pyrasolones and etc.; the presence of the β-DCC function as structural element in many biological important polymers permit to be investigated the influence of the radiation - UV-light, γ-ray, X-ray on their behaviour and to make conclusions about many biological processes; excited-state proton transfer - in this case: ***enol → keto*** have been employed as mechanistic tools and in technological applications in photolithography, as probes of the environment around proteins, micelles, β-cyclodextrins and films, in chemical lasers, energy storage devices at a molecular level, high-energy radiation detectors, and polymer stabilisers[25].

The result permits to be made conclusions of the specific properties of similar systems in the direction: **solution-monolayer-multilayers-films** and to be predicted unusual effects[21-24].

The presence of a polymeric β-DCC chain can be rate-limiting proton motion and it is an important type of reaction in organic chemistry and biochemistry.

UV-spectra of polymeric β-DCC in different solvents show an intensive bands in 245-260 nm and 270-300 nm region. The high intensity band is associated with the π-π^* transition in the carbonyl conjugated ethylene fragment (***enol form***). The enolic fraction in the polymers was measurement on NMR spectrometer and found to be in the interval from 34% to 96% depending on the structure. The absorption bands at ≈1725 cm^{-1} (***diketo***) and 1600 cm^{-1} (***enol***) are characteristics of the high degree of enolization – ***cis***-chelate form is dominant. There is not direct evidence for the presence of ***trans*** enol form in the

IR spectra. Under the influence of UV-light the absorption maximum at 250 nm (***enol*** form) disappear with the irradiation.There is a process of quiely restoration after the irradiation to the initial correlation keto-enol. The light-induced spectral changes are clearly due to a conversion: ***enol→ keto***.

2. Processes connected with the sensitive parts: *cis-trans-*, valencc-, open-closed-isomerization, electrochemical oxidation and reduction, photochemical synthesis of dyes, photo-, electro-, thermo-, halo-, radio-, piezochromism, conductivity.

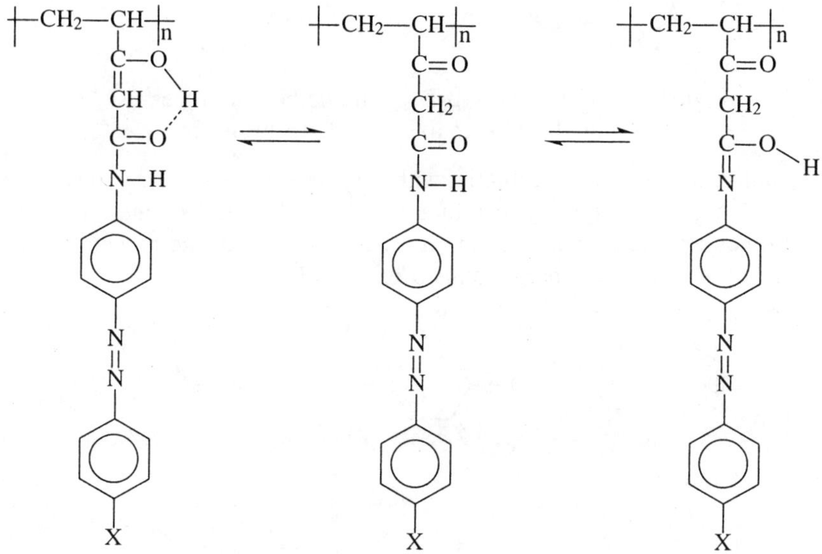

Figure 13. System poly(beta-DK)/azodyes.

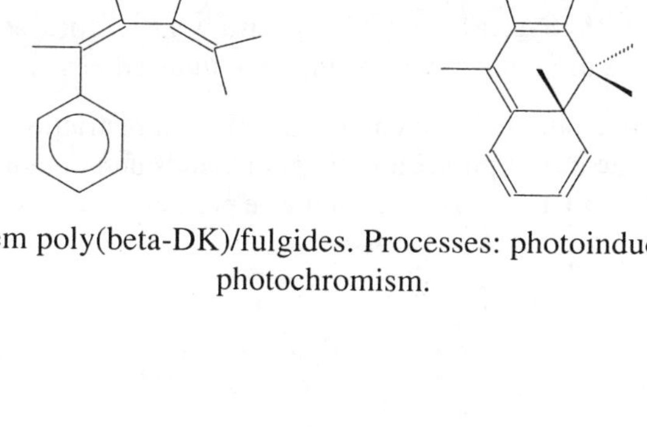

Figure 14. System poly(beta-DK)/fulgides. Processes: photoinduced ketonization, photochromism.

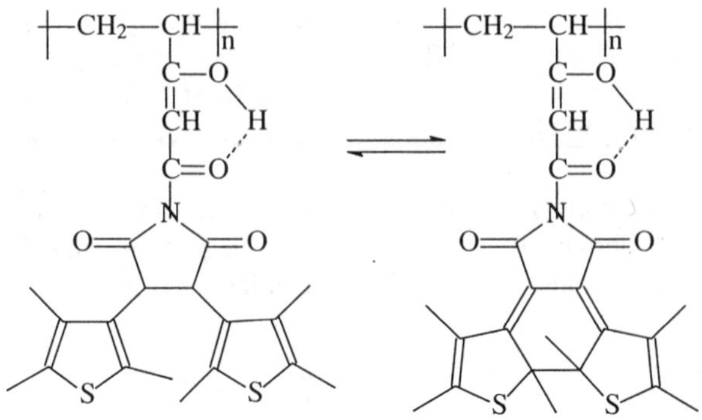

Figure 15. System poly(β-DK)/diarylethenes. Processes: photoinduced ketonization, photochromism

The combination of these possibilities with the photo-, radio-, electro-, piezo, thermo-, halochromic properties of spiropyrans, fulgides, diarylalkenes, azulenes, azodyes, stylbenes and other organic compounds and dyes can be base of the development of the sensitive multifunctional polymer systems (**Figures 13-16**) [26-31].

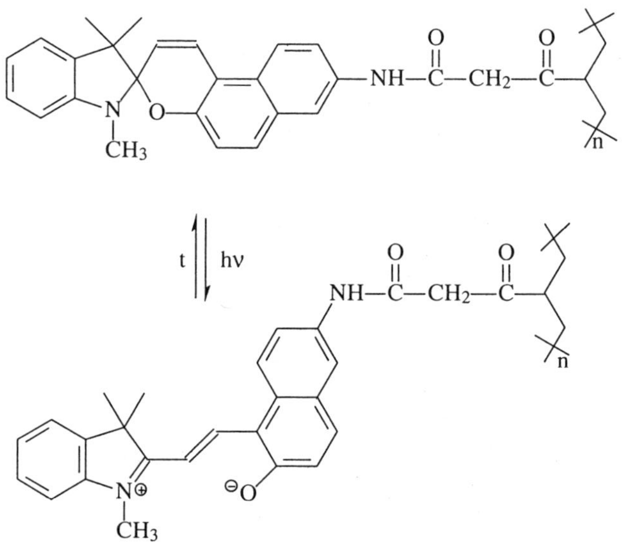

Figure 16. System poly(β-DK)/spiropyrans. Processes: photoinduced ketonization, photochromism, photoinduced *cis-trans* isomerization

Apart from economical, environmental and technical restrictions, the most important requirements for the use of photochromic compounds and organic dyes are: photochemical switching between the two forms should be possible; no thermal interconversion of the isomers should occur in a large temperature range(e.g.-20-80°C), allowing storage of information almost infinitely; the isomers should be fatigue resistant, with the possibility to perform the write/erase cycle many times; and no thermal or photochemical degradation to any side products should occur; both forms should be readily detectable; a non-destructive read-out procedure should be available; this read - out method should not interfere with or erase the written data.- high quantum yields should be achieved, allowing an efficient switching process and avoiding the necessity of long irradiation times – fast response times should be reached, leading to fast switching cycles; retention of all the properties is necessary when the switchable compound becomes part of a macromolecular structure.

3. Mixed processes – combination on the basis of antenna effect in the polymer systems. The experimental and theoretical data on the molecular reorientation and interaction under the radiation in various systems containing light absorbing chromophores showed that the most important factors, which control the phenomenon are chemical changes and the photoselection of the chromophores. Even in diluted solutions, they provide, respectively, the necessary rotational mobility of the chromophores and the consequent optical anisotropy of medium. In condensed phases(LB films, polymers, host-guest systems, sol-gel materials) the shape of the intermolecular interaction potential is also of paramount importance(**Figures 17-18**).

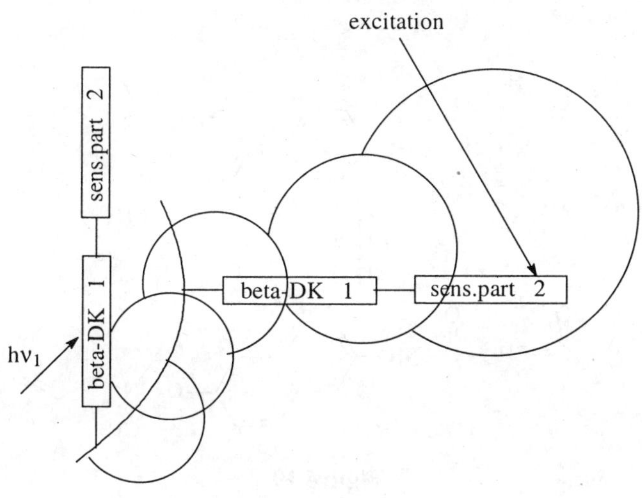

Figure 17. Antenna effect. Activity of polymer β-DK part.

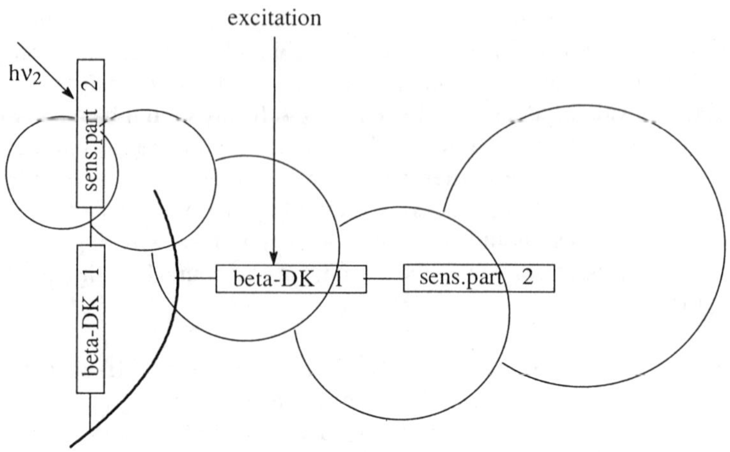

Figure 18. Antenna effect. Activity of sensitive part.

Model for application. Multifunctional polymer systems

<u>1 STAGE</u>. ACTIVITY OF POLYMER (β-DK) FRAGMENT (FIGURE 19).

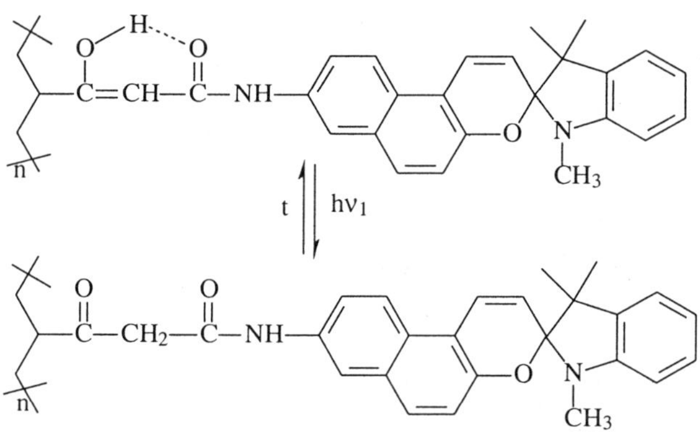

Figure 19

Processes: photoinduced proton transfer (keto-enol tautomerism, enamin-ketimin tautomerism, thiol-thione tautomerism, enhydrazine-hydrazone tautomerism); photoinduced transfer of groups, cations and anions (acylotropic tautomerism,cationotropic tautomerism, anionotropic tautomerism); photoinduced metallotropic tautomerism; photoinduced izomerization (pyrazolone derivatives).

2 STAGE. ACTIVITY OF THE SENSITIVE PARTS (FIGURE 20)

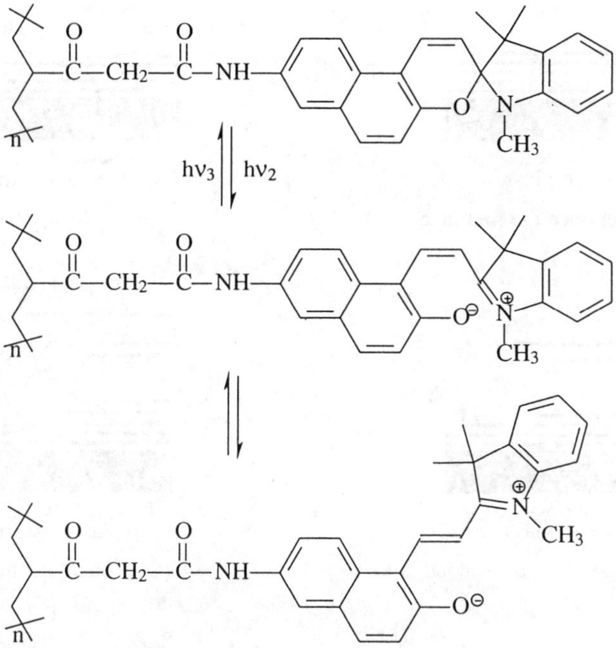

Figure 20.

Compounds: azo dyes, spiropyrans, fulgides, diarylethenes, azulenes, stilbenes, viologens, polymeric liquid crystals, donor-acceptor molecules, conducting polymers, host-guest system.

Processes: *cis-trans* izomerization, valence izomerization, open-closed izomerization, electrochemical oxidation and reduction, photochemical synthesis of dyes, photochromism, electrochromism, thermochromism, halochromism, radiochromism, piezochromism, conductivity.

Monolayer, bifunctional systems

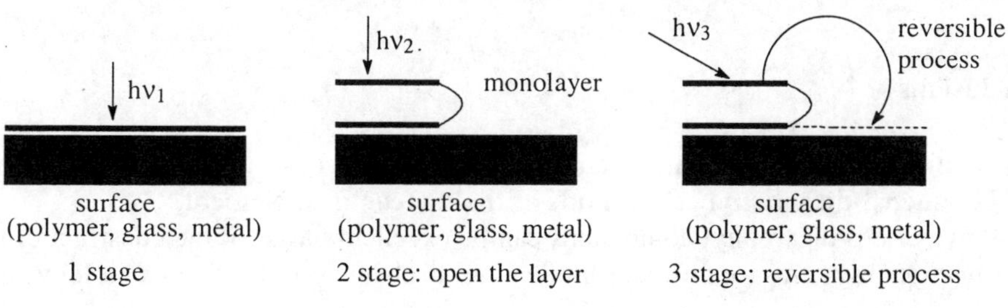

Multilayers, bifunctional systems

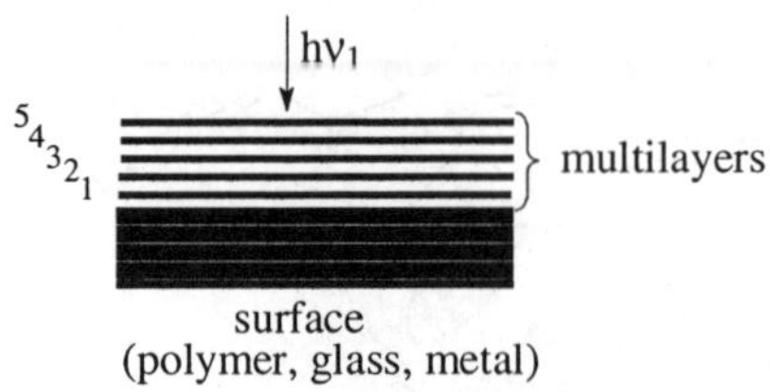

1 stage: open the first layer

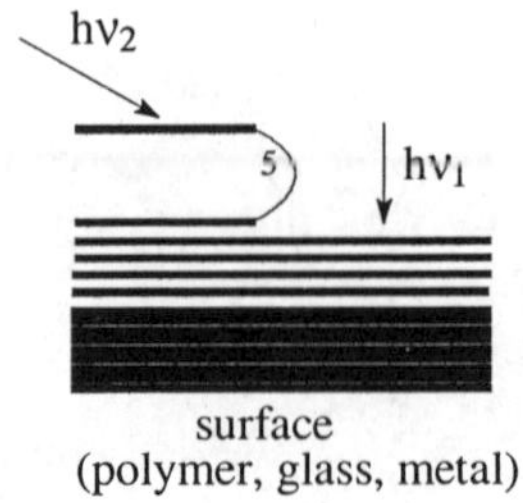

2 stage

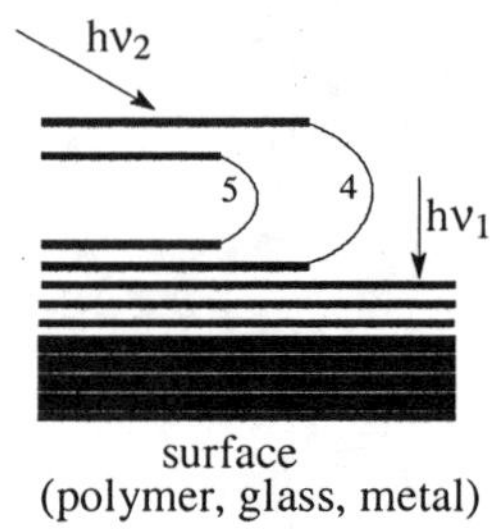

3 stage: open the second layer

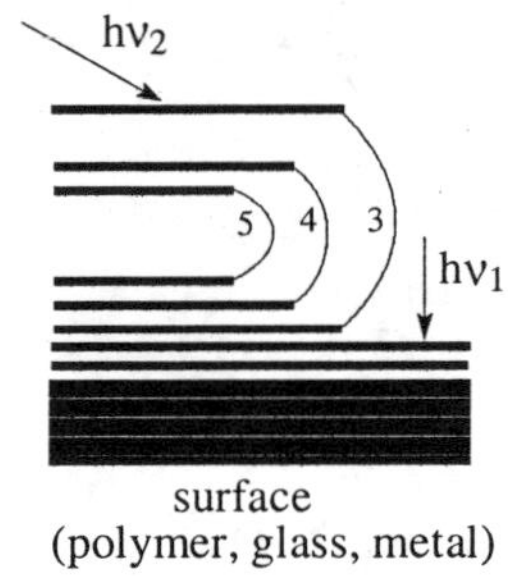

4 stage: open the third layer

Multilayers, multifunctional systems Compositions

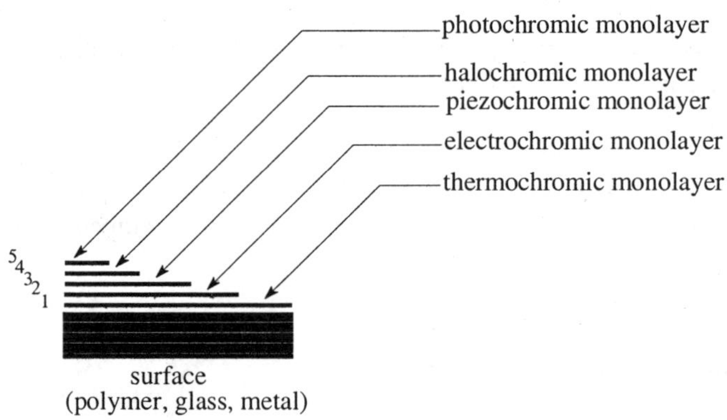

Conclusions

New multifunctional intelligent systems can be development on the basis of two parts:

– **Polymer β-dicarbonyl compounds as main structural element.**

Polymeric β-dicarbonyl compounds can be excellent basis for the construct of the multifunctional systems for different applications in chemistry, physics and biology. Specific interactions with different organic and inorganic compounds, metals permit to be developed devices on that base with big selectivity and sensitivity. The synthesis of the polymeric β-dicarbonyl compounds is simple chemical process. There is not need from heavy experimental equipment, difficult methods. Excellent possibility give the fact that can be used commercial polymers. There are many low molecular compounds, which can be incorporated into different medium. On the other side the polymer film or polymer fibers with these compounds will permit the construct of the easy analytical devices. Keto-enol equilibrium is mobile, the two functional groups are active and the system is multifunctional. On the basis of specific interactions they can be used for biochemical identification as high selectivity and sensitivity probes.

– **Modified with organic dyes and photochromic compounds β-dicarbonyl chains as sensitive parts.**

The main change in these materials will be the change of the colour. The presence of added colour or fluorescence will permit high level of selectivity and sensitivity. The investigations of the base groups of compounds permit to be directed attention to specific functional groups, specific compositions. The using of different ways for registration: spectral methods; conductivity; chemical methods; side reactions on the base of different kinds of chromism permit on the basis of the investigated systems to be build materials with multifunctional properties; the chemical reactions are simple and they give guarantee for high percentages of registration; can be realized orderly fragments, new compositions with new, unknown properties; the choice of the polymer can permit application in different area - from physics through chemistry to the biology and the medicine; can be eliminated some unwanted processes, which are characterization of the low molecular compounds; the price of the materials and the devices on their basis can be lower; can be used sol-gel method for the modified of the glass surfaces; can be prepared polymer, glass, carbon fibers, modification of paper.

REFERENCES

1. Malkin, J., Dvornikov, A. S., Straub, K. D., and Rentzepis, P. M. (1993) Photochemistry of molecular systems for optical 3D storage memory, *Research on Chemical Intermediates* **19**, 159-189.
2. Feringa, B. L., Jager, W. F., and Lange, B. (1993) Organic materials for reversible optical data storage, *Tetrahedron* **49**(37), 8267.
3. Lewis, T. W., and Wallace, G. G. (1997) The basis for development on intelligent materials, *Journal of Chemical Education* **74**(6), 703.
4. Markov, P., and Petkov, I. (1977). On the photosensitivity of dibenzoylmethane, benzoylacetone and ethyl benzoylacetate in solution, *Tetrahedron* **33**, 1013.
5. Markov, P., Petkov, I., and Jeglova, D. (1978) Photochemistry of enolizable β-dicarbonyl compounds: Study on the photoketonization of some esters of aroylacetic acids, *J. Photochem.* **8**, 277.
6. Markov, P., Petkov, I., et Ivanov, C. (1978) Evidence spectrale d'une cationotropie photo-activee dans des solutions des derives cuivres des esters acetylacetique et benzoylacetique.*C. R. Acad., Sc. Paris* **286**, ©, 505.

7. Markov, P., Petkov, I., and Ivanov, C. (1979) Influence of ultraviolet light on some copper chelates of β-dicarbonyl compounds. Spectral evidence for a photoactivated metallotropy, *J. Organomet. Chem.* **173**, 211.

8. Markov, P., Petkov, I., and Arnaudov, M. (1979) Influence of ultraviolet light on 3-aminocrotonic acid-ethylester and its N-substituted derivatives. Spectral evidence for a photoinduced hydrogen transfer, *Mol. Photochem.* **9**(3&4), 295.

9. Petkov, I., Markov, P. (1984) Photoinduced ene-hydrazine-hydrazone interconversion of N-substituted hydrazine derivatives of β-ketoesters in solution, *Comt. Rend. de l'Acad. Bulg. Sci.* **37**(8), 1051.

10. Yankov, P. D., Saltiel, S. M., Tomov, I. V., Markov, P. J., and Petkov, I. K. (1985) Measurements of the relaxation kinetics of photo-excited enol form of dibenzoyl-methane, *Laser Chem.* **5,** 107.

11. Yankov, P., Saltiel, S., Petkov, I., Markov, P. (1985) Study of the relaxation rate of the transient form of the photoexcited enol form of DBM, *Bulg. J. Phys.* **12**(6), 629.

12. Yankov, P., Saltiel, S., Petkov, I. (1986) Photoketonization of DBM in polar solvents, *Chem. Phys. Lett.* **128**(5,6), 517.

13. Yankov, P., Saltiel, S., Petkov, I. (1988) Photoketonization and excited state relaxation of dibenzoylmethane in non-polar solvents, *J. Photochem. and Photobiol. A. Chem.* **41**, 205.

14. Petkov, I., Dodov, N., Markov, P. (1990) Photoinduced tautomerism of some benzoylacetanilides in solution and polymer film, *J. Photochem. Photobiol. A: Chem.* **54**, 119.

15. Stojanov, St., Petkov, I., Antonov, L., Stoyanova, T., Karaginnidis, P., and Aslanidis, P. (1990) Spectral (UV-VIS) studies of the thione-thiol tautomeric equilibrium in 2- and 4-mercaptopyridines and 2-mercaptopyrimidines, *Can. J. Chem.* **68**, 1482.

16 Petkov, I., Stoyanov, S., Stoyanova, T., Karagiannidis, P., Akrivos, P. (1998) Studies of the photoinduced processes between thione, thiol forms Pof 2- and 4-mercapto-pyridines, 2-mercaptopyrimidines and their disulfides in solution, *Ann. de l'Universite de Sofia "St. Kl. Ohridski", Faculte de Chimie* **90**, 27.

17. Emelina, E. E., Ershov, B. A., Gribanov, A. V., Koltsov, A. I., Petkov, I., Petrov, A. A., Ushakova, I. L., Veber G., Zheglova D. Kh. (1999) Intramolecular hydrogen bond with two potential minima in some 1,3-keto enols in solutions and crystaline state, *J. Molecular Structure* **475**, 287-293.

18. Petkov, I., Sertova, N., Grigorov, L., and Masuda, S. (1995) Photoinduced ketonization of poly(ethylacryloylacetate) in solution, *J. Photochem. and Photobiol., A: Chemistry* **85**, 191.

19 Petkov, I., Masuda, S., Sertova, N., Grigorov, L. (1996) On the photosensitivity of poly(acryloylacetone) in solution, *J. Photochem. and Photobiol., A: Chemistry* **95**, 189.

20. Masuda, S., Sertova, N., Petkov, I. (1997) Photochemical behaviour of poly(ethylacryloylacetate) and poly(acryloylacetone) films, *J. Pol. Sci. Part A: Polym. Chem.* **35**, 3683-3688.

21. Balashev, K., Panaiotov, I., Ivanova, M., Petkov I., Proust, J. E., Boury, F., Masuda, S. (1997) Influence of pH and presence of Cu^{2+} ions on the properties of polyacryloylacetone (PAA) monolayers, *J. of Dispersion Science and Technology* **18**(6&7), 661-681.

22. Balashev, K., Proust, J. E., Boury, F., Petkov, I., Panaiotov, I., Masuda S. (1997) Comparative study of PAA - monolayers at Dichloromethane-Water and Air-Water interfaces, *Langmuir* **13**, 5362-5367.

23. Balashev K, Panaiotov I., Petkov I. (1998) Interfacial photochemical tautomerization in polyacryloylacetone monolayers, *Colloid&Polymer Sciences* **276**, 984-991.

24. Balashev, K., Panchev, N., Petkov, I., and Panaiotov, I. (2000) Photochemical behaviour of polyacryloylacetone and polyethylacryloylacetate monolayers at air-water interface, *Colloid&Polymer Sciences* **278** , 301-311.

25. Petkov, I., Sertova, N., Tersieva, A., and Ivanova, M. (1995) UV- and IR-study of γ-ray and fast electron beam initiated prototropic and metallotropic tautomerization of some β-dicarbonyl compounds and their Cu chelates in solution and in polymer (PVC) film, *Radiat. Phys. Chem.* **45**(1), 121.

26. Tchapkanov, A., Petkov, I. (1998) Spectral and photochemical behaviour of some pyrazolones in solutions, micelles and polymer films. *Analytical laboratory* **7**(3),

27. Sertova, N., Geffroy, B., Nunzi, J.-M., Petkov, I. (1998) PVC as photodonor of HCl for protonation of polyaniline, *J. Photochem. and Photobiol. A. Chemistry* **113,** 99-101.

28. Sertova, N., Petkov, I., Grigorov, L., Daiev, Ch., and Deligeorgiev, T. (1998) Photo-induced transport of HCl trough polymer (PVC) membranes, *ANALES DE QUIMICA International edition* **4-5,** 302-307.

29. Lessard, R. A., Lafond, Ch., Petkov, I., and Barachevsky, V. (1999) Use of Photo-chromic molecules as optical power limiters, *Nonlinear Optics* **21**, 115-124.

30. Philipova Tzvetanka, Petkov Ivan (1999) Synthesis and spectral properties of new 1,8-naphthalimide fluorophores applicable to modified epoxy oligomers. The Faculty of Technology and Metallurgy, Skopje, Republic of Macedonia, *Tom 1*, **1**, 115.

31. Petkov, I. (2000) Design and synthesis of photochromic systems, *Multiphoton and Light Driven Multielectron Processes in Organics:New Phenomena. Materials and Applications. NATO Science Series. 3. High Technology* **79**, 387-402.

CHARGE TRANSFER IN MOLECULAR LOGICAL DEVICE POSSESSING DIFFERENT MOLECULAR BRIDGES

J. TAMULIENE[a], M. L. BALEVICIUS[b], A. TAMULIS[a], V. TAMULIS[a]

[a]*Institute of Theoretical Physics and Astronomy, A. Gostauto 12, 2600 Vilnius, Lithuania*

[b]*Vilnius University, Sauletekio al. 9, III rumai, 2600 Vilnius, Lithuania*

ABSTRACT: The design applying the density functional theory time dependent (DFT-TD) method of molecular OR logical functions from the dithieno[3,2-b:2',3'-d]thiophene and azo-dye Disperse Orange 3 (DO3) molecules joined with various electron-insulator bridges are performed. The obtained results of the geometry optimization indicate the similarity of the spatial structures of these supermolecules, while the charge distribution in the various excited states are different. Additionally, the isomerization of the DO3 molecule is present despite this molecule connection to supermolecules. However, the twisting conformation should be obtained in the supermolecules consist on the above molecule joined with -N=N- or -CH_2-CH_2-bridge fragments. These results indicate influence of bridge fragments to properties of the supermolecules.

1. Introduction

Design and use of organic molecules for electronic circuits in signal processing is in initial stage now. It responds to the demands for miniaturization of electronic components, which are becoming ever smaller. However, it is not yet clear which of the variety of proposed new device technologies will ultimately prove most useful.

In the past few years the molecular chemical, photochemical and electrochemical switchings are designed [1]. Photorefractive materials are applied for optical applications including image processing, optical storage, programmable optical interconnects and simulation of neutral networks [2]. Molecular-level machines, based on pseudorotaxanes, rotaxanes and catenanes, have been proposed as possible components of chemical computers, and are currently being exploited for the construction of practical devices [3]. T. M. Tour et al. presented the several convergent synthetic routes to conjugated molecules with various potential digital device applications including a two-terminal molecular wire with a tunnel barrier, a molecular wire with a quantum well to serve as a resonant-tunneling diode (RTD), three -terminal systems with switch -like possibilities, and four-terminal systems that could serve as logical gates without the use of multipole transistors [4]. On the other hand, the above

F. Charra et al. (eds.), Organic Nanophotonics, 73–79.

molecular systems are constructed from the covalent attachment of donor and acceptor chromophores to rigid saturated bridges continue to provide important and fundamental insights into the characteristics of intermolecular electron transfer processes. Experimental studies on these systems have revealed that the bridges not only serve as a scaffolding that fixes the chromophores at well-defined separations and orientations with respect to each other, but they also play a crucial role in the mediation of the electron transfer process by a superexchange mechanism that could be applied in the design of molecular devices [5].

One of the advanced purposes in electron transfer is the design of molecular system that are capable of transducing photon energy. These supermolecules are designed from the photodonor and photoacceptor molecules joined with molecular electron-insulator bridges. Remarkable, the above molecular bridges should be chosen by such way that the quantum yield for the charge separation process would be as high as possible.

Thus, we intend to design the OR molecular logical function consist of the dithieno[3,2-b:2',3'-d]thiophene (thiophene) and azo-dye DO3 molecules joined by different molecular bridges in order to understand the above bridges influence to charge transfer processes and investigate the spectral properties of the designed compounds.

2. Results and Discussions

The design and investigation of the light-driven supermolecular OR logical devices based on the DO3 and thiophene molecules was performed applying DFT-TD B3PW91/6-31G model [6,7]. The above molecules are joined with following molecular bridges: -N=N-; -CH=CH-; $-CH_2-CH_2-$ and -CH=N- (Fig. 1).

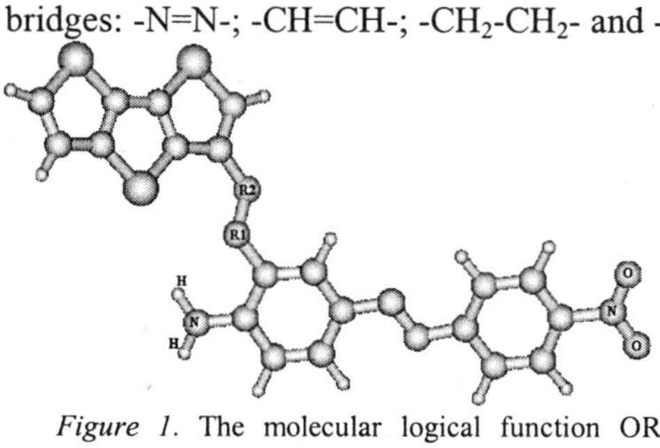

Figure 1. The molecular logical function OR designed from the DO3 and dithieno[3,2-b:2',3'-d]thiophene molecules where R1-R2 bonds correspond to molecular wires: -N=N-; -CH=CH-; $-CH_2-CH_2$ and -CH=N-.

TABLE I. The truth table of OR logical functions. In our investigated case the inputs are phenyl-NH_2 (I) and the thiophene molecule with bridge fragments (II), while output is phenyl-NO_2.

I input	*II input*	*Output*
0	0	0
1	0	1
0	1	1
1	1	1

Let us assume the phenyl ring of DO3 with the $-NO_2$ substitution and -N=N- bridge is output while the inputs are the other phenyl ring with the $-NH_2$ substitution (I) or the thiophene molecule with the bridge fragment (II). If any input is excited the additional photo-induced negative charge should be present on output. The combinations of all possible excitations and charge transfer should correspond to the results that is exhibited in Table I. Then the designed supermolecule is the OR logical device.

To show that our assumption is correct and to check the influence of the bridge attached thiophene and DO3, the charge distribution in the various excited states of the supermolecules were investigated.

It is well know that in the B3PW91 (the same in all generalized gradient approximation (GGA's) functionals) possesses some density inhomogeneity while PW91 retaining many of best features of local spin density (LSD) [8]. In this study the charge changes during excitation are obtained taking the atomic charge differences in the excited state and the ground state. Therefore, the obtained atomic charge changes describe electron excitation process qualitatively and allow us to compare electron density redistribution in the various excited states.

Firstly, it is necessary to pay attention to geometry search results that exhibit thiophene and the -CH=CH-, CH=N-, -CH_2-CH_2- bridge fragments in the same plane, while dihedral angle between that and DO3 is equal to 33.9 degrees. Meanwhile, the supermolecule consisting of thiophene and DO3 joined with -N=N- bridge is completely planar. This simple results confirm our assumption that the designed device input is thiophene within the bridge fragment, and the influence of the bridge fragment to the geometry of the supermolecule is exhibited also.

Before we proceed any further, let us introduce main results of our investigations. The essential point here is that the amount of transferred charge and its distribution are completely different in spite of similarity geometry that are confirmed by the results presented in Table II. It is evident, that the bridge fragment influents to charge transfer process appreciable.

In the case when the above molecules are joined with the -CH_2-CH_2- bridge the excitation by 424.49 nm (II excitation) and 314.10 nm (VII excitation) leads to charge transfer from the phenyl-NH_2 fragment to -N=N-phenyl-NO_2, while that is transferred from thiophene-CH_2-CH_2- fragment when the supermolecule is excited by 404.00 (III excitation) and 364.82 nm (V excitation) (Table II, Fig. 2). It is necessary to pay attention, that the localized charge on the output of this molecular device is large enough (more than 0.66) to be measured. However, the energy is absorbed mostly in the second excited state that lifetime is equal to 4.3 $*10^{-9}$ s. This clearly means, that experimentally observation of the additional charge on output is complicate enough. The lifetime of the VII excited state and the transferred charge amount are larger than that of the II one, what results this excitation as more reasonable to be apply.

The lifetime of the III and V states is comparable with that of the VII excited state and the transferred charge amount is equal to 0.8 and 0.9 (Table II.) Hence, these states are acceptable to adopt for OR logical functions.

In keeping with this described results one can state that non-conjugated bridge -CH_2-CH_2- should be applied to join thiophene and DO3 in order to design supermolecule that could be applied as OR logical function. It is evident that the 314.10 nm and 404.00 nm (or 364.82 nm) wave length serve the excitation of the above device that charge is transferred from the inputs and measured on output.

It is interesting that the joining of the above molecules via -CH=CH- bridge is not suitable because the charge that could be obtained experimentally is mostly transferred from thiophene with this bridge (Table I). It is not surprise, because the thiophene molecule is better photodonor than phenyl-NH_2. It implies, that attachment with the -CH=CH-bridge fragment should be possible if the thiophene molecule is changed, i. e. instead this photodonor should be joined the other molecule which molecular orbitals are closer to that of phenyl-NH_2 fragment, or the phenyl-NH_2 fragment should be changed with respect to the thiophene molecule also.

The bridge of -N=CH- is not suitable to join the thiophene molecule and DO3 to supermolecule that could be applied as OR logical device. In the case of this supermolecule the 455.98 nm (II excitation) or 356.78 nm (VII) and 387.73 nm (III) excitations should be present (Table II). In the II state the charge transfers from phenyl-NH_2 fragment mostly while that is from thiophene during the III and VII excitations. However, the lifetime of the II excited state could complicate to observe the additional charge as it was obtained in the case with the -CH_2-CH_2- bridge fragment.

TABLE II. The difference between charge located on the various fragments of the investigated molecules in excited states and the ground state. The results are obtained applying B3PW91 TD method within the 6-31G basis set .

State	-CH_2-CH_2-				-CH=N-			
	thiophene	*Ph-NH₂*	*Ph-NO₂*	*Lifetime (s)*	*thiophene*	*Ph-NH₂*	*Ph-NO₂*	*Lifetime (s)*
I	0	-0.08	0.08		0	-0.08	0.08	
II	0.09	0.56	-0.66	$4.30*10^{-9}$	0.18	0.49	-0.67	$4.78*10^{-9}$
III	0.92	-0.08	-0.84	$4.53*10^{-8}$	0.83	-0.04	-0.8	$2.07*10^{-8}$
IV	0	0	0		-0.51	0.43	0.08	$1.21*10^{-8}$
V	0.98	-0.12	-0.86	$2.63*10^{-7}$	-0.17	-0.09	0.26	$1.20*10^{-6}$
VI	0	-0.03	0.03		-0.51	-0.12	0.63	$1.10*10^{-6}$
VII	0.09	0.63	-0.72	$2.40*10^{-8}$	0.83	-0.06	-0.82	$3.21*10^{-8}$

State	-CH=CH-				-N=N-			
	thiophene	*Ph-NH₂*	*Ph-NO₂*	*Lifetime (s)*	*thiophene*	*Ph-NH*	*Ph-NO₂*	*Lifetime (s)*
I	0	-0.08	0.1		-0.08	-0.07	0.08	
II	0.56	0.18	-0.75	$1.33*10^{-8}$	0.15	-0.12	-0.02	
III	0.61	0.1	-0.71	$6.47*10^{-9}$	0.12	0.46	-0.58	$4.20*10^{-9}$
IV	0.67	0.05	-0.73	$7.80*10^{-9}$	-0.64	-0.25	0.9	
V	0.08	0.04	-0.12	$2.44*10^{-7}$	0.89	-0.1	-0.79	
VI	0.43	0.18	-0.61	$1.05*10^{-10}$	-0.17	0.28	-0.11	$1.32*10^{-8}$
VII	0.03	0.24	0.27	$1.17*10^{-8}$	0.57	0.1	-0.66	$1.39*10^{-8}$

Furthermore, in the above case the several excitation when the charge is transferred from the phenyl-NO_2 fragment take place also what is not observed in the cases of the -CH_2-CH_2-and -CH=CH- bridges. This clearly confirm the influence of the bridge fragments to the charge separation process.

A most interesting situation arises when the thiophene and DO3 molecules are joined with -N=N- bridge. There are excitations (III, V, VII) when charge is transferred from the thiophene -N=N- to the -N=N-phenyl-NO_2 fragment and one (IV) when that is transferred vice versa. Remarkable, that in this case the transfer charge amount is comparable with that of the supermolecules within -CH_2-CH_2- bridge. In keeping with this results, we can state that -N=N- bridge is not adopted for connection thiophene and DO3 to design OR device. A very interesting, and for the bridge influence to charge transfer exhibition quite relevant generalization follows if the transferred charge distribution are investigated (Fig. 2). In the I excited state the charge is transfer from the -N=N-bridge fragment of DO3 molecule while in the V excited state when one is transfer from the other bridge fragment.

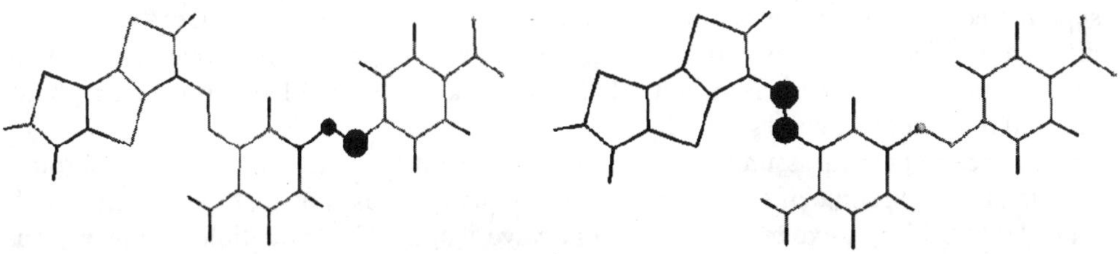

Figure 2 . The transferred charge distribution in the I and V excited states of the supermolecule consist on thiophene and DO3 joined with the bridge -N=N- fragment. In the first excited state the charge is transferred from the bridge that is joined phenyl ring of DO3 molecule, while that in the V excited state is transferred from the bridge that joined thiophene and DO3.

Furthermore, according to our investigations the different molecular bridges results to the blue shift of the absorption spectra in comparison with that of DO3 measured in hexane solution [9] (Fig 3). So, these simple examples also confirm that the bridge fragment provide important and fundamental insights into the characteristics of intermolecular electron transfer processes.

However, it is important that the first forbidden $\pi \rightarrow \pi^*$ type excitation of this device does not depend on molecular bridge.

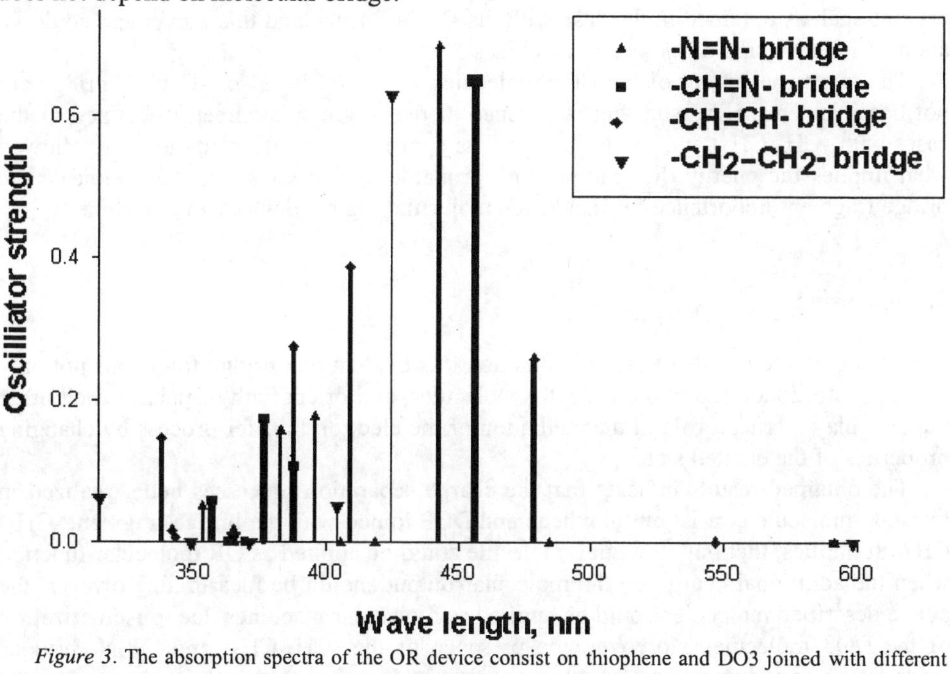

Figure 3. The absorption spectra of the OR device consist on thiophene and DO3 joined with different bridge fragment. This spectra is obtained applying B3PW91 TD method within the 6-31G basis set

It is interesting, that in all investigated cases in the I excited state the negative charge is transferred from the -N=N- molecular bridge that joins the phenyl rings of the DO3 molecule fragment. The excitation and similar charge distribution is obtained in the isolated DO3 molecule [9]. Additionally, the second excited state (III-th in the supermolecule with two -N=N- bridges) is the most intensive as is obtained in the isolated DO3 molecule also. Thus, the presence of these states of the investigated supermolecules leads to possibility of the *trans↔cis* isomerization of DO3 despite the different isomerization ways [10]. The results allow us to predict the movement of these supermolecules on the surface and all the supermolecule could be applied as photo-induced molecular machines [11,12].

Another important assuming arises from the result investigations of the absorption spectra and the lifetime of states. As example, let us again assume that the supermolecule with -N=N-bridge is excited by 395.51 nm wave length. The excitation is allowed, and lifetime time of that is $1.32*10^{-8}$s. Thus, the supermolecule should be gone to the lower excited state (as example V or IV) that are relatively long living. It implies, that in the above states, the energy is dissipated due to rotation and vibration, and the movement of the supermolecule due to presence of the bridge attached thiophene and DO3 should be present. In keeping with this results it is possible foresee two different path-ways of movement of the above mentioned molecule. When the molecule is excited by the wave length smaller than 442.39 nm the movement of the molecule is possible due to *trans↔cis* isomerization of DO3, while the excitation by more than 442.39 nm leads to transformation of the thiophene -N=N-DO3 molecule.

The similar situation is obtained in the cases when the above thiophene and DO3 molecules are joined with -CH_2-CH_2- bridge. It is allow us to foresee that the supermolecules with the -CH_2-CH_2- bridge fragments could differently move due to reasons such as obtained in the case with the -N=N- bridge and this movement could be adopted in the design of molecular machines.

The twice movement of the supermolecules within -CH=CH-or -CH=N- bridge are not possible. One of reason that is absence of the fragment symmetry obtained in the cases with -CH_2-CH_2- and -N=N- bridge thus approximately all excitations are allowed what implies the energy dissipation more probable due to emission. That indicate the bridge fragment importance for the design molecular logical devices and machines.

3. Conclusion

Basing on the results of our investigation it is evident that bridge fragments not only fix the photo donor and photo acceptor molecules, and orient with respect to each other but also play a crucial role in the mediation of the electron transfer process by changing properties of the excited state.

The obtained results indicate that the charge separation processes better realized in the supermolecule consist on thiophene and DO3 joined with the bridge fragment -CH_2-CH_2-. It implies, that only this supermolecule could be applied as OR molecular function when the additional charge on the molecular output should be measured. However, the above described molecules could be applied as molecular machines due to isomerization of the DO3 molecule. Moreover, the presence of the -CH_2-CH_2- and -N=N- bridges leads to the two different transformations that realize due to the above isomerization, vibration and rotation around these bridges.

4. Acknowledgement

Authors thanks Poznan's Supercomputing and Networking Center for allowance to perform calculations with installed Gaussian 98 A.7 package and Prof. A. Graja for the assistance.

REFERENCES:

1. Balzani, V., Gomez-Lopez, M., Fraser Stoddart, J. (1998) Molecular Machines , Acc. Chem. Res., **31**, 405-414.

2. P. Cheben, P., del Monte, F., Worsfold, D. J, Carlsson, D. J., Grover, C. P, Mackenzie, J. D. (2000) A photorefractive organically modified silica glass with high optical gain, *Nature*, **408**, 64-67.

3. Balzani,V., Credi, A., Marchioni, F., Stoddart, J. F. (2001) Artificial molecular - level machines. Dethreading-rethreading of a pseudorotaxane powered exclusively by light energy, *Chem. Commun.*, 1860-1861.

4. Tour, J. M., Kozaki, M., Seminario, J. (1998) Molecular scale electronics: a synthetic/computational Approach to digital computing, *J. Am. Chem. Soc.*, **120**, 8486-8493.

5. Jollife, K. A, Langford, S. J., Oliver, A. M., Shepard, M. J., Paddon-Row, M. N. (1999) A new class of giant tetrads for studying aspects of long-range intermolecular electron transfer processes;synthesis and computational studies, *Chem. Eur. J.*, **5**, 2518-2529.

6. Heinze, H. H., Gorling, A., Rosch, N. (2000) An efficient method for calculating molecular excitation energies by time-dependent density-functional theory, *J. Chem. Phys.*, **113**, 2088-2099.

7. Perdew, J. P., Burke, K., Ernzerhof, M. (1996) Generalized Gradient Approximation Made Simple, *Phys. Rev. Lett.*, **77**, 3865.

8. Foresman, J. B., Frish, A. *Exploring Chemistry with Electronic Structure Methods*, (Gaussian, Inc., Pittsburgh, PA, USA 1995-96)

9. Private talk with prof. J. M. Nunzi.

10. Balevicius, M. L., Tamuliene, J., Tamulis, A. **(**2000) Role of excited states on cis-trans isomerization of disperse orange -3 molecule, *Lithuanian Journal of Physics (Lietuvos Fizikos Zurnalas)*, **40**,. 387-393.

11. Balevicius, M. L., Tamuliene, J., Tamulis, A. (2001) Influence of various substituents for isomerization processes of Azo-Dye,, *Lithuanian Journal of Physics (Lietuvos Fizikos Zurnalas)*, **41**,83-88.

12. Natansohn, A., Rochon, P., Meng, X., Barrett, C., Buffeteau, T., Bonenfant, S., Pezolet, M., (1998) Molecular Addressing Selective Photoinduced Cooperative Motion of Polar ester groups in copolymers containing Azobenzene groups, *Macromolecules*, **31**, 1155-1161.

TUNNELLING INDUCED FLUORESCENCE AS A PROBE OF ELECTROMAGNETIC INTERACTION AT NANOMETRE PROXIMITY

GERMAR HOFFMANN
Institut für Experimentelle und Angewandte Physik,
Christian-Albrechts-Universität zu Kiel,
D-24098 Kiel, Germany

JAVIER AIZPURUA
National Institute of Standards and Technology,
Gaithersburg, Maryland 20899-8423, USA

S. PETER APELL
University Outreach, Kristianstad University,
SE-291 88 Kristianstad, Sweden

RICHARD BERNDT
Institut für Experimentelle und Angewandte Physik,
Christian-Albrechts-Universität zu Kiel,
D-24098 Kiel, Germany

1. Abstract

Fluorescence induced by the tunnelling current of a scanning tunnelling microscope is used to investigate the electromagnetic coupling of a metal tip and a metal sample. Sub-atomic scale modifications of the tunnelling junction geometry cause spectral shifts of the fluorescence. Such shifts are observed when the tip is vertically displaced relative to a flat sample surface. Similarly, monatomic sample steps close to the tip position shift the fluorescence. These experimental results are consistent with model calculations of the electromagnetic response of an appropriate tip-sample geometry. We find that for sharp tips the electromagnetic coupling of the tip and the sample is confined to a lateral range of a few nm.

F. Charra et al. (eds.), Organic Nanophotonics, 81–91.

2. Introduction

The electromagnetic interaction of objects in nanometre proximity plays a central role in surface enhanced Raman scattering [1] and it is vital to scanning probe techniques such as non-contact atomic force microscopy [2] or scanning near-field optical microscopy [3]. In these microscopies, the electromagnetic coupling of the tip and a sample leads to the formation of new, localized electromagnetic modes. In the scanning tunnelling microscope (STM), these modes can be excited by inelastic electron tunnelling and give rise to the emission of characteristic fluorescence [4, 5, 6, 7, 8, 9]. Owing to the ability of the STM to precisely control the vertical and lateral position of the tip spectral analysis of this fluorescence can be used to study the electromagnetic coupling on an atomic scale.

Here, we report on such experiments, first described in Ref. [10]. In the vicinity of a monoatomic step we observe gradual shifts of fluorescence spectra when crossing the step in lateral direction. Related shifts occur when the tip is retracted from an atomically flat metallic surface. These results are quantitatively compared to model calculation of the electromagnetic coupling where the tip is assumed to be a revolution hyperboloid with a radius of curvature of a few nm. From these analyses the lateral extension of the coupled surface plasmon mode, which is induced at the cavity by the proximity of the tip, is estimated. Understanding the implications of sub-atomic modifications of the cavity in these simple cases may give us a better approach for topographically more complex systems such as adsorbed molecules.

3. Instrumentation

The experiments were performed with a custom built STM in ultra-high vacuum (UHV) at low temperature (5 K). A lens system collects and transfers the light to a grating spectrometer and finally to a CCD camera [12]. Au and Ag (111)-surfaces as well as Au and Ag-covered W tips were prepared in UHV by standard Ar-ion sputtering and annealing cycles. Surface structures were verified before and after all experiments with STM to rule out surface modifications

4. Experimental observations

Figure 1b shows a fluorescence spectrum recorded from a defect-free Ag(111) surface at a distance of 55 Å from the next step. Circles indicate raw data, dots are corrected for the detection efficiency which is shown separately in Figure 1a. The acceptance angle, chromatic aberrations, reflectivities of

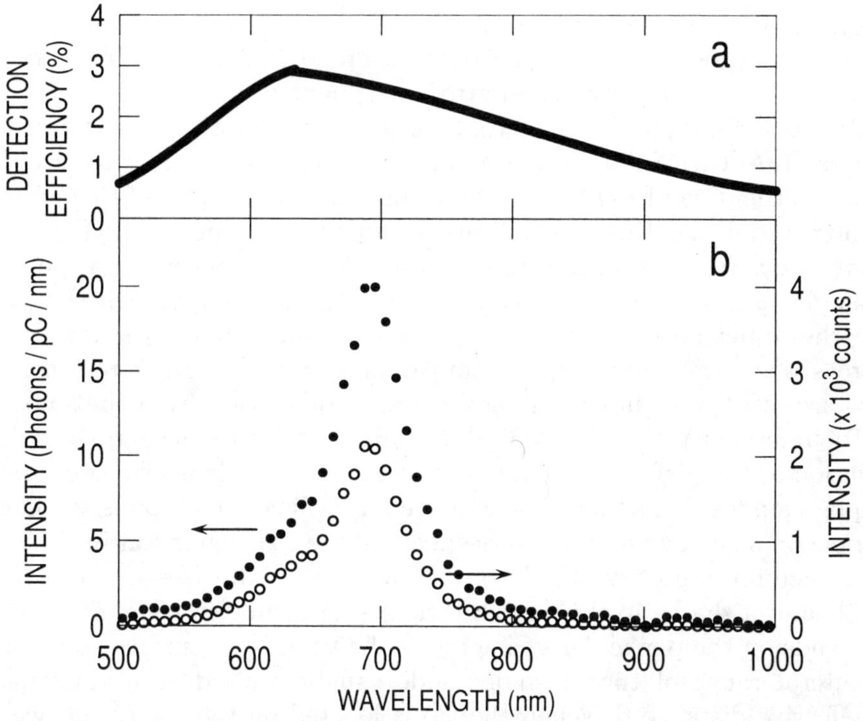

Figure 1. (a) Wavelength dependent detection efficiency of the optical setup. (b) Fluorescence spectrum of a flat Ag(111) surface and a Ag-covered W tip at a sample voltage $V = 2.9$ V and a tunnelling current $I = 5$ nA as measured in 0.1 seconds (circles) and corrected for the detection efficiency (dots).

the optical setup as well as the detection efficiency of the CCD camera are taken into account [15]. An intensity peak occurs at a wavelength $\lambda \approx 690$ nm with a broad tail to the short wavelength side. The spectrum is similar to previously reported and calculated spectra from Ag tips on Ag samples [6, 13]. It is due to the plasmon resonance for the given tip-sample distance and tip shape. We note that $V = 2.9$ V was chosen sufficiently large to have the peak fully developed.

At this point it is useful to introduce the definition of a mean emission wavelength $\overline{\lambda} = \int \lambda P(\lambda)\, d\lambda / (\int P(\lambda)\, d\lambda)$, where P is the spectral density of the measured fluorescence. Using this definition, the wavelength resolution of a shift is improved. This procedure is acceptable since the spectral shape does not change significantly over the range of distances probed here. Uncertainties due to counting statistics are indicated in the following graphs by error bars. The offset between the peak position and $\overline{\lambda}$ is ≈ -20 nm. The uncertainty of the peak position is the root mean square or combined standard uncertainty of the statistical uncertainty (less than ± 1 nm), and the systematic uncertainty (± 4 nm).

Figure 2 shows experimental data recorded from a monoatomic step on Ag(111). This step was chosen to be straight over a distance of ≈ 300 Å. The solid line represents a cross-sectional profile of a constant current topograph. At each STM image point a fluorescence spectrum was recorded, similar to the data of Fig. 1. From the spectra overall shifts of the emission are observed while the spectral shape remains unaffected. In Figure 2b (dots) we evaluate the shift of the mean wavelength (relative to a spectrum on the flat terrace) versus the tip position. As the tip approaches the step from the upper atomic terrace, the emission shifts to shorter wavelength whereas a red-shift occurs when approaching from the lower terrace. The shifts are discernible up to a lateral distance of ≈ 50 Å from the step.

To understand this spectral shift it is useful to consider the lateral extension of the surface mode induced at the tip-sample cavity and the tip-sample distance d. The lowest-energy mode, which corresponds to charges of opposite signs on the boundaries, causes the light emission in STM [5]. For a vacuum gap of width d between a semi-infinite free-electron metal sample and a sharp tip this mode disperses according to $\hbar\omega \sim d^{1/n}$, where n depends on the particular geometry of the tip [17]. A surface step within the lateral range of the plasmon mode can be viewed as a variation of the effective distance d. When the tip is located on the upper terrace the distance between tip and sample averaged over the lateral extension of the plasmon is effectively increased yielding a blue-shift. When the tip approaches the step the tip-sample distance effectively increases further as well as the blue-shift. This shift of the emission peak is a consequence of the interaction of the surface mode with the step. Estimating the extension of the localized mode as the distance where the shift becomes noticeable we find a value of ≈ 50 Å. This value is consistent with previous model calculations for the extension of the mode at the cavity [16].

The above concept of an effective tip-sample d can be tested experimentally. To this end we varied d over a flat area of a Au(111) sample, i.e., in absence of any steps, and simultaneously recorded some 680 fluorescence spectra. The spectra are similar to previously reported spectra of this material combination [13, 18]. The integrated intensity vs. the current is plotted in Figure 3a. As expected, an exponential variation is found. Figure 3b displays the shift of mean emission wavelength $\overline{\lambda}$ versus the tunnelling current and the associated tip excursion. An approximately linear blue-shift of $\overline{\lambda}$ with the tip excursion Δd is observed with a slope of 3 nm/Å.

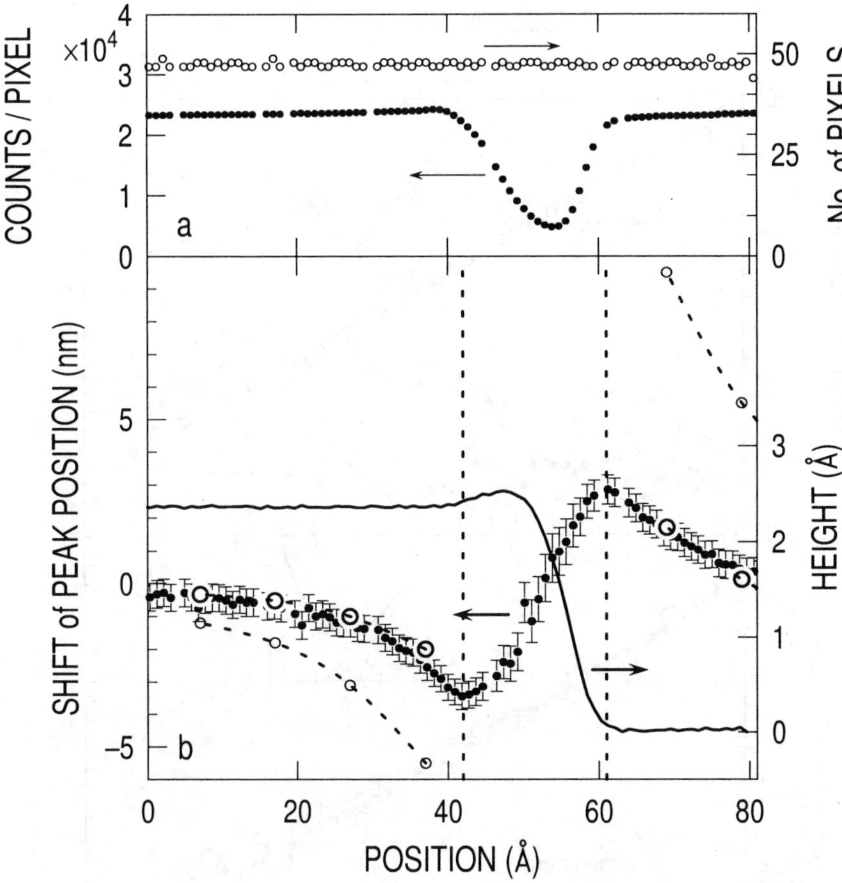

Figure 2. (a) Profile of the integrated intensity (dots) across a monoatomic Ag(111) step. To improve signal-to-noise ratio spectra from consecutive linescans recorded at identical distances from the step were integrated. Since the step was not exactly perpendicular to the scan direction each linescan is shifted by a fraction of pixel relative to the previous scan. Therefore, the average intensity at each data point and the number of pixels used per data point are indicated. Data points polluted by cosmic rays are removed. Directly at the step edge the intensity is reduced. (b) Cross-sectional profile of the constant current topograph (line), along with the shifts of the mean wavelength (dots with error bars) from fluorescence spectra recorded at each STM image point. Model calculations for an axially symmetric step (hole and island) are indicated by small circles ($\Delta\lambda_{\rm island}$). Corrected values for a one dimensional step are plotted as large circles ($\Delta\lambda_{\rm step}$). Vertical dashed lines delimit the immediate proximity of the step where significant vertical tip motion occurs.

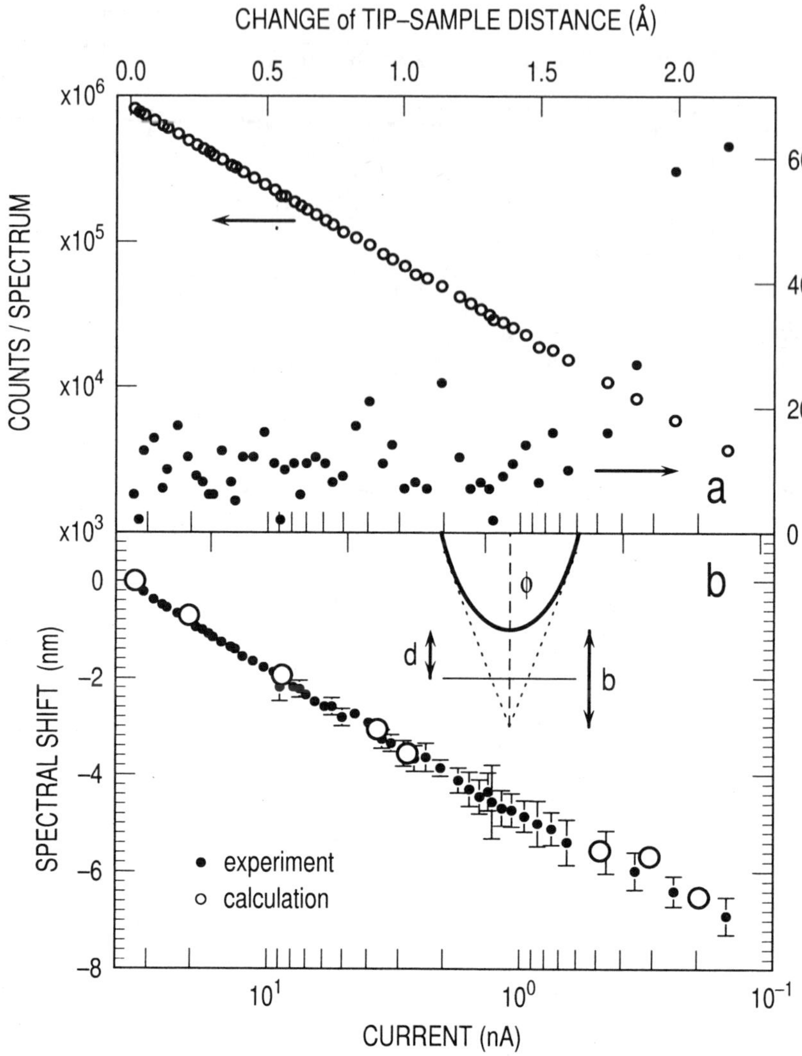

Figure 3. (a) Total intensity per spectrum vs. current (circles) and number of spectra used for averaging (dots). (b) Shift of mean emission wavelength $\overline{\lambda}$ vs. tunnelling current and associated vertical tip displacement, as measured (dots) and calculated (circles). The inset shows the tip geometry used for modelling the data.

5. Numerical Calculations

The experimental data can be analysed further by numerical model calculations within the model of Ref. [16]. In this model the tip is represented by a hyperboloid with an aperture angle ϕ and a tip apex parameter b located at a distance d above the flat surface (see inset in Fig. 3). The geometric parameters ϕ and b of the tip correspond to a radius of curvature of the tip $R = b\ tan(\phi)$. The light emission emission is calculated as an incoherent sum of the inelastic transitions of energy $\hbar\omega$ between tip and sample states at energies E_t and E_s which lead to the current density $\vec{j}_{ts}(\vec{r'}, \omega)$, weighed by the field enhancement in the tip-sample gap $G(\vec{r'}, \omega)$. In this model, the radiated power per photon energy and solid angle is:

$$\frac{d^2P}{d\Omega\ d(\hbar\omega)} = \frac{\omega^2}{8\pi^2\varepsilon_o c^3} \sum_{t,s} \left| \int d^3\vec{r'}\ G(\vec{r'}, \omega)\ \vec{j}_{ts}(\vec{r'}, \omega) \right|^2 \times \delta(E_t - E_s - \hbar\omega), \qquad (1)$$

where c is the velocity of light. The field enhancement factor $G(\vec{r'}, \omega)$ determines the electromagnetic modes at the cavity. It is determined with a boundary element method using bulk dielectric functions [19]. Tersoff and Hamann's model extended to high energy losses [16, 20] is used to calculate the tunnelling current.

The overall aperture of the tip ϕ mainly controls the position of the emission peak. Small changes of this parameter can lead to wavelength changes of several ten nm. The other geometrical parameters, the tip-sample separation distance d and the apex length parameter b, enable a fine tuning of the peak position. Moreover, b determines the shift of the peak with vertical tip excursion. We find shifts of ≈ 3 nm/Å for sharp tips ($R = 10$ Å) up to ≈ 10 nm/Å for blunt tips ($R > 150$ Å). In modelling a particular experiment, ϕ and b are first chosen to fit calculated spectra to the experimental data. Then d is varied as in the experiment. The resulting calculated shifts of the peak positions are shown in Fig. 3 (circles). The observed shifts are due to the geometry-controlled variation of the field enhancement $G(\vec{r'}, \omega)$. A quantitative agreement with the experiment is achieved for sharp tip apex ($R = 10$ Å) along with a fairly wide aperture of the tip ($\phi = 40°$).

From the above experimental and calculated results it is clear that the tip-sample separation d and the fluorescence spectrum are directly related. We now attempt to adapt the theoretical concept to the case of an atomic step. To reduce the numerical expense of the problem, in a first step, we preserve the axial symmetry by modelling the step as a hole (or island) with the tip placed in the center as indicated in Fig. 2a. Within this calculation different island (or hole) radii represent different distances from the step. Clearly, this geometry overestimates the influence of a step. This will be corrected for later on. Next, we restrict the calculations to the range where no topographical contrast in the linescan can be observed, i.e., outside of the close proximity of the step. Then we can safely assume that the energy dependent current density $\vec{j}_{ts}(\vec{r'}, \omega)$ is constant and does

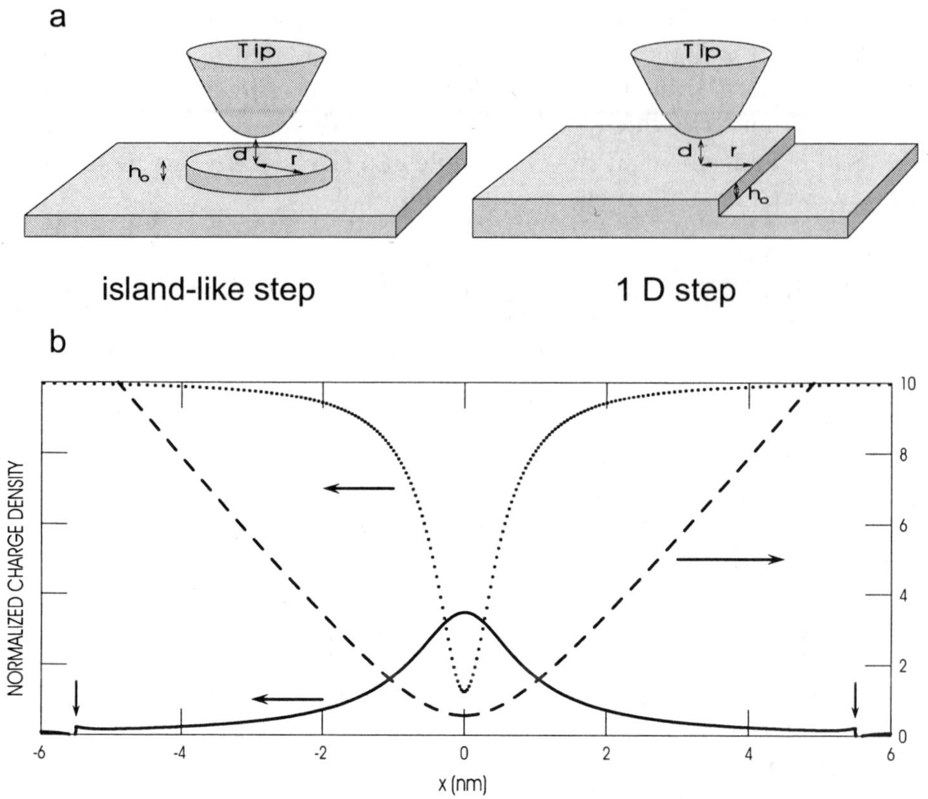

Figure 4. (a) Geometries used to model light emission near a step. Left: Step showing azimuthal symmetry to obtain Δ_{island} in Fig. 1. Right: One dimensional step used for obtaining the corrected Δ_{step} in Fig. 1. (b) Cross section of the tip-sample geometry (dashed line) and the calculated charge densities on the tip (dots) and sample (line) at a lateral tip-step distance of $\rho = 55$ Å.

not contribute to additional variations of the fluorescence. We note that quantitative modelling very close to the step would require information on the atomic tip structure and the precise electronic structure at the step. As in the experiment, the calculated spectral shape does not change as ρ is varied while there are overall shifts. These shifts are evaluated from the spectra and indicated by dots in Fig. 2. In agreement with the experiments we find a blue-shift when the tip is located on the top of an island (which corresponds to a tip position on an upper terrace) and a red-shift on the bottom of a hole (corresponding to the tip on a lower terrace).

Fig. 4b shows the calculated charge densities of the mode relevant for light emission and the tip geometry. Owing to the tip curvature the mode is localized. We define the extension of electromagnetic coupling to be the spatial extent of the induced surface plasmon on the sample side. For a tip with radius $R = 10$ Å, we find a lateral extension of only ≈ 20 Å as determined by the full width at half maximum of the plasmon charge density. Note, however, that the tail of the plasmon mode extends up to more than twice this value, as shown in Fig. 4b. At this distance the presence of the step only causes a minor redistribution of the charge density (indicated by two vertical arrows). The localized mode at the sample surface suffers a stronger modification when the center of the mode (which is identical to the tip position) is closer to the monoatomic step. The lateral distance where the shift becomes discernible is connected with the range of the interaction of the plasmon mode with the step. In the present case, the experimental extension is ≈ 50 Å which agrees with the model calculations for the sharp tips used in Fig. 4b.

As expected, the calculated shifts from the axially symmetric model are larger than the experimental data. Above it was shown that a change Δd of the tip-sample distance leads to a linear shift of the spectral maximum: $\Delta\lambda \sim \Delta d$. A variation of the surface height $h(\vec{r})$ within the lateral range of the plasmon mode will effectively change the average tip-sample distance. Therefore, we define an effective distance $d_{\text{eff}} = d + \Delta d_{\text{eff}}$ where

$$\Delta d_{\text{eff}} = \int \sigma_s(\vec{r})\, h(\vec{r})\, d^2\vec{r}. \tag{2}$$

σ_s is the normalized ($\int \sigma_s(\vec{r}) d^2\vec{r} = 1$) charge density of the plasmon mode on the sample surface as calculated. This convolution of height change and surface charge density accounts for any surface variation over the area where the surface plasmon extends. For a hole (or island), $h = 0$ within a circle of radius ρ, and $h = \pm h_0$ outside (see left side of Fig. 4). Similarly, for the case of a one-dimensional ascending or descending step at $x = \rho$, we have $h = \pm h_0$ for $x > \rho$ and $h = 0$ elsewhere (right of Fig. 4). The shift $\Delta\lambda_{\text{island}}$ originally calculated for the case of an axially symmetric step is then corrected by the convolution with the more realistic geometry:

$$\Delta\lambda_{\text{step}} = \frac{\Delta d_{\text{step}}}{\Delta d_{\text{island}}}\, \Delta\lambda_{\text{island}}. \tag{3}$$

The resulting data (squares in Fig. 2) are close to the measured spectral shifts. We note that the presence of a second descending step at the position 125 Å is also taken into account in calculating $\Delta\lambda_{\text{step}}$.

6. Molecular Systems

The role of an organic molecules on metal surfaces in light emission has repeatedly been investigated [21, 22, 23, 24, 25]. While molecular-scale contrast was observed in maps of the photon intensity, for the systems

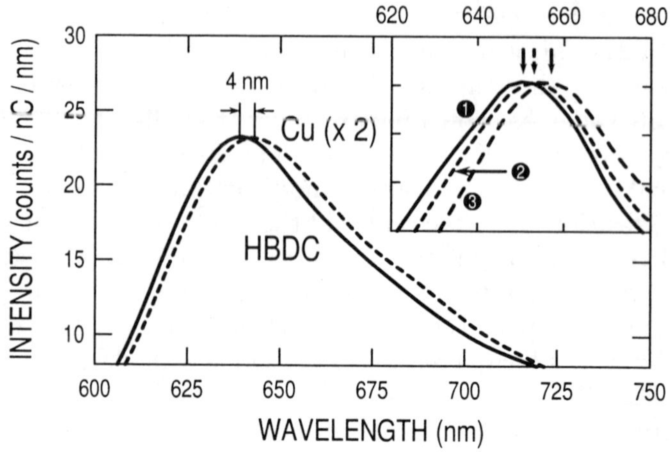

Figure 5. Fluorescence spectra from HBDC and the surrounding metal surface acquired with a W tip ($I = 5$ nA, $V = 2.3$ V). Spectra are scaled to same maximum intensity. They reveal a blue shift of the emission on HBDC. The inset shows three spectra from Au(111) recorded at different currents, i.e., different tip-sample distances. Spectrum (1) was recorded at $V = 3$V and $I = 0.3$ nA. Spectrum (2) at $I = 2.8$ nA is red-shifted by $\Delta\lambda = 2$ nm, spectrum (3) at $I = 33.7$ nA is shifted further with $\Delta\lambda = 6$ nm.

studied to date, no compelling evidence of molecular fluorescence has been obtained. A recent example is the organic molecule HBDC (hexa-tert.-butyl-decacyclene) adsorbed on Cu(111). HBDC, which appears as a ≈ 2 Å protrusion in STM images and causes prominent contrasts in photon maps on a sub-molecular scale [11]. Fluorescence spectra recorded on top of individual molecules and for comparison on the bare Cu(111) substrate are shown in Fig. 5. Except for a ≈ 4 nm red-shift of the fluorescence spectra on top of molecules the emission spectrum remain nearly unaffected. This is clear indication that the emission process is unchanged. For illustration the inset of Fig. 5 shows results from clean Au(111) where the tip-sample distance was varied. In both cases virtually identical characteristics are observed. When the tip is forced to retract from the metal substrate either by changing the set tunnelling current (inset) or by the presence of a molecule, the fluorescence is blue-shifted by similar amounts.

We are delighted to thank A. Baratoff for discussions. This work has been supported by the European Commission via the TMR network EMIT and the Deutsche Forschungsgemeinschaft.

References

1. *Surface Plasmons*, H. Raether (Springer Tracts in Modern Physic 111, Berlin 1988) and *Optical properties of Metal Clusters*, U. Kreibig and M. Vollmer (Springer Series in Materials Science 25, Berlin 1995).
2. H.-J. Güntherodt, D. Anselmetti, E. Meyer (Eds.), *Forces in Scanning Probe Methods*, NATO ASI Series E, Vol. **286** (Kluwer, Dordrecht, 1995).
3. D. W. Pohl and D. Courjon (eds.), *Near Field Optics*, NATO ASI Series E, Vol. **241** (Kluwer, Dordrecht, 1993).
4. J. K. Gimzewski, J. K. Sass, R. R. Schlittler, J. Schott, Europhys. Lett. **8**, 435 (1989).
5. P. Johansson, R. Monreal, S. P. Apell, Phys. Rev. B **42**, 9210 (1990).
6. R. Berndt, J. K. Gimzewski, P. Johansson, Phys. Rev. Lett. **67**, 3796 (1991).
7. B. N. J. Persson and A. Baratoff, Phys. Rev. Lett. **68**, 3224 (1992).
8. M. Tsukada, T. Schimizu, K. Kobayashi, Ultramicroscopy **42-44**, 360 (1992).
9. S. Ushioda, Y. Uehara, M. Kuwahara, Appl. Surf. Sci. **60/61**, 448 (1992).
10. J. Aizpurua, G. Hoffmann, S. P. Apell, R. Berndt, Phys. Rev. Lett. **89**, 156803 (2002).
11. G. Hoffmann, L. Libioulle, R. Berndt, Phys. Rev. B, **65**, 212107 (2002).
12. G. Hoffmann, J. Kröger, R. Berndt, Rev. Sci. Instrum. **73**, 305 (2002).
13. R. Berndt, J. K. Gimzewski, P. Johansson, Phys. Rev. Lett. **71**, 3493 (1993).
14. R. Berndt and J. K. Gimzewski, Phys. Rev. B **48** 4746 (1993); V. Sivel, R. Coratger, F. Ajustron, J. Beauvillain, Phys. Rev. B **50** 5628 (1994).
15. We use 'counts' to indicate the number of detected photons while 'photons' refers to the estimated number of emitted photons, i.e., the data is corrected for the detection efficiency.
16. J. Aizpurua, S. P. Apell, R. Berndt, Phys. Rev. B **62**, 2065 (2000).
17. R.W. Rendell, D.J. Scalapino, B. Mühlschlegel, Phys. Rev. Lett. **41**, 1746 (1978).
18. R. Péchou *et al.*, J. Phys. III **6**, 1441 (1996).
19. *Handbook of Optical Dielectric Constants of Solids*, D. Palik (ed.) (Naval Research Laboratory, Washington D.C., 1985).
20. K. Stokbro, U. Quaade, F. Grey, Appl. Phys. A: Mater. Sci. Process. **66**, S907 (1998).
21. R. Berndt *et al.*, Science **262**, 1425 (1993).
22. I. I. Smolyaninov, Surf. Sci. **364**, 79 (1996).
23. S. Evoy, F. D. Pardo, P. M. S. John, H. G. Craighead, J. Vac. Sci. Technol. A **15**, 1438 (1997).
24. D. Fujita *et al.*, Surf. Sci. **454-456**, 1021 (2000).
25. G. E. Poirier, Phys. Rev. Lett. **86**, 83 (2001).

STM-INDUCED LIGHT EMISSION : EXCITATION AND TIME-RESOLVED SPECTROSCOPY

FABIEN SILLY† AND FABRICE CHARRA
CEA-Saclay
Service de Physique et Chimie des Surfaces et Interfaces
DSM-DRECAM, F-91191 Gif-sur-Yvette Cedex France.
† present address :
Université de Lausanne IPMC CH-1015 Lausanne Swiss

Abstract

We review some spectroscopy experiments applied to the luminescence of nanostructured systems, excited highly locally by the tip of a Scanning Tunneling Microscope (STM). We first present briefly the principles of this STM-induced luminescence. The we present a review of selected results in STM-induced luminescence from the point of view of nano-scale photonics. We illustrate various contrast mechanisms with different examples of nanostructured systems.

Optical spectroscopy techniques are essential tools for probing the electronic and photophysical properties of condensed matter. A wide range of spectroscopic methods is available, including absorption-, emission-, excitation-, polarization- and ultrafast time-resolved-spectroscopy. However, the wavelength-limited spatial resolution in far-field optical techniques prevents local nanometer-scale photophysical studies (see figure 1-a). This is a severe limitation considering the new trends for nanoscale-structured functional opto-electronic materials such as advanced molecular materials. Several approaches have been adopted to circumvent this problem in order to excite locally nanostructured objects. A first *all-optical* approach takes advantage of near-field optical scanning probe techniques originally developed for high-resolution optical microscopy. Typically, the light scattered through a sub-wavelength sized aperture at the free end of an optic fiber is used as a secondary light source, scanned close to the studied surface (see figure 1-b). The resolution is then limited by the optical skin depth of the aperture material, typically a few 10 nm. This paper deals with an alternative approach which combines an optical detection system and a Scanning Tunneling Microscope (STM, see figure 1-c). It consists in taking advantage of the extreme localization of the tunnel current for exciting locally nano-structured objects with energies in the optical range. The locally-excited luminescence can then be analyzed in the far-field. As a matter of fact, when the STM junction is biased with a voltage V_T between 1 and a few volts, the

F. Charra et al. (eds.), Organic Nanophotonics, 93–102.

current at the tunnel junction behaves as a highly localized electro-magnetic source at frequency $\omega \leq eV_T/\hbar$ in the optical range.[1, 2] The resulting luminescence can provide photophysical information with an unparalleled spatial resolution, below 1 nm. However, the analysis is complicated since contrast, i.e. spatial variation of luminescence characteristics, not only occur through local variations of optical parameters but also from electronic structure which may influence the spectral properties of the exciting source, or the mechanisms of electron-photon conversion. Interpretations of photon maps must thus include a careful analysis of contrast origins.
In this review we present selected results of STM-induced luminescence from the point of view of nano-scale photophysics. We first present briefly the principles of STM-induced luminescence. Then we illustrate various contrast mechanisms with different examples of nanostructured systems.

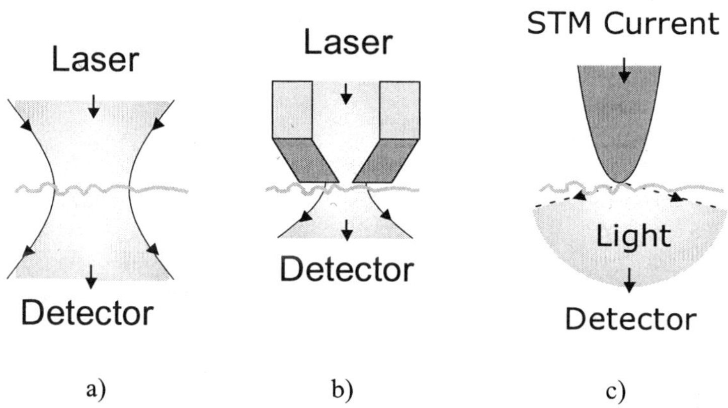

Figure 1 : Schematic representations of (a) far-field optical microscopy (e.g. confocal microscopy), (b) scanning near-field optical microsopy (SNOM) based on metal-coated fiber-tip probes, and (c) STM-induced luminescence.

Light emission from tunnel junction was first observed in metal-oxide–metal systems.[2] Following the advances in scanning probe microscopy, photon emission was observed in a STM with relatively high quantum efficiencies [3] and with molecular-scale imaging resolution [4]. The possibility of achieving a chemical contrast through this technique has been demonstrated for carbon clusters on silicon [5] and tip-induced surface chemical modifications [6]. Despite the weakness of the emission, generally detected using photon-counting techniques, the optical nature of this detection permitted local spectroscopic studies of rough metal surfaces [7,8] or single silver clusters [9].
An elementary electromagnetic source term consists in a current vector $\mathbf{j}(\omega, \mathbf{M})$ at frequency ω highly localized around a given point $\mathbf{M}_0$. The localized DC tunnel current in a STM arises from the overlap between the evanescent wave functions of filled initial tip electronic state φ_i with energy E_i and empty final sample state φ_f at the same energy. However, each tip state overlaps also with sample states at energies

E_f different from E_i, giving rise to AC currents at frequency $\omega=(E_f-E_i)/\hbar$ as sketched in Fig. 2. From a quantum point of view this corresponds to inelastic tunneling from φ_i to φ_f, with simultaneous excitation of an electromagnetic eigenmode. The intensity of those AC currents is a function of the local density of contributing φ_i and φ_f states at energies E_i and E_f and of the tunneling matrix element T_{if} between them in the tunneling Hamiltonian formalism [1]. The cutoff frequency ω_C is determined by the bias V_T as $\omega_C=eV_T/\hbar$. In metals, both density of states and matrix elements can be approximated as constants over the energy range thus defined. The spectrum of AC currents is fully determined by the DC resistance R of the tunnel junction [1] which is usually kept constant during the scan in STM imaging (so-called constant-current mode or Z mode). However, if other materials are involved, the spectrum of AC currents can be strongly affected by highly structured density of states.

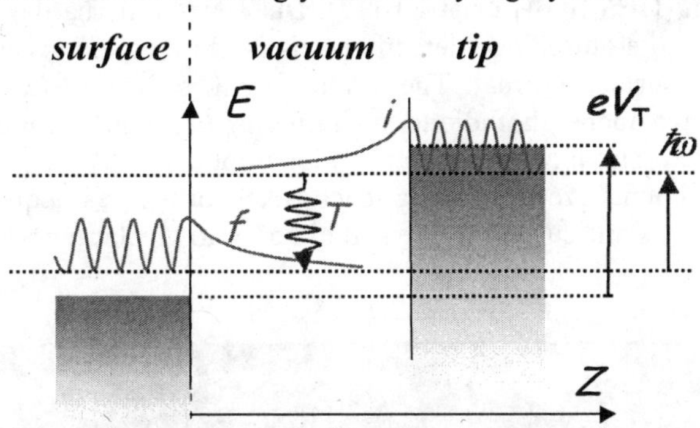

Figure 2 : Energy diagram for the inelastic tunneling process of electrons from a metal STM tip (right) to metal-sample surface (left) through a vacuum gap (middle). The maximum energy $\hbar\omega$ of the excited electromagnetic mode is imposed by the bias, eV_T.

For example, in semiconductor samples there are no available empty states φ_f at energy E_f below the gap. The inelastic tunneling process is thus not efficient for moderate biases. However, for *p*-type direct-gap semiconductors, the tunneling from tip to the conduction band can still generate a luminescence from subsequent electron-hole recombination. The localization of this two-step process may be affected by diffusion of charge carriers prior to recombination.

Then, the elementary source $\mathbf{j}(\omega,\mathbf{M})$ located around a point $\mathbf{M}_0$ will radiate energy through coupling with a given electromagnetic eigenmode at the same frequency. This process will thus depend on the amplitude of the electric field of the excited mode at point $\mathbf{M}_0$.[10]. The proximity of highly polarizable systems such as metal aggregates or conjugated molecules will thus strongly affect the efficiency of electron-photon conversion. On metal surfaces, the tip itself strongly favors light emission efficiency through the formation of a tip-induced plasmon mode (also called "gap-mode" [11]), highly localized at the tip-surface junction. When the tip scans the sample, the photon emission variations reflects additional variations of the excited electromagnetic mode

due to local perturbations. The first example presented below illustrates this phenomenon.
Enhancements by several orders of magnitude of photophysical processes, such as surface-enhanced raman scattering (SERS) [12], have been observed at rough noble-metal surfaces. The amplification role played by local field enhancements, especially in plasmon oscillation modes, is now clearly established [13]. To gain further insights in the microscopic mechanisms of plasmon-mediated processes, we have applied STM-induced luminescence on model systems with well-controlled geometry.[14] A particularly interesting system consists in thiol-molecule coated silver nanoparticles.[15] For low size dispersities, these nanoparticles self-assemble at room temperature as two-dimensional monolayers on an atomically flat (111) gold surface. [16]
To measure STM-induced measurements, a light collection optics must be adapted to the STM. In our laboratory, it consists in a high-numerical-aperture lens either under Ultra-High Vacuum (UHV) [14] or at air [17]. But several alernative geometries have been reported in the literature. The detection axis is close to the grazing incidence, at about 75° from the surface normal. The emitted light is focused (in UHV through a view port) onto an avalanche photodiode, operating in a photon-counting mode. Photon maps were acquired simultaneously with topographic maps using a home-made electronics. Photon counts are recorded at each pixel for a fixed acquisition time. The STM operates in a constant-current mode and either gold or platinum-iridium STM tips are used.

a) b)

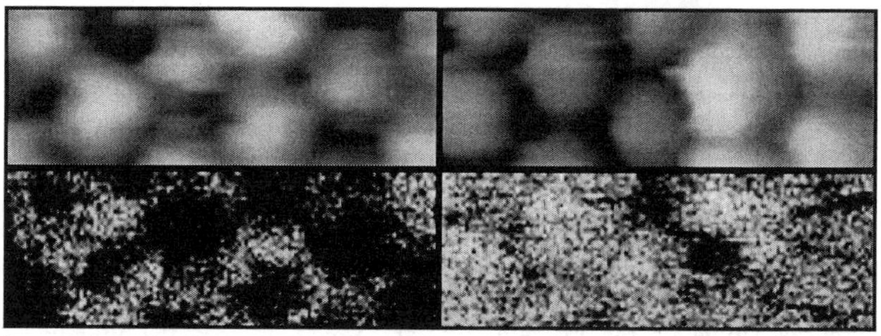

c) d)

Figure 3 : Simultaneously recorded STM topography (a,b) and photon map (c,d) for self-assembled thiol-coated silver particles. Scan area 26 × 10 nm, sample bias Vs=2.0 V (a,c) and 2.5 V (b,d). Set-point current It = 3.5 nA.

A closeup (26 × 10 nm^2) of a self-assembled domain is presented in Fig. 3. The topography (fig. 3a,b) and the photon maps (fig. 3c,d) were recorded simultaneously. Images 3a-d were recorded on the same area using different biases: V_T =2.0 V (a,c) and V_T =2.5 V (b,d). In both cases, the tunnel current was set to 3.5 nA. Although apparent topographies (fig. 3a,b) are similar, the photon maps (fig. 3c,d) exhibit drastic differences in photon emission efficiencies: at the lower bias (2.1 V, fig. 3a,c), light emission is below the detection limit when the tip is located above the top of a particle whereas a maximum is reached between particles. In contrast, when a bias of 2.5 V (fig.

3b,d) is applied, the maximum of photon emission is detected on the top of the particle and the minimum is attained when the tip is above a junction between two particles.
Electrons always tunnel toward silver. We can thus assume low variations of electronic structure during the scan. The formation of tip-induced mode results from the interaction of the tip apex with sample modes. Hence, the observed changes in photon emission rate during constant-current scanning mainly map the sample modes which, participate to the excited modes at a given bias. Then, the observed variations can be understood in terms of the collective plasmon modes of the hexagonal silver-particle network.[14] In particular the enhancement between particles at lowest bias reflects the lowest-energy mode of a simple two-particle model, which mainly consists in a gap-mode localized between them.[18] Analysis of STM-induced luminescence thus permits to study the effects of local geometry on photonic processes. Tuning the STM bias makes possible an excitation spectroscopy. The understanding of the collective excitations in such a system may lead to a control of field distribution and optimization of photonic properties at the nanoscale.

Despite a good ordering into a hexagonal network, the quantum efficiencies of STM-induced luminescence of single nanoparticles appears highly site dependent.[19] As shown in the cross-sections in figure 4, the STM-induced luminescence of some specific particles appear completely quenched. This behavior has been correlated with an increased mobility of such "dark" particles within their hexagonal site.[19] In the particular case represent in figure 4, moreover, a "dark" particle is located next to a vacant site. The STM tip can also be used to manipulate individual nano-scale objects at the single atoms [20] or single molecule [21] level at low temperature. Room temperature manipulation and control of the nanoparticle assembly at the single-particle scale can be realized through direct mechanical interactions with the sharp STM tip by increasing tunnel-current set-point.[22].

The switching of the luminescence of the "dark" particle through STM manipulation is reported in figure 4, presenting the tip height curves during moving of a single particle with simultaneously recorded STM induced light emission. The particle is first located at site A at time T_0. It is then moved toward the vacant neighbor site B by a right-to-left scan of the STM tip at T_1. As a consequence of this displacement the luminescence of the particle is nearly completely switched off. It is then switched on and off successively by moving the particle into site A and B at times T_2 and T_3. The process has been successively repeated about 10 times before the image was lost, due to drift in piezoelectric motion. These observations demonstrate that the luminescence quenching does not depend on the particle itself but on the site where it lies. To mechanisms of this site-dependent quenching may be invoked. The first one involves variations in the height of a second tunnel-barrier between tip and substrate. This could be caused by an inhomogeneous coverage of the gold surface by some alkanethiols molecules separated from silver particles. The reduced tip-particle junction resistance would then result in an increased mechanical influence of the tip on the particle, consistent with the observations. A second possibility involves the electron-phonon dynamics in the metal nanoparticle network. An increased particle mobility could open additional inelastic-tunneling channels with excitation of local phonon modes instead of the radiative optical mode. These results provide evidence for the realization of a light-emission switch with high reversibility, at nanometer scale and at room temperature. Here, STM was used for both manipulation and excitation of the modified luminescence of single nanoparticle.

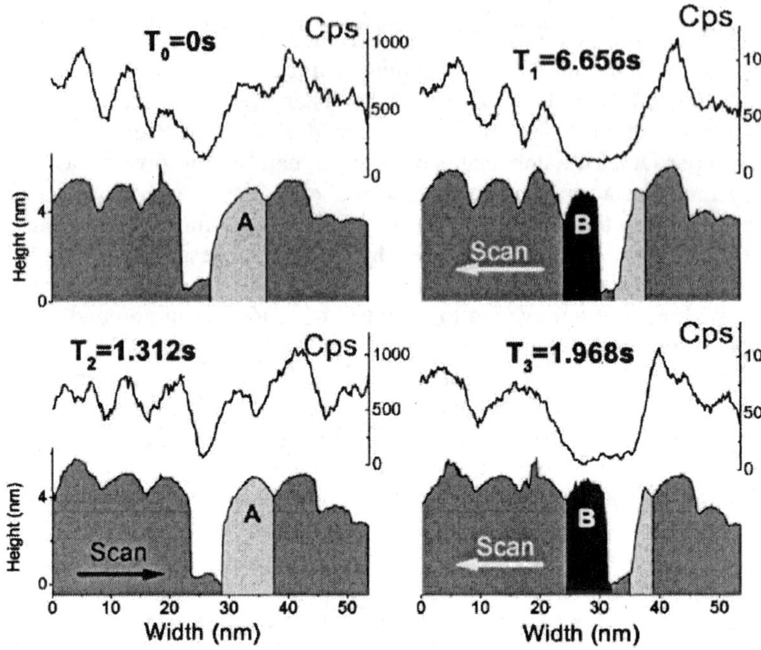

Figure 4 : Cross section of STM topography and corresponding photon emission curve during three successive particle manipulations, sample bias: V_T=2.5 V; tunnel-current set-point: I_T=5.3 nA. Particle was alternatively shifted through action of the STM tip between luminescent site A (light gray area) and site B (black area) where the light emission is quenched.

The next challenge in this technique is the excitation of individual molecules. A first demonstration of molecular-scale contrasts in photon maps was achieved on C60 molecules deposited on Gold.[4] The influence of a the organization of a monomolecular layer on STM-induced luminescence has been demonstrated recently on self-assembled alkane-thiol molecules attached to gold surface.[23] At first sight, the use of intrinsically luminescent molecules such as conjugated luminophores is an attractive route. However the quantum efficiency of such luminophores close to conducting materials is low, and large current densities are still needed. Such current density very rapidly damage the conjugated layer. To illustrate this effect, we report on figure 5 the effect of scanning under luminescence conditions on a terthiophene-functionalized thiol monolayer. The continuity of gold monoatomic step edges crossing the damaged square region shows that the underlying gold substrate has not been affected. The surface changes, as imaged by STM, thus result from damages to the molecular layer, most probably to a bleaching of the conjugated moiety of the self-assembled molecules.

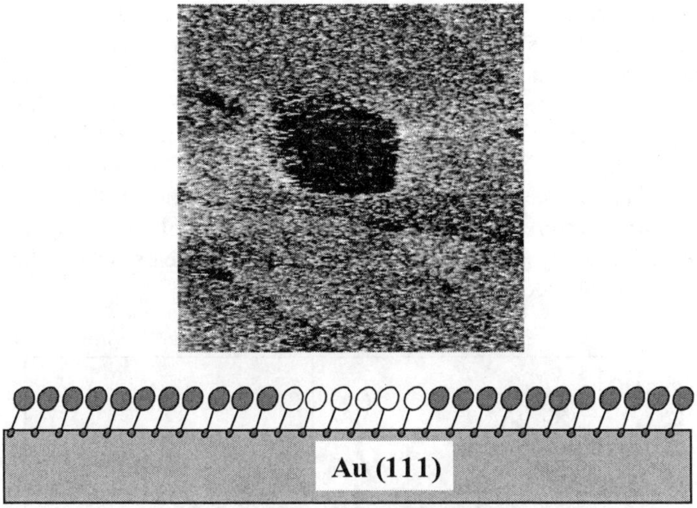

Figure 5 : STM image of a 500 × 500 nm self-assembled terthiophene-substituted alkane-thiol monolayer on gold (111) epitaxially grown on mica (low bias and current). A 90 × 90 nm area in the center of the image was previously scanned in ligh-emission conditions (V_T=3.2V, I_T=200pA)

One unique opportunity offered by optical detection is time-resolution down to ultrafast time-scales. According to the simple probabilistic mechanism of STM-luminescence described above, for stable samples, individual photon emission events are independent one from another, leading to randomly distributed photon-emission times. However, in more complex systems, dynamic effects could take place. Moreover the electronic mechanisms of STM induced photon emission may deviates from this simple model. Then the time statistics of single-photon emissions can be completely different. As an example, a time-resolved detection of STM-excited luminescence of polymer light-emitting diodes has permitted to evidence photon bunches in the millisecond time scale attributable to high-field avalanche processes.[24] Thus, in contrast to extended light sources, the highly local nature of the STM-induced luminescence makes possible strong correlations between successive single-photon emission events, as, for example, bunching or antibunching phenomena. Such photon correlations are intimately connected to the dynamics of the source that produced them.[25] The analysis of time correlations in single-photon detection can then give valuable information on the dynamic processes taking place after tip-induced excitation.

Time-autocorrelated two-photon counting measurements have been successfully applied to the study of time correlations in laser-excited systems like, e.g., few-molecule fluorescence.[26] We have used this technique to study STM-induced photon emission, with time resolutions down to the nanosecond time scale. For this purpose, the collected light is split symmetrically between two detectors connected to the start and stop inputs of a time-to-amplitude converter (TAC) Time intervals between single-photon detection can be measured with an accuracy better than 200 ps. For each event, which consists in the detection of a valid start-and-stop pulse pair, the electronic system records the

measured time interval, the current X– Y coordinates of the scanning tip. These data are stored for each individual event for subsequent off-operation statistical treatments. This permits to generate histograms of start-stop time intervals on a given area as well as maps of mean time intervals acquired simultaneously with STM topographic maps.
In the case of a gold surface at air, light emission exhibits a strong bunching behavior at the nanosecond scale for a bias of 2.5 V as reported on the histograms of figure 6a,b. The bunching peak width corresponds to a time constant of the order of 12 ns. Moreover this two-photon time correlation appears strongly bias dependent in the voltage range 2.0–2.5 V, since no such correlation appears when the voltage is decreased to 2.0 V (figure 6c,d).

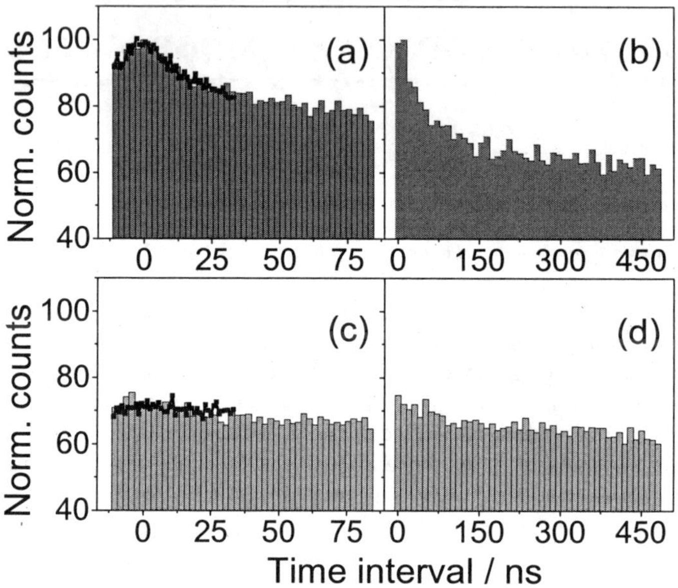

Figure 6 : Histograms of recorded time intervals between consecutive pairs of detected photons for two different bias voltages applied on the substrate: 2.5V (a, b) and 2.0V (c, d). The results obtained with TAC time ranges of 50nm (solid squares) and 100ns (bars) are superimposed for each bias: 2.5V (a) and 2.0 V (c). The histograms obtained with a range of 500 ns are represented on a larger scale for each bias: 2.5 (b) and 2.0 V (d).

The relationship between such photon correlation and the local topography is expected to give crucial information on both electronic and structural fast dynamic processes taking place locally at surfaces. The map of two-photon auto-correlations on a 50-ns time scale is represented on figure 7, together with the topography, in the case of a rough gold surface at air. Time correlations appear as a new contrast mechanism for scanning probe microscopy. As a matter of fact the bunching phenomenon present at biases near 2.5V appears strongly spatially dependent. More precisely it is present exclusively at specific less emissive bumps of the gold surface, pointed to by arrows on both topography and correlation maps. The similarity between the conditions for observation of bunching and some tip-induced surface modifications observed previously [27,28] suggest that this effect could be a manifestaton of a diffusion-controlled reorganisation of a tip-induced transient atomic-scale defect. Optical diagnostics then permits measurements of the dynamics of tip-induced defects at unprecedented time resolutions compared with other surface-characterization techniques. Moreover, observations of time-correlations in such down-scaled light sources opens opportunities in nanophotonics for addressing and triggering single photon sources which exhibits non-classical behaviors, like single molecules or quantum dots.

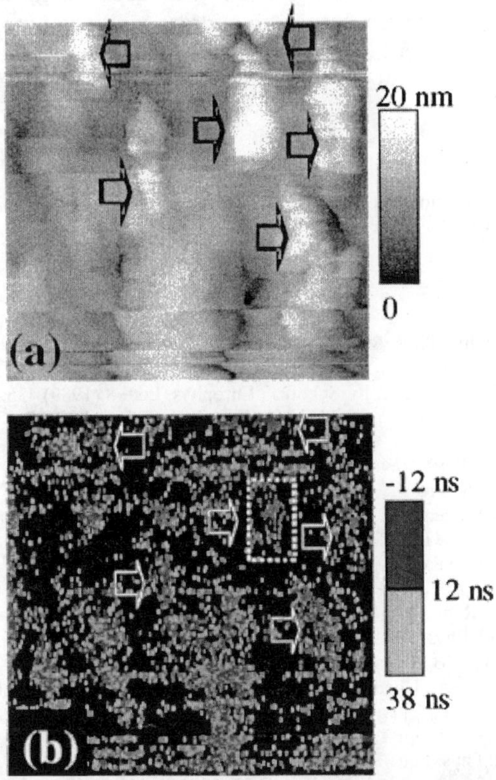

Figure 7 : Simultaneously recorded topography of a rough gold surface (a) and map of Time-autocorrelated two-photon events (b). Time-interval belonging to [-12ns,12ns] are represented in green whereas those belonging to [+12ns,38ns] are represented in pink. Image dimensions: 250nm x 250nm. (V_T=2.0V, I_T=1.5nA)

In conclusions, in STM-induced luminescence experiments, the STM tip is used simultaneously as a tunnel-current probe to map the topography of the studied sample and as a highly-localized excitation source generating a luminescence. A third possible simultaneous action of the tip is the modification of the surface. We have illustrated this possibility here for the reversible reorganization of self-assembled nanoparticle monolayer, the local bleaching of self-assembled dye-functionalized monomolecular layers, or the transient effects of a biased tip at air on gold surfaces. The simultaneous influence of such tip-induced effects on both nano-scale topography and locally-excited luminescence gives very valuable informations on the fundamental photonic processes taking place at molecular-scale level. As compared with other surface characterization techniques, the optical detection coupled to the STM offers a very powerful local diagnostic tool through the possibility of optical spectroscopy. We have illustrated these new possibilities here in the case of time resolution.

The next challenges are a better understanding of fundamental processes leading to luminescence at the nanoscale and the excitation of individual quantum objects like semiconductor quantum dots or conjugated molecules thus forming highly-correlated nanoscale light-sources. The observation of STM-induced luminescence at liquid-solid interface [29] may offer new perspectives to address individual molecules. As a matter of fact some conjugated molecules can form stable self-assembled monolayers at liquid-solid interface which can be probed by STM with intramolecular resolution.[30]

Acknowledgement

The results on terthiophene-functionalized thiol self-amonolayers were obtained in collaboration with A. Yassar and D. Fichou, from LMM, CNRS Thiais, France.

References

[1] Rendell R.W., D. J. Scalapino, Phys. Rev. B 24 (1981) 3276.
[2] Lambe J., McCarthy S. L., Phys. Rev. Lett. 37 (1976) 923
[3] Gimzewski J.K., Sass J.K., Schlitter R.R., Schott J., Europhys. Lett. 8 (1989) 435.
[4] Berndt R., Gaisch R., Gimzewski J.K., Reihl B., Schlitter R.R., Schneider W.D., Tschudy M., Science 262 (1993) 1425.
[5] Downes A., Welland M.E., Phys. Rev. Lett. 81 (1998) 1857.
[6] Thirstrup C., Sakurai M., Stokbro K., Aono M., Phys. Rev. Lett. 82 (1999) 1241.
[7] Bischoff M.M.J., van der Wielen M.C.M.M., van Kempen H., Surf. Sci. 400 (1998) 127.
[8] Ito K., Ohyama S., Uehara Y., Ushioda S., Surf. Sci. 324 (1995) 282.
[9] Dumas P., Syrykh C., Makarenko I.V., Salvan F., Europhys. Lett. 40 (1997) 447.
[10] Johansson P., Monreal R., Apell P., Phys. Rev. B 42 (1990) 9210.
[11] Mal'shukov A. G., Phys. Report 194 (1990) 343.
[12] Moskovits M., Rev. Mod. Phys. 57 (1985) 783
[13] García-Vidal F. J., Pendry J. B., Phys. Rev. Lett. 77 (1996) 1163
[14] Silly F., Gusev A. O., Taleb A., Charra F., Pileni M.-P., Phys. Rev. Lett. 84 (2000) 5840
[15] Taleb A., Petit C., Pileni M. P., Chem. Mater. 9 (1997) 950
[16] Taleb A., Petit C., Pileni M. P., J. Phys. Chem. 102 (1998) 2214
[17] Silly F. Charra F., Appl. Phys. Lett. **77** (2000) 3648
[18] Xu H., Aizpurua J., Käll M., Apell P. Phys. Rev. E 62 (2000)
[19] Gusev A. O., Silly F., Taleb A., Charra F., Pileni M. P., Adv. Mater. 12 (2000) 1583
[20] Eigler D. M., Lutz C. P., Rudge W. E., Nature 352 (1994) 600
[21] Kulzer F., Kummer S., Matzke R., Brauchle C., Basche Th., Nature 387 (1997) 688
[22] Silly F., Gusev A. O., Charra F., Taleb A., Pileni M.-P., Appl. Phys. Lett. 79 (2001) 4013
[23] Poirier GE, Phys. Rev. Lett. 86 (2001) 83
[24] Charra F., Bouchel S., Plessis O., Nunzi J.-M., Raimond P., Denis C., Gautier-Thianche E., Opt. Mater. 12 (1999) 249
[25] Loudon R., Quantum Theory of Light, Oxford University Press, Oxford (1983) Chaps. 3 and 6.
[26] Kitson S. C., Jonsson P., Rarity J. G., Tapster P. R., Phys. Rev. A 58 (1998) 620
[27] Sivel V., Coratger R., Ajustron F., Beauvilain J., Phys. Rev. B 51 (1995) 14598
[28] Uehara Y., Fujita T., Ushioda S., Phys. Rev. Lett. 83 (1999) 2445
[29] Nishitani R, Kasuya A Surf. Science 435 (1999) 283
[30] Charra F., Cousty J., Phys. Rev. Lett. 80 (1998) 1682

DIFFERENTIAL REFLECTION SPECTROSCOPY OF ULTRATHIN HIGHLY ORDERED FILMS OF PTCDA ON AU(111)

R. NITSCHE, H. PROEHL, AND T. FRITZ
Institut für Angewandte Photophysik
Technische Universität Dresden, D-01062 Dresden, Germany

1. Introduction

Over the last decades organic molecules have gained considerable interest because they show a functionality not common to inorganic semiconductors. With the help of organic chemistry they can be designed to exhibit desired properties like strong and wavelength-selective absorption, photo- and electroluminescence and even semiconducting properties. The construction possibilities of such molecules are almost unlimited so that they are ideal candidates for future molecular electronic devices.
Thin organic films on arbitrary substrates can easily be prepared by means of different techniques and can be utilized in potential applications like light emitting diodes or solar cells, the former one being already realized in commercially available devices. However, their electronic and optical properties strongly depend on the film morphology. During the last years, special attention has been paid to the optical properties of highly ordered organic films because these are the ideal systems to study the correlation between morphology and intrinsic physical properties and to relate these properties to existing theories.
There are a few indications in literature that the structure of organic thin films might indeed have an impact on their optical properties. Leonhardt et al. [1] describe the appearance of different fluorescence peaks in PTCDA films by varying the substrate temperature during evaporation. They attribute this to the coexistence of the α- and β-phase of PTCDA in their films, which is also in agreement with the structural characterizations they performed.
Another indication of the dependence of the optical properties on the structure of the crystal is given by Fritz et al.[2]. They perform fluorescence measurements of highly ordered and polycrystalline thin films of PTCDA on gold, which show a strong blueshift of the excimer peak in the highly ordered films compared to the polycrystalline samples. Furthermore, the appearance of a new feature at even higher energies is described, which only shows up in the highly ordered films.
Another important issue is that many organic crystals stack in a layer by layer fashion, making the distance for adjacent molecules in stacking direction much shorter than for neighboring molecules within a single layer. This gives rise to a quasi-one-dimensional behavior, since due to the large overlap of the π-systems of the adjacent molecules in the stacking direction the most relevant electronic interactions for the organic crystal

F. Charra et al. (eds.), Organic Nanophotonics, 103–117.

will occur within the stack rather than in the plane. This is the model of the one-dimensional linear chain [3]. Upon irradiation with light, excitons can be created whose coherence length extends over several molecules of the chain. The interesting question now is whether the optical properties of the organic film are influenced when the chain length (i.e. the film thickness) is shorter than the intrinsic coherence length of the excitons. This effect is also referred to as exciton confinement. Such thickness dependent experiments have been carried out previously on a variety of organic materials [4,5,6,7]. The authors all observe a blueshift of the optical transition with decreasing film thickness which can be as large as 150 meV [7]. This shows that it is not only important to know about the morphology, one also has to take care of possible exciton confinement effects when discussing the electronic and optical properties of organic thin films.

It is also important to realize that it is predicted by exciton theory that the strongest effects of a change length variation will already appear in the range from 1 to 4. Therefore, meaningful experiments will require a controlled thickness variation in the monolayer range, and it is clear that those ultrathin films will require special care in measuring the optical properties due to the small signals expected from such thin films.

In this contribution we will therefore address the following issues: *i)* Do ultrathin highly ordered organic films show an unusual absorption behavior as compared to polycrystalline films, and, *ii)* do their optical properties depend on the film thickness? To answer this we perform thickness dependent experiments of both, highly ordered (*ho*) and polycrystalline (*pc*) organic films.

PTCDA (3,4:9,10-perylenetetracarboxylicdianhydride) was chosen as the candidate molecule, since its structural data is well known [8,9], optical data of thick *pc* films are available for comparison [10], and a theoretical model already exists [3]. The substrates are a massive Au(111) crystal which enables to grow *ho* films by means of Organic Molecular Beam Epitaxy (OMBE), and polycrystalline gold on glass for the *pc* PTCDA films.

For the optical characterization of the films, *ex situ* Differential Reflection Spectroscopy (DRS), a variant of Reflection Absorption Spectroscopy, was applied and is described in greater detail in the following section.

2. Experimental

Ultrathin, highly ordered layers of PTCDA were grown on a massive Au(111) crystal by means of OMBE. Prior to evaporation, the gold crystal was sputtered (Ar-ion energy: 600 eV) and annealed (T approx. 500 °C) for 30 minutes, which produces large (100-200 nm) and flat (111)-terraces suitable for epitaxial growth of the PTCDA films. A thorough study of the growth of PTCDA on such substrates has been given elsewhere [8,9].

The PTCDA layers were grown in ultrahigh vacuum (UHV) at a substrate temperature of approx. 25 °C by evaporation from a Knudsen cell, which has been loaded with sublimation purified PTCDA (Aldrich). During deposition the pressure was in the range of 10^{-9} Torr and the Knudsen cell was held at 329 °C, resulting in a rate of 0.3 monolayers per minute (checked with a quartz balance). The relative thickness of the

PTCDA films was also confirmed by the *ex situ* optical measurements. They show that the ratio of the magnitude of the strongest absorption peak for varying layer thickness corresponds to the ratio of the layer thickness as estimated by the quartz balance.
The LEED (Low Energy Electron Diffraction) pictures were taken after the sample has been optically investigated under ambient conditions and transferred into the UHV chamber again in order to characterize the film structure actually present during the optical measurements (structural changes upon exposure to air can not be excluded *a priori*). Afterwards, STM pictures were recorded under UHV conditions, using an OMICRON UHV STM/AFM at room temperature.

For the polycrystalline PTCDA films, gold on glass was used as a substrate. The gold was evaporated under high vacuum (HV) conditions (10^{-6} Torr) at a substrate temperature of approx. 25 °C and a high rate of roughly 1.5 nm/s for 3-5 minutes, resulting in an opaque gold substrate.
Disordered polycrystalline layers of PTCDA were prepared in a HV-chamber (ca. 10^{-5} Torr) at a deposition speed equivalent to 8 monolayers per minute. The substrate temperature was held at approx. 25 °C and the thickness was again checked with a quartz balance and confirmed by the *ex situ* optical measurements.
STM pictures were taken after the optical investigations under ambient conditions, using a NanoScope II from Digital Instruments.

The reflectance measurements were performed using a commercially available spectrometer UV 3100 by Shimadzu. It is a two-beam spectrometer, which means that the main beam is split into two beams, which are used as sample- and reference beam, respectively. To be able to measure both, reflection and transmission, the spectrometer consists of two separate chambers, the transmission and reflection chamber. However, with the standard setup for measuring reflectance it was not possible to detect film thicknesses below 4 monolayers (ML) of PTCDA on gold. This is basically due the poor signal-to-noise ratio in the reflection chamber, caused by the much longer light paths and a series of additional optical mirrors as compared to the transmission chamber.

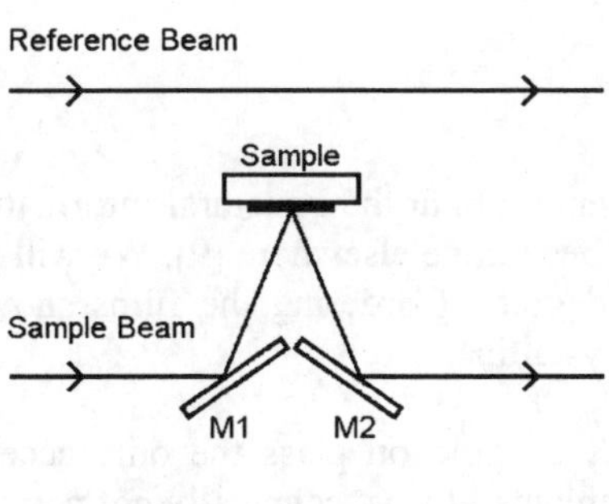

Fig. 1: Schematic sketch of the home-built setup to measure reflectance signals in the transmission chamber of the Shimadzu UV-3100.

Therefore, a setup was designed which enables to measure reflectances in the transmission chamber, where the signal to noise ratio is much better. This can be achieved by a rather simple arrangement of two optical mirrors, as shown in Fig. 1. The measured quantity does not correspond to the absolute values of the reflectance of the sample since the two mirrors give a considerable contribution to the shape and absolute height of the reflection spectrum.

However, to perform DRS measurements the absolute values are not important since the quantity one is interested in is calculated by

$$\frac{\Delta R}{R} = \frac{R_{sample} - R_{sub}}{R_{sub}} \qquad (1)$$

Here, R_{sample} is the reflectance of the film on the substrate while R_{sub} is the reflectance of the bare substrate only. By building this ratio one therefore eliminates the importance of the absolute values of the reflectance.
In practice, the following steps had to be carried out to measure the *ex situ* DRS of a single PTCDA film on a gold substrate:

1. Prepare the gold substrate (either reconstructed Au(111) under UHV, or gold on glass under HV conditions)
2. Measure the reflectance curve of the substrate with the home-built setup under ambient conditions
3. For the case of Au(111): re-prepare the substrate to get surface reconstruction (important to enable the growth of *ho* films)
4. Grow the film under UHV/HV conditions
5. Measure the reflectance curve of the sample (substrate + film) with the home-built setup under ambient conditions
6. Build the ratio (1).

Since it is absolutely crucial to have the substrate and the sample (substrate + film) at the same absolute position within the reflection setup when performing the optical measurements, a special sample holder was designed which ensures a good reproducibility of the reflection spectrum, even after repeated mounting and demounting of the sample (or substrate). Also, a set of lenses were introduced to focus the optical beam onto the sample.

3. Results

3.1. STRUCTURAL CHARACTERIZATION

This section is not intended to give a detailed report about the structural information of our films (especially for the *ho* films). This has been done elsewhere [9]. We will rather focus our investigation on the question what degree of ordering the films show, i.e. whether they are highly ordered or rather polycrystalline.

In the case of polycrystalline films of PTCDA on gold on glass the only accessible structural information stems from STM measurements. It was technically not possible to perform LEED experiments since no reliable electrical contact between the sample and grounding could be established.
Fig. 2 shows a STM picture of 3 ML of PTCDA on gold on glass. The film forms small clusters of PTCDA on top of the larger gold clusters. It was not possible to get

Fig. 2: STM image of 3 ML PTCDA on gold on glass, 500 x 500 nm^2, V = 0.6 V, I = 0.1 nA, raw data. Inset: 100 x 100 nm^2 - scan of the same sample. Large features correspond to gold clusters with smaller PTCDA clusters on top.

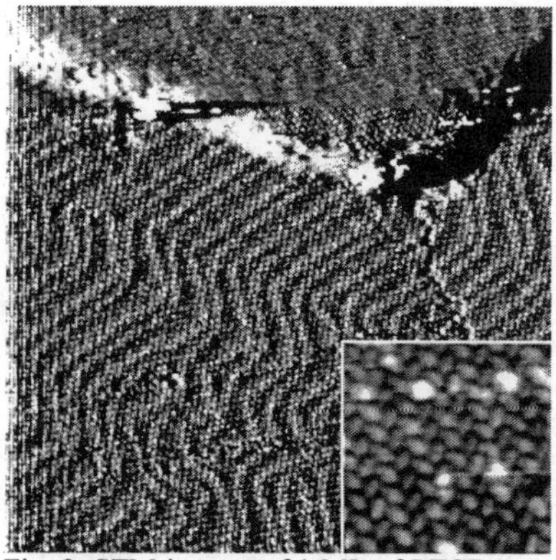

Fig. 3: STM images of 1 ML of PTCDA on Au(111), 90 x 90 nm, V = 1.2 V, I = 0.2 nA, raw data. inset: 7 x 7 nm zoom. The herringbone structure (inset) as well as the surface reconstruction can clearly be recognized.

molecular resolution of the PTCDA clusters. When zooming in on such clusters the picture started to smear out which is an indication that the tip picked up molecules and altered the surface. It is therefore reasonable to assume that the molecules are only loosely bound as it can be expected for polycrystalline films. STM measurements of different film thicknesses produced similar results.

STM images of 1 ML of PTCDA on Au(111), grown by OMBE are shown in Fig. 3. The well known herringbone structure with two molecules per unit cell [9] can be recognized (see inset), extending over two terraces of the gold substrate (large picture). The large scan also reveals the surface reconstruction of the gold crystal as clearly visible zigzag patterns underlying the molecules.

Similar pictures were taken for coverages of 2 and 4 ML, although in the latter case the film was already too thick to maintain stable tunneling conditions. For 6 ML and above no reliable picture could be recorded, so that one has to rely on LEED measurements.

Fig. 4 shows the LEED images of highly ordered PTCDA films of different film thickness on reconstructed Au(111) surfaces after they had been optically investigated.

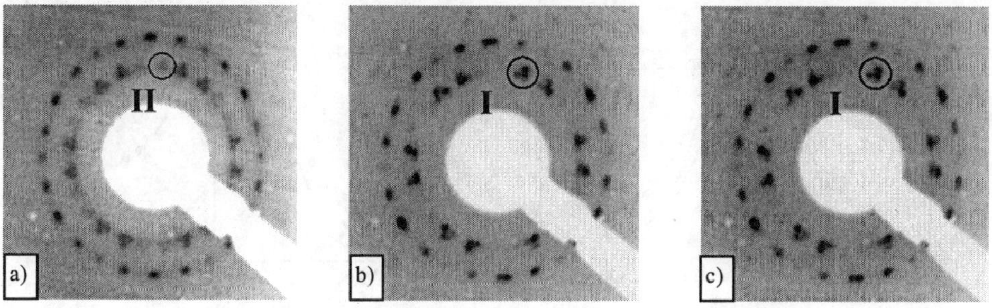

Fig. 4: LEED images of PTCDA on Au(111) for different coverages: a) 1 ML, E_0 = 20.6eV b) 2 ML, E_0 = 20.7eV c) 6 ML, E_0 = 16.1eV. For a description of structures I and II see [9].

For a coverage of 1 ML (Fig. 4a) two structures are found, labeled I and II, although the diffraction pattern of structure II (marked by a circle in Fig. 4a) is only of weak intensity, indicating that its volume fraction is rather low. These two structures correspond to findings of Mannsfeld et al. [9], who also give the lattice parameters as lying in between the values for the α- and β-polymorph of PTCDA. It was only at this coverage that both structures could be clearly resolved, for all other film thicknesses structure II was either barely to be seen or completely disappeared. Fig. 4b and c show LEED pictures taken at coverages of 2 and 6 ML, respectively. Although the triangles of structure I (marked by a circle in Fig. 4b,c) in both images are clearly visible, there is no sign of structure II. Summarizing the results above, one can conclude that our OMBE films are indeed highly ordered, showing the well known herringbone structure and that their ordering do not change with varying film thickness. In contrary, PTCDA films on polycrystalline gold exhibit a polycrystalline character with small PTCDA clusters on top of larger gold clusters.

3.2. OPTICAL CHARACTERIZATION

Fig. 5 shows the differential reflection spectrum of the polycrystalline PTCDA films on gold as a function of layer thickness. To enhance the signal-to-noise ratio and to give the measurements more reliability, the curves in Fig. 5 correspond to an average of five single scans. However, no further smoothing operation was applied. Although conclusions about the position of the optical transitions of the films can only be drawn after a thorough analysis of the spectra (this is discussed in section 4), qualitative differences of the shape of the spectra for varying film thickness can already be seen from the bare data (phenomenological discussion).

Two peaks A and B are visible throughout the spectra which seem not to change their energetic position. The overall shape of the spectra is very similar to thick *pc* films, whose calculated DRS is plotted as the dashed line, using the known optical constants of PTCDA, determined from thick PTCDA films (25...250 nm) on glass [10]. This is not unexpected if we recall that the *pc* PTCDA films build clusters instead of growing layer by layer.

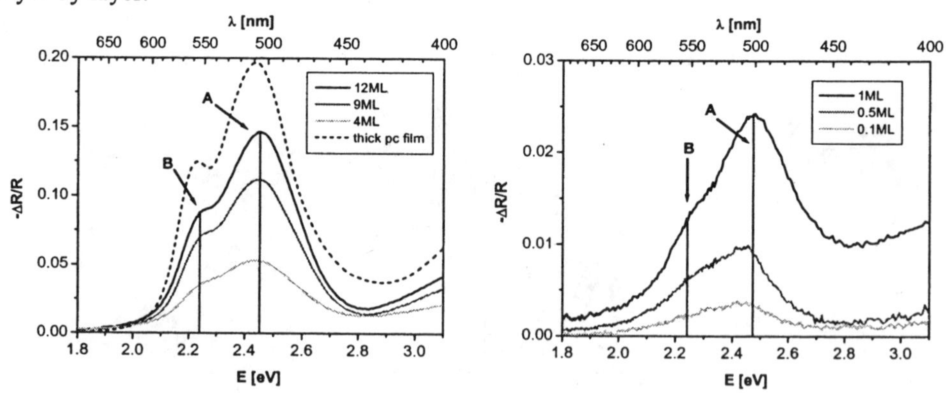

Fig. 5: *Ex situ* Differential Reflection Spectra of polycrystalline films of PTCDA for different film thickness, raw data. Dashed line: Simulated DRS of 12 ML polycrystalline PTCDA on gold, optical constants of polycrystalline thick PTCDA films, taken from [10]. The spectrum was calculated using the optical thin film program FILMWIZARD by SCI.

The situation changes remarkably for the *ho* films of PTCDA on Au(111). Their differential reflection spectra are plotted in Fig. 6. While for 12 monolayers (ML) the spectrum is very similar to both the *pc* case on gold and the thick *pc* films, already for 6 ML an additional feature C, whose energetic position is in between peaks A and B, shows up. For smaller film thicknesses this new peak C gets even more pronounced and finally becomes the dominating feature of the spectra, together with peak A, while peak B gradually disappears with decreasing film thickness. For coverages of 1 to 3 ML one also recognizes a broad, structureless feature at lower energies. It is intriguing that the absolute height of the spectra do not scale with the film thickness anymore for coverages of 2 ML and below. This is the reason why we labeled the prominent peak for submonolayer coverage A' instead of A, although it exists at the same energetic position like the latter one. We believe that this peak is of different origin which will be discussed in the next section.

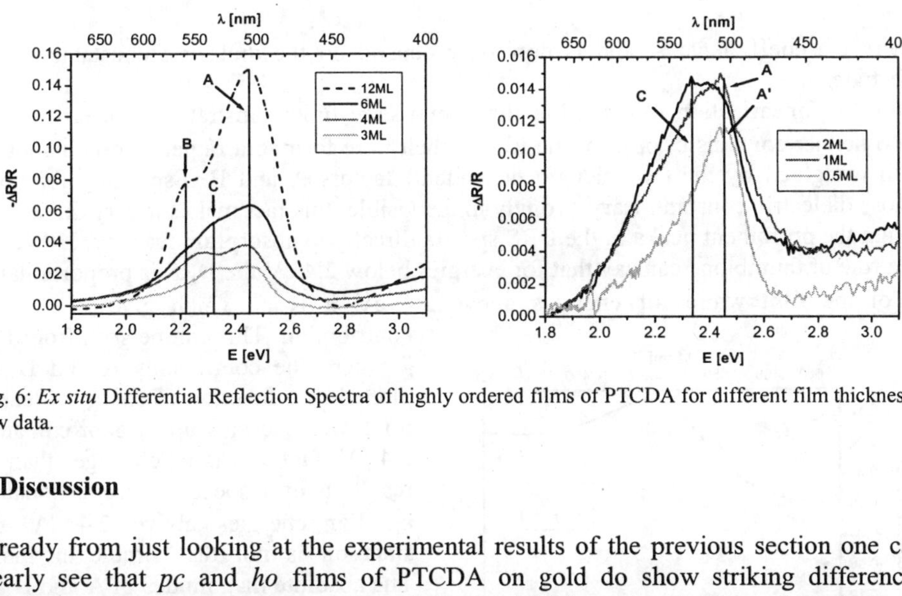

Fig. 6: *Ex situ* Differential Reflection Spectra of highly ordered films of PTCDA for different film thickness, raw data.

4. Discussion

Already from just looking at the experimental results of the previous section one can clearly see that *pc* and *ho* films of PTCDA on gold do show striking differences concerning their optical spectra. Especially for highly ordered films of PTCDA on Au(111) the following questions arise:

- What is the origin of peaks C and A'?
- Why is there no thickness dependence of the spectra for coverages below 3 ML?
- Do we observe any peak shift, i.e. exciton confinement?

But before we discuss these issues, a general remark concerning the DRS has to be given, because up to now it was not discussed clearly what kind of information such spectra contain. Do they correspond more or less to the absorption index *k* like in transmission experiments or is it something else?

To answer this question, McIntyre and Aspnes [11] describe a linear approximation for ultrathin layers. They expand the Fresnel coefficient of the differential reflection for ultrathin layers (ratio of film thickness t to wavelength $\lambda << 1$) and yield the following formula:

$$\frac{\Delta R}{R} = -\frac{8\pi t}{\lambda}\left(\frac{\varepsilon_2^s\left(\varepsilon_1^f - 1\right) - \varepsilon_2^f\left(\varepsilon_1^s - 1\right)}{\left(\varepsilon_1^s - 1\right)^2 + \left(\varepsilon_2^s\right)^2}\right) \quad (2)$$

with $\varepsilon^{s/f}$ being the real part (subscript 1) and imaginary part (subscript 2) of the complex dielectric constants of the substrate and film, respectively. Equation (2) can be written in a more compact form:

$$\frac{\Delta R}{R} = -\frac{8\pi t}{\lambda}\left(A(\varepsilon_1^f - 1) + B\varepsilon_2^f\right) \quad (3)$$

where the coefficients A and B now only depend on the dielectric constants of the substrate.

From this linearization for ultrathin films on a semi-finite substrate it is clear that the DRS neither contains ε_1 or ε_2 of the film in their pure form, but rather a combination of both, weighted by the (wavelength dependent) factors A and B. Especially for gold, whose dielectric constants vary strongly in the visible, this fact makes it very difficult to assign the prominent peaks in the DRS spectra directly to absorption features. However, as a rule of thumb one can say that for energies below 2.4 eV the DRS is proportional to ε_2 of the film while for energies above 2.4 eV it is ε_1 which gives the major contribution. This can be seen from Fig. 7 where the coefficients A and B are plotted as a function of energy. Clearly, for lower energies up to approximately 2.4 eV, factor B is much larger than A, resulting in a spectrum proportional to ε_2. For energies above 2.4 eV the situation is not that simple anymore. The absolute magnitudes of A and B are on the same order of magnitude but with opposite signs. Using this fact, (3) can be written as

$$\frac{\Delta R}{R} \cong -8\pi t A\left(\varepsilon_1^f - \varepsilon_2^f - 1\right) \quad (4).$$

Substituting the dielectric constants by the index of refraction n and absorption coefficient k by means of

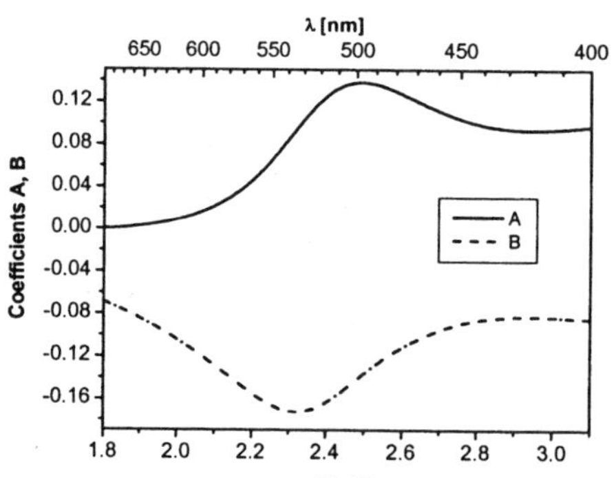

Fig. 7: Plot of the factors A and B of formula (3). For energies below 2.4 eV, factor B dominates A (A<<B) which results in a spectra proportional to ε_2. For energies higher than 2.4 eV we have $|A| \approx |B|$, but with opposite signs. For this situation, the spectra is proportinal to $(n-k)^2$. Since we know that for our films k is much smaller than n in this energy range, the DRS for this region is in its simplest approximation dominated by n^2.

$$\varepsilon_1^f = n^2 - k^2 \text{ and } \varepsilon_2^f = 2nk \tag{5}$$

yields:

$$\frac{\Delta R}{R} \cong -8\pi t A\left((n-k)^2 - 1\right) \tag{6}.$$

Since we know that for higher energies the absorption coefficient k of our films is usually much smaller than n, the DRS is for this energy region in first approximation dominated by the square of the index of refraction, n^2, which corresponds to ε_1 ($k^2 << n^2$). We will keep that issue in mind when discussing the results in the following.
As a further insight provided by the approximation for ultrathin films, formulae (3) clearly indicates that a DRS signal directly proportional to the layer thickness t is to be expected.
Coming back to the questions of interest, i.e. what causes the *ho* films to show such distinctive differences of their optical spectra when compared to the *pc* films, one can think of several reasons, which can be divided into extrinsic mechanisms like surface plasmons, interaction with the substrate, or dielectric effects, and intrinsic matters like exciton confinement, surface excitons, monomer absorption, or changes in the interaction of molecules due to structural differences of the film.

4.1. EXTRINSIC EFFECTS

It has been pointed out by several authors [12,13,14] that it is possible to excite surface plasmons on metals when the surface exhibit a periodic structure. A surface plasmon is a fundamental electromagnetic excitation on a metal-dielectric interface (see e.g. [15]). On a smooth metal surface it can not be optically excited, since its dispersion relation lies to the right of the dispersion line of light. In order to excite the surface plasmon an additional k-vector has to be provided, which can be done by a surface grating with appropriate periodicity. The question now is, can the Au(111) surface reconstruction cause the creation of a surface plasmon on the gold substrate? At a first glance it seems not important whether there is a surface plasmon or not, since we measure *differential* reflection, i.e. the difference between a film on a gold substrate and the bare substrate. If in both cases the surface reconstruction would be present it should cancel out in the DRS. However, we know from our STM measurements that the surface reconstruction vanishes after the bare gold substrate has been exposed to air[1], while in contrary, PTCDA films on top of the gold reconstruction preserves it even in air. Since we perform our optical measurements *ex situ* under ambient conditions, we not only measure the difference between a film on a substrate and the bare substrate, but also the contribution of a possible surface plasmon, caused by the preserved surface

---

[1] It has been reported by other authors that they were able to see by STM the gold surface reconstruction even in air. However, for our massive gold crystals we could only observe the surface reconstruction when imaging the crystal surface *in vacuo*.

reconstruction of the gold substrate underneath the PTCDA film. Therefore we conclude that a surface plasmon should be considered in our data analysis.
Another issue which is also closely connected to this is the possible interaction of the film with the metal substrate. It is known that molecules in close proximity to a metal can give rise to the formation of hybrid states. This contribution should be especially visible for the first layer. Could this cause additional absorption features?
Both questions can be answered by a rather simple experiment. Provided we have a material which preserves the surface reconstruction, has its absorption band at higher energies than PTCDA, and do not hinder the growth of *ho* films of PTCDA, we could use this material as an interlayer between the gold substrate and the PTCDA film and measure the DRS of the heterosystem PTCDA/interlayer/Au(111) taking interlayer/Au(111) as the reference signal rather than the bare substrate. By doing so, the influence of a possible surface plasmon is omitted and the interlayer serves as a spacer to bring the PTCDA molecules further away from the substrate to avoid direct interactions.
A material which fulfills these criteria is HBC (peri-hexaperibenzocoronene, provided by the group of K. Müllen, MPI für Polymerforschung, Mainz). Its growth on Au(111) has been studied by Sellam et al. [16] in detail. They also reported that PTCDA on top of HBC still has the same degree of ordering as if directly grown on Au(111). The absorbance spectra of a thick HBC film on glass has been measured by Proehl et al. [17] and shows that the onset of the optical absorption band is around 2.6 eV. We therefore use 3 ML of HBC as a intermediate layer between the gold substrate and the PTCDA film to check whether or not we observe the creation of a surface plasmon and/or interactions with the substrate.

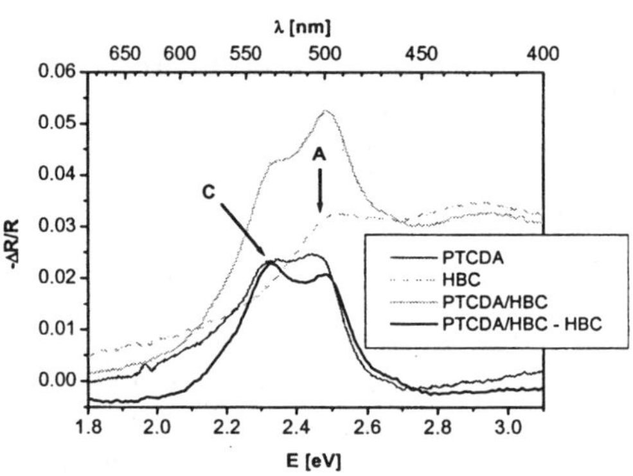

Fig. 8: *Ex situ* Differential Reflection Spectra for 3 ML of PTCDA on top of 3 ML of HBC on Au(111) (light grey curve) as well as for 3 ML PTCDA (grey curve) and 3 ML HBC (dash-dotted curve) directly evaporated on Au(111). The difference between the heterosystem PTCDA/HBC/Au and HBC/Au is plotted as black curve.

The dash-dotted curve of Fig. 8 represents the DRS of 3 ML HBC on Au(111). There is no absorption peak at the position of peak C but an unexpected feature at the position of peak A. In a next step, 3 ML PTCDA are evaporated on top of the 3 ML HBC which produces the light grey curve. If we now build the difference between the light grey and the dash dotted curve, i.e. eliminate the influence of a possible surface plasmon and compare the resulting black curve with 3 ML PTCDA *directly* grown on Au(111) (grey curve), which could contain the contribution from a surface plasmon, the following is observed:

- peak C remains unchanged
- peak A is significantly modified.

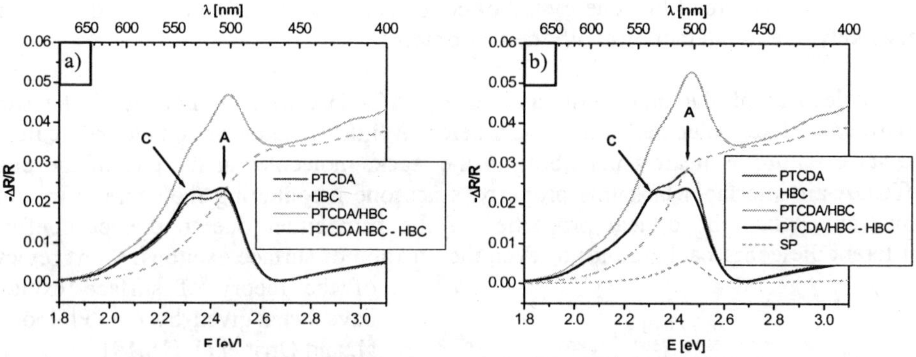

Fig. 9: Model calculation (program FILMWIZARD by SCI) of 3 ML PTCDA on top of 3 ML HBC on gold. The optical constants of 3 ML PTCDA have been extracted from the experimental DRS of 3 ML by means of a Lorentz Oscillator Model (LOM). Optical constants of HBC from transmission measurements [17] plus application of the LOM. a) without Surface Plasmon, b) an additional surface plasmon peak at 2.45 eV is assumed.

This clearly shows that peak C is neither due to a surface plasmon nor to interaction with the substrate. On the other hand, one can conclude that peak A' of Fig. 6 is most probably caused by a surface plasmon, whose absolute intensity is in the order of the difference between the heights of the grey and black curve of Fig. 8 at the position of peak A. It also explains why there is no thickness dependence of the absolute height of the DRS spectra of *ho* films of PTCDA for coverages of 2 ML and below, since the intensity of the surface plasmon should not depend on the film thickness. This is also confirmed by two model calculations with the program FILMWIZARD by SCI, shown in Fig. 9. 3 ML PTCDA on top of 3 ML HBC on gold without a surface plasmon are plotted in Fig. 9a, Fig. 9b includes an additional peak caused by a surface plasmon. The overall agreement of Fig. 9b with the experimental data of Fig. 8 is seen to be excellent. We therefore conclude that peak A' is indeed caused by a surface plasmon which is most likely created by the surface reconstruction of Au(111).

Let us now take a closer look at the low energy tail between 1.90 and 2.25 eV of the spectra of PTCDA/Au(111) (grey curve) and PTCDA/HBC/Au(111)-HBC/Au(111) (black curve) in Fig. 8. Comparing the two curves for this portion of the spectrum it seems that for PTCDA *directly* deposited on Au(111), a broad feature can be recognized which shows not up for the same amount of PTCDA deposited on HBC/Au(111). This is an indication for interactions of the molecules of the film with the substrate. Since we bring these molecules further away from the substrate by adding an interlayer like HBC, the interactions weaken, resulting in the disappearance of the broad feature at the low energy site. It is reasonable to assume that only molecules of the first few layers of the PTCDA film interact with the substrate so that these interactions are especially visible for 1 or 2 ML of PTCDA on Au(111) (see black and grey curve of Fig. 6) where its contribution for the overall spectra is rather high. Ongoing experiments with interlayers and smaller film thicknesses of PTCDA confirm this assumption. We therefore conclude that the broad low energy tail of coverages of 3 ML of PTCDA and below are very likely due to interactions with the substrate.

The molecules of the outermost layer of the PTCDA film do not 'feel' the same environment like molecules in the layers below. While the latter ones have molecules of the same nature as nearest neighbors in the stack, molecules at the end of the chain (PTCDA-air interface) lack this property, since one neighboring molecule is missing. This can change the optical properties of the film either due to dielectric effects (different dielectric background) or even the creation of surface excitons. Short reviews of the theory of surface excitons have been given by e.g. Philpott et al. and Orrit et al. [18,19].

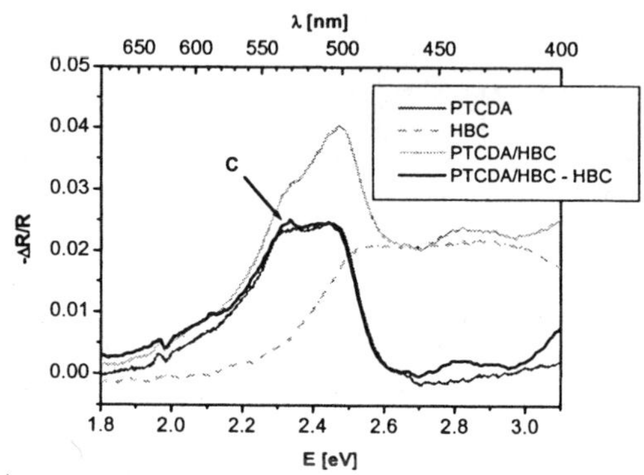

Fig. 10: *Ex situ* Differential Reflection Spectra for 2 ML of HBC on top of 3 ML of PTCDA (light grey curve) as well as for 3 ML PTCDA (grey curve) and 2 ML HBC (dash dotted curve) directly evaporated on Au(111). The difference between the heterosystem HBC/PTCDA/Au and HBC/Au is plotted as the black curve.

A simple test to check whether peak C can be assigned to a surface exciton or to dielectric effects is to alter the environment of the top layer, i.e. evaporate a thin film of HBC molecules on top of the PTCDA film, because as stated already HBC has no significant absorption at the position of peak C and its static dielectric constant is very similar to PTCDA. If peak C is indeed created by such mechanisms one should see a shift or intensity change of this transition.

Fig. 10 shows the differential reflection spectrum of 2 ML HBC on 3 ML PTCDA on Au(111) as well as the spectra of 3 ML PTCDA and 2 ML HBC directly on Au(111). The difference between the heterosystem HBC/PTCDA/Au and HBC/Au is plotted as the black curve of Fig. 10. Obviously, already the simple subtraction of these spectra reproduces almost exactly the measured spectrum of 3 ML of PTCDA directly evaporated on gold (grey curve). It is therefore highly unlikely that peak C stems from surface exciton transitions or dielectric effects since its shape or position did not significantly change upon evaporation with HBC molecules.

4.2. INTRINSIC EFFECTS

Since it was not possible to explain peak C in the *ho* films of PTCDA by means of extrinsic effects, let us turn to the questions what kind of intrinsic reasons could cause the observed behavior.

Exciton confinement effects are one possible candidate. In this case peak C would not really be a new or additional absorption feature but rather a peak which exists in thicker films as well and its energetic position would be shifted by a certain amount with decreasing film thickness. Looking at the series of spectra in Fig. 6, at the first glance one could think that with increasing film thickness peak C develops into peak B. This peak is well pronounced for thicker films (6 and 12 ML) but seems to disappear when decreasing the film thickness while at the same time peak C starts to show up. Such a blue shift of peak B to peak C is seen to be on the order of 100 meV, i.e. rather large. However, it is not clear why for 6 ML both peaks, B and C, are simultaneously visible. This could be explained if one assumes the PTCDA films to consist of islands of different thickness instead of closed layers. Our STM measurements do, however, not confirm this assumption. It is therefore highly questionable if we indeed observe a blueshift of peak B as proposed above or if peak C is rather an additional absorption feature.

In the structural characterization of chapter 3.1. it was shown that for certain coverages the PTCDA films can simultaneously consist of two structures I and II. These two structures exhibit different lattice parameters as described by Mannsfeld et al. [9]. Could this structural change give rise to the creation of peak C? As already described in the introduction the major contribution concerning the optical spectra stems from interactions of neighboring molecules within the stack rather than in the plane. Therefore, such a small variation of the lattice constants are expected to have only a small impact on the electronic properties of the film, if any. Leonhardt et al. [1] show optical spectra of the α- and β-polymorph of PTCDA, which do not contain a peak at the position of our peak C. Furthermore, their spectra are rather similar exhibiting only small energetic differences concerning the peak positions. Since we know that the lattice parameters of our films lie in between α- and β-polymorph, the creation of an additional peak C due to a transition between the two observed structures seems highly unlikely.

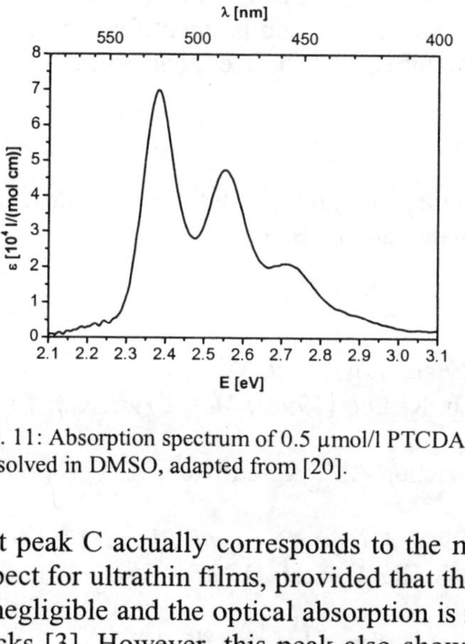

Fig. 11: Absorption spectrum of 0.5 μmol/l PTCDA dissolved in DMSO, adapted from [20].

As explained before, the DRS is proportional to the imaginary dielectric constant of the PTCDA film for energies below 2.4 eV. Due to this, we estimate the energetic position of peak C to be around 2.33 eV. Comparing this value to the monomer absorption of PTCDA in solution (see Fig. 11), which lies at 2.38 eV, the similarity of the two values is intriguing. It gives rise to the assumption that peak C actually corresponds to the monomer absorption of PTCDA as one would expect for ultrathin films, provided that the in-plane interaction of the PTCDA molecule is negligible and the optical absorption is readily explained by the interaction within the stacks [3]. However, this peak also shows up as a small shoulder for a thickness of 6 ML, which stands in contrary to the fact that monomer absorption can only occur for single molecules or monolayers of PTCDA. To account for this experimental observation but still being able to assign peak C to the monomer absorption, one has to assume either island growth (not confirmed by our STM measurements) or the collection of single molecules at the boundaries of the Au(111)-terraces. Since we are not able to observe single molecules at the terrace edges by means of STM measurements, the latter assumption still remains to be clarified. It is also not clear why we do not observe a vibronic progression and why the DRS for 2 ML and below have this very broad and unstructured character. Both matters could be connected to the fact that we deal with a metal substrate which could give rise to interactions with the molecules.

5. Summary

The optical properties of ultrathin films of PTCDA on reconstructed Au(111) and polycrystalline gold on glass were investigated by means of differential reflection spectroscopy. LEED and STM measurements confirmed the high degree of ordering for PTCDA films on Au(111), and the polycrystalline nature of PTCDA films on polycrystalline gold. For polycrystalline films, the well known optical behavior is observed with no significant change in the form of the spectra with decreasing film thickness.
In contrast, highly ordered films show a striking thickness dependence of the spectral shapes. The appearance of a new absorption peak at 2.33 eV was found, who does not show up for the polycrystalline films. All known extrinsic effects concerning the creation of such a peak could be ruled out, however, the obvious intrinsic reason remains unclear. A broad feature at the low energy site for coverages of 3 ML of PTCDA and below could be found which is very likely due to interactions with the substrate. More over, a thickness independent feature A' could be identified as surface plasmon, most likely caused by the periodic structure of the surface reconstruction.

Acknowledgement

The authors gratefully acknowledge stimulating and helpful discussions with Prof. Dr. V. Agranovich, Prof. Dr. K. Leo, Dr. M. Hoffmann, and K. Schmidt.

References

1. M. Leonhardt, O. Mager, H. Port (1999), *Chem. Phys. Lett.* **313**, 24.
2. T. Fritz, M. Hoffmann, T. Schmitz-Hübsch, K. Leo (1998), *Mol. Cryst. Liq. Cryst.* **314**, 279.
3. M. Hoffmann, K. Schmidt, T. Fritz, T. Hasche, V.M. Agranovich, K. Leo (2000), *Chem. Phys.* **258**, 73.
4. H. An, B. Chen, J. Hou, J. Shen, S. Liu (1998), *J. Phys. D - Appl. Phys.* **31**, 1144.
5. E. I. Haskal, Z. Shen, P. E. Burrows, S. R. Forrest (1995), *Phys. Rev. B - Condensed Matte*r **51**, 4449.
6. Z. Shen, S.R. Forrest (1997), *Phys. Rev. B - Condensed Matte*r **55**, 10578.
7. L.-K. Chau, C. D. England, S. Chen, N. R. Armstrong (1993), *J. Phys. Chem.* **97**, 2699.
8. T. Schmitz-Hübsch, T. Fritz, F. Sellam, R. Staub, and K. Leo (1997), *Physical Review B* **55**, 12, 7972.

9. S. Mannsfeld, M. Törker, T. Schmitz-Hübsch, F. Sellam, T. Fritz, and K. Leo (2001), *Organic Electronics* **2**, 121-134.
10. A.B. Djurisic, T. Fritz, K. Leo (2000), *Opt. Comm.* **183**, 123.
11. J.D.E. McIntyre, D.E. Aspnes (1971), *Surface Sci.* **24**, 417.
12. M.B. Sobnack, W.C. Tan, N.P. Wanstall, T.W. Preist, J.R. Sambles (1998), *Phys. Rev. Lett.* **80**, 5667.
13. W.C. Tan, T.W. Preist, J.R. Sambles, N.P. Wanstall (1999), *Phys. Rev. B - Condensed Matte*r **59**, 12661.
14. T. Lopez-Rios, D. Mendoza, F.J. Garcia-Vidal, J. Sanchez-Dehesa, B. Pannetier (1998), *Phys. Rev. Lett.* **81**, 665.
15. V.M. Agranovich, D.L. Mills (1982) *Surface Polariton*s, North-Holland Publishing Company.
16. F. Sellam, T. Schmitz-Hübsch, M. Törker, S. Mannsfeld, H. Pröhl, T. Fritz, K. Müllen, and K. Leo (2001), *Surface. Sci.* **478**, 113 - 121.
17. H. Proehl, M. Toerker, F. Sellam, T. Fritz, and K. Leo (2001), *Physical Review B* **63**, 205409.
18. M.R. Philpott, J.M. Turlet (1976), *J. Chem. Phys.* **64**, 3852.
19. M. Orrit, J. Bernard, J.M. Turlet, Ph. Kottis (1983), *J. Chem. Phys.* **78**, 2847.
20. K. Schmidt (1999), *Diploma Thesi*s, Technical University of Dresden, Germany.

STM-INDUCED PHOTOEMISSION AT SOLID-LIQUID INTERFACE

KAREN PERRONET and FABRICE CHARRA
CEA-Saclay
DSM-DRECAM-SPCSI, F-91191 Gif-sur-Yvette Cedex, France

Abstract. We extended STM-induced photoemission from metal surface to solid-liquid interface. A Au(111)-Au junction immersed in a liquid droplet was studied. We observed a decrease of the voltage threshold for photoemission when adding liquid compared to air medium, a strong decrease of the apparent tunnel barrier height and red-shifted emission spectra with two distinct modes of emission. This is attributed to the refractive index of the liquid. Our observations permit to investigate the influence on STM-induced photoemission of a self-assembled molecule monolayer formed on a Au(111) substrate.

1. Introduction

Photoemission of a metal sample excited by a Scanning Tunnelling Microscope (STM) tip has been studied under UHV and in air for several years. Several teams have worked on Au(111)-Au junctions in air or under UHV. R. Nishitani and A. Kasuya [1] studied the same junction at solid-liquid interface. They observed red-shifted spectra in liquids and attributed this to the dielectric constant at $\omega = 0$ of the liquids. They did not achieve measurements for most of the solvents chosen because their $\varepsilon(\omega = 0)$ were too large.

In this paper, we report our investigations of the STM-induced photoemission of a Au(1 1 1) / Au junction immersed in a liquid. We measured the voltage emission threshold, the apparent height of the tunnel barrier and the spectra of the light emitted. We compare our results with those obtained at air or under UHV and conclude on the influence of the dielectric constant $\varepsilon(\omega = 0)$ and of the refractive index of the liquids.

2. Experiment

A Au(1 1 1) / Au junction immersed in a liquid droplet was studied with a home made STM. Liquids were chosen to be hydrophobic to avoid problems due to a contamination with water, and with a good electrochemical

F. Charra et al. (eds.), Organic Nanophotonics, 119–126.

stability to achieve the spectroscopic measurements. Gold samples were prepared by evaporating pure gold onto freshly cleaved mica. Gold tips were made of gold wire and mechanically cut. Four liquids were considered : phenyloctane, dodecan-4-ol, tetradecane and perfluorooctane (cf. Table I). Phenyloctane and tetradecane were chosen because they are very commonly used in STM structural studies of self assembled monolayers, perfluorooctane had an usually low refractive index ans dodecanol had a higher dielectric constant at $\omega = 0$.

TABLE I. Characteristics of studied solvents. [1] value reported for perfluorohexane. [2] estimation according to the decanol serie figuring in Handbook. [3] value reported for dodecan-1-ol.

Medium	dielectric constant (for $\omega = 0$)[4]	refractive index (for Na D ray)[4]	emission threshold (V)	apparent barrier height (eV)
phenyloctane	2,26	1,49	$< 1,25$	0,5
perfluorooctane	1,76[1]	1,28	$< 1,25$	0,8
tetradecane	2,03	1,43	$< 1,25$	0,9
dodecan-4-ol	≈ 4[2]	1,44[3]	$< 1,25$	1,2
air	1	1	1,3	1,9

2.1. EXCITATION SPECTROSCOPY

The light emitted at the tunnel junction is collected by a large aperture lens and focused onto a single photon counting module (EGG, SPCM-AQ-15, with 60 cps dark count). The spectral range of detection is 400-1000 nm. Tunnel current characteristics $I(V)$ and photon counting rates $N(V)$, as well as $I(s)$ and $N(s)$ (s being the tip height along the tip axis) were recorded on different points of the sample.

2.2. EMISSION SPECTROSCOPY

We also observed the wavelength distribution of the light emitted by the tunnel junction immersed in tetradecane. In this case, the light was collected by a microscope objective, then passed through a dispersing prism (flint glass) and is finally focused on a CCD camera (Andor, DU401-BR-DD) (cf. Fig. 1). The spectral range of detection is 400-1100 nm. The spectral resolution is about 5 nm, due to the low dispersive power of the prism. Each spectrum is integrated for 2 min, tunnelling current is set to 3nA.

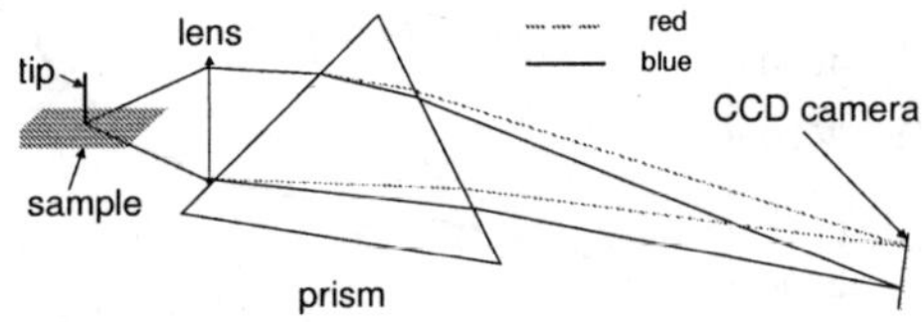

Figure 1. Spectrometer

3. Results

3.1. VOLTAGE SPECTROSCOPY

Fig 2 shows a typical set of characteristics of the studied junctions. General shape of the curves are similar to those obtained in air or under vacuum. $I(V)$ relationship is linear between -1V and 1V and superlinear above. $N(V)$ curves are symmetrical, at the dark noise level for low tunnel biases and increasing above a threshold bias. In air, the threshold value is 1,3 V. For all the studied liquids, we recorded a signal for biases larger than 1,25 V. As it is known, the mechanism of excitation of the plasmon modes is through inelastic tunnelling, so that the energy of the photons emitted is less or equal to eV_{tunnel} where V_{tunnel} is the tip-sample bias. Due to the limited spectral range of the detector, we are not able to detect photons with a wavelength larger than 1000 nm, ie. with an energy less than 1,25 eV. We can therefore deduce that the emission threshold is less than 1,25 V.

3.2. TIP HEIGHT SPECTROSCOPY

On Fig. 3(a) are plotted current and emitted photons versus tip retraction s for tetradecane. For all the liquids we observed an exponential decay with the same decay constant for both current and photons. In air also, both current and photon emission have an exponential decay with the same slope. However, this decay is weaker in liquids (cf. Fig.3(b)). We could deduce the apparent height of the tunnel barrier using the formula [2]

$$\Phi(\mathrm{eV}) = 0,952 \left[\frac{\mathrm{d} \ln I}{\mathrm{d}s(\mathring{A})} \right]^2 \quad (1)$$

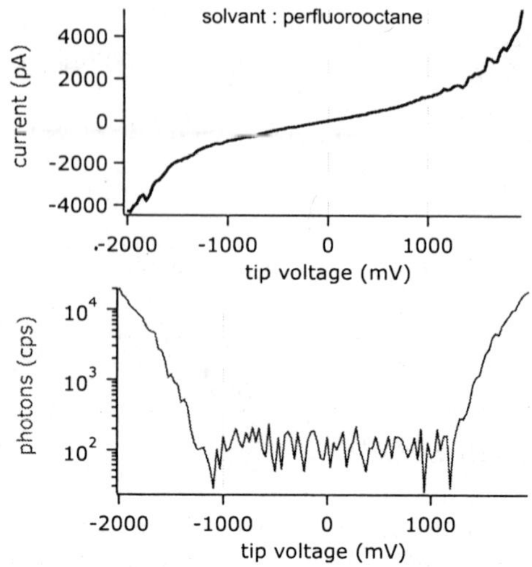

Figure 2. Voltage spectroscopy for Au(111)/Au junction in perfluorooctane

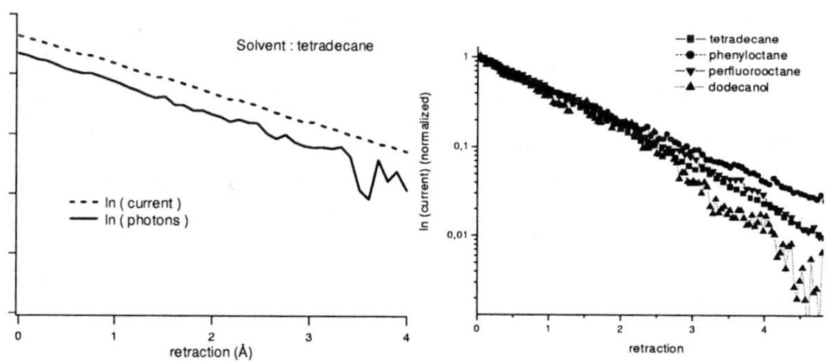

Figure 3. (a) Tip height spectroscopy obtained for tetradecane. Current and photon emission. (b) Comparison of the decay constants for the different liquids

The calculated values are reported in Table I. For perfluorooctane, we observed two regimes with distinct decay constants and the value given in Table I corresponds to the second part of the curve (cf. Fig 4).

3.3. EMISSION SPECTROSCOPY

Spectra of the emitted light were recorded only for tetradecane (cf. Fig.5. We observed two different modes of emission : for voltages less than 1.6V,

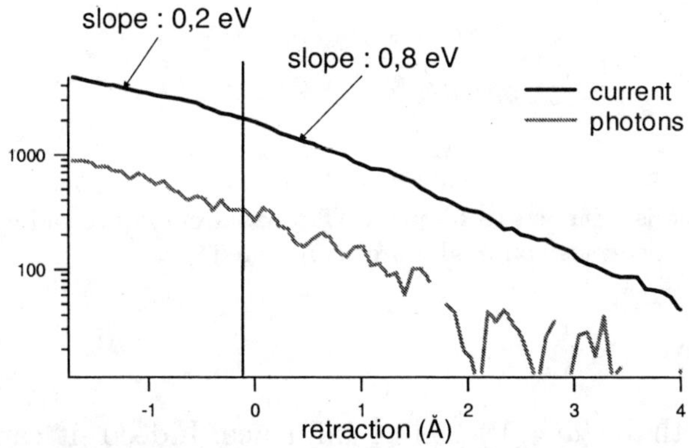

Figure 4. Existence of two domains with different barrier characteristics for perfluorooctane

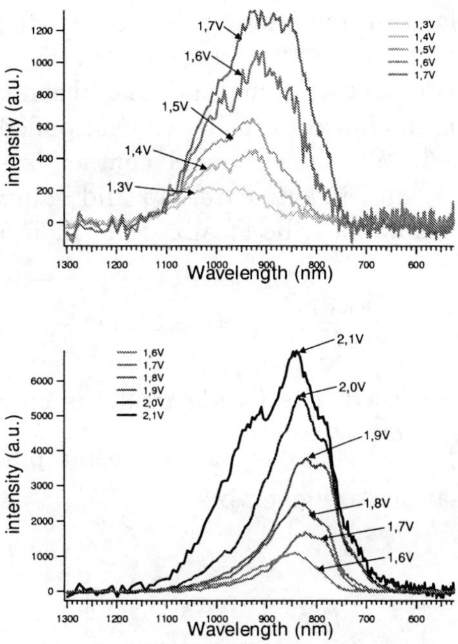

Figure 5. Emission spectra in tetradecane for different tunnel voltages. Tunnel current is set to 3 nA.

spectra are centered at about 960 nm while for voltages more than 1.7V, they are centered at about 820 nm. For 1.6V and 1.7V, we saw both modes, but not simultaneously. The shape of the peaks is rather bumpy. However, in most cases, we can clearly see a substructure and, at least two components.

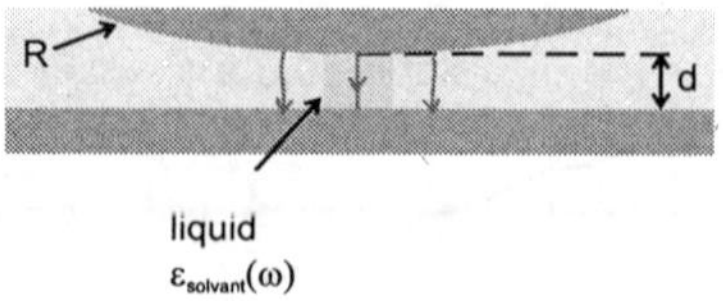

Figure 6. Sample is assumed to be plane. Tip has a curvature radius R. Tip-sample distance is d. Localized plasmon modes are represented.

4. Discussion

First, it seems that $\varepsilon(\omega = 0)$ has no influence. Indeed, it can only modify the static electric field. However we measured an emission threshold lower than in air. This means that the applied voltage was totally transmitted to the photon, despite a $\varepsilon(\omega = 0)$ of the liquid above 2. More specifically, an intermediate mechanism involving electrons of the solvent molecules in the coupling with the plasmon mode may exist but is not preponderant.

On the contrary, the refractive index of the liquid plays here a key role. The dielectric constant at ω modifies the plasmon mode frequencies. We can quantify this modification using the simplified junction geometry represented on Fig. 6. When the tunnel contact is established, we can assume $d \ll R \ll \lambda$, and in our case, tip and sample are made of the same metal. We can show that the localized plasmon modes exist for [3]:

$$\frac{\varepsilon_{\text{solvent}}(\omega)}{\varepsilon(\omega)} = -(l + 1/2)\sqrt{\frac{d}{2R}}$$

For the frequencies used, the Drude model is correct to describe the metal and then $\varepsilon(\omega) = \frac{\omega_p^2}{\omega}$ where ω_p is the plasma pulsation of the metal. So the plasmon pulsations are given by :

$$\omega^2 = \sqrt{\frac{d}{2R}}\,(l + 1/2)\,\frac{\omega_p^2}{\varepsilon_{\text{solvent}}(\omega)} \tag{2}$$

As a consequence, the higher the refractive index is, the more the emission spectrum is red-shifted and the less the detector is sensitive. We can estimate the first plasmon mode energy (for $l = 0$). Using $d \simeq 10$ and $R \simeq 100$ nm (typical values for a gold tip) we get : $\varepsilon(\omega) \simeq -30\varepsilon_{\text{solvent}}(\omega)$. The calculated values are reported in Table II. They are consistent with the red-shifted threshold observed. Furthermore, if the refractive index

TABLE II. Estimation of the first plasmon mode energy

medium	refractive index	plasmon mode energy (eV)
vacuum	1	1,4
perfluorooctane	1,28	1,1
tetradecane	1,43	1
dodecanol	1,44	1
phenyloctane	1,49	1

increases, the difference between the plasmon modes decreases. So an electron tunnelling can excite more modes and consequently, the probability to excite the first (dipolar) mode, ie. the one with a radiative decay, decreases.

Besides, we observed a lower barrier height for all the liquids compared to the value in air or under vacuum. This is probably due to the electrons of the solvent molecules reducing the barrier height to the tunnelling electron. We can also notice that even with a liquid, $I(s)$ and $N(s)$ curves have the same exponential decay constant. This again lets us think that the preponderant mechanism for tip induced plasmon excitation has not been drasticly modified by the presence of the liquid.

The wavelength distribution of the light emitted, for tetradecane, shows two distincts modes of emission : the first occurs when the tunnel bias is less than 1.7V and the second above 1.6V. We could first think that these two modes corresponds to two different plasmon modes, but in this case, the second mode would correspond to $l = 1$, that is to say to a quadrupolar mode and it would not be so intense. Furthermore, these two modes don't coexist, except for 1.6V and 1.7V. As a consequence, these modes correspond to two different junctions : at 1.7V, the structure of the junction is strongly modified. There is no sudden change of the geometry. On the other hand, either the optical properties or the mechanism of the electron - photon conversion can be modified.

We can however notice that for all spectra, experimental high-energy cut-offs are in good agreement with calculated ones ($\lambda_{\text{cut-off}} = \frac{hc}{eV_{tunnel}}$). This indicates again that a direct tunnelling from tip to surface occurs at all voltages. On the contrary, as V_{tunnel} increases while tunnel current is fixed, the tip-sample distance increases. Consequently, above a given bias value, an additional molecular layer can be located between the tip and the sample and this strong change in the structure of the solvent at the junction level modifies significantly the value of $\varepsilon(\omega)$. This change is confirmed by a tip height spectroscopy obtained with perfluorooctane. The decay plotted on Fig. 4 shows a non linearity which is enhanced if we increase the set point value, ie. if we go closer to the sample. We can even observe two different domains with two distinct decay constants. This proves again that the barrier characteristics change with the tip-sample distance and that he molecular organization of the liquid right under the tip changes with s and modifies the exact (microscopic) value of $\varepsilon(\omega)$ at the junction.

5. Conclusion

We have investigated the conditions of STM induced light emission at solid-liquid interface for several liquids and the characteristics of the tunnel barrier in these conditions. We have shown that the refractive index of the liquid is responsible for the lower emission voltage threshold but may also decrease the emission efficiency because more plasmon modes (and in particular the non radiative ones) can be excited. The presence of the liquid doesn't modify the main mechanism of light emission, ie. the coupling between the inelastic tunnel electrons and the localized plasmon modes but the microscopic refractive index value can vary with the tip-sample distance and so affect the plasmon mode frequency. Finally, the electrons of the solvent molecules contribute to reduce the apparent height of the tunnel barrier because their presence gives to the tunnelling electron more pathes to go through the barrier. Our observations permit to investigate the influence on STM-induced photoemission of a self-assembled molecule monolayer formed on a Au(111) substrate.

References

1. R. Nishitani and A. Kasuya. : 1999, 'STM induced photon emission at the solid-liquid interface', *Surf. Sci.*, **1**, pp. 433-435
2. N.D. Lang : 1988, Apparent barrier height in scanning tunneling microscopy, *Phys. Rev. B*, **37**, pp. 10395
3. A.G. Mal'Shokov : 1990, Surface-enhanced raman scattering : the present status, *Physics Reports*, **194**, pp. 343
4. *Handbook of Chemistry and Physics 80th edition.* 2000.

XYZ on a Chip: Nanoscale fabrication, fluidics, and optics directed toward applications within biology and medicine.

ZHONGLIANG TANG,[1] GRACE CHAO,[2] AUREA TUCAY,[3] ERICA TAKAI[2], DJORDJE DJUKIC[3], MARY LAURA LIND[1], CLARK HUNG,[2] EDWARD GUO[2], ALAN WEST,[1] RICHARD OSGOOD,[3] AND JAMES T. YARDLEY[1]

[1]Department of Chemical Engineering,
[2]Department of Biomedical Engineering,
[3]Department of Electrical Engineering,
Columbia University, New York, NY 10027

Abstract.

We have developed techniques for fabrication of micro-scale and nano-scale structures in polymeric media and we have developed methodology for control of fluid flow within these systems. We have also developed methods for modeling fluid flow in these systems under both electrokinetically driven and hydrostatically driven conditions. These capabilities offer new opportunities for exploration of biological function in many systems. We have illustrated these capabilities for *in vitro* studies by examination of the influence of medium composition and flow rate on the growth of cells by studying cell volume and shape changes of chondrocyte cells and ACL fibroblast cells at regulated osmotic loadings close to physiological frequencies. We have also demonstrated the development of an artificial medium for study of the kinetics and biological function of biological cells with specific application to understanding the function of osteocyte and osteoblast cells within the bone structures. These experiments demonstrate the efficacy of new generations of bio-chips for studies of biological function.

1. Introduction.

Microfluidic devices are suitable for biomedical and biological *in vitro* studies, where *in vivo* investigations are either extremely difficult or costly[1, 2]. We have begun to develop technologies suitable for creation of such devices. These technologies include fabrication of polymeric micro fluidic devices through a low-cost replication process, modeling tools for describing quantitatively fluid flow in microscale and nanoscale devices, capabilities for experimental examination of flow parameters driven by both hydrostatic forces and electro osmotic forces, optical examination of biological processes, and processes for rendering these

F. Charra et al. (eds.), Organic Nanophotonics, 127–138.

devices biocompatible. These tools can be used to provide *in vitro* environments in which living cells may be isolated and then subjected to fluid environments that may be controlled at will. Thus for example while static osmotic loading has been used as a tool for investigating cell mechanotransduction [1] as well as cell material properties [2] for cells including osteocytes, chondrocytes and intervertebral disk cells, the response to dynamic osmotic loading is still vastly unknown. Here we demonstrate a microfluidic device designed to apply automated dynamic osmotic loading up to 0.1 Hz on cultured cells. The gas permeable and transparent polymer provides a sterile tissue culture environment and permits visualization of cells using fluorescence microscopy.

2. Device Fabrication.

We have chosen to fabricate our devices by creating a master with the inverse of the desired structure created in a negative photoresist material SU-8 using conventional photolithography. The SU-8 material provides for accurate production of high aspect ratio geometrical shapes with thicknesses ranging from a few hundred nanometers to several hundred microns [3]. In this process, commercially available SU-8 photoresist (MicroChem Corp., Newton, MA) is spun onto silicon wafer substrates, yielding films with thicknesses ranging from 500 nm to 200 μm. The films are then subjected to a pre-exposure bake at 95°C for a time specified by the manufacturer according to the film thickness. After lithographic exposure with a conventional mask aligner, the films are then subjected to a post-exposure baked at 95°C for 15 min before developing in a SU8 developer (MicroChem Corp.) for 5-15 minutes depending on film thickness. Figure 1 shows a typical master device structure on a 10-30 micron dimensional scale.

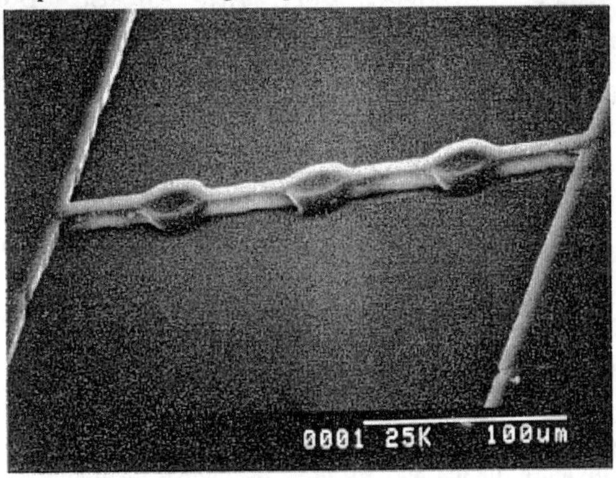

Figure 1. Scanning electron micrograph of a portion of a biochip master fabricated in SU-8 photoresist.

We then create the desired nanostructure or microstructure through molding into polydimethylsiloxane (PDMS). PDMS has been chosen because it is capable of high-resolution replication, is highly optically transparent, can be surface-treated to achieve biocompatibility, and because surfaces of PDMS form quality liquid seals reversibly. Our PDMS is prepared from Sylgard 184 (Dow Corning) by mixing the precursor and curing agent at 10:1 weight ratio. This material is then poured over the SU-8 master, which has been previously coated with a thin layer (~50nm) of gold to improve the release of the PDMS replica after curing. Before curing, the mixture is placed in a desiccator to vacuum out the bubbles formed during mixing.

Typical curing conditions are 70°C for 12 hours[4]. Figure 2 shows a scanning electron micrograph of the replica of the structure shown in Figure 1. The features are accurately replicated including slight undercut in the master for this particular example.

3. Fluidics for polymeric microstructures and nanostructures.

For flow modulation, driven by either electrokinesis or pressure differential, our approach is to introduce a composition modulated fluid flow into the microsystem of interest. The ability to modulate at a given frequency or in a given time domain is determined through two primary factors. First, the characteristic length of the composition gradient must be greater than the diffusion length characteristic of the process involved. This characteristic diffusion distance increases in proportionality to the square root of distance downstream of the inlet junction. The maximum modulation frequency may also limited by fundamental characteristics of the rate of convection. In short, to create an electrolyte with a spatially defined compositional structure, the region of interest must be completely filled with solution from one fluid source before switching to another source. This is achieved at higher fluid velocity or lower switching frequency. These conditions are approximated with the requirement [5]

Figure 2. Scanning electron micrograph of a replica of the master shown in Figure 1 prepared in PDMS.

$$f < \frac{U}{W} \tag{1}$$

where U is the fluid flow velocity, W is the channel width and f is the modulation frequency. If we define a dimensionless frequency as

$$f^* = \frac{fW}{U} \tag{2}$$

then equation (1) corresponds to $f^* < 1$. Of course, equation (1) is only an order of magnitude approximation. For example, computer simulations show that the actual transition frequency changes slightly with inlet junction geometry[5] for geometries similar to those explored here.

4. Electrokinetic Flow Modulation.

We are interested in creating fluid flow within our microstructures with controlled flow rate and fluid composition. Electro-osmotic flow control is very attractive for microstructures because high flow rates can be achieved even through highly constricted structures with application of very modest electric fields [6]. In this scheme, the flow velocity is proportional to the applied electric field E (or gradient of the potential Φ) through the electroosmotic mobility μ_{eo}.

$$\vec{u} = -\mu_{eo}\vec{E} \tag{3}$$

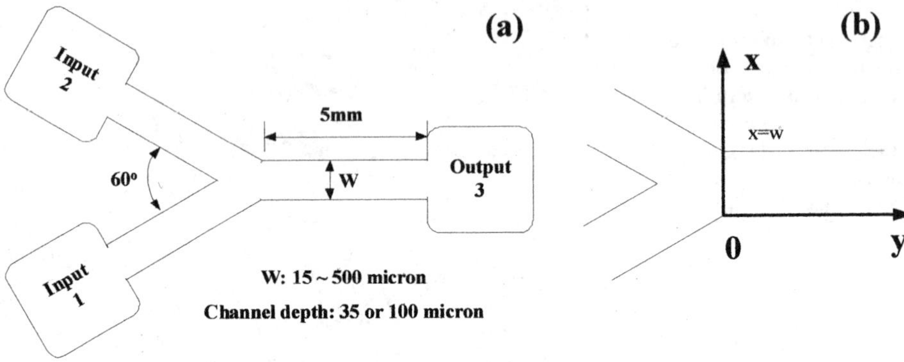

Figure 3. (a) Illustration of the chip used (drawn not to scale); (b) Coordinate system.

The mobility μ_{eo} depends on the type of fluid and the nature of the interface with the solid. For flow of a solution of 0.2 mM KNO_3 in our PDMS structures for example, we have found $\mu_{eo}=2.8\times10^{-8}$ $m^2\ V^{-1}\cdot s^{-1}$ by direct measurement of flow velocity in a single channel [7]. For prototypical flow modulation studies we examined the geometries depicted in Figure 3. In the experiments described here, the junction angle is 60° and the inlet and outlet channel lengths are 5 mm. The channel-width W varies between 15 μm and 500 μm, and a typical channel depth is 35 μm. Where flow is measured with fluorescence imaging, we also need to consider the effect of dye diffusion. The influence of diffusion on these experiments can be estimated through a dimensionless parameter known as Peclet Number (Pe) which depends on the diffusion coefficient D_{ab} through the relationship Pe = UW/D_{ab} [8]. So long as the diffusion coefficient is chosen so that Pe >> 1, effects of diffusion will be dominated by the convective phenomena of interest here.

We have fabricated many "biochips" based on the geometry of Figure 3 as described above. Typical experimental procedures are as follows. Prior to use, the chip is cleaned and assembled as described previously[4]. A 60 μL aliquot of 0.2 mM KNO_3 electrolyte is dispensed into each of the two input reservoirs and a vacuum is applied at the output reservoir to flood the channel. After flooding, the input and output reservoirs are then cleaned, and the input reservoir 2 and the output reservoir 3 are refilled with 60 μL of 0.2 mM KNO_3 solution. Input reservoir 1 is refilled 60 μL 0.2m M KNO_3 with 50 ppm calcein dye. This fluorescent dye is used as an imaging probe. A platinum wire electrode is dipped into each of the reservoirs and the circuit is connected as illustrated in figure 4 (a).

The assembled chip is then placed under a fluorescence microscope on a precisely leveled stage and a voltage is applied across the channel. An HP 3312A function generator is used to control a relay (SPDT) to switch voltage between the two input reservoirs. A CCD camera (Sensys0401E, Roper Scientific) is used to record a time-lapse fluorescence image sequence. The fluid flow modulation in the outlet channel is analyzed by processing the image sequence. For example, Figure 5 shows relative fluorescence intensity profiles at one width down stream ($y=W$) at three different positions x/W=0.25, 0.50, 0.75 across the channel width for several example dimensionless frequencies. Detailed conditions are: (a) W = 500μm, f = 0.07 Hz, and E = 200V/cm (Pe = 509, f^* = 0.063); (b) W = 500μm, f = 0.12Hz and E = 200V/cm (Pe = 509, f^*=0.11); (c) W = 135μm, f = 0.96Hz and E = 200V/cm (Pe = 137, f^*=0.23); (d) W = 500μm, f = 0.53Hz and E = 200V/cm (Pe = 509, f^*=0.47). These data demonstrate that the intensity profiles predominantly depend on the dimensionless frequency[5]. At low dimensionless frequencies, different composition packets completely separate from each other, thus form regions of defined composition, while at higher dimensionless frequencies, flow modulation diminishes and different composition regions oscillate in half of the channel width. The dimensionless frequency that separates those two regimes is about f^*=0.2[5].

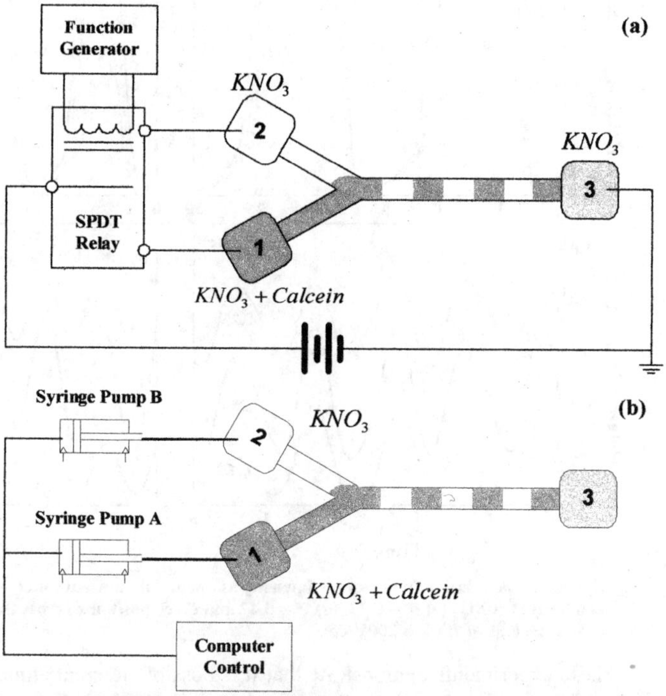

Figure 4. (a) Setup of microfluidic flow modulation by electrokinesis; (b) Setup of microfluidic flow modulation by hydrostatic pressure differential.

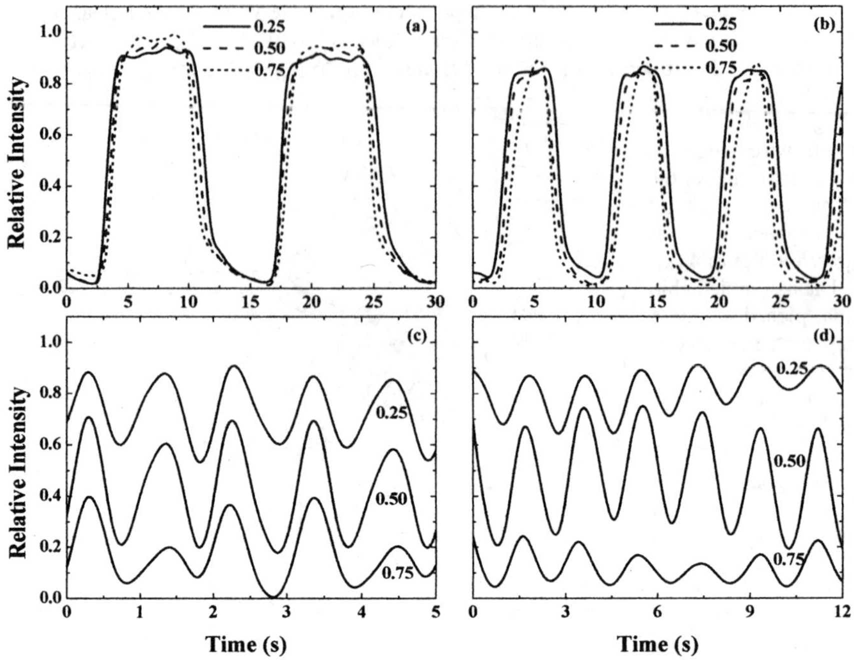

Figure 5. The relative fluorescence intensity at one width downstream (y=W) for (a) f* = 0.063, (b) f* = 0.11, (c) f* = 0.23, (d) f* = 0.47 and three positions across the channel-width x/W=0.25, 0.50, 0.75; E = 200V/cm.

These experiments demonstrate that it is possible to create modulated electro-kinetic flow within microfluidic structures that can be modulated at frequencies higher than 10 Hz, depending upon device geometry. This flow matches well that expected from conventional fluid flow modeling.

5. Flow control by hydrostatic flow modulation

It is also possible to deliver modulated fluid flow in microfluidic systems through hydrostatic pressure differential. Although there are severe limitations on the modulation frequencies that can be obtained, there are some advantages, particularly for biological applications. For example, due to high conductivity of the cell media, considerable current is associated with electro kinetic flow, which may cause hydrolysis and corresponding bubble formation. Also, the electric field used in the electrokinetic flow and the corresponding current flow may impose complications to cells and other living tissues in the biological and biomedical studies. Thus we have also explored flow modulation due to hydrostatic pressure differential. For these experiments the biochips are cleaned, assembled, flooded and the reservoirs are refilled as described in the previous section. A fluorescent dye is also used as an imaging probe. A syringe pump, controlled by a computer, is connected to each of the input wells to modulate the hydrostatic height in the well through a needle (Becton-Dickenson, 21G), as illustrated in figure 4 (b). The assembled chip is then placed under a fluorescence microscope on a precisely leveled stage, and the syringe pumps are programmed to control the hydrostatic pressure level as illustrated in Figure 6. Images are acquired and processed in the same way as described in the previous section. Typically, only intensity profiles at the middle of the channel across the width (x/W=0.50) are analyzed due to the low frequencies operable for hydrostatic pressure differential driven flow modulations. At low frequencies, intensity profiles across the entire channel width are similar as seen also in the case of electrokinetic flow (Figure 5a).

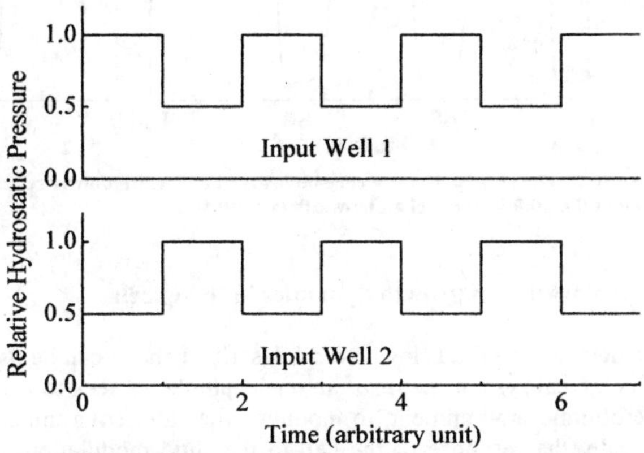

Figure 6. Hydrostatic pressure changes driven by syringe pumps.

From the point view of forming spatially composition modulated fluid flow in the outlet channel, no characteristic difference exists between flow driven by electrokinesis and flow driven by hydrostatic pressure differential. The basic principal illustrated in equation (1) holds for both. There are advantages of employing hydrostatic pressure differential as driving force including (1) provision of an electric field free, biologically friendly environment to cells and living tissues and (2) avoidance of electrolysis (bubble formation due to decomposing of water) at the high conductivity levels required to sustain living cells. The disadvantage of this mechanism is that it is more difficult to go to high modulation frequencies at small geometries due to the limitation of syringe pump's response time.

Figure 7 shows relative fluorescence intensity profiles at one width downstream ($y=W$) at the middle of the channel (x/W=0.50) across the channel at two different modulation frequencies. The hydrostatic pressure differential applied in those studies is about 10 Pa/mm, which induces a flow of about 220 μm/s in the outlet channel. Detailed conditions are: (a) W = 300 μm and f = 0.027 Hz (Pe = 120, f^* = 0.037); (b) W = 300 μm and f = 0.015 Hz (Pe = 120, f^*=0.021). The calculated wall shear stress was about 1.22×10^{-2} Pa (an order of magnitude smaller than observed to elicit a biological response in most cells). Theoretical calculation of the hydrostatic pressure inputs in figure 7 also predicts that the volumetric flow rate in the output channel is constant due to chip symmetry [9].

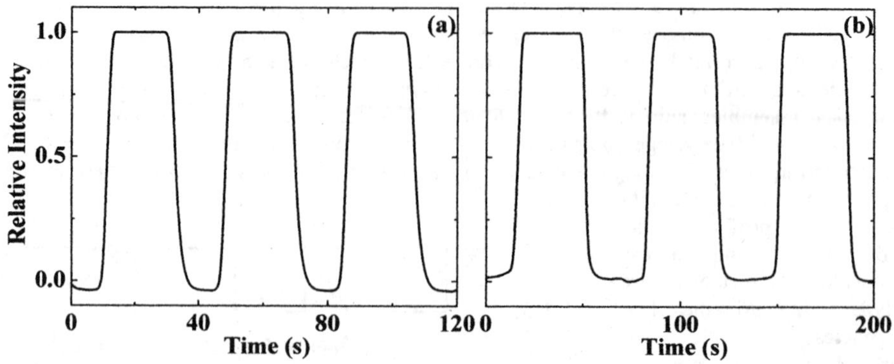

Figure 7. The relative fluorescence intensity at one width downstream (y=W) for (a) f* = 0.037, (b) f* = 0.021 and at the middle of the channel-width (x/W=0.50).

6. Exploration of growth dynamics in living cells.

The devices and fluid flow control described above can be used to explore dynamical processes in biological systems *in vitro*. In one example we seek to isolate individual living cells within a microfluidic environment, to modulate the fluid environment of the cell, and then to observe optically the response of the cell to the fluid modulation. Primary bovine chondrocytes are isolated via enzymatic digestion from young calf cartilage and cultured in DMEM with 10% FBS. Primary anterior cruciate ligament (ACL) fibroblasts are harvested by explant culture from young calf ligaments. Prior to experiments, cells are isolated with trypsin EDTA and labeled with CellTracker-CMFDA dye (Molecular Probes) for 20 minutes and plated in the microchannel. A hyperosmotic medium is prepared by adding sucrose to serum-free DMEM and the resultant osmolarity is determined by an osmometer (Advanced Instruments) to be 580 mOsm (injected through well 1), while the serum-free DMEM is 360 mOsm (injected through well 2). The hyperosmotic medium is loaded with 5 mg/L Texas Red-conjugated-dextran (70 kDa, Molecular Probes) for fluorescence detection. For isotonic loading, one of the isotonic channels is labeled with the same dye. Dynamic osmotic loading is applied at frequencies 0.1, 0.05, and 0.025 Hz. Fluorescence images of the cells (n=3-6/group) and the medium are captured with MetaFluor software (Universal Imaging) at four times the applied frequency. In these experiments we employ hydrostatic pressure differential driven flow modulation to avoid electrolysis due to the high conductivity of the cell culture media. The cell osmotic loading is done under hydrostatic pressure differential modulation at 0.1, 0.05 and 0.025 Hz. Under these flow conditions, chondrocyte cells demonstrate significant cell size changes (33%) with dynamic osmotic loading at 0.025 as demonstrated in Figure 8.

In order to examine the cell responses more quantitatively, we have developed an image processing algorithm including: smoothing with a Gaussian filter, computation of bimodal histogram, identification of the 'best' gray level threshold to isolate cells and binary segmentation [10]. The cell osmotic loading is done under hydrostatic pressure differential modulation at 0.1, 0.05 and 0.025 Hz. Flow conditions are specified in the previous section. Under such flow conditions, ACL fibroblasts demonstrate significant cell size changes with dynamic osmotic loading at both 0.025 (30%) and 0.05 Hz (27%) (Figure 9 (a)). Chondrocytes at 0.025 Hz also show cell size changes up to 33%. At 0.05 Hz, however, chondrocytes only exhibit size changes up to 4% (Figure 8, 9 (b)). Similar results are observed at 0.1 Hz, whereas isotonic control of all groups shows no correlation with the applied medium changes. These dynamic hyperosmotic loading studies reveal that chondrocyte cell size does not change significantly at applied frequencies above 0.05 Hz, whereas ACL fibroblast cell size continues to change. The physiological implications of these findings are at present unknown, and may suggest disparate material properties between these cell types. Application of optical probes (e.g., fluorescently labeled ion sensitive dyes) will permit additional biological changes of the cell to be studied in response to dynamic osmotic loading. With the channel geometry afforded by microfluidic technology, this versatile system can be used to apply different conditions to cultured cells with minimal flow-induced shear.

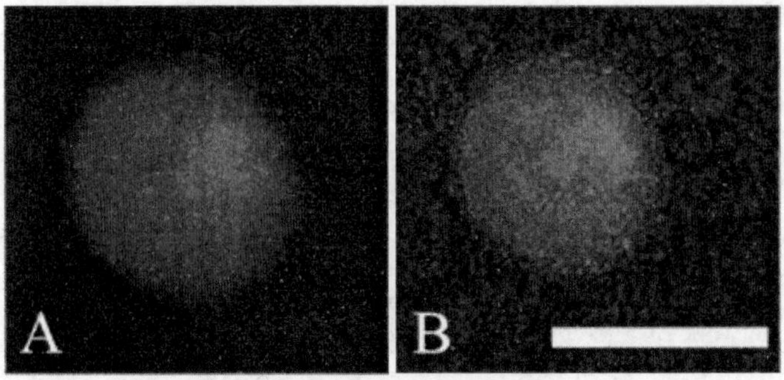

Figure 8. Chondrocyte size change at 0.025 Hz in (A) isotonic and (B) hyper-osmotic medium. The cell is labeled with CellTracker dye and hyperosmotic media labeled with Texas Red. Scale bar = 10μm.

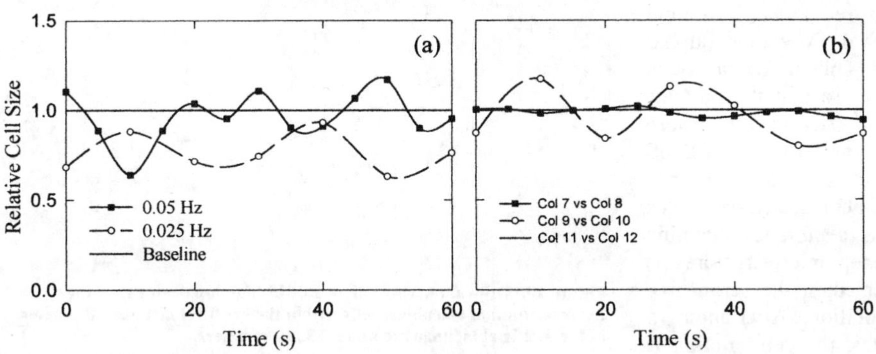

Figure 9. Mean cell size changes in response to applied dynamic osmotic loading in (a) ACL fibroblasts and (b) chondrocytes.

7. Placement of cells within polymeric microstructures for in vitro studies.

The results demonstrated here suggest that polymeric microstructures and nanostructures can provide new and exciting means for explorations of biological responses under controlled conditions. As a specific example we can consider understanding dynamics of cells found in human bone structures. Important cells include osteocytes and osteoblasts that are known to be critical for the management of calcium levels within the body. These cells normally reside in the bone within "pores" or lacunae of dimension approximately 10-15 microns. These lacunae are interconnected through a network of nanoscalar "channels" or canaliculae of diameter approximately 500 nm with interconnection distances typically a few hundred micrometers. The cells are capable of extending the cell wall into the channel to form "pseudopods", the precise function of which is not well understood at present. Since the bone structure is generally opaque and since network of interconnections is complex, it is difficult to examine directly processes and responses in these important systems. Our proposition is to create an *optically transparent artificial bone structure* in which we can introduce specific biological cells into structures of well-defined geometry. We can use the power of microfluidics to control the rate of flow and the flow composition of the fluid environment and to analytically examine changes in composition induced by the cell. In addition, we can optically observe the cell and its environment and we can use optical techniques such as fluorescence to monitor concentrations of specific species (such as Ca^{++} ions for example).

For our initial investigations into this concept we have fabricated in PDMS through the techniques described here simple structures about 15 microns thick with 25 micron diameter pools to simulate a "pore" interconnected through 10 micron wide channels over a distance of 30 microns to simulate the canaliculae as indicated for example in the structures shown in Figure 2. We then fill the internal regions of the devices with a solution of fibronectin (FN) and incubate for 90 minutes at 37°C to provide a surface suitable for cell attachment. Following this treatment we treat the entire structure with bovine serum albumen for 60 minutes at 37°C to provide a surface that will reject cell attachment. The osteoblast cells (MC3T3) are suspended in serum-free alpha Minimum Essential Medium (alpha-MEM, GibcoBRL) to avoid the chance that the proteins found in serum will absorb onto the treated PDMS surface and decrease cell attachment selectivity. The PDMS membrane containing the open microstructures is immersed in the serum-free cell solution. After about 15 minutes, the cell solution is removed and the microstructures are rinsed with phosphate buffer solution. Until needed, the PDMS

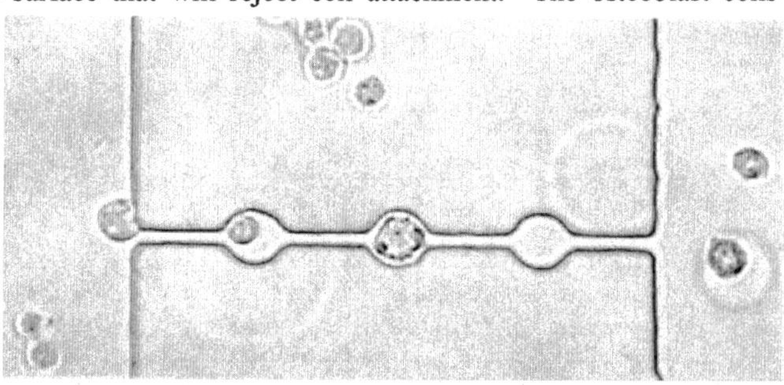

Figure 10. Optical micrograph of PDMS bio-chip device showing incorporation of osteoblast cells within the artificial lacunae. Diameters of the artificial lacunae are about 15 micrometers.

microstructures with attached cells are incubated at 37 degrees C in alpha-MEM with 10% fetal bovine serum (FBS). Figure 10 shows that it is possible to capture individual cells within the artificial "pore" structures. When these cells are allowed to grow for 1 day the cell walls extended to meet, creating pseudopod structures as indicated in Figure 11. These experiments clearly show that individual cells can be placed within artificial pore structures suitable for microfluidic control, that these cells can live, and that cell morphological changes to form pseudopods can be accomplished in artificial nanostructures. We are now producing geometries in which the pore geometries and the channel geometries can be independently controlled. This should set the stage for a number of very exciting investigations.

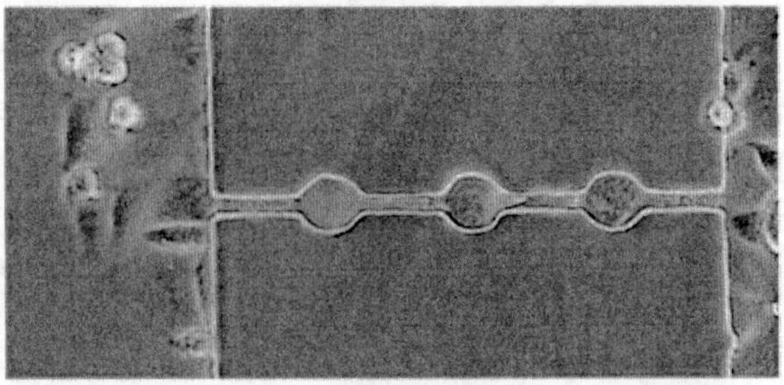

Figure 11. Optical micrograph of a bio-chip device containing osteoblast cells after incubation for 24 hours.

Acknowledgements

This work was supported by the National Science Foundation under award ECS-99-80828, the National Institute of Health under award AR48791 and the Whitaker Foundation.

References

1. Beebe, D., Wheeler, M., Zeringue, H. , Walters, E., and Raty, S. (2002) *Microfluidic technology for assisted reproduction* Theriogenology **57**, 125-135.

2. Erickson, G.R., Alexopoulos, L.G., and Guilak, F. (2001) *Hyper-osmotic stress induces volume change and calcium transients in chondrocytes by transmembrane, phospholipid, and G-protein pathways* Journal of Biomechanics **34**, 1527-1535.

3. Lorenz, H., Despont, M., Fahrni, N., Brugger, J., Vettiger, P., and Renaud, P. (1998) *High-aspect-ratio, ultrathick, negative-tone near-UV photoresist and its applications for MEMS* Sensors and Actuators a-Physical **64** (1), 33-39.

4. Hong, S., Tang, Z., Djukic, D., Tucay, A., Bakhru, S., Osgood, R., Yardley, J., West, A.C., and Modi, V. (2001) *Simulation and experimental validation of electroosmotic flow in a microfluidic channel* 2001 Microelectromechanical Systems Conference, California, 73-76.

5. Tang, Z., Hong, S., Djukic, D., Modi, V., West, A.C., Yardley, James T., and Osgood, R. (in press) *Electrokinetic Flow Control for Composition Modulation in a Microchannel* Journal of Micromechanics and Microengineering.

6. Ocvirk, G., Munroe, M., Tang, T., Oleschuk, R., Westra, K., and Harrison, D. J. (2000) *Electrokinetic control of fluid flow in native poly(dimethylsiloxane) capillary electrophoresis devices* Electrophoresis **21** (1), 107-115.

7. Locascio, L. E., Perso, C. E., and Lee, C. S. (1999) *Measurement of electroosmotic flow in plastic imprinted microfluid devices and the effect of protein adsorption on flow rate* Journal of Chromatography A **857** (1-2), 275-284.

8. Bird, Byron R., Stewart, Warren E., and Lightfoot, Edwin N. (1960) *Transport Phenomena.* Wiley, New York.

9. Chao, Grace, Tang, Zhongliang, Angelini, Elsa, West, Alan, and Hung, Clark (Accepted) *A novel microfluidic device to study the real-time response of cultured cells to applied dynamic osmotic loading* 49th Annual Meeting of the Orthopaedic Research Society, New Orleans, LA, Feb 2003.

10. Palmer, G. D., Chao, P. H. G., Raia, F., Mauck, R. L., Valhmu, W. B., and Hung, C. T. (2001) *Time-dependent aggrecan gene expression of articular chondrocytes in response to hyperosmotic loading* Osteoarthritis and Cartilage **9** (8), 761-770.

NEW ORGANIC DENDRIMERS WITH GREATLY ENHANCED MULTI-PHOTON ABSORPTION FOR PHOTONICS APPLICATIONS

C. W. SPANGLER, Z. SUO
Department of Chemistry and Biochemistry
Montana State University

M. DROBIZHEV, A. KAROTKI , A. REBANE
Department of Physics
Montana State University
Bozeman, MT 59717

1. Introduction

Tree-like dendritic structures based on identical repeat units and a variety of branching patterns are an exciting new class of materials with a myriad of potential applications at the nanophotonics level. Dendrimers offer advantages over traditional linear polymers containing a mixture of chain lengths in that they are monodisperse macromolecules amenable to detailed structure property relationship studies. For the study of multi-photon absorption (MPA), dendrimers can be designed using chromophore repeat units with known photonic properties, and the resulting highly branched three-dimensional structures allow a very high packing density per unit volume, with the number of chromophore repeat units increasing exponentially with dendrimer generation.

2. Structure-Property Relationship Studies of MPA

During the past five years there has been a dramatic increase in interest in the design of new organic chromophores with enhanced multi-photon absorption. Chromophores with enhanced two-photon absorption (TPA) have been shown to be useful in a variety of potential applications including such diverse fields as optical power limiting[1-4], two-photon photodynamic therapy[5,6] and three-dimensional microfabrication[7]. Some important design parameters for enhancing TPA have emerged through detailed structure-property relationship studies from several research groups including Reinhardt and coworkers[8-10], Marder and Perry[13-14], as well as our group at Montana State University[15-17]. Although originally proposed for linear chromophore structures, we shall see that the important design parameters devolved from these studies are equally valid for a variety of dendrimer constructs. A series of general rules for increasing the intrinsic TPA cross-section of a given chromophore type have gradually emerged, and are now generally accepted for the design of new chromophores with enhanced TPA:

F. Charra et al. (eds.), Organic Nanophotonics, 139–153.

1. Extend the conjugation length
2. Change the identity of the conjugated bridge
3. Increase the π-donor strength of e-donating substituents
4. Increase the π-acceptor strength of e-withdrawing substituents
5. Increase the chromophore planarity

Of the more simple chromophore types that are amenable to this structure-property relationship studies of this type are stilbene and PPV dimer structures:

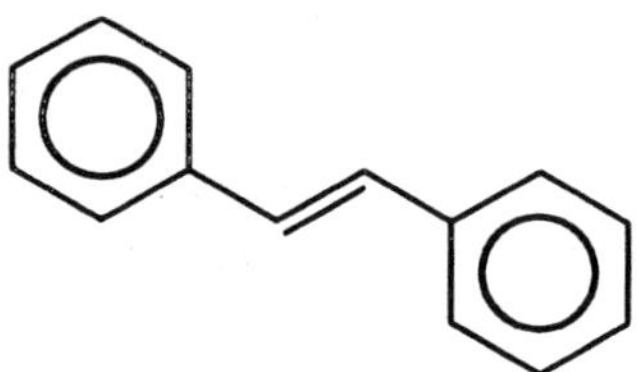

Stilbene-type chromophore

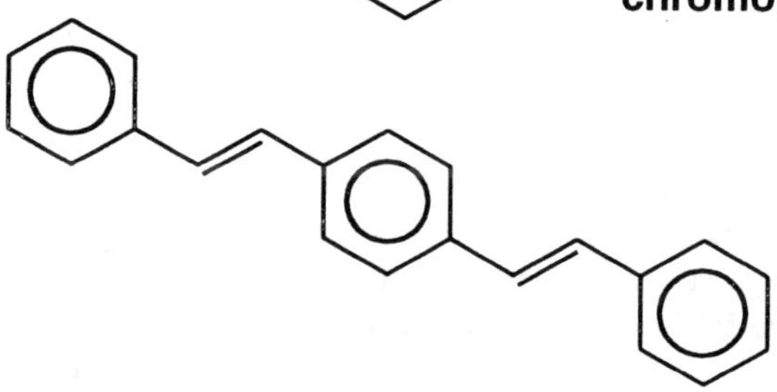

PPV-dimer type chromophore

Perry, Marder and coworkers[11-14] have particularly noted the effect of strong donor substitution on the ring termini in these systems, noting, for example, that bis(di-n-butylamino)-E-stilbene has a TPA cross-section almost 20 times that of E-stilbene itself. They have also observed similar effects for the PPV-dimer chromophore type. Substituents efects in materials with enhanced TPA seem to fall into two distinct types:

Type I: D-A-D

Type II: A-D-A

Thus one can design a chromophore with electron donation from the ends of the π-system, with electron withdrawal in the interior, or vice versa.

3. Design of Dendrimers with Stilbene Repeat Units

Our particular approach to the design of new chromophores has been based on bis-(diphenylamino)stilbene and α,ω-bis(diphenylamino)-PPV-dimer structures[3,4]. We have proposed that dendrimers based on these repeat units might exhibit enhanced cross-sections due to cooperativity effects and enhaced delocalization in highly branched π-electron structures structures. We have defined two distinct series of dendrimers based on bis-(diphenylamino)stilbene repeat units which we refer to as the 3-arm and 4-arm series. We have reported picosecond fluorescence decay rates in the G-0, G-1 and G-2 family of dendrimers based on this 4-arm branched structural motif[16], that is consistent with fast energy transfer from initially created localized states to more delocalized ones. The G-2 dendrimer in this series has a record high intrinsic two-photon cross-section (11,000 GM) for femtosecond pulses for a monodisperse macromolecule, and it was demonstrated that the maximum σ_2 value in the series increases faster than the total number of stilbene chromophores[17]. In this present study, we have examined the possibility that even greater TPA enhancement might be observed in dendrimer structures with e-donor and e-acceptor groups incorporated into the structure.

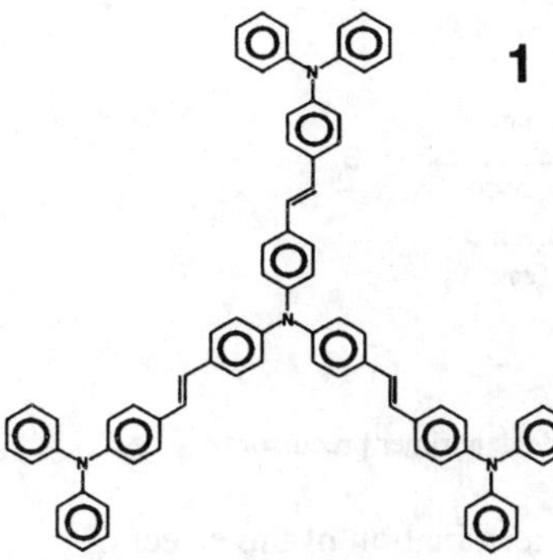

1

3-arm dendrimer

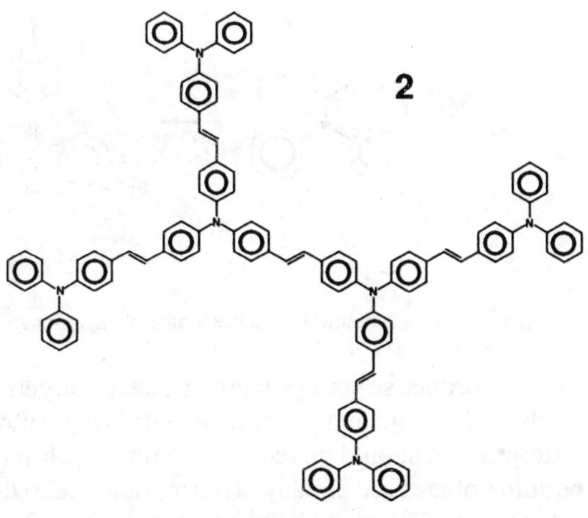

2

4-arm dendrimer

In addition, it has not yet been established by detailed structure-property relationships whether there are inherent advantages to having either a 3-arm structural motif radiating from a central nitrogen, or a four arm motif radiating from a stilbene core. In this present study we will attempt to provide preliminary evidence that addresses these two key questions.

4. Synthesis of Donor and Acceptor Derivatized 3-Arm Dendrimers

The syntheses of the parent and substituted 3-arm dendrimers were carried out by Wittig methodology developed over the past six years by our group. The key intermediate for the syntheses of these symmetrically substituted dendrimers is 4, 4', 4''-triphenylaminetricarbaldehyde. We have developed two separate approaches to the synthesis of this key intermediate, as shown below in Scheme 1.

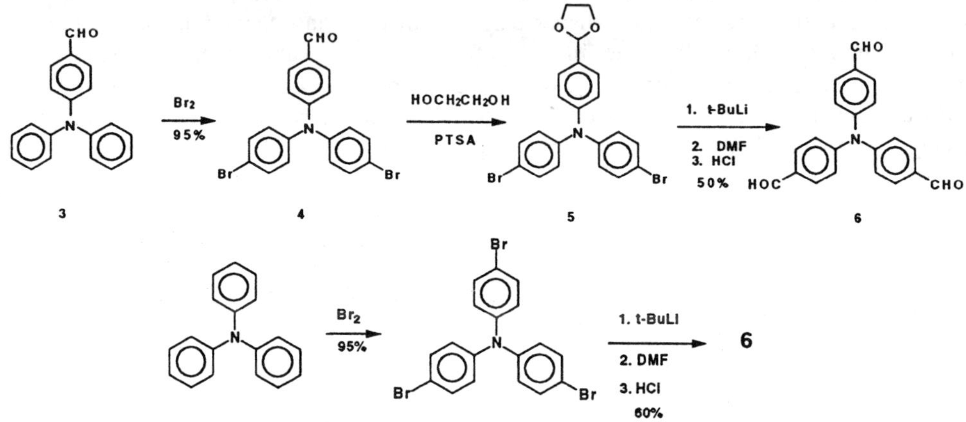

Scheme 1. Synthesis of triphenylaminetricarbaldehyde dendrimer precursor

Two dendrimer structures were chosen as targets for the determination of the effect of electron-donating and electron-withdrawing substituents on the intrinsic two-photon absorption. It should be recognized that dendrimers based on bis-(diphenylamino)stilbene building blocks are already electron rich due to the presence of the diphenylamino substituents. The amino nitrogens thus act as electron donors at each branch point. We chose long-chain alkoxy substituents ($OC_{10}H_{21}$) as our donor group, substituted on the para position of each terminal phenyl ring. The electron withdrawing substituent was a cyano group, CN. The synthesis of the dendrimers required the synthesis of an appropriate Wittig reagent, followed by coupling with the precursor trialdehyde **6**. The synthetic route to these new dendrimers is shown in Scheme 2. The dendrimers are processible, and can be purified by conventional column chromatography.

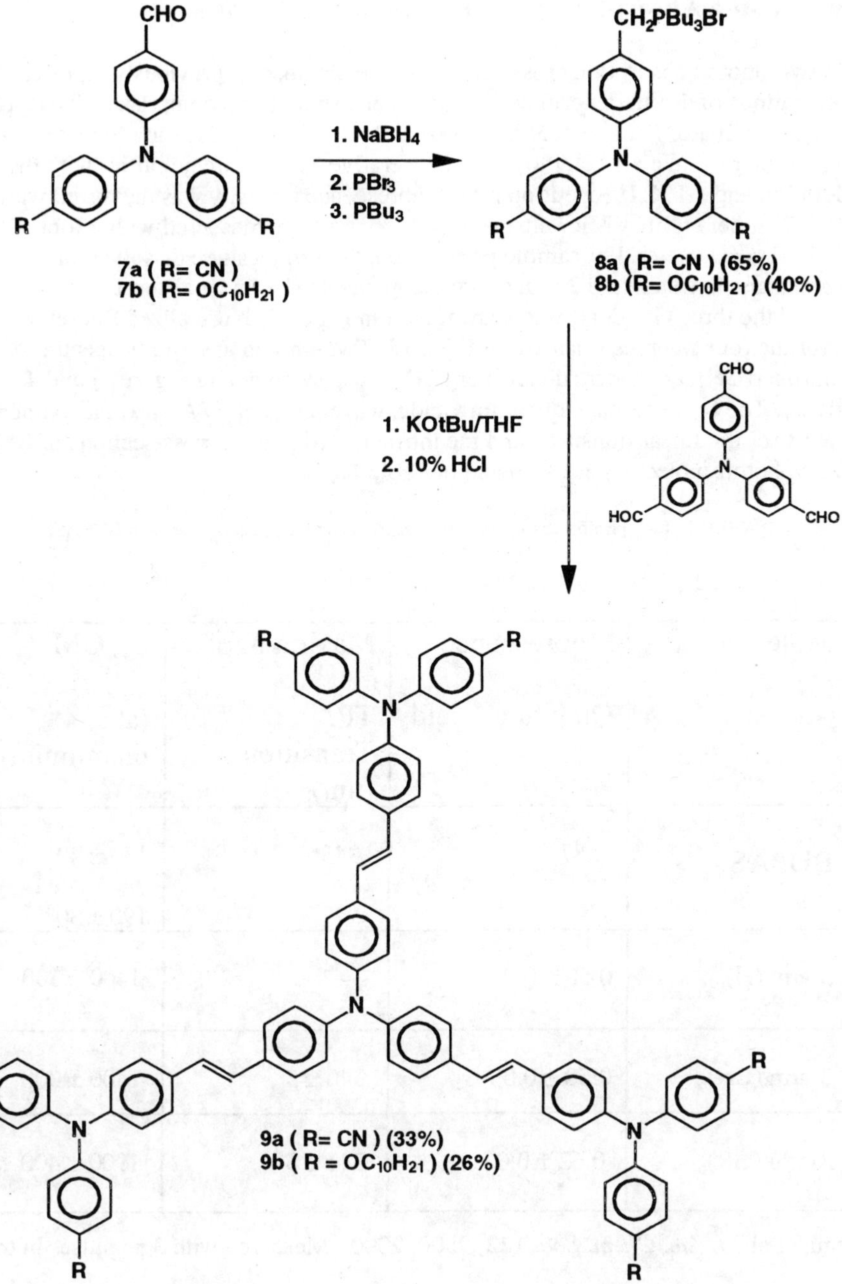

Scheme 2. Synthesis of substituted 3-arm dendrimers

5. Two-Photon Absorption in 3-Arm Dendrimer Models

Intrinsic two-photon absorption cross-sections were obtained as previously reported for the three generations of 4-arm dendrimers[17]. Our experimental setup comprised a Ti:sapphire femtosecond oscillator (Coherent MIRA 900) with 100-200 fs pulse duration and a 76 MHz repetition rate. We also employed a regenerative Ti:sapphire amplifier with the same pulse duration and a 1 KHz repetition rate. Fluorescence intensity was measured with a digital CCD camera (Xillix Microimager), and spectra were measured with a Jobin-Yvon TRIAX 550 spectrometer. For sample preparation, the samples wre dissolved in dichloromethane in a standard 2 mm spectroscopic cell. The linear absorption spectra for BDPAS and the three G-0 dendrimers are shown in Figure 1. Normalized fluorescence spectra for the four samples is shown in Figure 2. Two-photon absorption spectra for the 3-arm dendrimer$(CN)_6$ and 3-arm dendrimer $(OC_{10}H_{21})_6$ are shown in Figures 3 and 4 respectively. The fluoresecence quantum yields, wavelength of TPA maxima, extinction coefficients for the linear transition and the intrinsic two-photon cross-sections in GM units ($10^{-50}cm^4s$/photon-molecule) are all listed in Table 1.

TABLE 1. Two-photon absorption cross-sections of 3-arm dendrimers and BDPAS

Molecule	Fluorescence Quantum Yield	Maximum of TPA Transition (nm)	σ_2, GM (at mamimum)
BDPAS	0.87	345*	130 ± 40 190 ± 30[a]
3-arm $(H)_6$	0.58 ± 0.03	---	1400 ± 300
3-arm $(CN)_6$	0.58 ± 0.03	340 ± 5	1400 ± 300
3-arm $(OR)_6$	0.62 ± 0.03	341 ± 5	1900 ± 400

*M. Rumi, et al., *J. Am. Chem. Soc.* **122**, 9500, 2000. [a]Measured with 3 ps pulses in toluene.

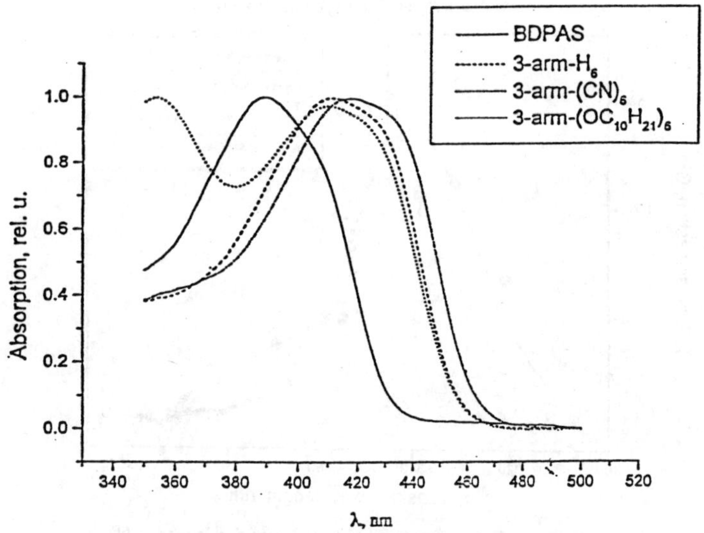

Figure 1. Absorption spectra of BDPAS and 3-arm dendrimers (CH_2Cl_2)

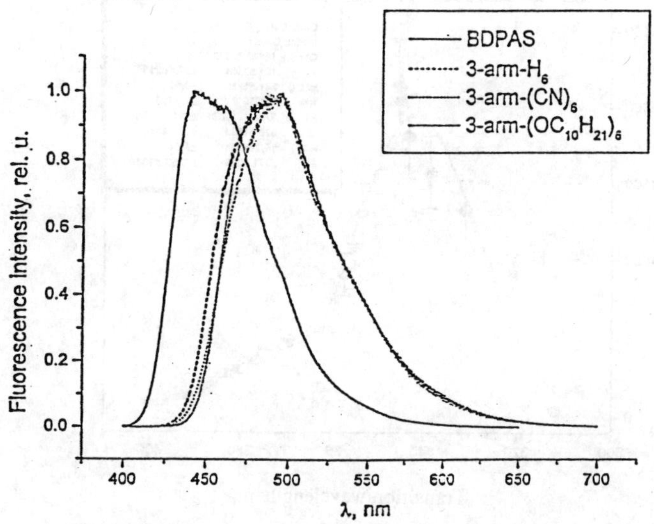

Figure 2. Fluorescence spectra of BDPAS and 3-arm dendrimers (CH_2Cl_2)

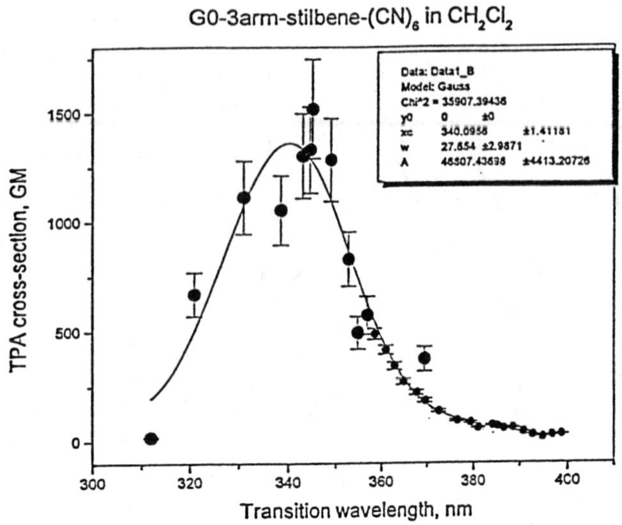

Figure 3. Two-photon spectrum of G-0 3-arm dendrimer $(CN)_6$

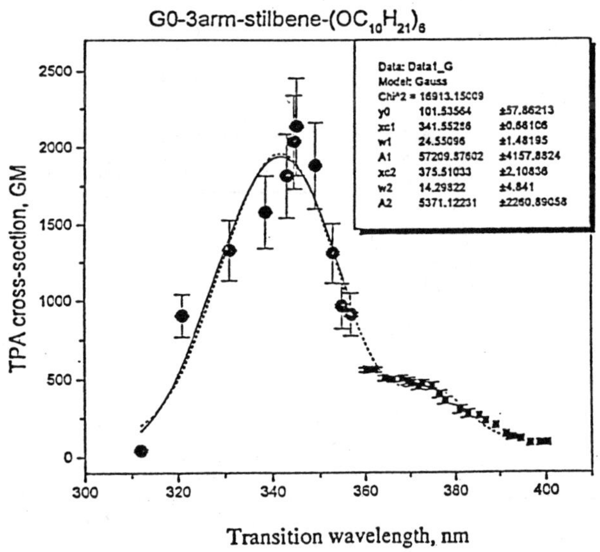

Figure 4. Two photon spectrum of G-0 3-arm dendrimer $(OC_{10}H_{21})_6$

6. Synthesis of Donor and Acceptor Derivatized 4-Arm Dendroimers

The syntheses of the parent and substituted 4-arm dendrimers were carried out in a manner similar to that previously described for the 3-arm series. However, the 4-arm series employs a different structural motif in that the dendrimer structure radiates from a bis-(diphenylamino)stilbene core rather than from a triphenylamine core. The syntheses of the 4-arm dendrimer $(OC_{10}H_{21})_8$ and 4-arm dendrimer $(CN)_8$ are shown respectively in Schemes 3 and 4 below.

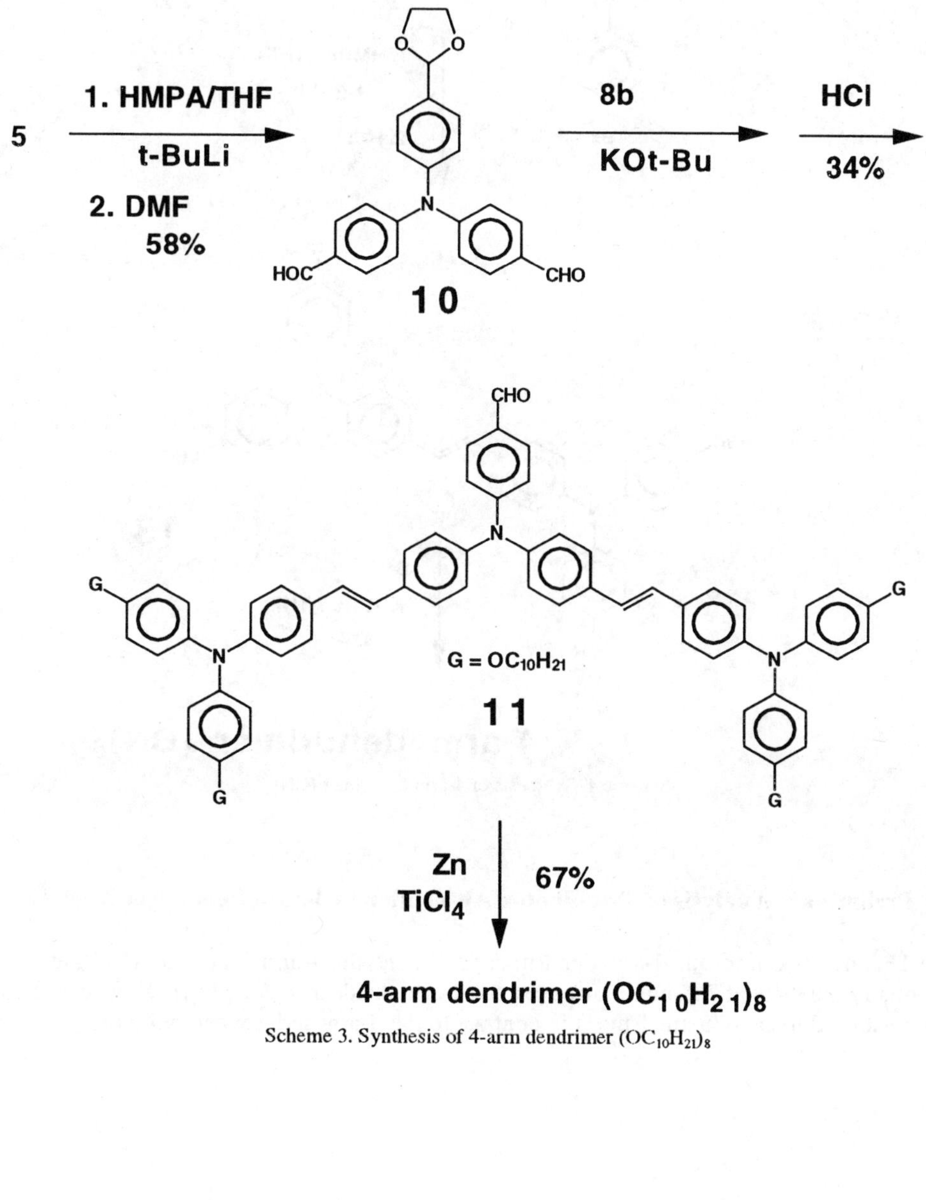

Scheme 3. Synthesis of 4-arm dendrimer $(OC_{10}H_{21})_8$

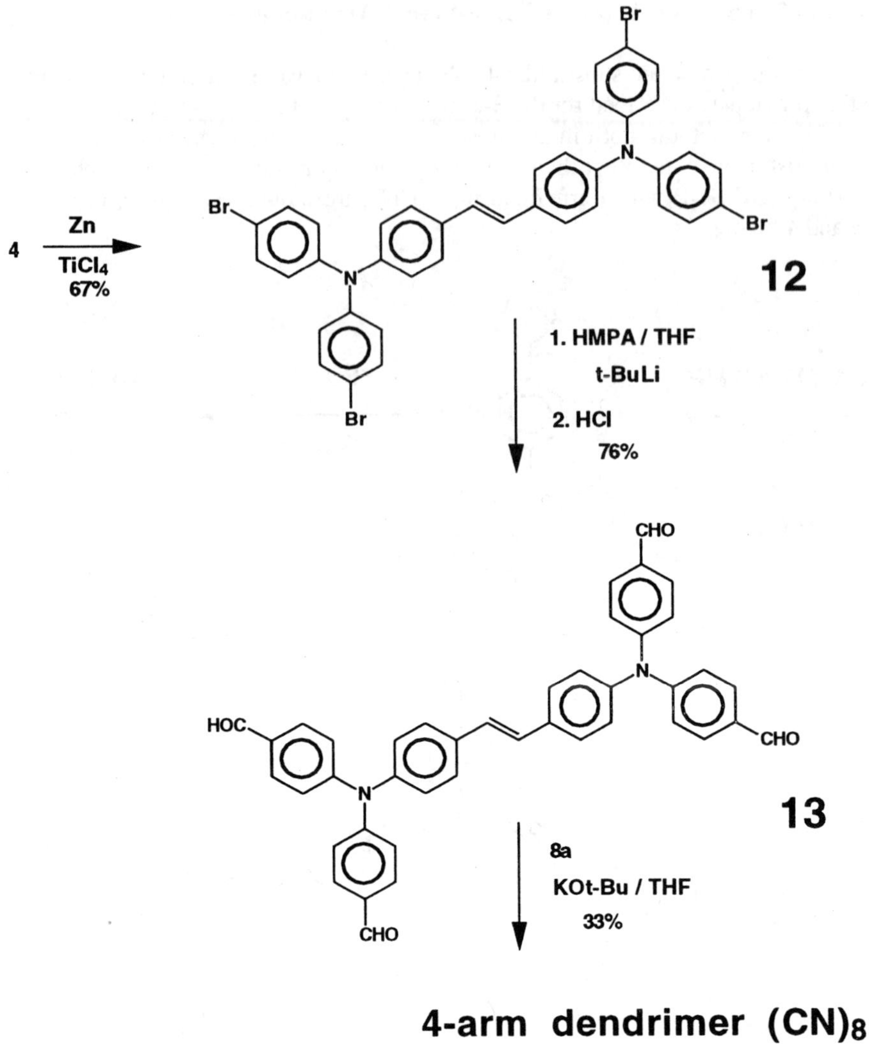

Scheme 4. Synthesis of 4-arm dendrimer $(CN)_8$

7. Preliminary Analysis of Two-Photon Absorption in 4-Arm Dendrimer Models

The TPA maximum for the 4-arm dendrimer parent and the 4-arm dendrimer $(CN)_8$ are essentially identical at 705 nm. Howeever, the 4-arm dendrimer $(OC_{10}H_{21})_8$ TPA maximum is blue-shifted to ca. 675 nm. This is in contrast to the 3-arm series where all three dendrimers display essentially the same TPA maximum. The 4-arm dendrimer parent and 4-arm dendrimer $(CN)_8$ are also red-shifted by ca. 30 nm with respect to their 3-arm counterparts. One can rationalize this difference by the greater potential for delocalization in the 4-arm series. As is the case for the 3-arm series, the relative magnitudes of the two-photon cross-sections are : $(OR)_8 > (CN)_8 > (H)_8$ (parent). The values we have obtained for the intrinsic two-photon cross-sections for femtosecond pulses should be regarded as preliminary, and more detailed measurements are in progress. The maximum TPA cross-sections show dramatic cooperativity effects in the same manner as the 3-arm series. The G-0 4-arm dendrimers are constructed from 5 BDPAS chromophore units. Thus if there were no cooperativity effects, to a first approximation one would expect the dendrimer two-photon cross-sections to be five times larger than BDPAS. The three 4-arm dendrimers in this study range from 2.3 to 4.6 times larger than the expected value. Since we realize that there are still large error bars on the crude preliminary data, we are using the lower value in all cases. We expect that the cross-sections are, in fact, much greater than the numbers we have included below in Table 2. At this point in time it is premature to say whether the 3-arm structural motif is better than the 4-arm motif, but it is clear that both series show large cooperativity effects. We are encouraged to extend our current program to both G-1 and G-2 dendrimer generations, which we expect will have the largest two-photon cross-sections yet measured.

TABLE 2. Two-photon absorption cross-sections of 4-arm dendrimers

Molecule	TPA maximum, nm	σ_2, GM at TPA maximum
4-arm$(H)_8$	705	2000
4-arm$(CN)_8$	705	2300
4-arm$(OR)_8$	675	3000

It is again interesting to note that the largest cross-sections for both 3-arm and 4-arm series are for the dendrimers containing all donor groups, including the amine nitrogens. The addition of electron acceptor groups on the dendrimer periphery does not affect the cross-section values. We anticipate that we will pursue only all-donor systems as we extend our synthetic efforts to the G-1 and G-2 generations.

8. Discussion

As can be seen in Tables 1 and 2, the measured two-photon cross-sections for the 3-arm and 4-arm dendrimers show significant enhancement over the simple additive effect of the number of BDPAS repeat units in the dendrimer structure (3 repeat units/dendrimner) irrespective of whether the dendrimers are unsubstituted, or substituted in the six terminal ring para positions by electron donor or acceptor groups. Depending on the cross-section one accepts for the BDPAS parent, the enhancement factor is between 3 and 5. The presence of acceptor groups does not appear to enhance the cross-section compared to the unsubstituted parent. This is somewhat surprising, considering other recent reports in the literature in which 3-arm dendrimers are constructed from arms that can be represented as D-π-A polyenes. Chung and coworkers[18] have reported cooperative enhancement of two-photon absorption in multi-branched (3-arm) structures. The individual arms of this branched structure were composed of a D-π-A moiety, with a central triphenyl amine moiety serving as the donor group for all three arms. This structure is similar to one of the dendrimers reported in our current study, G-0 3-arm dendrimer $(CN)_6$, where we also observe cooperative enhancement of the intrinsic σ_2. The observed enhancement (ca. 2 fold for 173 fs pulses) is lower than what we observe in the dendrimers discussed in this study. More recently, Adronov and coworkers have reported[19] studies in which a dendron similar to that reported by Chung, et al.[18], except that two D-π-A chromophores are incorporated in a dendron, that is then coupled to a central AB_2 core molecules to yield a dendrimer incorporating 4 D–π–A chromophore units. Iteration to a G-1 and G-2 levels produces dendrimers incorporating 8 and 16 chromophore units in the dendrimer structure. Thus the number of chromophores in the four samples contained 2, 4, 8 and 16 chromophore units (1:2:4:8). The relative values for σ_2 for these three constructs for 160 femtosecond pulses were, respectively, 1:3:6:9, thus showing a slight enhancement of TPA. This enhancement is also considerably lower than we have observed for either the 3-arm or 4-arm dendrimers in this current study. Our studies seem to indicate that D-π-A type structures are not required for large intrinsic TPA. In fact, our largest intrinsic TPA was observed for the 3-arm dendrimer with only donor substituents. This type of system, wherein each unit is resonantly coupled to every other unit, is much easier synthetically to extend the interacting cooperative subunits to higher dendrimer generations. This is quite difficult to do for D-π-A chromophore repeat units, particulary from the point of building in AB_2 functionality into the dendrimer structure. As we have shown[17], this is quite easy, although tedious, to do with all D-π-D repeat units.

9. Conclusions

In this paper we have reported the syntheses of new 3-arm and 4-arm dendrimers built from bis(diphenylamino)stilbene repeat units, and incorporating all-donor or all-acceptor substuituent groups attrached to the peripheral phenyl rings. These new dendrimers exhibit greatly enhanced TPA with respect to the parent BDPAS, showing a 3-5 fold enhancement compared to simple addition of the individual number of chromophores. To the best of our knowledge, this is the largest cooperative effect in dendrimer TPA observed to date. It is also interesting to note that the largest enhanced TPA cross-section was obtained for the dendrimers containing only donor substituents. This implies that D-π-D subunits may be better than D-π-A subunits in the design and construction of dendrimers for enhanced TPA. We are continuing these detailed studies to evaluate the effect of increased branching in these systems, as well as the effective conjugation length of each repeat unit, on the TPA for higher dendrimer generations.

10. Acknowledgments

This research was supported in part by the Air Force Office of Scientific Research (F49620-01-1-0324) and the Army research Office (49075-CH-DPS).

11. References

1. Spangler, C. W. (1999) Recent development in the design of organic materials for optical power limiting, *J. Mater. Chem.* **9**, 2013-2020.
2. Perry, J. W., Barlow, S., J. E. Erlich, Heikal, A. A., Hu, Z.-Y., Lee, L.-Y. S., Mansour, K, Marder, S. R., Rockel, H., Rumi, S., Thayumanavan, S. and Wu, X. L. (1999) Two-photon and higher-order absorptions and optical limiting properties of bis-donor substituted conjugated organic chromophores, *Nonlinear Optics* **21**, 225-241.
3. Spangler, C. W., Elandaloussi, E. H., Ozer, B., Ashworth, K., Madrigal, L., and Reeves,, B. (2000) Design, synthesis and characterization of new bimechanistic optical power limiters based on reverse saturable and two-photon absorption, *Mat. Res. Soc. Symp. Proc.* **597**, 369-379.
4. Elandaloussi, E. H., Spangler, C., Casstevens, M., Kumar, D., Weibel, J., Burzynski, R., He, G. and Prasad, P. N. (2000) Optical power limiters for nanosecond pulses: design of new dendritic chromophores with exceptionally large two-photon cross-sections, *Mat. Res. Soc. Symp. Proc.* **597**, 339-344.
5. Bhawalker, J. D., Kumar, N. D., Zhao, C. F. and Prasad, P. N. (1997) Two-photon photodynamic therapy, *J. Clin. Laser Med. Surg.* **66**, 201-204.

6. Karotki, A., Kruk, M., Drobizhev, M., Rebane, A., Nickel, E. and Spangler, C. W. (2001) Efficient singlet oxygen generation upon two-photon excitation of new porphyrin with enhanced nonlinear absorption, *IEEE J. Selected Topics Quantum Electronics* **7**, 971-975.
7. Zhou, W., Kuebler, S. M., Braun, K. L., Yu, T., Cammack, J. K., Ober, C. K., Perry, J. W. and Marder, S. R. (2002) An efficient two-photon generated photoacid applied to positive tone 3D microfabrication, *Science* **296**, 1106-1109.
8. Reinhardt, B. A., Brott L. L., Clarson, S. J., Dillard, A. G., Bhatt, J. C., Kannon, R., Yuan, L., He, G. S. and Prasad, P. N. (1998) Highly active two-photon dyes: design, synthesis and characterization toward application, *Chem. Mater.* **10**, 1863-1874.
9. Reinhardt, B. A., Brott, L. L., Clarson, S. J., Kannon , R. and Dillard, A. G. (1997) The design and synthesis of new organic molecules with large two-photon absorbing cross-sections for optical limiting applications, *Mat. Res. Soc., Symp. Proc.* **479**, 3-8.
10. Reinhardt, B. A., Brott, L. L., Clarson, S. J., Kannon, R. and Dillard, A. G. (1997) Optical power limiting in solution via two photon absorption: new heterocyclic dyes with greatly improved performance, *Proc. SPIE Int. Soc. Opt. Eng.* **3146**, 2-11.
11. Albota, M., Beljonne D., Bredas, J.-L., Erlich, J. E., Fu, J.-Y., Heikal, A. A., Hess, S. E., Kogej, T., Levin, M. D., Marder, S. R., McCord-Maughon, D., Perry, J. W., Rockel,, H., Rumi, M., Subramanian, G., Webb, W. W., Wu, X. L., and Xu, C. (1998) Design of organic molecules with large two-photon absorption cross-sections, *Science* **281**, 1653-1656.
12. Erlich, J. E., Wu, X. L, Lee, I.-Y., Heikal, A. A., Hu, Z.-Y., Rockel, H., Marder, S. R. and Perry, J. W. (1997) Two-photon absorbing organic chromophores for optical limiting,*Opt. Lett.* **22**, 1843-1847.
13. Erlich, J. E., Wu, X. L., Lee, I.-Y., Heikal A. A., Hu, Z. Y., Rockel, H., Marder, S. R. and Perry, J. W. (1997) Two-photon absorbing organic chromophores for optical limiting, *Mat. Res. Soc. Symp. Proc.* **479**, 9-15.
14. Rumi, M., Erlich, J. E., Heikal, A. A., Perry, J. W., Barlow, S., Hu, Z., McCord-Maughon, D., Parker, T. C., Rockel, H., Thayumanavan, S., Marder, S. R., Beljonne, D. and Bredas, J.-L. (2000) Structure-property relationships for two-photon absorbing chromophores: bis-donor diphenylpolyene and bis(styryl)benzene derivatives, *J. Am. Chem. Soc.* **122**, 9500-9510.
15. Spangler, C. W., Elandaloussi, E. H., Casstevens, M. K., Kumar, D. N., Weibel, J. F. and R. Burzynski, Design and synthesis of new optical power limiting chromophores with enhanced two-photon absorption, *Proc. SPIE Int. Soc. Opt. Eng.* **3798**, 117-122.
16. Drobizhev, M., Rebane, A., Sigel, C., Elandaloussi, E. H. and Spangler, C. W. (2000) Picosecond dynamics of excitations studied in three generations of new 4,4'-bis(diphenylamino)stilbene-based dendrimers, *Chem. Phys. Lett.* **325**, 375-382.
17. Drobizhev, M., Karotki, A., Rebane, A. and Spangler, C. W. (2001) Dendrimer molecules with record large two-photon absorption cross-section, *Opt. Lett.* **26**, 1081-1083.

18. Chung, S. J., Kim K.-S., Lin, T. C., He, G. S., Swiatkiewicz, J. and Prasad, P. N. (1999) Cooperative enhancement of two-photon absorption in multi-branched structures, *J. Phys. Chem. B* **103**, 10741-10745.
19. Adronov, A., Frechet, J. M., He, G. S., Kim, K.-S., Chung, S. J., Swiatkiewicz, J. and Prasad, P. N. (2000) Novel two-photon absorbing dendritic structures, *Chem. Mater.* **12**, 2838-2841.

DENDRITIC POLYMERS FOR OPTICAL APPLICATIONS

Anders Hult, Claire Pitois, Mikael Lindgren[a]

The Royal Institute of Technology,
Dept. of Fiber- andPolymer Technology,
100 44 Stockholm, Sweden
[a]*Swedish Defence Research Agency*
Department of Sensor Technology, P.O. Box 1165
SE-581 11 Linköping, Sweden

1. Introduction

There is significant interest for polymers in electronics and/or opto-electronics. This is due to the fact that polymeric materials are relatively easy to give certain properties. From a processing point of view polymers are also advantageous over inorganic materials.

Dendritic macromolecules (dendrimers[i] and hyperbranched polymers[ii]) are synthesized from AB_x-type monomers (where x is two or greater) resulting in macromolecules with a highly branched architecture and a multitude of end-groups (figure 1). Dendrimers are symmetrically branched structures while hyperbranched polymers are irregularly branched. Both types have a multitude of end groups.

The field of dendritic polymers is still rather young and unexplored. Many groups worldwide work on the synthesis of macromolecules having a highly branched architecture. However, not many focus on dendritic polymers as a *material* – most groups deal with pure synthetic challenges.

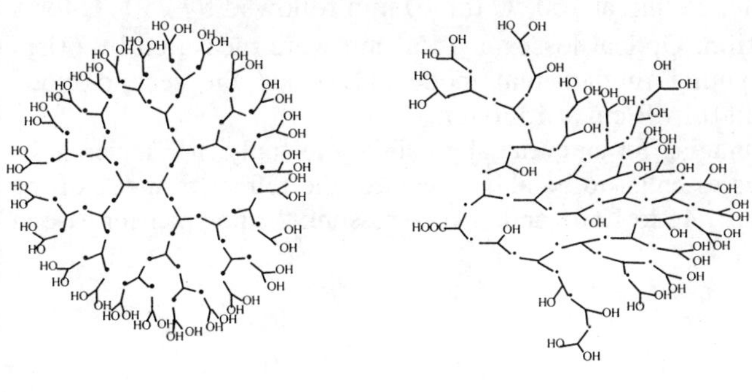

Figure 1. Schematic representation of a hydroxy-functional dendrimer (left) and a hydroxy-functional hyperbranched polymer (right).

F. Charra et al. (eds.), Organic Nanophotonics, 155–161.

This paper will present the use of functional dendritic polymers to build complex materials for use in photonic devices. The controlled shape, size and differentiated functionality as well as their ability to form isotropic and anisotropic assemblies, compatibility with many other types of nanometer scale building blocks (DNA etc), their potential for self-assembly, their capacity to form surfaces and interfaces and their ability to either encapsulate or be engineered into unimolecular devices make dendrimers uniquely versatile amongst existing nanometer scale building blocks.

2. Optical waveguides

The continual trend toward high transmission speed, data capacity and data density in integrated circuits demands a solution to the bottleneck resulting from the limited data rate of electrical interconnects. One approach to this problem is the use of optical interconnection operating with polymer waveguides. This alternative has been widely reported and shows advantages for large-system applications (a few cm to 1 m). It reduces electromagnetic interference and has lower power requirements. Optical polymer waveguides are potentially useful for low-level hierarchies, such as board-level or backplane-level interconnections, and flexible manufacturing techniques have been reported that can be applied on a large variety of substrates. Polymer materials exhibit high absorption or high optical losses compared to silica fibers in the range of wavelengths used for telecommunication transmission, the near-infrared (NIR). The loss can be limited by shifting the high absorption signal toward longer wavelengths by replacing hydrogen atoms with heavier atoms like halogens, essentially chlorine, fluorine or deuterium.

A hyperbranched fluorinated polymer based on fluorinated polyaryl ether was synthesized according to the procedure worked out by Y. M. Miller *et al* (*J. Am. Chem. Soc.* **1992**, *114*, 1018-1025.), *Scheme 1*. This polymer has a low hydrogen content (C-H is only 8.8 bond %), amorphous structure, and posses high thermal stability (more than 95 % of its mass is retained up to 500 °C under nitrogen at 10 °C/min). It is obtained in high yield by a one-sep polymerization of the AB_2 monomer, 3, 5-dipentafluorophenyl-1-(hydroxy)benzene. The obtained polymer exhibits molecular weights of Mn = 25 100, Mw = 43 170, Mw/Mn = 1.72, and a glass transition temperature at 10 °C/min under nitrogen of 140 °C.

Slab waveguides were made by dissolving **1** in 2, 4-pentanedione, spin-coating the solution onto SiO_2-silicon plate, baking at 160 °C for 10 min followed by 250 °C for 5 min, to give a 4.6 Πm thick film. Optical losses at 1550 nm were measured by prism coupling method. The propagating fundamental mode (TE0) and the second mode (TE2) have optical loss at about 0.1 dB/cm at 1550 nm.

This result was very encouraging for an eventual possibility to apply such material for optical devices. However it was essential to be able to control the refractive index of the polymer over a wide range (from 1.46 to 1.60), and to UV-crosslink the polymer for ease in processing.

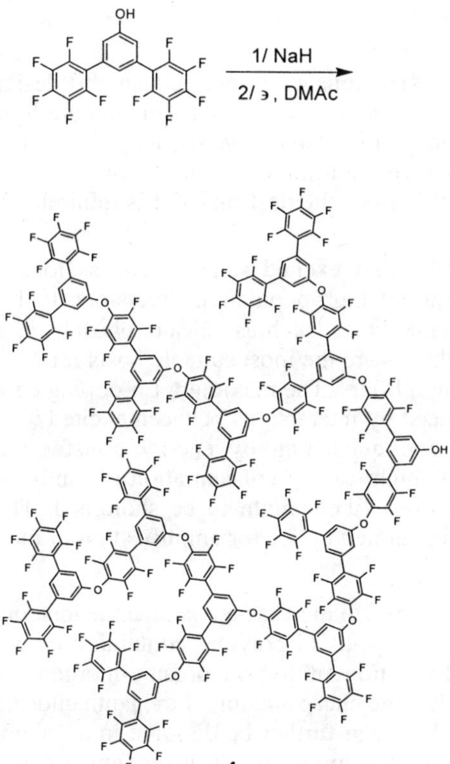

1

Scheme 1.

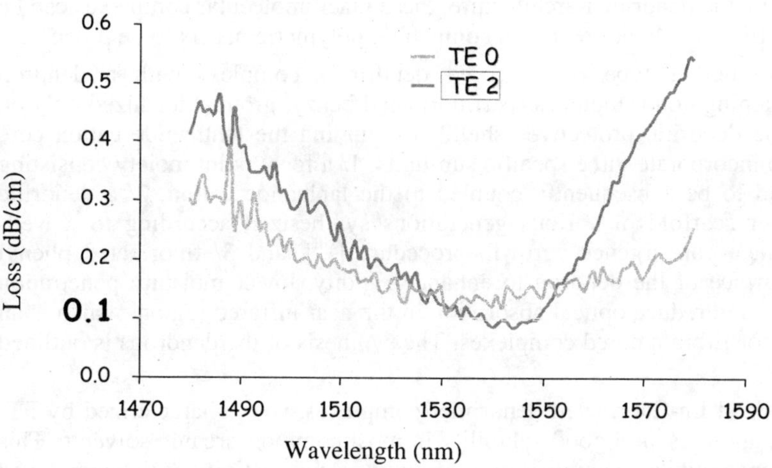

Figure 2. Optical loss measurements on fluorinated hyperbranched polyethers.

3. Amplifying waveguides

The large interest for trivalent lanthanide ions as fluorescers in the design of light amplifiers for optical fiber communications has risen from their unique luminescence characteristics from the electronic transitions between the 4f energy levels [iii]. Because their partially filled 4f shells are shielded from the environment by the outer filled 5s and 5p shells, the effect of the crystal field of the host material is minimized and leads to rather sharp emission bands [iv].

The luminescence lifetime of the first excited state can be as long as several milliseconds, which permits the required high-population inversions to be obtained using modest pump power [v]. Inorganic hosts, such as silicate, phosphate, borate and more recently germanate [vi]-based glasses are the most suitable hosts for Nd^{3+} and Er^{3+} ions commonly used in optical fiber amplifiers. The maximum Ln-doping concentration is limited due to parasite effects induced by interactions between excited ions, notably, cooperative up-conversion process and quenching by energy transfer with micro-clustering in the host glass matrix. A high doping concentration of lanthanide ions is desirable because it allows the Ln-doped fiber length to be shortened. This issue is particularly important in the area of integrated optics for amplification in planar optical waveguides.

Several approaches have been reported to increase the lanthanide ion concentration in optical media, such as dispersion Ln^{3+} salts in polymer containing carboxylic acid and sulfonic acid ligands [vii], or formation of hybrid organic-inorganic lanthanide complex via a sol-gel process [viii]. Also, the encapsulation of the lanthanide cation in an organic ligand to form a complex, which can further be dissolve in a polymer matrix, has been extensively studied [ix, x, xi]. In this approach, sensitizer-functionalized ligands were complexed to lanthanide ions to enhance their luminescence. Finally, site isolation of the lanthanide ion and energy transfer due to an "antenna effect" was demonstrated by self-assembled lanthanide-cored dendrimer complexes [xii]. Beside the "antenna effect" provided by the dendrimer architecture, these macromolecular complexes can be directly cast as a film and do not require a compatible polymeric matrix to be processed.

Here we report another type of Ln-cored dendrimer complex, with a dendron architecture containing donor moieties (perfluorinated benzyl groups) localized only on its periphery. The dendritic protective "shell" surrounding the lanthanide cation core were designed to incorporate three specific sub-units: 1/ a focal point moiety consisting of a benzoic acid to be subsequently coupled to the lanhanide cation, 2/ a dendritic aliphatic polyester scaffold of various generations synthesized according to a well-known double-stage convergence growth procedure [xiii], and 3/ fluorinated phenyl groups on the surface of the dendron to enhance rigidity, lower moisture penetration into the core unit, and reduce optical absorption in the near infrared region, which is an essential feature for Erbium cored complexes. The synthesis of the dendrons is outlined in **Scheme 2**.

The synthesized Ln-fluorinated dendrimer complexes, were characterized by FT-IR spectroscopy, and present a good solubility in most common organic solvents. This class of dendrimers exhibits an amplification of luminescence activity of the complexed Eu^{3+} and Tb^{3+} ions. The control over the dendrimer architecture and possibility to modify of end-group functionality make the dendritic shell an interesting approach for

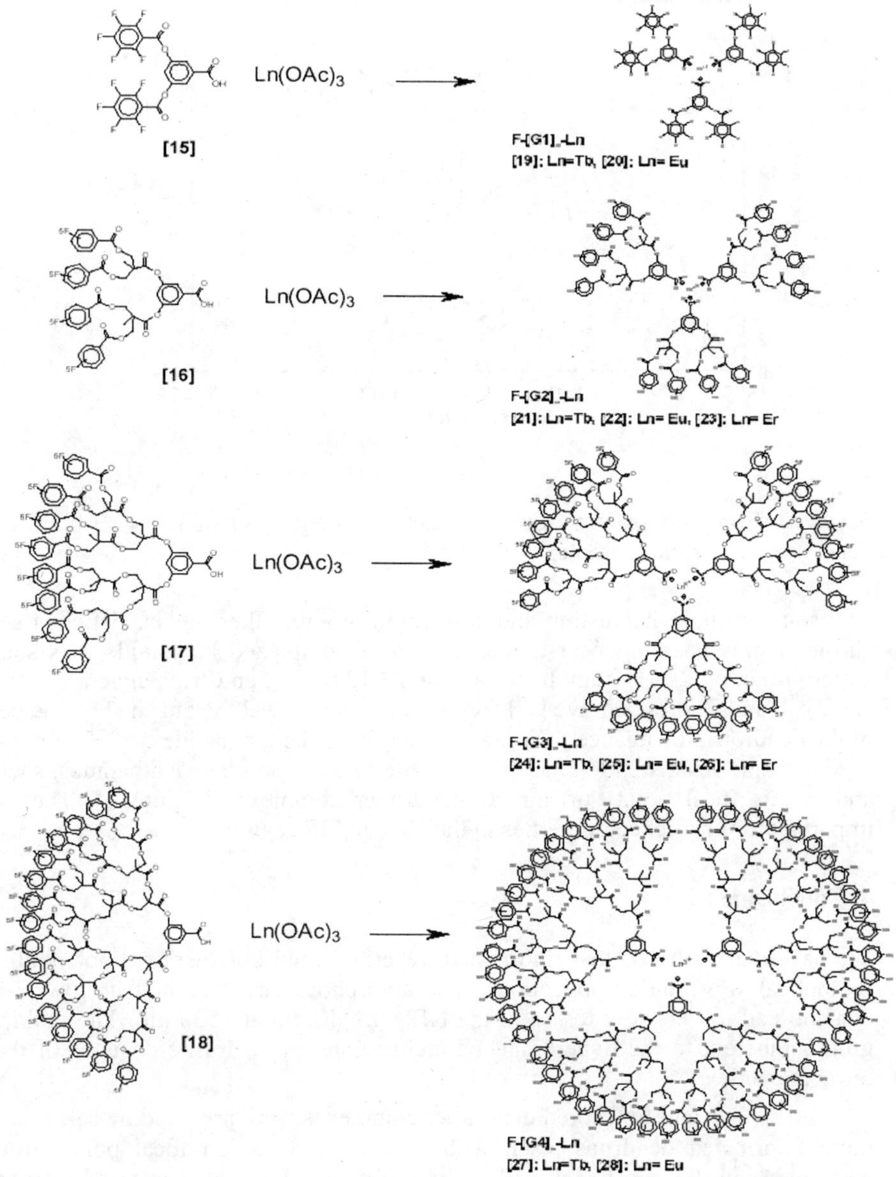
[15]
Ln(OAc)3
F-[G1]3-Ln
[19]: Ln=Tb, [20]: Ln= Eu
[16]
Ln(OAc)3
F-[G2]3-Ln
[21]: Ln=Tb, [22]: Ln= Eu, [23]: Ln= Er
[17]
Ln(OAc)3
F-[G3]3-Ln
[24]: Ln=Tb, [25]: Ln= Eu, [26]: Ln= Er
[18]
Ln(OAc)3
F-[G4]3-Ln
[27]: Ln=Tb, [28]: Ln= Eu

Scheme 2.

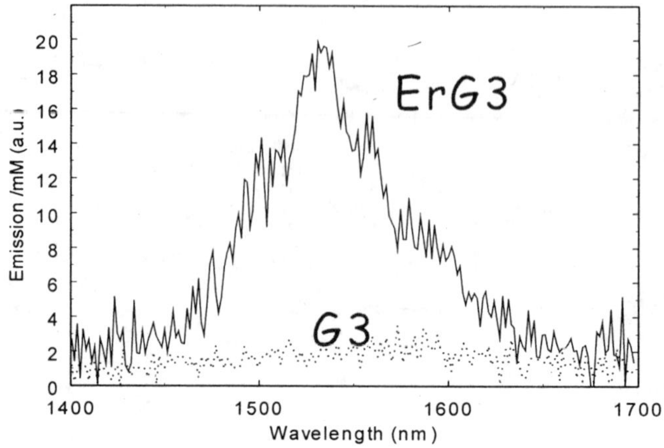

Figure 2. Emission spectra from 3rd generation dendrimer (G3) and Erbium cored 3rd generation dendrimer complex (ErG3).

site isolation, light harvesting and compatibility with other media. Efficient sensitizer chromophores enabling near-UV excitation of complexed lanthanide ions such as 7-amino-4-methyl-2(1H)-quinoline (carbostyril 124) [xiv], and triphenylene [xv] for Eu^{3+}, and Tb^{3+}, and fluorescein dye [xvi] for Er^{3+} ion, may as well be attached on the periphery of the dendrons, to enhance their energy transfer to the lanthanide core. Our preliminary spectroscopic results show that it is possible to use also other lanthanides, such as Nd and Er, as focal points of similar dendrimer complexes (Figure 2). These exhibit important luminescence properties in the NIR and IR regions.

4. Conclusions

We have demonstrated that fluorinated hyperbranched polymers are potentially useful for optical waveguide applications. The amorphous structure and its low C-H bond content lead to very low losses in the NIR, 0.1 dB/cm at 1550 nm. The multiple end-groups inherent to the hyperbranched architecture allow drastic changes of the initial refractive index.

Eu^{3+}, Tb^{3+} and Er^{3+} cored dendrimer complexes were prepared by self-assembly of three fluorinated dendrons, each with a carboxylate anion focal point, around the lanthanide ion. Energy transfer from the peripheral fluorinated phenyl moieties of the dendrons to the lanthanide cation was evidenced spectroscopically for Eu^{3+}- and Tb^{3+}-cored dendrimer complexes, in solution. The luminescence yield of the lanthanide ion was found to be enhanced by the combined light harvesting and site-isolation provided by the dendrimer structure.

References

[i] Inoue, K. *Prog. Polym. Sci.* "Functional dendrimers , hyperbranched and star polymers" *Prog. Polym. Sci.* 2000, **25**, 453

[ii] Vögtle, F., Gestermann, S.; Hesse, R.; Scwierz, H., Windisch, B. "Functional Dendrimers" *Prog. Polym. Sci.* 2000, **25**, 987

[iii] (a) S. Hufner, "Optical Spectra of Transparent Rare Earth Compounds" Academic press, New York, **1978**. (b) L. R. Matthews, E. T. Knobbe, *Chem. Mater.* **1993**, *5*, 1697.

[iv] (a) "Handbook on the Physics and Chemistry of Rare-Earths", edited by K. A. Gschneidner, Jr. and L. Eyring, Chap. 35, M. J. Weber, "Rare Earth Lasers", 1979. (b) "Optical fiber Amplifiers, Materials, Devices, and Applications", edited by Shoichi Sudo, 1994. (c) "Erbium-Doped Fiber Amplifiers, Principles and Applications", E. Desurvire, edited by J. Wiley & Sons, inc, 1994.

[v] W. J. Miniscalco, *Journal of Lightwave Technology*, **1991**, *9*, (2), 234-250.

[vi] (a) J. E. Canale, R. A. Condrate, Sr. K. Nassau, and B. C. Cornilsen, "Characterization of various glasses in the binary PbO-GeO2 and Bi2O3-GeO2 systems," *J. Can. Ceram. Soc.* **1986**, *55*, 50-56. (b) R. Balda, M. Sanz, J. Fernández, J. M. Fdez-Navarro, *J. Opt. Soc. Am. B.,* **2000**, *17*, (10), 1671-1677.

[vii] Y. Okamoto, Y. Ueba, N. F. Dzhanibekov, and E. Banks, *Macromolecules*, **1981**, *14*, 17-22.

[viii] (a) T. Jin, S. Tsutsumi, Y. Deguchi, K. Machida, G. Adachi, *Journal of Alloys and Compounds*, **1997**, *252*, 59-66. (b) D. Dong, S. Jiang, Y. Men, X. Ji, and B. Jiang, *Advanced Materials*, **2000**, *12*, (9), 646-649.

[ix] S. I. Weissman, *J. Chem. Phys.*, **1942**, *10*, 214.

[x] C. Koeppen, S. Yamada, G. Jiang, A. F. Garito, L. R. Dalton, *J. Opt. Soc. Am. B*, 1997, *14*, (1), 155-162.

[xi] **(a)** L. H. Sloff, A. Polman, M. P. Oude, Wolbers, F. C. J. M. van Veggel, D. N. Reinhoudt, and J. W. Hofstraat, *J. Appl. Phys.* **1998**, *83*, (1), 497-503. **(b)** F. J. Steemers, W. Verboom, J. W. Hofstraat, F. A. J. Geurts, and D. N. Reinhoudt, *Tetrahedron Letters*, **1998**, *39*, 7583-7586. **(c)** S. I. Klink, G. A. Hebbink, L. Grave, F. C. J. M. van Veggel, D. N. Reinhoudt, L. H. Sloff, A. Polman, J. W. Hofstraat, *Journal of Applied Physics,* **1999**, *86*, (3), 1181-1185. **(d)** L. H. Sloff, A. Polman, S. I. Klink, G. A. Hebbink, L. Grave, F. C. J. M. van Veggel, D. N. Reinhoudt, J. W. Hofstraat, *Optical Materials*, **2000**, *14*, 101-107.

[xii] M. Kawa, and J. M. J. Fréchet, *Chem. Mater.*, **1998**, *10*, 286-296.

[xiii] H. Ihre, A. Hult, J. M. J. Fréchet, and I. Gitsov, *Macromolecules*, **1998**, *31*, (13), 4061-4068.

[xiv] M. Lin and P. R. Selvin, *J. Am. Chem. Soc.* **1995**, *117*, 8132-8138.

[xv] (a) M. P. Oude Wolbers, F. C. J. M. van Veggel, J. W. Hofstraat, F. A. J. Geurts, and D. N. Reinhoudt, *J. Chem. Soc., Perkin Trans. 2* **1997**, 2275. (b) S. I. Klink, L. Grave, D. N. Reinhoudt, F. C. J. M. van Veggel, M. H. V. Werts, F. A. J. Geurts, J. W. Hofstraat, *J. Phys. Chem. A.* **2000**, *104*, 5457-5468.

[xvi] M. H. V. Werts, J. W. Hofstraat, F. A. J. Geurts, J. W. Verhoeven, *Chem. Phys. Lett.* **1997**, *276*, 196-201.

MAGNETIC AND MORPHOLOGIC CHARACTERISATION OF SOME SUBSTITUTED FERRITES SYNTHESISED BY A NON-CONVENTIONAL METHOD

ELENA CRISTUREAN[a], DANA MARINESCU[a], RODICA OLAR[a], MIHAELA BADEA[a], N. STANICA[b], ANA EMANDI[a] and MARIA BREZEANU[a]

[a] *Department of Inorganic Chemistry, Bucharest University, 23 Dumbrava Rosie St., Bucharest, Romania*

[b] *Institute of Chemical Physics, 202 Spl. Independentei, Bucharest, Romania*

Substituted ferrites with the chemical composition $M_xFe_{1-x}Fe_2O_4 \cdot nH_2O$ (M: Mn(II), x:0.35, n: 0.4; M: Ni(II), x:0.38, n: 1; M: Zn(II), x:0.45, n: 0.35) have been prepared by a soft chemical method using as starting materials magnetite and suitable acetates. These compounds have been characterised X-ray diffraction on powder and magnetic measurements at room temperature.

Keywords: Substituted ferrites, saturation magnetization

1. Introduction

One of the main challenges in the oxidic materials synthesis remains to be the obtaining of the samples that have dimensions within the nanoparticles range. These kinds of materials have electric, magnetic and catalytic properties[1-5] that vary with the particle dimension.

The non-conventional methods as sol-gel method[6], thermal decomposition of suitable complexes[7], coprecipitation[8] and hydrothermal[9] method have been widely used in the last years in order to obtain substituted ferrites with different morphologies.

This paper deals with the magnetic and morphologic characterisation of manganese, nickel and zinc ferrite obtained by a soft chemical method using as starting materials magnetite and suitable acetates in aqueous medium.

F. Charra et al. (eds.), Organic Nanophotonics, 163–168.

2. Experimental part

The substituted ferrites were obtained by reaction of magnetite and $M(CH_3COO)_2 \cdot nH_2O$ (M: Mn, Ni, Zn) as reported elsewhere[10].
X-ray powder diffraction patterns were measured on the Philips diffractometer using Cu-Kα radiation source with the wavelength λ=1.54 Å.
Magnetic measurements were done by alternating current permeameter, at room temperature, using Ni-powder as standard.

3. Results and discussions

In this paper, we report the morphologic and magnetic characterisation of the substituted ferrites $M_xFe_{1-x}Fe_2O_4 \cdot nH_2O$ (M: Mn(II), x:0.35, n: 0.4; M: Ni(II), x:0.38, n: 1; M: Zn(II), x:0.45, n: 0.35) obtained by a soft chemical method.

The X-ray powder diffraction patterns for magnetite and nickel ferrite are shown in Figure 1. The d values and the mean diameter of particles (D_{XR}) determined with the Scherrer's formula[11] are summarised in Table 1; these values show that the substituted ferrites are nanocrystals.

As shown in figure 1, the magnetite has a cubic structure, the substituted ferrites preserving this structure. The substitution of Fe(II) (r = 0.74Å) with M(II) (M = Mn (r = 0.91Å, Ni (r = 0.69Å) and Zn (r = 0.83Å)) leads to a modification of the distances between atomic planes, as shown in table 1.

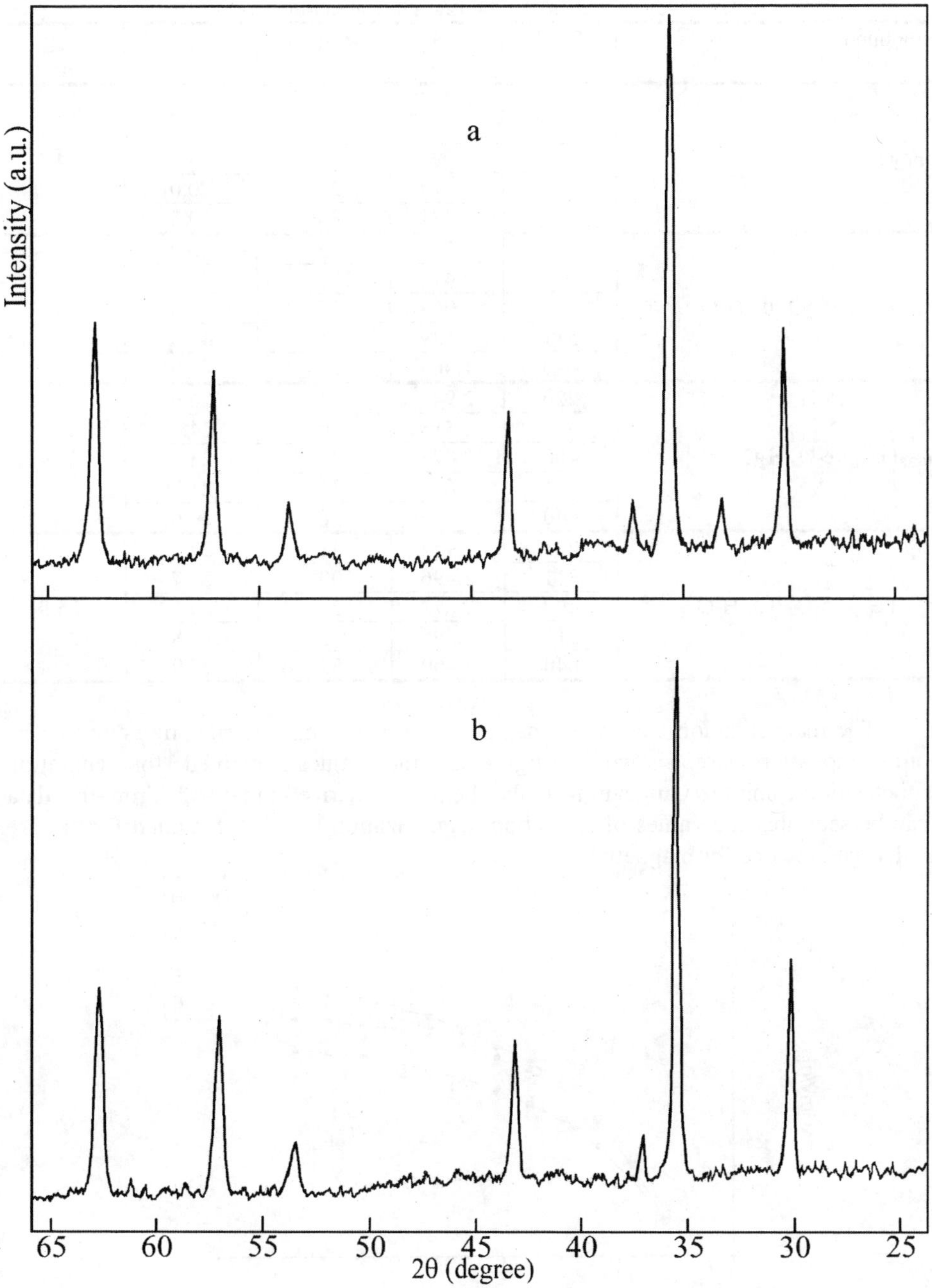

Figure 1. The diffractogram pattern for $Ni_{0.38}Fe_{0.62}Fe_2O_4{\cdot}H_2O$ (a) and $FeFe_2O_4$ (b)

Table 1. X-ray diffraction data for magnetite and substituted ferrites

Compound	[hkl]	d [Å]	I/I_0 [%]	D_{XR} [nm]	D_{medXR} [nm]
$FeFe_2O_4$	220	2.964	34	27.6	24.4
	311	2.561	100	29.1	
	400	2.095	22	25.6	
	511	1.614	29	20.0	
	440	1.483	33	19.5	
$Mn_{0.35}Fe_{0.65}Fe_2O_4 \cdot 0.4H_2O$	220	2.879	32	40.7	43.7
	311	2.478	100	41.5	
	400	2.063	25	43.0	
	511	1.589	31	44.6	
	440	1.463	52	48.9	
$Ni_{0.38}Fe_{0.62}Fe_2O_4 \cdot H_2O$	220	2.942	34	26.6	23.4
	311	2.518	100	24.1	
	400	2.075	23	23.9	
	511	1.610	30	21.7	
	440	1.479	38	20.6	
$Zn_{0.45}Fe_{0.55}Fe_2O_4 \cdot 0.35H_2O$	220	2.907	32	27.1	26.8
	311	2.496	100	23.7	
	400	2.072	27	28.6	
	511	1.595	33	26.8	
	440	1.469	54	27.9	

The magnetization curves for magnetite and substituted ferrites measured at the room temperature are shown in figure 2, the values obtained for saturation magnetization using the Langevin formula[12] being summarized in table 2. From this data it can be seen that the values of saturation magnetization for all substituted ferrites are smaller than the one for magnetite.

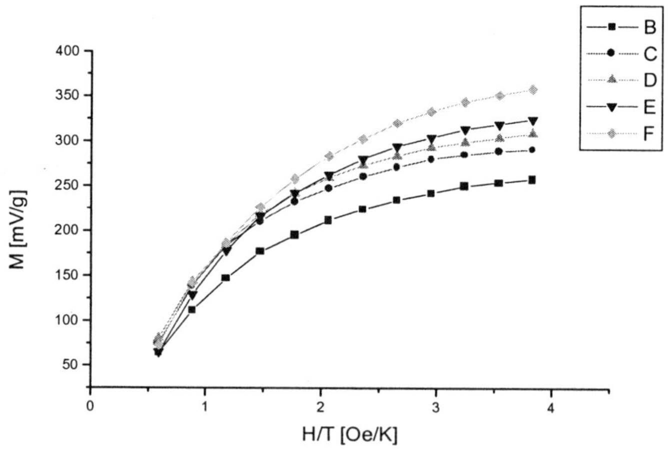

Figure 2. Magnetization curves for Ni (B); $Ni_{0.38}Fe_{0,62}Fe_2O_4$ (C); $Zn_{0.45}Fe_{0,55}Fe_2O_4$ (D); $Mn_{0.35}Fe_{0,65}Fe_2O_4$ (E); $FeFe_2O_4$ (F)

Table 2. The saturation magnetization values for nickel, magnetite and substituted ferrite

Sample	M (emu/g)
Ni	54,0
$Ni_{0,38}Fe_{0.62}Fe_2O_4$	59,4
$Zn_{0,45}Fe_{0.55}Fe_2O_4$	63,2
$Mn_{0,35}Fe_{0.65}Fe_2O_4$	70,1
$FeFe_2O_4$	78,6

Néel postulated that magnetic moment of ferrites is a sum of magnetic moments of individual sublattices, the sublattice A consisting of cations in tetrahedral positions and sublattice B with cations in octahedral positions. Usually, the interaction between paramagnetic ions of sublattices A and B (AB-interactions) is the strongest while AA-interaction is almost ten times weaker and BB-interaction is the weakest.

In the inversed ferrites, magnetic moments are mutually compensated and the resulting moment of the ferrite is due to the magnetic moment of divalent cations in the B-positions.

$$\begin{array}{ccc} A & B & \\ Fe^{3+} & [M^{2+} & Fe^{3+}]\,O_4 \\ \downarrow & \uparrow & \uparrow \end{array}$$

Partial substitution of Fe(II) ($3d^6$, four unpaired electrons) in this positions with Zn(II) ($3d^{10}$, diamagnetic) or Ni(II) ($3d^8$, two unpaired electrons), by maintaining the inversed spinel structure, leads to the decrease of the ferromagnetic interaction BB this behavior having as a final result the decrease of the saturation magnetization.

$$\begin{array}{cccc} A & & B & \\ Fe^{3+} & [Ni^{2+}_{0.38} & Fe^{2+}_{0.62} & Fe^{3+}]\,O_4 \\ \downarrow & \uparrow & \uparrow & \uparrow \end{array}$$

The partial substitution of Fe(II) ($3d^6$, four unpaired electrons) with Mn(II) ($3d^5$, five unpaired electrons) in B sites should lead to an increase of the saturation magnetization. The experimental value, that is lower than expected, could mean that this substituted ferrite adopts a mixed spinel structure. This behavior for manganese ferrite has already been reported for such a substitution degree.

4. Conclusion

The manganese, nickel and zinc ferrites possess the same cubic structure as magnetite with lattice parameter within 20-45 nm range. The size of particles determined from analysis of the diffraction peak indicated that these samples are nanocrystals.
The magnetic behaviour agrees with the substitution degree established by the chemical analysis.

Acknowledgements
The present work was partially supported by a Grant for Scientific Research from the Ministry of Education and Research of Romania.

REFERENCES

1. L. G. Hubert-Pfalzgraf, *New J. Chem*, **11** (1987) 663.
2. S. I. Swartz and V. E. Wood, *Cond. Matter New*, **1** (1992) 4.
3. M. Singh and S. P. Sud, Modern Phys. Lett., **14** (2000) 531.
4. A. Miyakosshi, A. Ueno and M. Ichikawa, *Appl. Cat. A: General*, **216** (2001) 137.
5. G. K. Boreskov, V. S. Muzikantov and T. T. Schulmester, *Kinet. Katal.*, **12** (1971) 979.
6. D. H. Chen and X. R. He, *Mat. Res. Bull.*, **36** (2001) 1369.
7. I. Mindru, L. Patron, D. Crisan and M. Brezeanu, *Rev. Roum. Chim.*, **42** (1997) 31.
8. N. Viart, G. Pourroy, J. M. Greneche, D. Niznansky and J. Hommet, *Eur. Phys. J.*, **12** (2000) 37.
9. J. H. Lee, C. K. Kim, S. Katoh and R. Murakani, *J. All. Comp.*, **325** (2001) 276.
10. R. Olar, M. Badea, L. Diamandescu, E. Cristurean, D. Marinescu, D. Tarabasanu-Mihaila, N. Stanica and M. Brezeanu, *J. Alloy and Compds.*, (in press).
11. A. Guinier, "Theorie et Technique de la Radiocristallographie", Dunod, Paris, 1964, 462.
12. P. Langevin, *Ann. Chim. Phys.*, **8** (1905) 70.

DENDRITIC POLYESTERS FOR OPTICAL APPLICATIONS

ROBERT VESTBERG, EVA MALMSTRÖM, ANDERS HULT
Dept. of Fiber and Polymer Technology
Royal Institute of Technology
SE-100 44 Stockholm, Sweden

CESAR LOPES, ANDERS ERIKSSON, MIKAEL LINDGREN
Swedish Defence Research Agency
Department of Sensor Technology, P.O. Box 1165
SE-581 11 Linköping, Sweden

Abstract

In the protection against lasers, optical power limiting (OPL) is an area of interest. We have explored the effect of using dendritic molecules, based on bis-MPA (2,2-bis(methylol)-propionic acid), as encapsulants of porphyrine and thiophene dyes having OPL-properties. Preliminary results indicate that the OPL-effect of the porphyrine is increased approximately by a factor 2 if the porphyrine is decorated with dendritic wedges. This is thought to be a result of isolation of the dye leading to extended life-times for the excited states, increasing the probability for multi-photon absorption. For the thiophenes no OPL-effect was observed. This was probably due to the small size of the dendrons (first generation).

Introduction

DENDRIMERS

Dendritic molecules[1], which are highly branched polymers, have showed a lot of interesting properties. The highly branched dendrimers are based on AB_x monomers, they are monodisperse and perfectly branched (i.e. each monomer contains a new branching point and it contains no linear segments). The dendrimer is based on layers of repeating units, the number of layers is defined as the generation of the dendrimer. They adopt a globular shape which becomes more pronounced with higher generations. Because of the high branching the dendrimers contain many end-groups, and about half the weight is made up by the end-groups. Therefore the properties of the dendrimers are highly dependent on the end-groups and hence it can also be altered to a great extent by changing the end-groups. Dendrimers have been proven to exhibit increased solubility and reduced viscosity in both melt and solution in comparison to its linear counterparts[2,3]. It has also been shown that dendrimers can isolate the core moiety from

F. Charra et al. (eds.), Organic Nanophotonics, 169–176.

the surroundings, suggesting that they can provide site isolation[4,5]. The high degree of freedom in the molecular design of dendrimers makes them very attractive for use in photonic applications. The aim of this study is to show the versatility of the bis-MPA (2,2-bis(methylol)-propionic acid) based dendrons when used in optical applications.

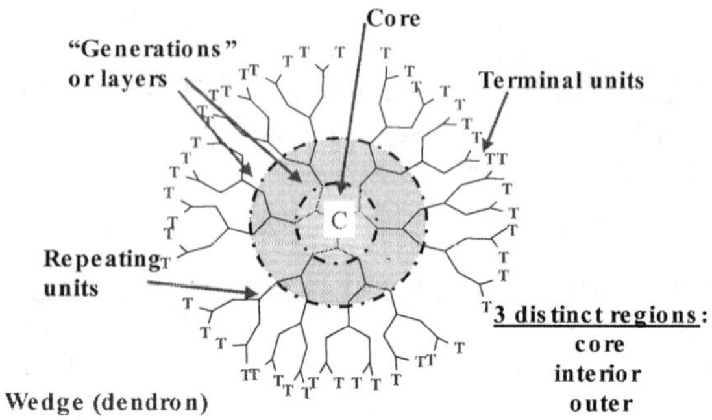

Figure 1. Schematic representation of a dendritic molecule.

OPTICAL POWER LIMITING

Optical equipment is used by the military personal for many purposes, e.g. observation, navigation, fire control, and reconnaissance. The laser, on the other hand, is a powerful countermeasure weapon against optical equipment, particularly the detectors. In direct vision, e.g. through a sight, the human eye (the retina) is the detector and can receive damaging laser doses. To shield the detectors from the dangerous laser light Non Linear Optical (NLO) materials can be used. NLO materials absorb high intensity light while they are transparent to low intensity light.

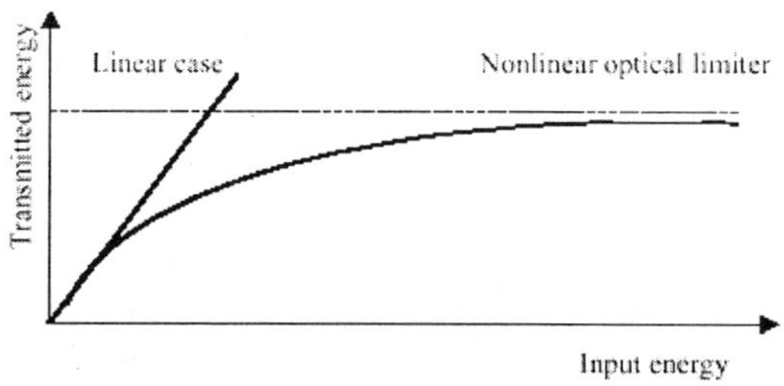

Figure 2. Graphic illustration of nonlinear optical limiting.

We have explored the effect of using dendritic molecules as encapsulants of dyes having OPL properties. The role of the dendritic part is to increase the solubility and to provide site isolation. The site isolation is thought to depress de-excitation by acting as a unimolecular bumper in collision with other molecules. This should increase the NLO behavior and hence the OPL properties.

In this study we have investigated a dendron-decorated porphyrine molecule. Some initial tests have also been carried out on a thiophene dye.

Synthesis

The dendrons were synthesized according to the divergent route employing anhydrides of bis-MPA[6].

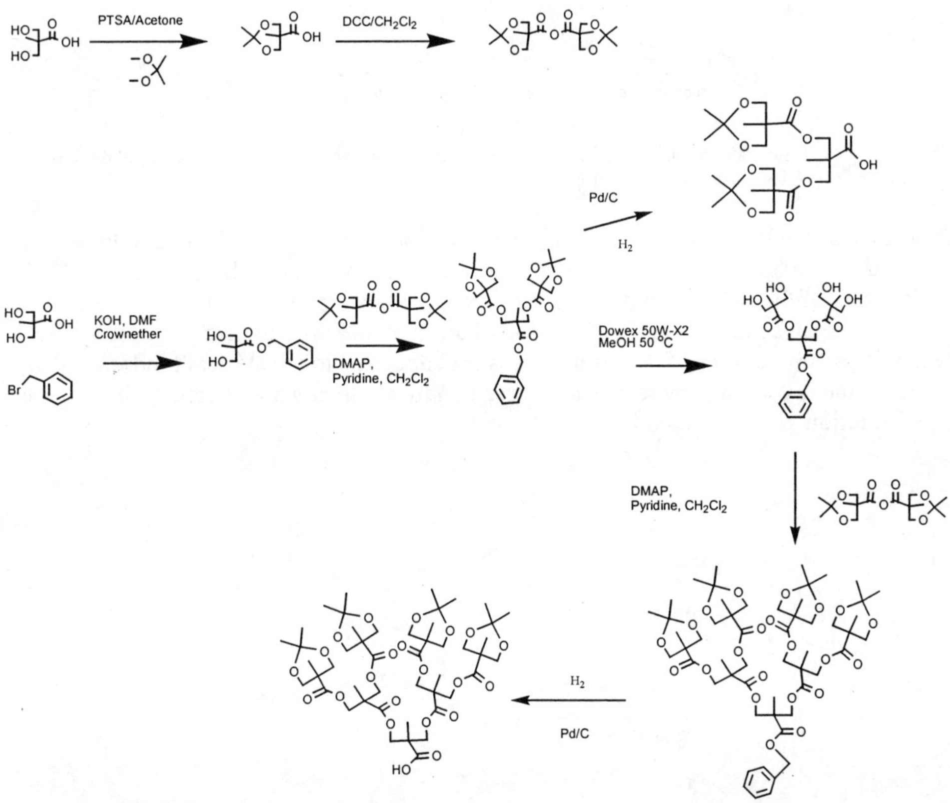

Scheme 1. General synthesis route to bis-MPA dendrons.

A central Zn-atom was added to the porphyrine core by reaction with zinc acetate[7]. The reaction was monitored with UV-Vis spectroscopy. In the spectra it can be seen that the reaction only takes a couple of hours for complete insertion. The absorption at around 520 and 650 nm decreases/disappears whilst the absorption around 600 nm increases when the metal atom is inserted.

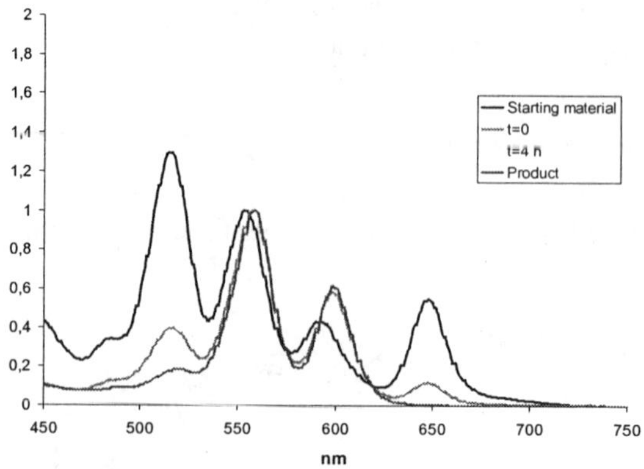

Figure 3. UV-Vis spectra of the insertion reaction of Zn into a porphyrine molecule. The curve at 4 hours is the same as the product curve showing complete conversion.

Synthesis of dendron-decorated porphyrines were performed through direct addition of the dendron wedges to the Zn-porphyrine core as described earlier by Fréchet *et al.*[4] Dendrimers of the second and third generation have been investigated. However it proved to be difficult to achieve 100% substitution on the third generation porphyrine even with a large excess of dendron. This is probably due to the sterical hindrance that occurs in the divergent synthesis when three dendrons already are attached. The degree of substitution was measured by NMR to 85 %.

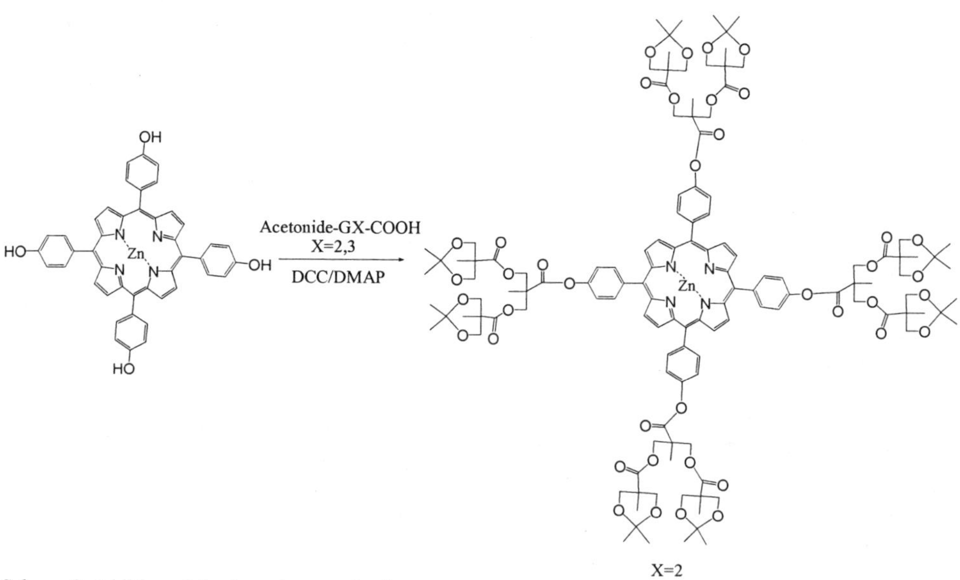

Scheme 2. Addition of dendrons to a porphyrine core.

One thiophene molecule has also been synthesized according to the procedure outlined in *Scheme 3*.

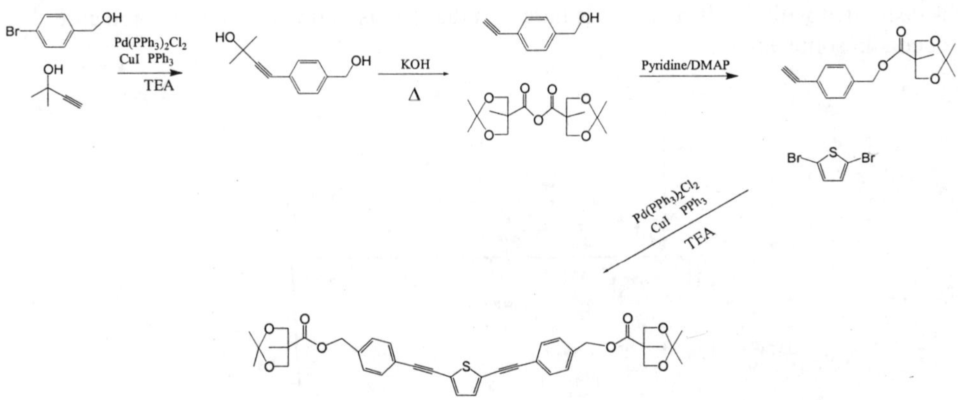

Scheme 3. Synthetic route to dendron decorated thiophenes.

Measurements

The assessment of the optical limiting performance of the compounds was investigated with the dyes in solution. The solutions were prepared with a photopic transmission of approximately 70%. The investigations were performed with a frequency doubled Nd:YAG laser delivering 5 ns pulses at 532 nm and with a repetition rate of 10 Hz. The laser-beam is focused in a 2 mm cuvette by an arrangement of lenses.

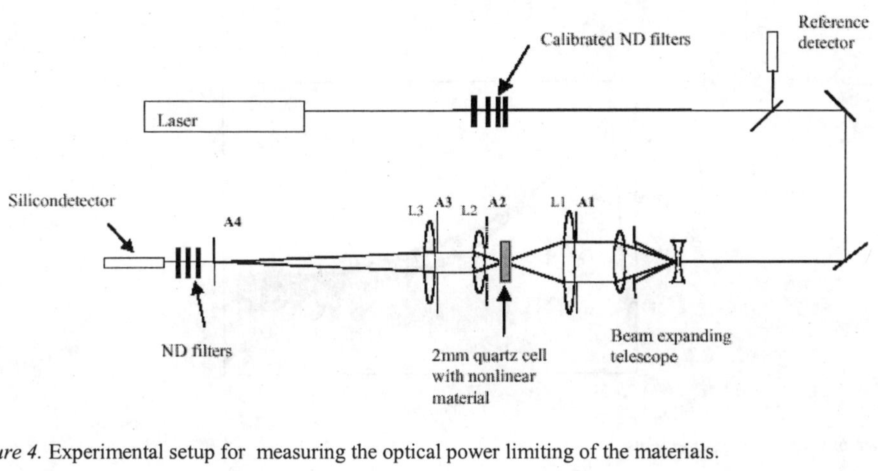

Figure 4. Experimental setup for measuring the optical power limiting of the materials.

Results and Discussion

The optical limiting of the porphyrine-cored dendrimer of the second generation is depicted below. It can be seen that the optical limiting is increased when the porphyrine is decorated with dendrimers. This indicates that site isolation of the core molecule occurs in some extent.

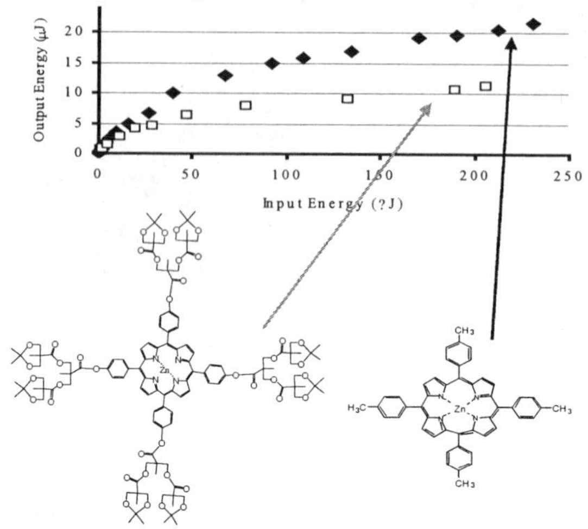

Figure 5. Optical limiting of a second generation porphyrine-cored dendrimer compared to a pure porphyrine molecule.

Further proof of the site isolation is given by the life time measurements, which showed extended lifetimes for the dendron decorated porphyrines.

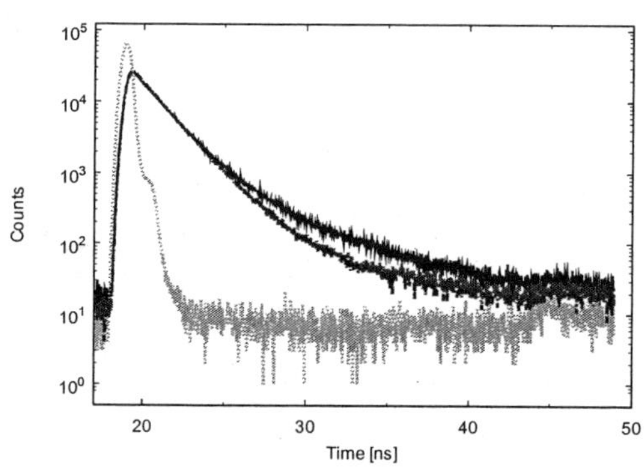

Figure 6. Lifetime measurements.

Measurements on the third generation dendron decorated porphyrine resulted in even higher optical limiting implying that the shielding of the core is better with larger dendrons.

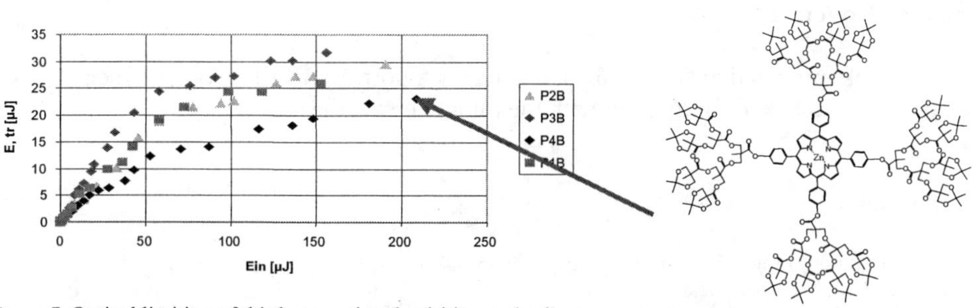

Figure 7. Optical limiting of third generation dendritic porphyrines.

For the thiophene samples this effect is not seen. The reference sample shows higher optical limiting than the first generation dendron functionalized, probably because the first generation dendron is small and the effect of the dendron is therefore thought to be very little. However the solubility of this molecule proved to be good and the optical limiting increased with increased concentration.

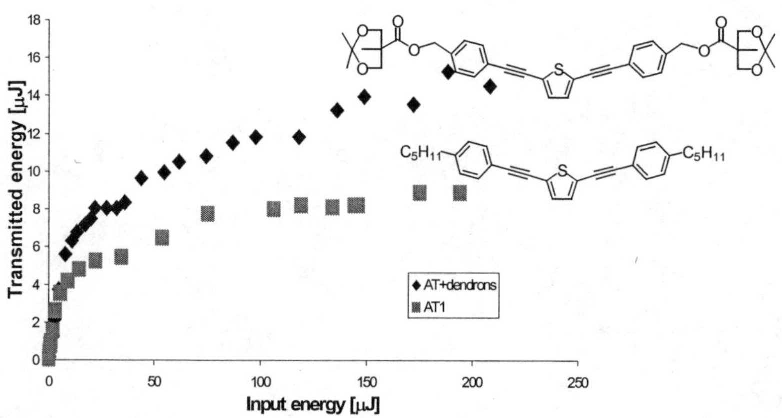

Figure 8. Optical limiting results for the thiphene molecules.

Conclusion

We have proven that the optical limiting of dendron decorated porphyrine-Zn samples increases compared to pure porphyrine-Zn. Also that increasing size of the dendrimer yields higher optical limiting. These results have led us to the conclusion that we have site-isolation of the porphyrine core..For the thiophene samples we have not seen the same effect. However, the lack of site isolation probably due to the small size of the dendrons. If the higher generation dendrons were used there is very probable that site-isolation could be seen.

Acknowledgement

FOI in Linköping is thanked for their financial support. Bertil Eliasson at Umeå University is thanked for helping with the thiophene synthesis.

References

1) Inoue, K. (2000) *Progress in Polymer Science* **25**, 453-571.
2) Fréchet, J. M. J. (1994) *Science* **263**, 1710-15.
3) Hawker, C. J.; Malmström, E. E.; Frank, C. W.; Kampf, J. P. (1997) *JACS,* **119**, 9903-9904.
4) Kawa, M.; Fréchet, J. M. J. (1998) *Chemistry of Materials* **10**, 286-296.
5) Hecht, S.; Vladimirov, N.; Frechet, J. M. (2001) *JACS* **123**, 18-25.
6) Malkoch, M.; Malmström, E.; Hult, A. (in press) Rapid and Efficient Synthesis of Aliphatic Ester Dendrons and Dendrimers, *Macromolecules.*
7) Pollak, K. W.; Sanford, E. M.; Fréchet, J. M. J. (1998) *Journal of Materials Chemistry* **8**, 519-527.

NONLINEAR OPTICAL STUDY OF FULLERENE-DOPED CONJUGATED SYSTEMS: NEW MATERIALS FOR NANOPHOTONICS APPLICATIONS

NATALIE V. KAMANINA
Vavilov State Optical Institute
12 Birzhevaya Line, St. Petersburg, 199034, Russia
E-mail: kamanin@ffm.ioffe.rssi.ru

Nonlinear optical properties have been studied in C_{60} and C_{70}-doped organic thin films (polyimide, 2-cyclooctylamino-5-nitropyridine, *N*-(4-nitrophenyl)-(*L*)-prolinol) and in the liquid crystal compounds based on them under a laser irradiation at wavelengths of 532, 805, and 1315 nm. The films can be applied as laser power attenuators at the laser energy density of more than 1.5-2 J cm^{-2} and at the laser energy density up to 0.8-1 J cm^{-2} in the visible and near infrared spectral ranges, respectively. The nonlinear refraction and the third order susceptibility have been determined to be n_2~10^{-10} cm^{2}W^{-1} and $\chi^{(3)}$ ~10^{-9} esu for the thin fullerene-doped films and n_2~10^{-9} cm^{2}W^{-1} and $\chi^{(3)}$ ~10^{-8} esu for polymer-dispersed liquid crystal compounds.

1. Introduction

Organic nanophotonics problems have stimulated research in designing new materials for effective optical limiting (OL) and holographic recording systems, as well as good modulated media for information processing. Great interest has been expressed in nonlinear optical properties of organic conjugated systems doped with fullerenes [1-9]. The optical properties, which are conditioned by excitation of π-electrons, demonstrate that a new charge transfer complex is formed between fullerene and a donor fragment of an organic molecule [10-13]. This complex is of a higher excited state absorption cross section than the ground state one. That is why, fullerene-doped systems are reverse saturable absorption (RSA) materials and may be used for a sensor and eye protection as a nonlinear optical absorbers. Moreover, a large value of the light-induced change in refractive index in these compounds contributes significantly to the OL effect (due to energy losses by diffraction) and explains holographic recording in these media because of the photorefractive effect even without an electrical control. It also predicts large nonlinear refraction n_2 and third order susceptibility $\chi^{(3)}$, which allow the materials to be used for conversion and modulation of a laser beam over broad spectral range.

In the present paper the nonlinear optical properties have been studied in the C_{60} and C_{70}-doped organic thin films (polyimide, 2-cyclooctylamino-5-nitropyridine, *N*-(4-nitrophenyl)-(*L*)-prolinol) and in the liquid crystal structures based on the materials

F. Charra et al. (eds.), Organic Nanophotonics, 177–192.

mentioned above under a laser irradiation at wavelengths of 532, 805, and 1315 nm. It has been shown that the systems studied can be applied as diffactive elements and laser power attenuators in the visible and near infrared spectral ranges.

2. Experiment

3 % solutions of conjugated materials in 1,1,2,2–tetrachloroethane (TClE) were used. Polyimide 6B (PI6B) systems, which general formula was previously described in paper [14], 2-cyclooctylamino-5-nitropyridine compounds (COANP), which linear and nonlinear optical properties for fullerene-free phase were studied in paper [15], *N*-(4-nitrophenyl)-(*L*)-prolinol (NPP) structures shown in the viability test by Zyss *et al.* [16] were used as photosensitive organic matrixes. The solutions were doped with fullerenes C_{60} or/and C_{70}. The fullerene concentration was varied from 0.2 wt.% to 5 wt.%. It should be noticed that fullerenes are of relatively high solubility (about 5.3 mg mL^{-1}) [17] in TClE. The 2.5–5 μm thick films were spun on the glass substrates coated with transparent indium-tin-oxide contacts. A general views of the films obtained and PDLC structure based on the materials treated are presented in Fig. 1. The film structures before and after drying are shown in Fig. 2.

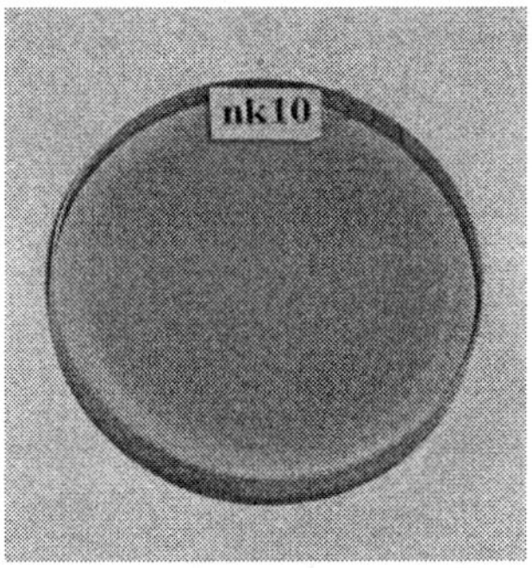

Figure 1a. A view of the fullerene-doped polyimide thin film on the glass substrate.

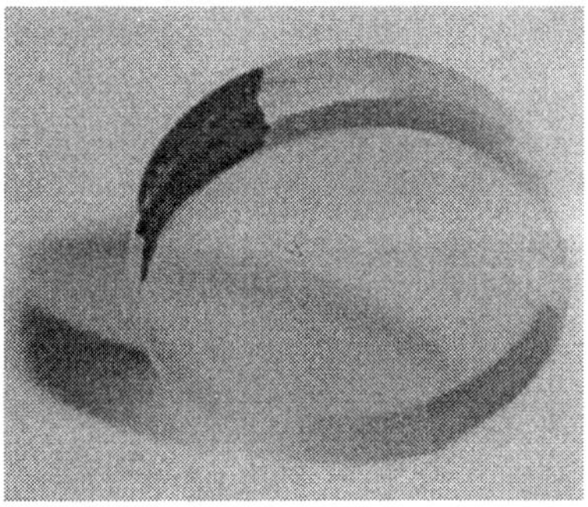

Figure 1b. A view of the fullerene-doped PDLC.

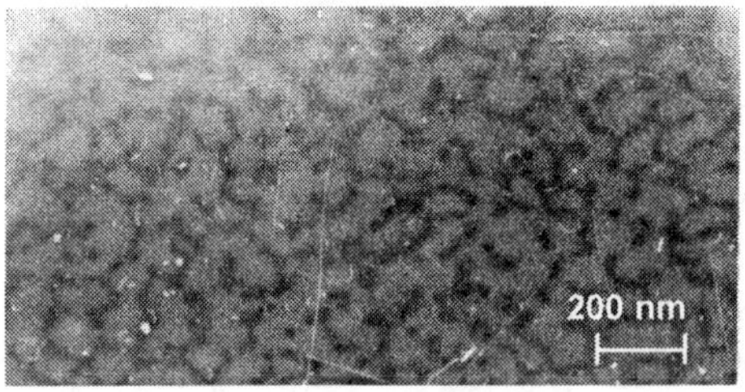

Figure 2a. SEM image of fullerene structures with different configuration, forming chains of quasi-pentagonal and hexagonal shapes.

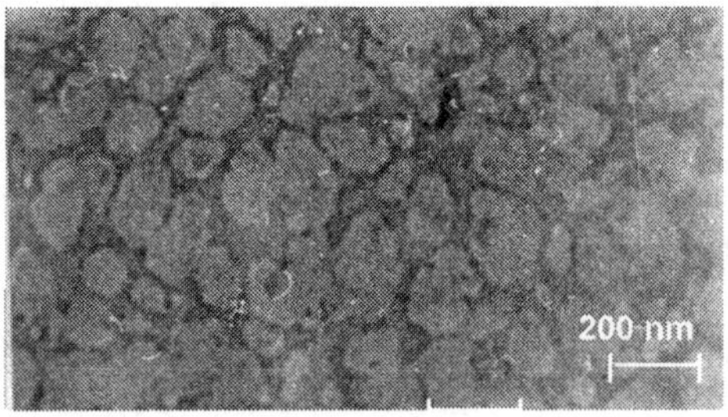

Figure 2b. SEM image of the 6B polyimide film formed on the fullerene "skeleton".

The films were investigated with a scanning electron microscope (SEM) HU-11B at the accelerating voltage of 75 kV. The multiplication was $12500^{\times}$. To enhance the contrast, the SEM images were obtained with partial focusing. Spectroscopic measurements were carried out using a Perkin–Elmer Lambda 9 instrument in the wavelength range 200-3000 nm.

10-12 μm polymer-dispersed liquid crystal structures (PDLCs) were made when the refractive index of LC ordinary wave n_o was close to refractive index of polymer n_p. ZK999, ZK1282, and E7(BDH) were used as a liquid crystal mesophase. The initial mixture of polymer PI6B and nematic liquid crystal (NLC) was prepared in the ratio of 3 polymer to 2 NLC. PDLC based on COANP and NPP compounds was made in the ratio of 1 photosensitive component to 2 NLC. A 2.5% solution of non-photosensitive polyimide 81A in TClE was used as a film-forming base for COANP and NPP. The relation between COANP and NPP compounds and the film-forming base was 2:1. A size of NLC drops in the photosensitive matrix was 2–4 μm.

The pulsed Nd-YAG laser at the wavelength of 532 nm was applied as a radiation source to study OL and to record thin holographic grating in the structures. Passive modulation was performed with a LiF crystal. A KDP crystal was used as a second-harmonic converter. A pulsewidth was 20 ns and a laser spot on the sample surface was 3-3.5 mm in diameter. Beam energies incident on and transmitted through the sample were measured. A set of neutral light filters was used to vary the incident beam energy. The OL scheme was analogous to that presented in paper [18]. The holographic experiments were performed at the spatial frequency of 100 mm^{-1}. In this case the pulsewidth was 20 ns and 400 ps and laser spot on the sample surface was 4-5 mm in diameter. The diffraction was observed in the self-diffraction mode under the Raman-Nath diffraction conditions. The scheme was the same to that shown in paper [19].

To estimate the nonlinear transmission at 1315 nm the dependence of the output energy on the input one was measured with a use of a photodissociative iodine laser. The first near-infrared OL results were published in paper [20]. The laser was pumped by a nonmagnetic coaxial Xe lamp, which an interior quartz tube was filled in by components of a laser mixture: n-C_3F_7I (RI) and SF_6. Partial pressure of n-C_3F_7I (RI) and SF_6 was 35 and 500 mm of Hg, respectively. The diameter and length of the active zone were 0.8 cm and 50 cm, respectively. A Xe flash-lamp had a broad spectral band. The flash preceded the laser pulse with the time interval of 5 μs . The pumping and laser pulsewidth was 8 μs and 50 ns, respectively. A spot on the sample surface was 2 mm. The input energy was measured with a calorimeter. The energy transmitted through a set of filters and the sample was measured with a pyroelectric photometer. It should be mentioned that to control the OL level in the near infrared, an additional scheme solution was proposed [21]. The holographic experiments at λ=1315 nm were made at the spatial frequency of 100 mm^{-1}.

Taking into account that the additional absorption band in the range 800-820 nm was preliminary found in the COANP compound sensitized by fullerene C_{70}, OL experiments have been carried out for PDLC based on COANP-C_{70} and E7 under femtosecond pulsed irradiation of a quasi-CW Ti-sapphire laser (λ = 805 nm). The pulsewidth was 70 fs, the repetition frequency was 80 MHz, the power was 150 mW. The spot diameter at the film surface was 0.2 mm.

3. Results and Discussion

3.1. OPTICAL LIMITING STUDY

One of the most important effect observed now in fullerene-doped media is OL. The study of this effect allows the laser-matter interaction to be investigated and the important technical problem, namely, search for the new materials for protection of human eyes and technical devices against high laser radiation to be solved.

The dependence of the output energy density W_{out} on the input one W_{in} for the polyimide and COANP thin films with different fullerene concentrations followed by different number of excited molecules is shown in Fig. 3. The same dependence for a 1 μm thick pure fullerene film is shown in the inset. On synthesis of thin films, preference has been given to C_{70} because the molecule is anisotropic. As the result, the compound with C_{70} has 3-10 times more absorption than the one with C_{60} in the range 400-700 nm [22]. The absorption spectra of the films with C_{70} are red-shifted 20-30 nm and 50-70 nm for PI [23] and COANP [24], respectively. Because the lifetime of C_{70} in triplet T_1 state is one order of magnitude longer than that of C_{60} [25], more pronounced responses and their clear detection have been obtained over visible spectral range.

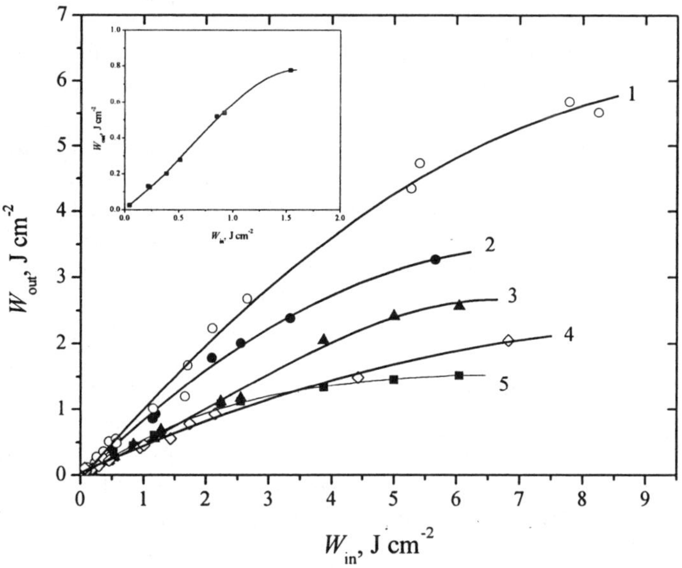

Figure 3. Dependence of the output energy density (W_{out}) on the input one (W_{in}) for films: 1 – COANP with 0.5 wt.% C_{70}; 2 – polyimide with 0.2 wt. % C_{70}; 3 – polyimide with 0.5 wt. % C_{70}; 4 –COANP with 5 wt.% C_{70}; 5 – polyimide with 0.7 wt. % C_{70}. In the inset: Dependence of the output energy density (W_{out}) on the input one (W_{in}) for 1 μm pure fullerene thin film. (λ =532 nm).

From the intersection of linear and nonlinear parts of the dependence $W_{out} = f(W_{in})$ (Fig. 3), the OL level could be estimated. It is corresponded to the input energy density of 1.1-1.3 J·cm^{-2} and 2.2-2.5 J·cm^{-2} for polyimide with 0.7 wt.% C_{70} and COANP with 5 wt.% C_{70}, respectively. 3-fold attenuation of the laser beam has been established for these structures that is close to that for 1 μm pure fullerene thin film (see the inset of the Fig. 3). It should be mentioned that the level of input energy density, at which the saturation of the transmission takes place, decreases as the fullerene concentration increases. This fact is in a good agreement with the results of papers [26,27].

The dependence of the output energy density (W_{out}) on the input one (W_{in}) for polymer-dispersed liquid crystal (PDLC) structures is shown in Fig. 4.

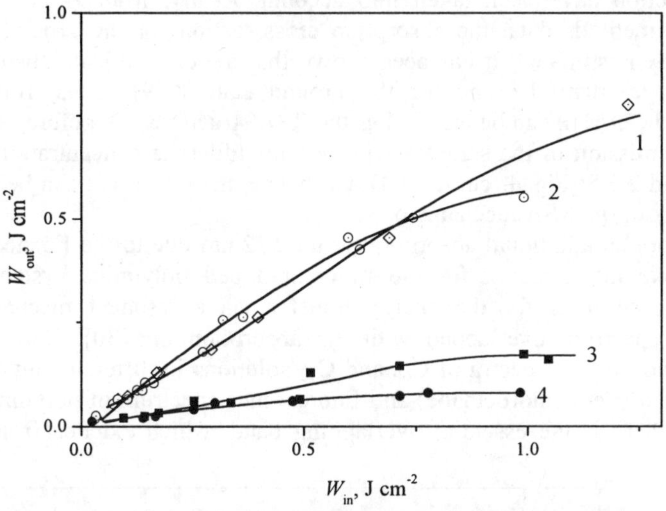

Figure 4. Dependence of the output energy density (W_{out}) on the input one (W_{in}) for PDLC structures: 1 – COANP-5 wt.%C_{70}-LC (E7); 2 – NPP-1 wt.%C_{60}-LC (E7);3 – polyimide-(0.25 wt.% C_{60}+C_{70})-LC (ZK999); 4 – polyimide-(0.5 wt.% C_{60}+C_{70})-LC (ZK999). (λ =532 nm).

More than 10-fold attenuation of the laser beam has been obtained for the PDLC cell based on polyimide 6B at the incident energy density of 0.15-0.2 J·cm^{-2}. Twice attenuation of the laser energy density has been found in the fullerene-doped PDLC based on COANP and NPP compounds that is close to the OL level for pure C_{60} film. The difference in the nonlinear transmission of PDLC based on COANP structure and NPP one is associated with different matching between COANP and NLC refractive indexes as compared with that between NPP and NLC refractive indexes. It should be noticed that OL for NPP compounds has been revealed for the PDLC cell based on NPP when NPP structure was doped with fullerene C_{60}. A little enhancement of absorption at the wavelength of laser irradiation (λ=532 nm) has been observed on doping NPP materials with fullerenes C_{60}. It may be caused by two-photon absorption mechanism because the resonance line for C_{60} (565 nm) is close to the experimental irradiation wavelength.

The first OL experiments have been carried out for PDLC based on COANP-C_{70} and E7 under femtosecond pulsed irradiation of a quasi-CW Ti-sapphire laser (λ = 805 nm). Taking into account the saturation effects in quasi-CW mode, the 10-fold attenuation of the laser beam has been detected. The experiment allowed dynamics of photo-induced processes to be investigated at upper excitation levels, which were characterized by the transition times of 1 fs, and the reverse saturable absorption to be revealed in the fullerene-doped organic structures based on COANP, that did not contradict the results of Ref. [28]. In that paper it was found in pure fullerene films that a photo-induced absorption increase occurred at the forbidden transition of 1.6 eV due to the electron state hybridization in the electric field of the laser beam.

To discuss the OL mechanisms, which are responsible for the OL effect in the fullerene-doped polyimide and COANP structures RSA and complex formation have been considered. Moreover, an additional Förster energy transfer process and a dipole-dipole interaction have been taken into account. Really, from Z-scan and electro-photographic methods data, the absorption cross-section for the ground and excited states have been estimated. It has been shown that the excited state absorption cross-section is larger than the one for the ground state [8,29]. Thus, fullerene-doped polyimide and COANP can be testified as the RSA structures. Therefore, the difference between transmission of the samples with various fullerene concentration (see Fig. 3, curves 1,4 and 2,3,5; Fig. 4, curves 3,4) and hence, the OL effect can be explained in the framework of the RSA mechanism.

Furthermore an additional absorption at the 532 nm due to the Förster mechanism should be take into account for the fullerene-doped polyimide system. T. Förster considered possibilities for the energy transfer via a resonant mechanism, if the fluorescence spectrum overlapped with the absorption one [30]. From comparison between the absorption spectra of C_{60} and C_{70} solutions in different solvents (hexane, toluene, benzene, tetrachloroethane) and fluorescence spectrum of polyimide matrix, it follows that there is the essential overlapping band, which extends from 500 nm to 700 nm [18].

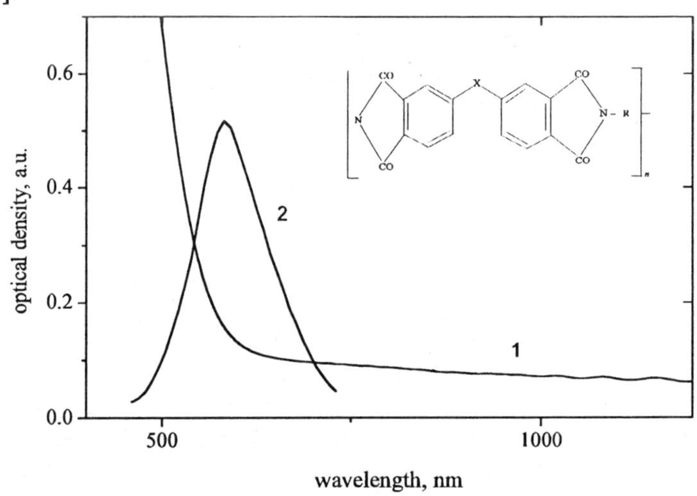

Figure 5. Absorption spectrum of the C_{70}-doped polyimide 6B film (1) and fluorescence spectrum of a polyimide matrix (2). In the inset: a general formula of polyimide 6B is shown.

In our case, the overlapping of the fluorescence spectrum of polyimide matrix and the absorption one of the C_{70}-doped polyimide is observed at the excitation wavelength of 532 nm. This spectra are shown in Fig. 5. As the result, polyimide doped with fullerenes C_{70} should attenuate the laser beam stronger due to resonance absorption at 532 nm via the Förster mechanism.

As for C_{70}-doped COANP compounds, one can note that electronic states of COANP, corresponding to the red edge absorption near 430 nm, are in the same spectral region as the C_{70} electronic states. The spectra are presented in Fig. 6. As the result, the theoretical model for the absorption spectra under consideration based on the electric dipole-dipole interaction between electronic degrees of freedom of C_{70} and COANP molecules has been proposed [24]. Due to this rather close energy splitting one deals here with a possibility of a quasi-resonant dipole-dipole interaction between COANP and C_{70} states. This model explains the absorption enhancement at 532 nm and do not contradicts with the complex formation in the COANP-C_{70} system.

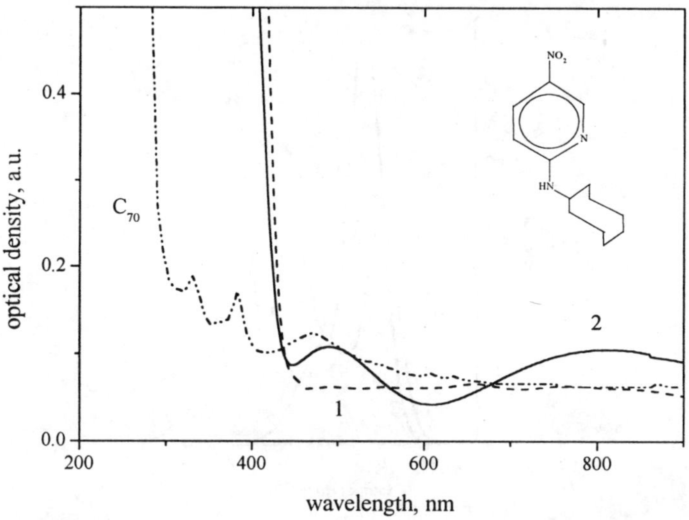

Figure 6. Absorption spectra of thin films: 1 – pure COANP, 2 – C_{70}-COANP. For comparison, the absorption spectrum of pure C_{70} is presented. In the inset: a general formula of COANP is shown.

It should be mentioned that the fullerene sensitization provokes the creation of the charge transfer complex between the donor fragment of photosensitive conjugated molecules and fullerene ones [8,11,13,29,31]. The charge-transfer complex formation should influence OL due to an additional absorption by the complex. The structural peculiarities of polyimide are favorable for the complex formation between the polyimide donor fragment (triphenylamine) and fullerene. The excitation causes the conformational transformation of the triphenylamine fragment from the neutral tetrahedral form to the ionized planar one. The parallel arrangement of these planes creates a necessary condition for the charge transfer.

The fullerenes have electron affinity of 2.65-2.7 eV. They are stronger sensitizers than the intramolecular acceptor fragment (diimide) and hence, they increase the charge phototransfer in that system. Really, fullerene doping of polyimide 6B results in a significant increase in photocurrent. The increase is at least one order of magnitude at potential of 70 V [8]. The acceptor fragment of the COANP and NPP molecule is an NO_2 group, which is bound with the donor fragment by a benzene ring. The electron affinity of an individual molecule or radical NO_2 is 2.3 eV, but the electron affinity of the NO_2 group bound with the benzene ring is only 0.54 eV [32] that is four times less than the one of fullerene. Thus, the fullerene molecule dominates the acceptor NO_2 group in COANP (or NPP) and changes the intramolecular donor-acceptor interaction. The photoconductivity measurements [13] and the mass spectroscopy investigations of the C_{70}-COANP system give additional arguments in favor of the charge transfer complex formation.

The mass spectroscopy data pointed to the charge transfer complex formation between fullerene and triphenylamine and between fullerene and the HN group in the C_{70}-polyimide and C_{70}-COANP systems are shown in Fig. 7. The curves for the C_{70} release rate on heating the films are presented.

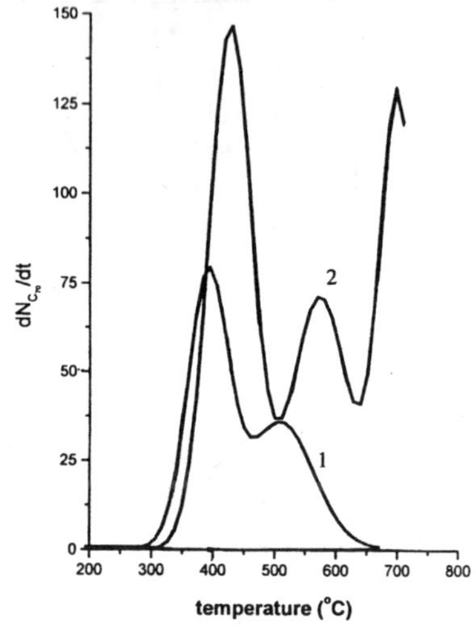

Figure 7. The release rate of C_{70} molecules on heating the films of polyimide 6B and COANP sensitized: 1 – 5 wt.% C_{70}-COANP and 2 – 0.5 wt.% C_{70}-polyimide 6B.

Curve 1 for the 5 wt.% C_{70}-COANP film contains two peaks. The first one at 400 °C corresponds to the release rate of fragments with free fullerene masses. The second one is shifted to the temperature range of 520 °C and associated with the decomposition temperature of a fullerene-HN group complex. Curve 2 for the 0.5 wt.% C_{70}-polyimide 6B film contains three peaks. The first one is observed also close to 400 °C. The second peak is located at 560 °C and associated with the decomposition of the fullerene-triphenylamine complex. It should be noticed that the melting temperature of the polyimides is 700-1000 °C [33,34], thus, the third peak at the temperature higher 700 °C corresponds to the total decomposition of polyimide materials.

Therefore, photoconductivity and mass spectrometry data confirm the complex formation in the polyimide and COANP structures.

Moreover, the drastic spectral dependence of photoconductivity has been revealed for fullerene-doped polyimide system. It should be noticed that photoconductivity of pure fullerenes (C_{60} and C_{70}) decreases drastically [35] in the near-infrared range (λ>700 nm), while the one of fullerene-doped polyimide 6B continues to rise. From this peculiarity, the OL effect has been studied for the structure of C_{70}-doped polyimide

6B with 0.5 wt.% of fullerenes using a pulsed iodine laser ($\lambda = 1315$ nm) with the pulsewidth of 50 ns. The results for the near-infrared range are presented in Fig. 8.

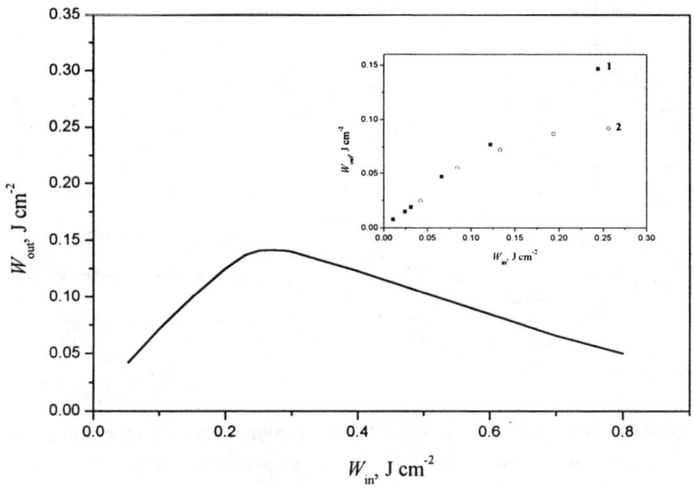

Figure 8. Dependence of W_{out} on W_{in} in polyimide 6B film with 0.5 wt.% of fullerene C_{70} ($\lambda = 1315$ nm). In the inset: Dependence of W_{out} on W_{in} in C_{70}-polyimide 6B films using an additional scheme solution [21].

The effect began at the input energy density of 0.25-0.3 J·cm^{-2}, while the 16-fold attenuation was observed when the incident energy density reached 0.8 J·cm^{-2}. Furthermore, the initial input energy density of 0.25-0.3 J·cm^{-2} can be reduced down to 0.1 J·cm^{-2} by using a Xe flash-lamp with a broad spectral band (see the inset in Fig. 8). The filter with the transmission peak at 470 nm, that corresponded to the resonant line of C_{70}, was applied. The flash-lamp irradiation was incident on the back surface of the film. The flash preceded the laser pulse with the time interval of 5 μs. Thus, the fullerene-doped polyimide film may be applied as an optical limiter operating in the infrared spectral range. The OL effect level can be controlled by the low-power irradiation of the flash-lamp.

3.2. HOLOGRAPHIC RECORDING STUDY

To estimate the laser-induced change in the refractive index and hence, the nonlinear coefficients n_2 and $\chi^{(3)}$ of the system the holographic grating was written. The grating was recorded in the self-diffraction mode. Self-diffraction in the fullerene-doped structures was investigated under nano- and picosecond pulsed laser irradiation at $\lambda = 532$ nm. Two crossed waves with the same frequency induce the grating due to the nonlinearity of the refractive index. The diffraction of the probe wave on this grating or the self-diffraction of each wave inducing the grating could be applied to determine optically-induced changes of the refractive index. The results for the nonsecond pulse regime are shown in Fig. 9. One can see that when fullerene is introduced in the organic matrix, the first diffraction order response increases drastically.

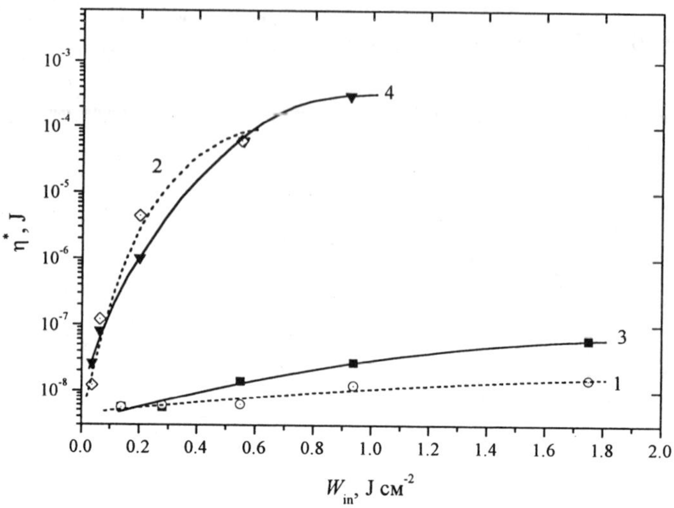

Figure 9 Dependence of the response η* in the first diffraction order on the incident energy density (W_{in}) for the films : pure polyimide (1) and COANP (3) ; C_{70}-doped polyimide (2) and COANP (4). (λ =532 nm).

From these data the change in the laser-induced refractive index has been found using the relationship [36]:

$$\eta = I_1 / I_0 = (\pi \Delta n_i d / 2\lambda)^2 \qquad (1)$$

where η is the diffraction efficiency, I_1 and I_0 are the laser intensity in first diffraction order and incident one, respectively, d is the film thickness, λ is the laser wavelength, and Δn_i is the laser-induced change of refractive index.

The essential increase in the laser-induced change of refractive index has been observed. Δn_i increases from 2.15×10^{-4} to 4.68×10^{-3} and from 3.16×10^{-4} to 6.89×10^{-3} as the incident energy density increases from 0.03 to 0.6 $J \cdot cm^{-2}$ and from 0.03 to 0.9-1.0 $J \cdot cm^{-2}$ for fullerene-doped polyimide and COANP, respectively. This result is associated with thermal nonlinearity, a conformation change of the polymer chain under the nanosecond pulsed laser irradiation, as well as with the photorefractive effect. Really, the field gradient is formed as the charge transfer path changes. In this case the charge is transferred from the intramolecular donor fragment not to the intramolecular acceptor fragment but to fullerene because the electron affinity (2.65 eV) of fullerene is twice and four times more than the one of the polyimide and COANP acceptor fragment, respectively. As the result, the photorefractive effect is possible in these structures even without electric field.

It should be noted that larger laser-induced change of refractive index in COANP is likely to be associated with two reasons: 1) higher fullerene concentration in the organic matrix based on COANP and 2) correlation with the electron affinity of acceptor fragments. The less electron affinity results in more redistribution of electron density between the COANP donor fragment and fullerene, a more field gradient, stronger photorefractive effect, and finally in larger laser-induced change of refractive index.

It should be noticed that the value of Δn_i obtained is more than that corresponding to the thermal nonlinearity (~10^{-5}-10^{-4}).

In spite of the fact that, in the nanosecond and longer time scales, the principal mechanisms for grating recorded nonlinearities are thermal and density fluctuations, the values of input energy density, at which the effect was observed, do not contradict to a high conformational rotation energy in the polyimide matrix. Really, it should be noticed that the values of the write energy density, which are applied in this work, make possible conformational modifications of the polymer chain. For example, for fullerene-doped polyimide 6B, the write energy density of 0.5–0.6 $J \cdot cm^{-2}$ corresponds to 350-400 $kcal \cdot mol^{-1}$ at molecular mass of a monomeric unit of 750 as well as at the sample thickness and the spot diameter mentioned above. The rotation threshold is 10-700 $kcal \cdot mol^{-1}$ for cyclic groups relative to ordinary bonds in polyimide [37]. Therefore, the write energy density applied suffices to the rotation of hinge-joint bonds in polyimide, making the conformational modifications of polymer chains easier. This effect is likely to result in a better arrangement of macromolecule donor fragment and fullerene planes and hence, in an efficient charge transfer between them.

From the holographic recording experiments in the near-infrared spectral range ($\lambda = 1315$ nm), the laser-induced change of refractive index was found for the fullerene-doped polyimide thin films. Its value was ~ 10^{-3} that was close to the one corresponded to the visible spectral range.

The dependence of the response in the first diffraction order η^* on the input laser energy density W_{in} for PDLC based on fullerene-doped COANP is shown in Fig. 10. From the data presented in Fig. 10, the induced change of refractive index could be estimated. The first diffraction order response was drastically changed as the incident energy density increased from 2.5 to 17.5 $mJ \cdot cm^{-2}$ (nanosecond irradiation) and from 4.2 to 35 $mJ \cdot cm^{-2}$ (picosecond irradiation), respectively. Therefore, the light-induced refractive index was changed from 7.3×10^{-4} to 1.4×10^{-3} and from 5.1×10^{-4} to 7.2×10^{-4} under nanosecond and picosecond laser irradiation, respectively. The values obtained for PDLC was more than thermal component of Δn typical of liquid crystals (~10^{-5}). The thermal component of Δn for the COANP compound can be calculated for the nanosecond regime as follows:

$$\Delta n_{therm} = \frac{dn}{dT}\Delta T \tag{2}$$

where ΔT is heating of the system in one laser pulse:

$$\Delta T = \frac{W_{in}\alpha_{eff}}{\rho c} \tag{3}$$

where W_{in} is the incident energy density, α_{eff} is the effective absorption, $\rho = 1.4$ $g\,cm^{-3}$ is the density, and $c = 2$ $J\,g^{-1}K^{-1}$ is the specific heat. Therefore for the present organic system, taking into account that $dn/dT \sim 10^{-4}$ and $\alpha_{eff} = 430$ cm^{-1} at the sample thickness of 10 μm and its transmission of about 0.65, $\Delta n = 2.7\times10^{-4}$. Thus the light-induced changes of refractive index determined in the present experiment for the nanosecond pulse region, namely, 1.4×10^{-3}, is more than the thermal component of Δn for both COANP and LC.

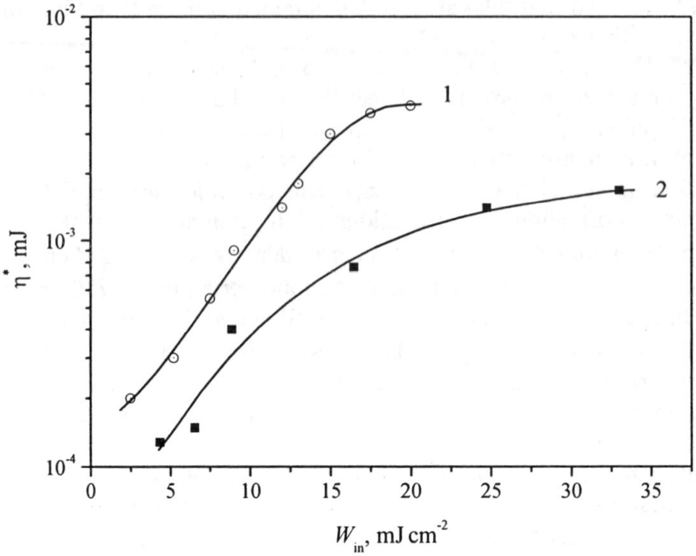

Figure 10. The dependence of the response in the first diffraction order η^* on the input laser energy density W_{in} for PDLC based on C_{70} (5 wt.%)-doped COANP at the laser pulsewidth: (1) 20 ns and (2) 400 ps. (λ =532 nm).

The large value of the light-induced refractive index in the fullerene-doped compounds contributes significantly to the OL effect (due to energy losses by diffraction), explains holographic recording in these media because of the photorefractive effect even without an electrical control, and also predicts the large values of nonlinear refraction n_2 and third order susceptibility $\chi^{(3)}$. These parameters have been estimated using the relationships [38]:

$$n_2 = \frac{\Delta n_i}{I} \tag{4}$$

$$\chi^{(3)} = \frac{n_2 n_0 c}{16\pi^2} \tag{5}$$

where Δn_i is the laser-induced change of refractive index, I is the input laser intensity, n_0 and n_2 is the linear and nonlinear refractive index, respectively, c is the velocity of light. The corresponding data are presented in Table 1.

It should be mentioned that the data obtained are close to those for silicon and gives an opportunity to estimate the place of fullerene-doped conjugated systems studied among other materials used traditionally for nonlinear optics [45].

Table 1. Nonlinear optical coefficients of materials

Materials	n_2, $cm^2\ W^{-1}$	$\chi^{(3)}$, $cm^3\ erg^{-1}$	T, K	Reference
CS_2	3×10^{-14}	10^{-12}	300	[38]
Silica	3×10^{-16}	10^{-14}	300	[38]
C_{60} film		0.7×10^{-11}		[39]
C_{60} film		8.7×10^{-11}		[40]
C_{60} film		2×10^{-10}		[41]
C_{70} film		1.2×10^{-11}		[42]
C_{70} film		2.6×10^{-11}		[40]
Polyimide + C_{70}	0.78×10^{-10}	2.64×10^{-9}		Present estimation
Polyimide + C_{70}	-1.2×10^{-9}	1.9×10^{-10}		[43]
COANP + C_{70}	$0,77\times10^{-10}$	$2,4\times10^{-9}$		Present estimation
PDLC* based on COANP + C_{70}	1.6×10^{-9}	4.86×10^{-8}		[44]
Si	10^{-10}	10^{-8}	300	[38]
Liquid crystal	10^{-4}	10^{-3}	300	[38]

* Polymer dispersed liquid crystal (PDLC) structure was made using 5 wt.%-doped COANP and LC E7(BDH) compound. The 10 μm cell was treated at the incident energy density of 17.5 mJ·cm^{-2} and Δn=1.4×10^{-3}.

The comparative scheme is shown in Fig. 11.

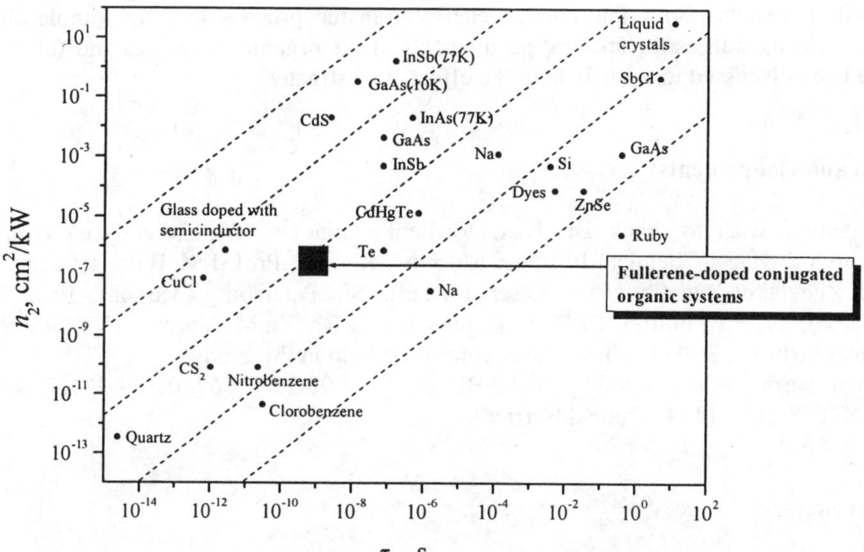

Figure 11. Values of nonlinear refraction n_2 in the plane (n_2, τ_{nl}) for different substances, where τ_{nl} is the build-up time of a nonlinear response [45]. The place of fullerene-doped conjugated structures studied in the present paper is shown by arrow.

4. Conclusion

Nonlinear optical properties of the fullerene-doped conjugated structures based on polyimide, 2-cyclooctylamino-5-nitropyridine and *N*-(4-nitrophenyl)-(*L*)-prolinol, as well as polymer-dispersed liquid crystal compounds have been studied to apply these materials as effective nonlinear absorber and holographic recording element in the visible and in the near-infrared spectral ranges. In the OL experiments 10-fold attenuation of the laser beam has been obtained for the PDLC cell based on polyimide 6B at the incident energy density of 0.15-0.2 $J \cdot cm^{-2}$ in the visible spectral range. The OL level in PDLCs is intermediate between those for liquid crystals (less than 0.09 $J\ cm^{-2}$) and for polymers (more than 1 $J\ cm^{-2}$). More than 10-fold attenuation of the laser beam was observed for the thin C_{70}-polyimide film with fullerene concentration of 0.5 wt.%. at the incident energy density of 0.8 $J \cdot cm^{-2}$ in the near infrared. It has been shown that the OL effect level can be controlled by the low-power irradiation of the flash-lamp.

From the induced changes of the refractive index, the nonlinear refraction n_2 and the third order susceptibility $\chi^{(3)}$ were determined. They were, respectively: $\sim 10^{-7}\ cm^2kW^{-1}$ and $\sim 10^{-9}$ esu for thin films of the fullerene-doped organic structures; $\sim 10^{-6}\ cm^2kW^{-1}$ and $\sim 10^{-8}$ esu for thin films of the fullerene-doped polymer-dispersed liquid crystals.

The place of the materials studied has been found among other nonlinear optical systems. It has been shown that the reverse saturable absorption, complex formation and high-frequency Kerr effect are responsible for the optical limiting and hologram recording results. Both the Förster energy transfer process and the dipole-dipole interaction as well as the spectral peculiarities of the organic molecules and fullerenes have been discussed to contribute to the effects investigated.

5. Acknowledgements

The author wish to thank Dr. N.A. Vasilenko and Dr. V.I. Berendyaev (Karpov Research Physical-Chemical Institute, Moscow, Russia), Prof. I.M. Belousova and Dr. A.P. Zhevlakov (Institute for Laser Physics, St. Petersburg, Russia), Prof. D.I. Stasel'ko, Dr. S.E. Putilin, Dr. L.N. Kaporskii and Dr. Yu.M. Voronin (Vavilov State Optical Institute, St. Petersburg, Russia) for their help in this study.

This work was supported by the RFBR grants ## 00-15-99067; 01-03-33162 and by the ISTC Project 1454 "Optical Barrier".

References

1. Tutt, L.W., and Boggess, T.F. (1993) A review of optical limiting mechanisms and devices using organics, fullerenes, semiconductors and other materials, *Prog. Quant. Electron.* **17**, 299-338.
2.. Kost, A., Tutt, L., Klein, M.B., Dougherty, T.K., and Elias, W.E. (1993) Optical limiting with C_{60} in polymetyl methacrylate, *Opt. Lett.* **18**, 334-336.
3. Khoo, I.C. (1995) Holographic grating formation in dye- and fullerene C_{60}-doped nematic liquid-crystal film, *Opt. Lett.* **20**, 2137-2139.
4. Kajzar, F., Taliani, C., Zamboni, R., Rossini, S., and Danieli, R. (1996) Nonlinear optical properties of fullerenes *Synth. Met.* **77**, 257-263.
5. Belousov, V.P., Belousova, I.M., Bespalov, V.G., Budtov, V.P., Volynkin, V.M., Grigor'ev, V.A., Danilov, O.B., Zhevlakov, A.P., Kalintsev, A.G., Ponomarev, A.N., Tul'skii, S.A., and Yutanova, E.Yu. (1997) Nonlinear optical properties of fullerene-containing media, *J. Opt. Technol.* **64**, 870-871.

6. Kamanina, N.V., Kaporskii, L.N., and Kotov, B.V.(1998) Absorption spectra and optical limiting of the fullerene-polyimide system, *Opt. Commun.* **152**, 280-282.
7. Bartkiewicz, S., Miniewicz, A., and Kajzar, F. (2000) Incoherent-to-coherent image converter based on hybrid liquid crystal-photoconducting polymer structure, *Synth. Met.* **109**, 105-108.
8. Kamanina, N.V. (2002) Mechanisms of optical limiting in π-conjugated organic system: fullerene-doped polyimide, *Synth. Met.* **127**, 121-128.
9. Ganeev, RA., Ryasnyansky, A.I., Kamalov, S.R., Kamanina, N.V., Kulagin, I.A.,.Kodirov, M.K, and Usmanov, T. (2002) Investigation of nonlinear optical characteristics of colloidal metals, semiconductors, fullerenes and organic dyes by Z-scan method and third harmonic generation of laser radiation, *Nonlin. Opt.*, **28**, 263-282.
10. Itaya, A., Sizzuki, I., Tsuboi, Y., and Miyasaaka, H. (1997) Photoinduced electron transfer processes of C_{60}-doped poly(N-vinylcarbazole) films as revealed by picosecond laser photolysis, *J. Phys. Chem.* **B101**, 5118-5123.
11. Konarev, D.V., Zubavichus, Y.V., Slovokhotov, Yu. L., Shul'ga, V.N., Drichko, N.V., and Lyubovskaya, R.N. (1998) New complexes of fullerenes C_{60} and C_{70} with organic donor DBTTF : synthesis, some properties and crystal structure of DBTTF·C_{60}·C_6H_6 (DBTTF=dibenzotetrathiafulvalene), *Synth. Met.* **92**, 1-6.
12. Du, Ch., Li, Yu., Wang, Sh., Shi, Zh., Xiao, Sh., and Zhu, D. (2001) Synthesis and characterization of [60]fullerene-substituted oligopyridines ruthenium complexes, *Synth. Met.* **124**, 287-289.
13. Kamanina, N.V. (2001) Peculiarities of optical limiting effect in π-conjugated organic systems based on 2-cyclooctylamino-5-nitropyridinedoped with C_{70}, *J. Opt. A: Pure Appl. Opt.* **3**, 321-325.
14. Dubenskov, P.I., Zhuravleva, T.S., Vannikov, A.V.,Vasilenko, N.A., Lamskaya, E.V., and Berendyaev, V.I. (1988) Photoconductive properties of some soluble aromatic polyimides, *Vysokomol. Soedin.* **A30**, 1211-1217 [in Russian].
15. Bosshard, Ch., Sutter, K., Günter, P., and Chapuis, G. (1989) Linear- and nonlinear-optical properties of 2-cyclooctylamino-5-nitropyridine, *J. Opt. Soc. Am.* **B6**, 721-725.
16. Zyss, J., Nicoud, J.F., and Coquillay, M. (1984) Chirality and hydrogen bonding in molecular crystal for phase-matched second-harmonic generation: N-(4-nitrophenyl)-(L)-prolinol (NPP), *J. Chem. Phys.* **81**, 4160-4167.
17. Ruoff, R. S., Tse, D. S., Malhotra, R., and Lorets, D. S. (1993) Solubility of C_{60} in a variety of solvents, *J. Phys. Chem.* **97**, 3379-3383.
18. Kamanina, N.V. (1999) Reverse saturable absorption in fullerene-containing polyimides. Applicability of the Förster model, *Opt. Commun.*, **162**, 228-232.
19. Danilov, V.V., and Kamanina, N.V. (1999) Self-diffraction and relaxation in a resonant liquid crystal medium, *J. Opt. A: Pure Appl. Opt.*, **1**, 37-40.
20. Kamanina, N.V., Bagrov, I.V., Belousova, I.M., Kognovitskii, S.O., and Zhevlakov, A.P. (2001) Fullerene-doped π-conjugated organic systems under infrared laser irradiation, *Opt. Commun.*, **194**, 367-372.
21. Kamanina, N.V., Bagrov, I.V., Belousova, I.M., Zhevlakov, A.P. (2001) The possibility of controlling the optical limiting effect by a weak light signal, *Opt. Spectrosc.* **91**, 1-2.
22. Bensasson, R.V., Hill, T., Lambert, C., Land, E.J., Leach, S., and Truscott, T.G. (1993), Triplet state absorption studies of C_{70} in benzene solution, *Chem. Phys. Lett.*, **206**, 197-202.
23. Kamanina, N.V., Kaporskii, L.N., Pozdnyakov, A., and Kotov, B.V., (2000) Optical limiting in organic polyimide systems doped with fullerenes and dyes, *Proceed. SPIE*, **3939**, 228-233.
24. Kamanina, N., Barrientos, A., Leyderman, A., Cui, Y., Vikhnin, V., and Vlasse, M. (2000) Effect of fullerene doping on the absorption edge shift in COANP, *Mol. Mater.*, **13**, 275-280.
25. Razbirin, B. S., private communication at the seminar in Ioffe Physico-Technical Institute, St. Petersburg, Russia, 2002.
26. Wray, J.E., Liu, K.C., Chen, C.H., Garrett, W.R., Payne, M.G., Goedert, R., Templeton, D. (1994) Optical power limiting of fullerenes, *Appl. Phys. Lett.*, **64**, 2785-2787.
27. Belousov, V.P., Belousova, I.M., Budtov, V.P., Danilov, V.V., Danilov, O.B., Kalintsev, A.G., and Mak, A.A. (1997) Fullerenes: Structural, physical-chemical, and nonlinear optical properties, *J. Opt. Technol.* **64**, 1081-1109.
28. Bazhenov, A.V., Gorbunov, A.V., Maksimuk, M.Yu., Fursova, T.N. (1997). Photoinduced absorption of C_{60} films at 0.08-4.0 eV, *J. Exp. Theor. Phys.* **85,** 135-145.
29. Kamanina, N.V., Plekhanov A.I. (2002) Mechanisms of optical limiting in fullerene-containing π-conjugated organic structures based on polyimide and COANP, *Opt. Spectrosc.* **93**, to be published.

30. Förster, T. (1959) Transfer mechanisms of electronnic excitation, *Disc. Farad. Soc.*, **27**, 7-17.
31. Kamanina, N.V. (2001) Light-induced variation in the refractive index in a polyimide–fullerene system, *Opt. Spectrosc.* **90**, 867–871.
32. Gurvich, L.V., Karachevtsev, G.V., Kondrat'ev, V.N., Lebedev, Yu.A., Medvedev, V.A., Potapov, V.K., and Khodeev, Yu.S. (1974) *Energies of Chemical Bond Breaking, Ionization Potentials And Electron Affinity*, Nauka, Moscow, [in Russian].
33. Vinogradova, S.V., Slonimskii, G.L., Vygodskii, Ya.S., Askadskii, A.A., Mzhel'skii, A.I., Churochkina, N.A., Korshak, V.V. (1969) On structure and properties of aromatic polyimides, *Vysokomol. Soedin.*, **A11**, 2725-2740 [in Russian].
34. Bessonov, M. I., Kuznetsov, N. P., Koton, M. M. (1978) On transition temperatures of aromatic polyimides and physical principles of their chemical classification, *Vysokomol. Soedin.*, **A20**, 347-354 [in Russian].
35. Hosoya, M., Ichimura, K., Wang, Z.H., Dresselhaus, G., Dresselhaus, M.S., Eklund, P.C. (1994) Dark conductivity and photoconductivity in solid films of C_{70}, C_{60}, and K_xC_{70}, *Phys. Rev.* **B49**, 4981-4986.
36. Collier, R.J., Burckhardt, C.B., and Lin, L.H. (1971) *Optical Holography*, Academic Press, New York and London,.
37. Bershtein, V.A. and Egorov, V.M. (1990) *Differential Scanning Calorimetry in Physical Chemistry of Polymers, Chemistry*, Leningrad [in Russian].
38. Akhmanov, S.A. and Nikitin, S.Yu. (1997) *Physical Optics*, Clarendon., Oxford.
39. Liu Huimin, Taheri B., Weiyi Jia (1994). Anomalous optical response of C_{60} and C_{70} in toluene, *Phys. Rev.* **B49**, 10166-10169.
40. Kaizar, F., Taliani, C., Muccini, M., Zamboni, R., Rossini, S., Danieli, R. (1994) Third order nonlinear optical properties of fullerenes, *Proceed. SPIE* **2284**, 58-68.
41. Li, J., Feng, J., Sun, J. (1993) Quantum chemical calculation on the spectra and nonlinear third order optical susceptibility, *J. Chem.Phys.* **203**, 560-564.
42. Krätschmer, W., Lamb, L.D., Fostiropoulos, K., Huffman, D.R. (1990) Solid C_{60}: a new form of carbon, *Nature* **347**, 354-358.
43. Ganeev, R.A., Ryasnyansky, A.I., Kodirov, M.K., Usmanov, T. (2000) Nonlinear optical characteristics of C_{60} and C_{70} films and solutions, *Opt. Commun.* **185**, 473-478.
44. Kamanina N.V. (2002) Optical investigations of C_{70} - COANP - liquid crystal system, *J. Opt. A: Pure Appl. Opt.* **4**, to be published
45. Akhmanov, S.A. Vorontsov, M.A. (eds.) (1990) *New Physical Principles of Optical Information Processing*, Nauka, Moscow [in Russian].

TRANSPORT AND OPTICAL PROPERTIES OF DNA

E. M. CONWELL and D. M. BASKO
Department of Chemistry
University of Rochester
Rochester, New York 14627 USA

1. Introduction

The structure of DNA is now very well known. The two helices that form the backbone are made up of alternating sugar and phosphate groups. Attached to the sugars are the bases, organic compounds consisting of one or two planar heterocycles and associated N, O, and H. There are four bases in DNA—guanine, cytosine, adenine, and thymine. In what follows these will be abbreviated G, C, A, and T, respectively. Hydrogen bonds connect one base from each helix with its complementary base from the other helix. Thus G is paired with C, A with T. The bases all have closed electronic shells. Guanine has the distinction of its highest occupied molecular orbital (HOMO) having the highest energy. Thus holes added to a DNA stack would tend to sit on a G.

Another important feature of DNA is the negative charge on the helices in solution. Because the phosphate groups are acidic, in vivo they give up protons to the surrounding solution (basically water) to become negatively charged. Overall neutrality is achieved by the presence of counterions, such as Na^+ or Ca^+, in the solution.

In the form of DNA that is found in living cells—B-DNA—the bases face each other. Thus their π orbitals overlap. Not long after the structure was understood it was suggested that electron or hole conduction is possible along the base stack. Transport in DNA is of interest because oxidation of a base and subsequent transport of the hole may result in mutation and carcinogenesis. Also, if DNA were a good conductor it might be useful for wires in nanocircuits, where its assembly properties could be useful

Conduction requires two features: (1) existence of free carriers and (2) ability of the carriers to move along the stack. Concerning these features a couple of general statements may be made. Free carriers may be introduced by thermal generation, injection at contacts or elsewhere, and light of sufficiently high frequency. Thermal generation is unlikely, at least in pure and perfect DNA; because of the large gap; the absorption edge of DNA is close to 4eV. The requirement that carriers be able to move makes it clear that conduction must depend on the base sequence. For the case of hole conduction, although the HOMO levels of the bases have not been measured for DNA in solution, measurements of

F. Charra et al. (eds.), Organic Nanophotonics, 193–206.

individual bases in solution indicate that for the most part the HOMOs differ by amounts much larger than kT at room temperature. Thus a hole is faced by large barriers to motion in an arbitrary sequence. Whether this is also true for electrons is not clear.

Recently there have been many measurements of conductivity and diffusion, leading to conflicting results: DNA was found to be metallic in some, insulating in others. In the next section, the current experimental and theoretical situation will be discussed, first for diffusion and then for the measurements involving an electric field. It will be seen that diffusion in DNA in solution is reasonably well understood with the exception of the mechanism for transport in long stretches of As or ATs. The same cannot be said for conduction. Still, there is agreement in a large part of the community that, despite experimental results to the contrary, DNA is not a metal in the absence of doping.

It has been suggested that for a long stretch of like bases, which is usually the case in DNAs synthesized to show good conductivity in air or vacuum, diffusion and conduction are carried out by means of large polarons--electrons or holes, dressed with chain distortion, whose wavefunction is spread out over a number of bases [1]. In section 3 we will present the calculated properties of the polarons. Our theory was tested by showing it can account for the measured difference in the depths of hole traps provided by a single G, a pair of Gs or three Gs together. We then describe the results of our calculations of the effect of water and counterions on the polarons. At the end of section 3 we discuss the role of the polarons in conduction.

Section 4 is devoted to the optical properties, notably absorption and emission, of DNA arising from the bases. We note here that the base sequence can be of great importance in these properties also.

2. Transport

2.1. DIFFUSION EXPERIMENTS AND THEORY

The diffusion experiments were carried out with DNA surrounded by water and counterions. Expecting no free carriers in the DNA due to the large gap, the experimenters arranged to introduce a carrier on the stack. In a frequently used technique, photoexcitation of an acceptor associated with the stack in some fashion, sometimes intercalated, allows transfer of an electron, usually from a G, leaving a hole on the stack. The progress of the hole along the stack was monitored by observing, on many DNA molecules in dilute solution, the percentage of holes penetrating successive traps. The hole traps were usually a sequence of two guanines (GG) or 3 guanines (GGG). The possibility of determining the percentage of holes being caught by a trap stems from the fact that, although these traps are relatively shallow, there is a possibility that a G^+ will undergo an irreversible reaction with water or oxygen. The relative amount of such reactions, thus the relative number of holes that reach a particular G in the trap sequence, can be determined by using further chemical treatment that leads to cleavage of the DNA at the site where the reaction of the trapped hole took place. Another technique used for determining how far a hole could travel was the observation of fluorescence quenching of a donor intercalated in the stack.

Using the techniques just described, many experimenters were able to show that an injected hole could travel 10nm or further on a DNA stack. Such experiments led to the description of DNA as a "molecular wire", implying, unfortunately, a metallic nature [2]. The very fact that a hole had to be introduced on the chain by relatively elaborate means indicated that there was not a large body of free carriers in DNA as there must be in a metal. Some clarity was brought into the situation by experiments of the Giese group in which G/C pairs were separated by a few A/T pairs. It was found that transmission decreased rapidly with the number of A/Ts, going down a factor of ~8 for each intervening A/T [3]. It was concluded from this strong decrease that the transport was by single-step tunneling between Gs [4,5], the hole population on the bridge between the Gs being always negligible.

These experiments suggested that a negligible number of holes would penetrate 4 or more A/Ts. However, Schuster and collaborators had shown that holes could penetrate 5 A/Ts and proceed beyond that [6]. In further experiments it was shown that a hole could penetrate as many as 10 A/Ts and continue beyond [7]. Finally, Giese and collaborators showed that the rate of charge transfer between two Gs decreases with increasing separation only if they are separated by no more than 3 A/Ts. If additional A/Ts intervene between the Gs the transfer rates change to a very weak distance dependence [8]. As this is inconsistent with the tunneling picture, Giese et al. concluded that after 3 A/Ts the hole hopped out of the G trap onto the bridge, i.e., onto an A, the change in mechanism occurring because the time for tunneling through 4 A/Ts is long enough so that the hole can with reasonable probability obtain the energy for thermal activation. Previously, thermal activation had not been considered because the difference in ionization energy between G and A was thought, on the basis of single base measurements in solution [9,10] and vapor [11,12], to be 0.4 or 0.5eV. However, it had been suggested and demonstrated in ab initio calculations of the ionization potentials of pairs or triples of bases, that the ionization potential of a base is much affected by its neighbors [13,14], and the difference in the ionization potentials can be significantly smaller, which makes thermal activation possible.

While the tunneling mechanism of charge transfer between two traps separated by no more than 3 bases is reasonably well understood and agreed on by now, the nature of the long-range transport, when the hole has hopped onto the relatively long A/T bridge, has not been agreed on. As noted above, a clear picture of the long-range transport is of paramount importance also for understanding conduction in dry DNA. Several models have been proposed so far. Jortner et al. [5] used perturbation theory in the hole nearest-neighbor transfer integral, thus assuming that the hole is localized on one base and proceeds by incoherent nearest-neighbor hopping. Another perturbative treatment of this problem was given by Bruinsma et al. [15], where the role of dynamical fluctuations was emphasized. Clearly, validity of the perturbation theory based on the transfer integral being a small quantity, although adequate for describing the tunneling because of significant difference in the hole energies in the trap and on the bridge, becomes questionable for describing hole transport between degenerate states on a long bridge consisting of the same base pair. Schuster and collaborators were the first to suggest that, to minimize its energy, the wavefunction of an extra charge in DNA, as in other essentially one-dimensional entities, would spread out over a number of bases [6]. This point of view agrees with the numerical calculations of Grozema et al. [16,17], who solved the equations of motion for the wavefunction of the hole coupled to thermally excited vibrations. Their work has given some evidence for delocalization of the hole on the bridge, but no clear physical picture was provided. Recent ab initio calculations [18] also indicated that the hole wavefunction is delocalized over several bases.

2.2. CONDUCTION EXPERIMENTS

Experiments without contacts or other source of carrier injection should be most revealing. Two such experiments have been performed, both on λ-DNA. In one it was ascertained that the dc conductivity of the DNA is very small, comparable to that of mica or glass [19]. This was to be expected because of the random distribution of bases. The other experiment measured microwave absorption, which was interpreted under the assumption that it was due to electronic processes. The frequency and temperature dependence found, the latter corresponding to an activation energy of 0.3eV, could be attributed to carrier hopping between a small number of strongly localized electron states [20,21]. It was suggested that these states were due to disorder, perhaps arising from the random base distribution, or the finite persistence length, or counter ion fluctuations [21].
Of course it is also possible that the states were due to an impurity.

The experiments in which a dc field was applied to DNA through contacts were carried out in air or vacuum. Paradoxically, the experiments that yielded the highest conductivity values were carried out on DNA with random base sequence. It has been concluded by many that the DNA was inadvertently doped in these experiments.. In the ingenious experiment of Fink and Schonenberger, where a λ-DNA "rope" was imaged in a modified low-energy electron point source microscope, a resistivity of ~1 mΩ cm was measured [22]. It was shown later, however, that the slow electron bombardment of the DNA causes a conducting layer to be set up on its surface [23]. In the case of Kazumov et al., who found the DNA to be metallic, showing proximity superconductivity at low temperatures [24], inadvertent doping may have been due to a glow discharge in pentylamine vapor carried out "to promote adhesion" of the DNA.

Based on the realization that DNA with all bases the same, i.e., a single base pair repeated, should have the largest conductivity, many of the later experiments were carried out on samples with all G/Cs or all A/Ts. It was noted that the former behaves as a p-type semiconductor, the latter as an n-type semiconductor [24]. Even with all bases the same there was a wide range of results, although the conductivities were in general characteristic of semiconductors. One may speculate that part of the variation may result from the fact that oxygen is a dopant for DNA [25]. The data most frequently quoted are those of Porath et al. on a single duplex strand consisting of 30 G/C pairs [26]. They found no current flow until the voltage exceeded a threshold value of ~2 to 3 volts, the value increasing with temperature. The existence of such a threshold is in accord with the idea that there are no thermal equilibrium carriers, all carriers being injected from the contacts. The authors suggest that the gap originates from the offset between the Fermi level of the electrode and the molecular energy levels of the DNA molecule. Yoo et al. [24] obtained a similar gap at low temperatures, but found that the gap decreases with increasing temperature, vanishing by 300K. Also at 300K they found much greater conductance than would be inferred from the data of Porath et al.

3. Properties of Polarons in DNA

Large polaron formation is expected when an electron or hole is added to a one-dimensional or quasi-one dimensional chain or stack [27]. The lack of constraint on the motion of atoms or molecules in one dimension compared with those bound in a three-dimensional configuration allows a distortion of the chain or stack that decreases the energy of the added carrier. The combination of free carrier and distortion, the polaron, has been studied extensively both theoretically and experimentally, particularly in conducting polymers.

3.1 STATIONARY POLARONS

The essential ingredients of a one-dimensional Hamiltonian that gives rise to polarons are: (i) wavefunction overlap of nearest neighbors, quantified by the transfer or resonance integral t. (ii) dependence of the transfer integral on the base displacements. Conveniently, t is broken up into the term t_0 representing the value for uniform spacing of the sites and a term representing the change in t due to nearest-neighbor displacements. The latter term plays the role of the electron-phonon coupling; (iii) an elastic restoring force that keeps the distortion finite. A simple tight-binding Hamiltonian embodying these elements, which has been used extensively to study large polarons in conducting polymers, is the Su-Schrieffer-Heeger, SSH, Hamiltonian [28]. Written for a situation in which there may be sites of different energies, and in a form allowing addition of an electric field, the Hamiltonian is:

$$H = H_{el} + H_{lat} \tag{1}$$

$$H_{el} = \sum_{n,s} d_n c^+_{n,s} c_{n,s} - (t_0 - \alpha(u_{n+1} - u_n))[e^{i\gamma A} c^+_{n,s} c_{n+1,s} + e^{-i\gamma A} c^+_{n+1,s} c_{n,s}] \tag{2}$$

$$H_{lat} = K/2 \sum_n (u_{n+1} - u_n)^2 + M/2 \sum_n \dot{u}_n^2. \tag{3}$$

Here u_n is a displacement parallel to the stack of the base pair at site n, c^+_n (c_n) is an electron or hole creation (annihilation) operator at site n, d_n are on site energies, including the polarization energy, α the electron-phonon coupling constant, $= \partial t/\partial u$, K the spring constant, M the mass of a base pair, $\gamma = 2\pi ea/hc$, a being the average distance between base pairs, c the light velocity and e the electronic charge, and A is the vector potential of the external electric field. The spin indices have been dropped because the quantities are not spin-dependent. Equation (3), embodying a classical treatment of the lattice vibrations, is justified by the large mass of the bases. Static solutions are found by minimization of the Hamiltonian above with respect to the nearest-neighbor distance change $y_n = u_{n+1} - u_n$ and electronic wavefunction, which yields the following equations:

$$(E_h - 2t_0)\,\psi_n = d_n\psi_n + (t_0 - \alpha y_n)\,\psi_{n+1} + (t_0 - \alpha y_{n-1})\psi_{n-1} \qquad (4)$$

$$y_n = (2\alpha/K)\,\psi_{n+1}\psi_n \qquad (5)$$

where $E_h - 2t_0$ is the energy of a hole relative to the bottom of a free hole band in a chain with d_n and u_n both equal to zero. Note that only the strain $u_{n+1} - u_n$ enters the equations, not the displacements themselves. The total energy of the system in the stationary state is taken to be the sum of the hole energy and the lattice deformation energy, the first term in Equation (3).

These equations must be solved numerically. The value of t_0 was taken in the range 0.1 to 0.3eV, the latter value obtained from an ab initio calculation of the energy of a pair of Gs as a function of the distance between them [29] for a separation of 3.4Å. Other calculations have given t_0 values in the range 0.1 to 0.2eV [30]. The electron-phonon coupling constant α, the derivative of t with respect to displacement, was obtained as 0.6evV/Å from the data of Ref.29. When a value of t_0 other than 0.3eV was used α was scaled accordingly. Although t_0 and α values depend on the particular pair of neighboring bases, the calculations were simplified by using the same value for all pairs of bases. The value of the elastic constant K was taken as 0.85eV/Å^2, derived [31] from the measured value of the sound velocity in DNA.

In Figure 1 we show the polaron calculated for the sequence shown at the bottom of the figure. The extent of the polaron, ~ 6 sites, was much the same for different sequences. It is seen that change in t_0 changes only the strain (change in nearest neighbor distance). The maximum strain is ~0.3-0.4Å.

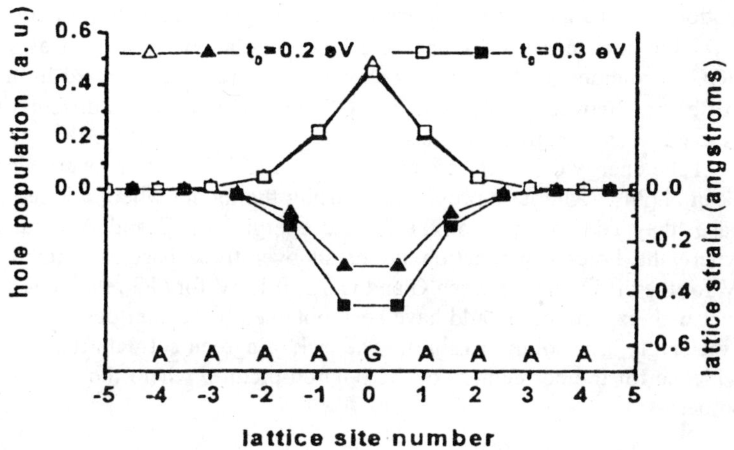

Figure 1. Hole population $|\psi_n|^2$ (open symbols) and lattice strain (filled symbols) for one guanine at n = 0 among adenines; t_0 = 0.2eV (triangles) and t_0 = 0.3eV (squares).

3.2 HOLE TRAPS IN DNA

As noted earlier, G, GG and GGG act as hole traps in DNA, with the trap depth expected to increase as the number of Gs increases. The difference in trap depths of GG and GGG relative to G, with the Gs in all cases surrounded by As, have been measured by Lewis et al [32]. They find that, relative to G, GG is deeper by 0.052eV, GGG by 0.077eV, both much shallower than expected from ab initio calculations.

To compare with the results of Lewis et al. we calculated the wavefunctions and energies, using the equations of the last section, for the same sequences as they measured. The parameters used were t_0 = 0.1, 0.2 and 0.3eV and the K and α values discussed in the last section. The only important quantity not known is the difference between the site energies of A and G; we took this as a parameter. We found the results were quite insensitive to the values of t_0, as is seen in Figure 1, but depended strongly on the difference between the site energies of A and G. The differences in trap depths measured by Lewis et al. between GG or GGG and G could be matched within a few % for this site energy difference equal to 0.17eV [33]. This was the value used for the calculation of Figure 1, whereas from measurements of single bases in solution [9-12] a value of 0.4eV or greater was expected. There is good evidence for a value of the site energy difference $\approx$ 0.2eV, notably, as discussed in section 2.1, that a hole can hop from G to a neighboring A with thermal energy. With 0.17 eV for the site energy difference the hole population, or $\psi\psi^*$, was very little different for GGG surrounded by As from that shown in Figure 1, as expected from the very small energy difference between them [33]. For GG the shape of $\psi\psi^*$is different because, by symmetry, the wavefunction must be the same on each of the two Gs [33].

Calculations of the energy differences between G, GG and GGG traps were also carried out by Voityuk et al.[34]. Although pressumably using the model where a hole is isolated on a single base, they used values for the on-site energies of G and A obtained from calculations where the hole wavefunction is spread over three bases. With these they obtained 0.3eV for the difference between G and GGG, 0.13eV for GG relative to G. Much worse agreement with experiment would have been obtained if the energies of G and A had been taken as the single base solution values. The lack of agreement with the experimental data [32] is evidence for the correctness of the polaron picture, where the wavefunction is spread over ~ 6 bases

3.3 EFFECTS OF SOLVATION ON HOLE MOTION IN DNA

In vivo DNA is surrounded by water and positive ions to neutralize the negative charge on the backbone. This environment could be neglected in the calculations of the last section because we were calculating differences between overall energies, which would not be much affected by replacement of one or two As by Gs. However, interaction with such a polarizable environment would clearly affect the binding energy of a polaron and its motion. In describing the interaction of a hole with the environment, it is only necessary to consider the change in the surrounding polarization due to the presence of the hole. This may be taken into account through an additional term in the electronic part of the Hamiltonian representing the change in the hole energy due to the change in polarization it causes. Reasoning from the classical form of the electrostatic energy change, we took this term to be $(1/2)\ \Sigma g(n-n')\ |\psi_n|^2\ |\psi_{n'}|^2$, where ψ_n represents the wavefunction of the hole on the nth site or base and $g(n-n')$ the coupling to the additional polarization. Quantitative calculations were carried out only for the case of the DNA stack consisting of the same base pair repeated.

To study the effect of the environment we considered the DNA molecule to be placed inside a cylindrical cavity with diameter R equal to that of the double helix, ~20A. Due to the hydrophobicity of the bases we took the cavity to be empty, neglecting the small polarizability of the backbone and the bases. It is quickly seen that the high dielectric constant of water causes the screening effect of the counterions to be negligible; the effect of the environment on the polaron is mainly due to water. To simulate a polaron, the hole charge was assumed to be concentrated on the axis of the cylinder at evenly spaced points corresponding to the base spacing. It is then useful to make a Fourier transform of the charge density in cylindrical coordinates. After numerical integration of the interaction energy between the polaron and the surrounding water it was determined that $g(0) = -0.87e^2/R$, and $g(n-n')$ decreases slowly as $n-n'$increases from 0 [35]. The latter suggests that the wavefunction of the polaron including solvation effects is not much different from that obtained neglecting solvation. This is seen in Fig.2 where $|\psi|^2$ for the stationary state is contrasted for DNA with and without the water and counterion surroundings. The parameters used were $t_0 = 0.2$eV and $\alpha^2/K = 0.2$eV.

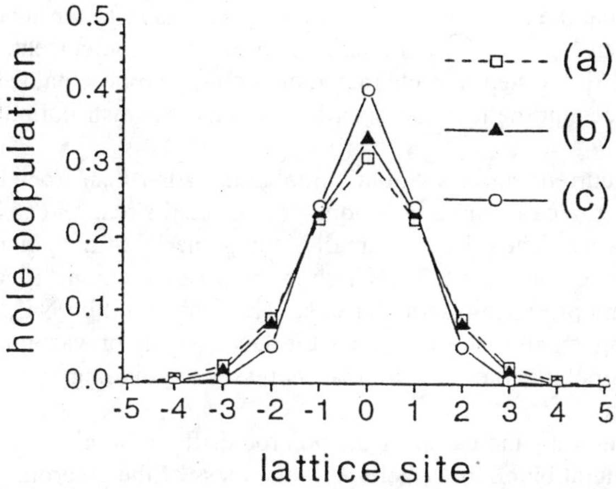

Figure 2. Hole population $|\psi_n|^2$ for a stationary state formed by interaction with (a) lattice only, (b) environment only, (c) both. $\alpha^2/Kt_0 = 1.0$, $e^2/Rt_0 = 7.0$.

The drag of the water on the polaron in motion results in its losing energy and achieving a steady state with a terminal drift velocity v_d. Knowing the interaction with water, $g(n - n')$, we can determine the rate of energy loss P and v_d. Conservation of energy stipulates that $P = e\, v_d^2/\mu$, where μ is the mobility, making it possible to calculate μ [35]. If there are significant scattering sources, such as the thermal vibrations of the stack, for example, this μ will be an upper limit. From μ, using the Einstein relation, we obtain an upper limit for the diffusion constant D, which could be measured experimentally. Our calculations give the upper limit on D for a hole diffusing in a DNA stack with all bases the same as 3×10^{-5}cm^2/s [35].

3.4 CONDUCTION IN DNA

As noted in section 2.2, the two sets of data on conduction vs. temperature in air or vacuum of a single DNA strand composed of G/Cs are quite in disagreement. Yoo et al. have attempted to fit their data to a small polaron (i.e., occupying a single site) hopping theory, but the temperature dependence they measure is not in agreement with the theory [24]. Hjort and Stafstrom [36] support an extended wavefunction and have set up a theory for the conduction based on resonant tunneling through narrow bands arising from HOMO and HOMO – 1 levels in guanine and cytosine. The theory results in a step-like current-voltage characteristic, which can be seen for some temperatures and some current directions in the data of Porath et al. [26]. A step-like characteristic could also be obtained with a polaron theory by taking into account the fact that disorder broadens the distribution in energy of the polaron states.

We believe that the current carriers would be polarons rather than free electrons or holes because, as noted earlier, a carrier in a quasi-one-dimensional situation can lower its energy by forming a large polaron. The relatively small binding energy of the polaron calculated in section 3.1 is not likely to apply for the DNA strand in air or vacuum. Electrical neutrality of the DNA requires the presence of counterions. The ions could be Na^+, protons, or ions from the air, depending on the way in which the DNA in air or vacuum was produced. These ions constitute a polarizable medium that increases the binding energy, as was found for the DNA in solution.

In the absence of scattering and damping the polaron drifts smoothly in an electric field of magnitude E with acceleration eE/m_p, where m_p is the mass of the polaron. This was shown in simulations based on equations 1-3 [37]. From the simulations it was deduced that m_p is ~5% of the mass of C/G or A/T.

4. Optical Properties

In discussing optical properties of DNA we will consider only the bases, and only properties that involve changes in electronic states. Because each base consists of one or two planar heterocycles, their optical properties bear a resemblance to those of polymers with phenyl rings in their backbone, such as poly(phenylene vinylene), PPV. A major source of differences is the fact that the bases contain, beside σ and π electrons, nonbonding electrons, denoted by n. Thus excited states may be formed by $n \rightarrow \pi^*$ as well as $\pi \rightarrow \pi^*$ transitions.

4.1 ABSORPTION

Monomers of DNA including any one of the four bases have strong optical absorption with an edge at 3×10^4cm^{-1} (~300nm or ~ 4eV). This is attributed to π to π^* transitions, and corresponds to the creation of singlet excitons. Native DNA, characterized by inclusion of monomers of all 4 bases in more or less random sequence, shows a strong absorption starting at ~ 3×10^4cm^{-1}. It has been shown by many researchers that this absorption is virtually identical to the sum of the absorptions of the constituent monomers This suggests that there is little interaction between the constituent bases. However, as will be discussed, some of the optical properties do show strong interaction between adjacent bases.

Lifetimes of optically excited states in nucleosides (monomers of DNA minus the phosphate groups; each of these is expected to have properties similar to those of the corresponding monomer) have recently been measured by fs pump-probe experiments. At room temperature in water plus ions, specifically a solution matching biological conditions, the lifetimes measured range from 290 fs for adenosine to 720fs for thymidine The short lifetimes are due to ultrafast internal conversion to the electronic ground state. The internal conversion is much faster than found in simple aromatics [38] or the polymers, for example. It has been suggested that this very short lifetime is due to interaction between $\pi\pi^*$ states and close-lying $n\pi^*$ states, which may cause the planar molecules to distort out of plane [38]. At 78K the lifetimes of the monomers have increased to a few ns [38].

4.2 FLUORESCENCE

The short exciton lifetimes result in very small emission at room temperature and it has not been much studied there. Fluorescence of the monomers has been studied in detail at 78K. For all the bases it starts at ~3.3×10^4cm^{-1} and, at least for monomers of C and T, the fluorescence band is a good mirror image of their main absorption band. In native DNA only A and T fluoresce. Excited singlets are found to be quenched in poly G/C [40]. Evidence that there is interaction between bases comes from the fact that at low temperatures the emission and yields of DNA may differ greatly from those of the monomers. The low temperature singlet emission spectra show the frequent occurrence of exciplexes.

An exciplex is formed when two non-identical molecules that do not interact in their ground states do so when one of the molecules is in an excited state. As a result of charge-transfer and exchange interactions of the overlapping π electrons of the two molecules on the one hand, and their mutual repulsion on the other, the molecules are drawn together into a potential minimum at a separation smaller than that in the ground state [41]. Emission from an exciplex has a lower frequency that that from the excited state of either molecule because its initial state is in this potential minimum, and the final state, being at a smaller separation of the two molecules than the ground state, is a repulsive state. These features also result in the emission being broad and structureless. Also, emission becomes less probable from the exciplex state because it usually has different symmetry from the ground state, resulting in a long lifetime for the exciplex [41]. Singlet exciplexes are more common than triplet, and the former usually consist of the two molecules arranged in a sandwich fashion. The separation of the two molecules in an exciplex is typically 0.3 to 0.35nm [41]. Thus the bases in B-form DNA, which are parallel and have a spacing of 0.34nm, are well situated to form singlet exciplexes.

From the study of DNA emission at 78K exciplexes have been identified and characterized in a number of dinucleotides, C-C, A-A, A-C and A-T [38]. They have also been found in poly C, poly A and poly (A-T)/(A-T), a duplex in which A alternates with T on both strands. At 78K the center of the luminescence of native DNA is shifted from ~310-320nm for the singlet to 350-370nm for the center of the exciplex. Exciplex emission from poly (A-T)/(A-T) has also been detected at room temperature with the DNA in solution. In this case the maximum emission shifted to 410nm [42]. The large shift for the exciplex emission indicates binding energies ~ 0.3 to 0.5eV, somewhat larger than excimers or exciplexes in the conducting polymers.

From the facts that the fluorescence spectrum of native DNA at 78K is broad and peaked at 355nm, about where the exciplex emission peaks at 78K, it has been concluded that it is an exciplex spectrum, due to A-A and A-T exciplexes [38]. As noted earlier, emission spectra at room temperature have not been available because of the very poor quantum yield. Recently, however, Lakowicz et al. [43] have shown that weak emission can be enhanced many fold by spatial proximity to silver island films. With this technique they found the DNA fluorescence spectrum at room temperature to be about as broad as that at 78K, with a peak at ~340nm. As noted above, A-T exciplex emission peaks at 410 nm at room temperature. It appears that the room temperature emission from native DNA (calf thymus) is more monomer than exciplex emission; the latter may be represented by a shoulder at long wavelengths seen in the data of Reference.43. However, no conclusions can be drawn without further investigation because of the different surroundings of the DNA at the two temperatures—rigid glass at 78K and liquid at 300K. There is some evidence for interaction of the DNA with the solvent [38].

In regard to polaron transport there is interest in exciplex formation in DNA. In the process of intrastrand exciplex formation the bases must move together, requiring rotation and translation; decrease of the intrastrand base distance is also required for polaron formation. To account for the non-tunneling transport along a string of As mentioned earlier we suggested polaron formation. The fact that exciplexes have been found for polyA indicates that the motion required to make polarons does occur.

4.3 PHOSPHORESCENCE

Phosphorescence is also observed for monomers of all the bases. All the monomers have a lowest $\pi\pi^*$ triplet to which intersystem crossing can take place. For some cases, notably thymine, the triplet cannot be excited directly and populating it requires a sensitizer with a higher triplet state, such as acetone. At 78K the phosphorescence spectrum for monomers of all the bases ranges from ~350 to 500nm. The lifetime of the phosphorescent state is ~1 second. The phosphorescence of DNA is found to be close to that of the T triplet state, which is the lowest triplet of the four. It is not clear how the T triplet is formed. One possibility is intersystem crossing on an exciplex involving T. It is also possible that it is populated by triplet-triplet transfer from the other, higher energy triplets [38].

5. References

1. Rakhmanova, S.V. and Conwell, E.M. (2001) Polaron Motion in DNA, *J. Phys. Chem.* **B105**, 2056-2061.
2. Turro, N.J. and Barton, J.K (1998) Paradigms, supermolecules, electron transfer and chemistry at a distance. What's the problem? The science or the paradigm? *JBIC, J. Biol. Inorg. Chem.* **3**, 201-209.
3. Meggers, E., Michel-Beyerle, M.E. and Giese, B. (1998) Sequence Dependent Long Range Transport in DNA, *J. Am. Chem. Soc.* **120**, 12950-12955.
4. Giese, B. (2000) Long Distance Charge Transport through DNA: The Hopping Mechanism, , *Acc. Chem. Res.* **33**, 631-636.
5. Jortner, J., Bixon, M., Langenbacher, T., and Michel-Beyerle, M. (1998) Charge transfer and transport in DNA, *Proc. Natl. Acad. Sci. USA* **95**, 12759-12765.
6. Schuster, G.B. (2000) Long-Range Charge Transfer in DNA: Transient Structural Distortions Control the Distance Dependence, *Acc. Chem. Res.* **33**, 253-260.
7. Williams, T.T., Odom, D.T. and Barton, J.K. (2000) Variations in DNA Charge Transport with Nucleotide Composition and Sequence, *J. Am. Chem. Soc.* **122**, 9048-9049.
8. Giese, B. et al. (2001) Direct observation of hole transfer through DNA by hopping between adenine bases and by tunnelling, *Nature* **412**, 318-320.
9. Seidel, C.A.M., Schulz, A. and Sauer, M.H.M. (1996) Nucleobase-Specific Quenching of Fluorescent Dyes. 1. Nucleobase one-Electron Redox Potentials and Their Correlation with Static and Dynamic Quenching Efficiencies, *J. Phys. Chem.* **100**, 5541-5553.
10. Steenken, S. and Jovanovic, S.C. (1997) How Easily Oxidizable is DNA? One-Electron Reduction Potentials of Adenosine and Guanosine Radicals in Aqueous Solution, *J. Am. Chem. Soc.* **119**, 617-618.
11. Hush, N.S. and Cheung, A.S. (1975) Ionization potentials and donor properties of nucleic acid bases and related compounds, *Chem. Phys. Lett.* **34** 11-13.
12. Lifschitz, C., Bergman, D. and Pullman, B. (1967) The ionization potentials of biological purines and pyrimidines, *Tetrahedron Lett.* **46**, 4583-4586.
13. Sugiyama, H. and Saito, I. (1996) Theoretical Studies of GG-Specific Photocleavage of DNA via Electron Transfer: Significant Lowering of Ionization Potential and 5' Localization of HOMO of Stacked GG Bases in B-Form DNA, *J. Am. Chem. Soc.* **118**, 7063-7068.
14. Giese, B. and Spichty, M. (2000) Long Distance Charge Transport through DNA: Quantification and Extension of the Hopping Model, *CHEMPHYSCHEM* **1**, 195-198.
15. Bruinsma, R. ,Gruner, G. ,D'Orsogna, M.R. and Rudnick, J. (2000) Fluctuation-Facilitated Charge Migration along DNA, *Phys. Rev. Lett.* **85**, 4393-4396.
16. Grozema, F.C., Berlin, Yu.A. and Siebbeles, L.D.A. (1999) Sequence-Dependent Charge Transfer in Donor-DNA-Acceptor Systems: A Theoretical Study, *Int. J. Quantum Chem.* **75**, 1009-1016.
17. Grozema, F.C., Berlin, Yu.A. and Siebbeles, L.D.A. (2000) Mechanism of Charge Migration through DNA: Molecular Wire Behavior, Single Step-Tunneling or Hopping? *J. Am. Chem. Soc.* **122**, 10903-10909.
18. Barnett, R.N. et al. (2001) Charge Migration in DNA: Ion-Gated Transport, *Science* **294**, 567-571.
19. Gomez-Navarro, C. et al. (2002) Contactless experiments on individual DNA molecules show no evidence for molecular wire behavior, *Proc. Natl. Acad. Sci. USA* **99**, 8484-8487.
20. Tran, P., Alavi, B. and Gruner, G. (2000) Charge Transport along the λ-DNA Double Helix, *Phys. Rev. Lett.* **85**, 1564-1567.
21. Helgren, E., et al. (2001) Electrons on the double helix: optical experiments on native DNA, *arXiv:cond-mat/***0111299 v1**, 1-5.
22. Fink, H.-W. and Schonenberger, C. (1999) Electrical conduction through DNA molecules, *Nature* **398**, 407-410.
23. de Pablo, P.J. et al. (2000) Absence of dc-Conductivity in λ-DNA, *Phys. Rev. Lett.* **85**, 4992-4995.
24. Yoo, K-H. et al. (2001) Electrical Conduction through Poly(dA)-Poly(dT) and Poly(dG)-Poly(dC) DNA Molecules, *Phys. Rev. Lett.* **87**, 19802-1 – 19802-4.
25.Lee, H-Y. et al. (2002) Control of electrical conduction in DNA using oxygen hole doping, *Appl. Phys. Lett.* **80**, 1670-1672.

26. Porath, D., Bezryadin, A., de Vries, S. and Dekker, C. (2000) Direct measurement of electrical transport through DNA molecules", *Nature* **403**, 635-638.
27.Emin, D. (1986) Self-trapping in quasi-one-dimensional solids, Phys. Rev. **B33**, 3973-3975.
28. Su, W.P., Schrieffer, J.R.,and Heeger, A.J. (1980) Soliton excitations in polyacetylene, *Phys. Rev.* **B22**, 2099-2111.
29. Sugiyama, H. and Saito, I. (1996) Theoretical Studies of GG-Specific Photocleavage of DNA via Electron Transfer: Significant Lowering of Ionization Potential and 5' Localization of HOMO of Stacked GG Bases in B-Form DNA, *J. Am. Chem. Soc.* **118**, 7063-7068.
30.Voityuk, A.A., Rosch, N., Bixon, M. and Jortner, J. (2000) Electronic Coupling for Charge Transfer and Transport in DNA, *J. Phys. Chem.* **B104** 9740-9745.
31. Conwell, E.M. and Rakhmanova, S.V. (2000) Polarons in DNA, *Proc. Natl. Acad. Sci.USA* **97**, 4556-4560.
32. Lewis, F.D. et al. (2000) Dynamics and Equilibria for Oxidation of G, GG, and GGG Sequences in DNA Hairpins, *J. Am. Chem. Soc.* **122**, 12037-12038.
33. Conwell, E.M. and Basko, D.M. (2001) Hole Traps in DNA, *J. Am. Chem. Soc.* **123**, 11441-11445.
34. Voityuk, A.A, Jortner, J., Bixon, M. and Rosch, N. (2000) Energetics of hole transfer in DNA, *Chem. Phys. Lett.* **324**, 430-434.
35. Basko, D.M. and Conwell, E.M. (2002) Effect of Solvation on Hole Motion in DNA, *Phys. Rev. Lett.* **88**, 098102-1 – 098102-4.
36. Hjort, M. and Stafstrsom,S. (2001) Band Resonant Tunneling in DNA Molecules, *Phys. Rev. Lett.* **87**, 228101-1-228101-4.
37. Rakhmanova, S.V. and Conwell, E.M. (2001) Polaron Motion in DNA, *J. Phys. Chem.* **B 105**, 2056-2061.
38. Gueron, M., Eisinger, J. and Lamola, A.A. (1974) Excited States of Nucleic Acids in P. O. P. Ts'o (ed.), *Basic Principles in Nucleic Acid Chemistry*, Academic Press, New York, pp. 311-398.
39. Pecourt, J-M., Peon, J. and Kohler, B. (2000) Ultrafast Internal Conversion of Electronically Excited RNA and DNA Nucleosides in Water, *J. Am. Chem. Soc.* **122**, 9348-9349.
40. Eisinger, J. et al. (1966) Excimer Fluorescence of Dinucleotides, Polynucleotides, and DNA, *Proc. Natl. Acad. Sci. USA*, **55**, 1015-1020.
41. Michl, J. and Bonacic-Koutecky, V. (1990) *Electronic Aspects of Photochemistry*, Wiley, New York.
42. Georghiou, S. Kubala, S. and Large, C.C. (1998) Environmental Control of the Deformability of the DNA Double Helix, *Photochem. and Photobio.* **67**, 526-531.
43. Lakowicz, J.R. (2001) Intrinsic Fluorescence from DNA Can Be Enhanced by Metallic Particles, *Biochem. and Biophys. Res. Commun.* **286**, 875-879.

DYNAMICS OF EOSIN Y ENCAPSULATED IN A FOURTH GENERATION FUNCTIONALISED POPAM DENDRIMER

GILBERTO TEOBALDI and FRANCESCO ZERBETTO
Dipartimento di Chimica 'G. Ciamician', Università di Bologna, V. F. Selmi 2, 40126 Bologna, Italy

KEYWORDS/ABSTRACT: molecular dynamics / dendrimers/ guest-host systems/ encapsulated dye / photophysical properties / excited state lifetimes / kinetics

Molecular dynamics is used to investigate the encapsulation of several eosins Y in a fourth generation dendrimer. The experimental limit of 6 eosins inside the dendrimer is reproduced. The dyes are distributed radially from the center of the dendrimer and their dynamics is accounts for the two fluorescence lifetimes observed experimentally.

1. Introduction

Dendrimers are well-defined macromolecules that radiate from a central core branches-upon-branches.[1] A series of repetitive synthetic steps forms successive shells that yields generations of perfectly mono-dispersed non-linear polymers.[2] Excellent control is now achieved over size, shape and terminal functionalisation of the branches, a feature that bears on surface chemistry.[1,2] An important potential application of dendrimers is the encapsulation of guest molecules. Balzani et al. [3,4] showed that poly(propylene amine), in short POPAM, dendrimers, functionalized in the periphery with dansyl units, can accept a varying number of guests. Here, we perform molecular dynamics, MD, calculations on a guest-host system made by the fourth generation, 4D, dansyl functionalised POPAM dendrimer with a varying number of Eosin Y guests, scheme 1.
The peripheral functionalization of a dendrimer can modify its encapsulation properties mechanically and sometimes electronically. For instance, the dendrimer considered here can encapsulate up to 6 Eosin Y and up to 25 molecules of Rose Bengal.[3,4] The guest-host interactions modify the photophysical properties of eosin itself and turn its mono-exponential decays in aqueous solution into a bi-exponential one in CH_2Cl_2, a solvent where eosin is not soluble without the presence of POPAM. The existence of two lifetimes was taken to indicate that there are two types, or two families, of sites in the dendrimer. The ability to tune and modify the electronic properties of one, or more, guest molecules can have practical consequences and be exploited practically, for instance in the field of chemical sensors.

F. Charra et al. (eds.), Organic Nanophotonics, 207–216.

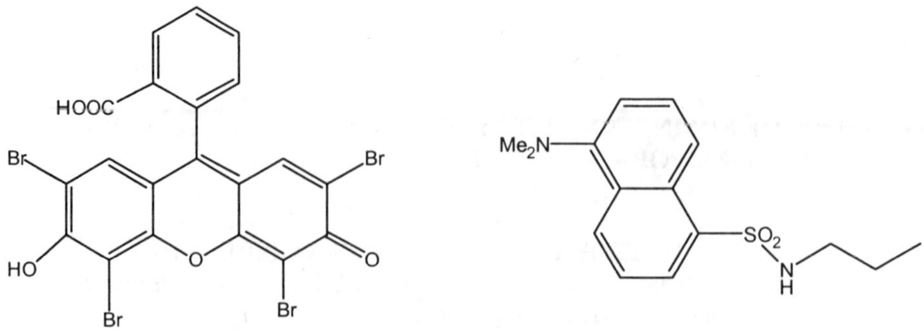

Eosin Y

Dansyl group

Scheme 1

The elucidation of the structural and dynamical properties of dendrimers and their guest-host systems is one of the crucial steps towards their practical exploitation. However, the flexible branches that characterize the majority of dendrimers, which can re-arrange to form a myriad of conformations, and the rapid exchange between co-conformers, which exist in the liquid and can be partly transmitted to the solid, make a structural characterization very difficult or impossible. Molecular modelling can assist the process of developing a quantitative understanding of the geometrical and dynamical properties of dendrimers.[1,5] It is perhaps surprising that we were able to find only one, recent and thorough computational investigation of organic encapsulation in dendrimers,[5] which is concerned with the interaction of Rose Bengal with the phenylalanine capped PAMAM dendrimer. Here, we use molecular dynamics to study the eosin@POPAM system.

2. Computational

The calculation were performed with the MM3 force field that has been found to be accurate for organic systems[6] and was parameterized explicitly to describe the $\pi-\pi$ stacking interactions that govern a sizeable part of the interactions in this system. The molecular dynamics (MD) calculations were run with the Tinker 3.8 program [7] using the approach of Berendsen *et al.*[8] with periodic boundary conditions (PBC), a cubic box with a linear dimension of 73 Å, at constant volume and at 300 K. The approach based on MM3 implemented in the Tinker program has been rather successful in a variety of applications carried out in our laboratory.[9-12]

Solvent molecules of CH_2Cl_2 were included explicitly. The fourth generation POPAM dendrimer was built in successive steps and the geometry was initially optimized in the vacuum. Eosin Y was also optimized in the vacuum and then inserted into the dendrimer in regions containing the largest cavities and then the guest-host structure was further optimized.

The dendrimer with 7 randomly placed eosins was then placed in the solvent box, and all solvent molecules within 1.4 Å of any of the dendrimer or the eosins atoms were excluded. This leads to a system of 2949 CH_2Cl_2 molecules plus 7Eosin@4D-

POPAM. Equilibration was performed in the NVT ensemble at 300 K until the potential energy converged.

To provide a more quantitative insight, a link between the MD simulations and the description of the excitation decay is needed. A simple kinetic equation describes the decay

$$-\frac{dC}{dt}=\sum_i k_i C \tag{1}$$

considering the single eosin molecules, C is the number of molecules that can decay and k_i is the probability of decay for a single channel (see below for definition of decay channel). Since the MD simulation consists of discrete steps, we discretize eq. 3 to read

$$\frac{C_{n+1}-C_n}{\Delta t}=-\sum_i k_i C_n \tag{2}$$

where n and n+1 are successive steps in the simulation. The number of surviving, i.e. still excited molecules at step n+1 is

$$C_{n+1}=C_n\left(1-\Delta t\sum_i k_i\right) \tag{3}$$

the number of molecules that remain excited at each step is obtainable from the MD if one can estimate k_i.

3. Results and discussion

Figure 1 shows a snapshot of the dendrimer in the simulations. Figure 2 displays the potential energy variation during the dynamics of the box with either 4D-POPAM or 7eosin@4D-POPAM embedded in solvent molecules. The small change in energy shows that the system is well equilibrated. Figure 3 shows the number of solvent molecules inside 4D-POPAM, both in the presence and absence of the guests. Two distances are used: the first is the dendrimer gyration radius, the second is the van der Waals distance, given by the sum of van der Waals radii, between an atom of the solvent and an atom of the macromolecule. The gyration radius roughly describes the inner part of the polymer so that the differences between the two numbers of CH_2Cl_2 estimates the solvent molecules on the surface. There are ~60 solvent molecules inside 4D-POPAM and a similar number on its surface. The presence of 7 eosins hardly affects the number of solvent molecules.

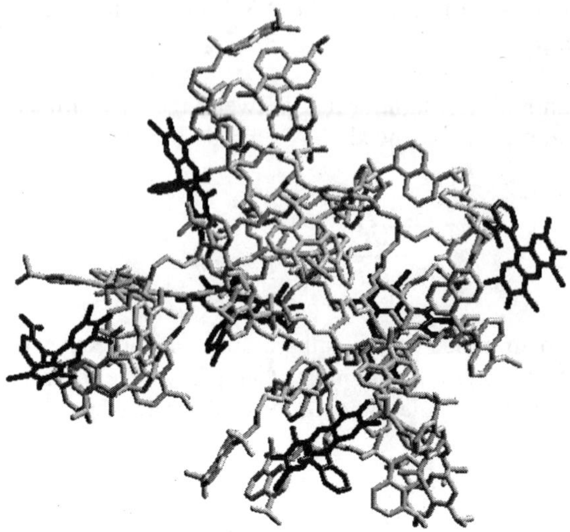

Figure 1. A snapshot of <u>6eosin@4D-POPAM</u>. The guest molecules are in van der Waals contact with the dendrimer, the 7th eosin and the solvent molecules are not shown.

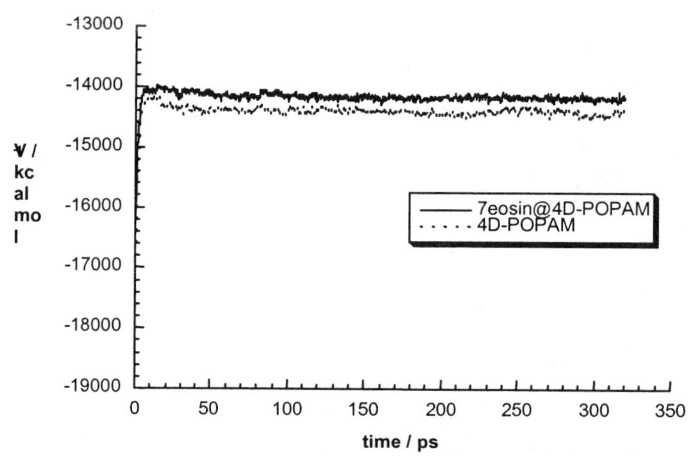

Figure 2. *Potential energy of 4D-POPAM and <u>7eosin@4D-POPAM</u> in CH_2Cl_2.*

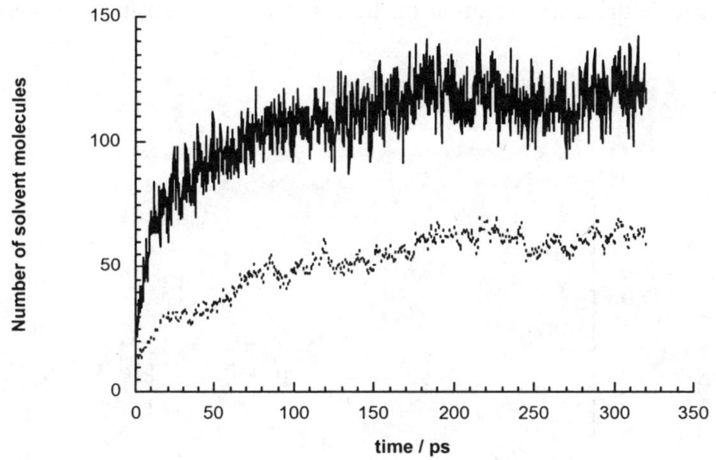

a)

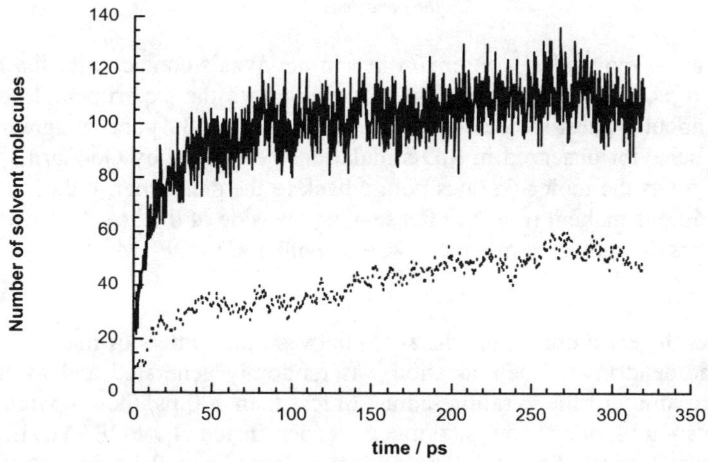

b)

Figure 3. *Number of solvent molecules a) 4D-POPAM, b) 7eosin@4D-POPAM. The top, solid line is the number of solvent molecules in van der Waals contact with the dendrimer, the bottom, dotted line is the number of molecules inside the gyration radius.*

Figure 4 illustrates the time evolution of the distance of the eosin molecules from the dendrimer.

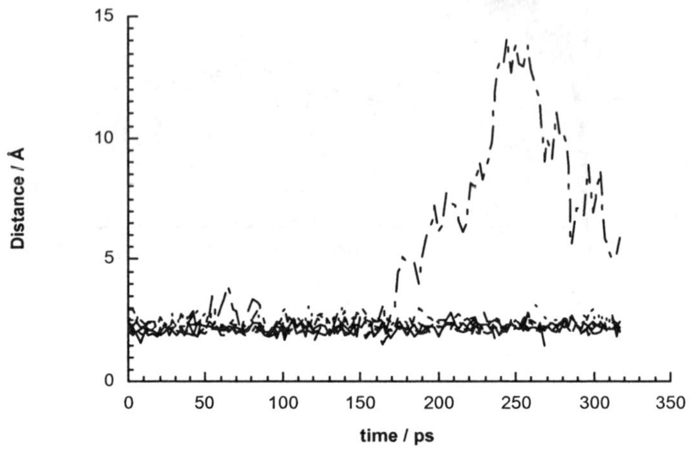

Figure 4. *Shortest atom-atom distance for each of the 7 eosins and 4D-POPAM: one guest molecule leaves the dendrimer.*

Of the seven guest molecules, six remain in van der Waals contact with the host, and, after about 170 ps, one leaves it. This in agreement with the experimental observation that there are about 6 guest molecules inside the dendrimer,[3,4] and in agreement with an analogous behavior observed in MD simulations performed by Goddard's group.[5] Notice that ~250 ps the molecule does not go back to the dendrimer, rather the periodic boundary conditions make it re-enter after leaving one side of the box. In a real solution, the event mimics the molecule moving towards another 4D-POPAM.

Figure 5 shows the evolution of the distances between the centers of mass of the eosins and that of the dendrimer. Their location was randomly generated and most of them occurred in proximity of the gyration radius. In less than 100 ps, they separate into two groups. Interestingly, one eosin sits much deeper inside 4D-POPAM. In different simulations, introduction of a second molecule was forced near the center of mass of the dendrimer but resulted in its removal from the 10 Å inner radius.

It is tempting to ascribe the two different experimental lifetimes (τ_1=2.2 ns, τ_2=0.6 ns) observed for eosins@4D-POPAM[3,4] to their radial distribution inside the dendrimer. In this picture, one lifetime would be ascribed to the eosins in the core and the other to the eosins residing in the periphery. Qualitatively, one could propose that the dye molecules in the core of 4D-POPAM are subject to more interactions with the host and therefore have a shorter lifetime.

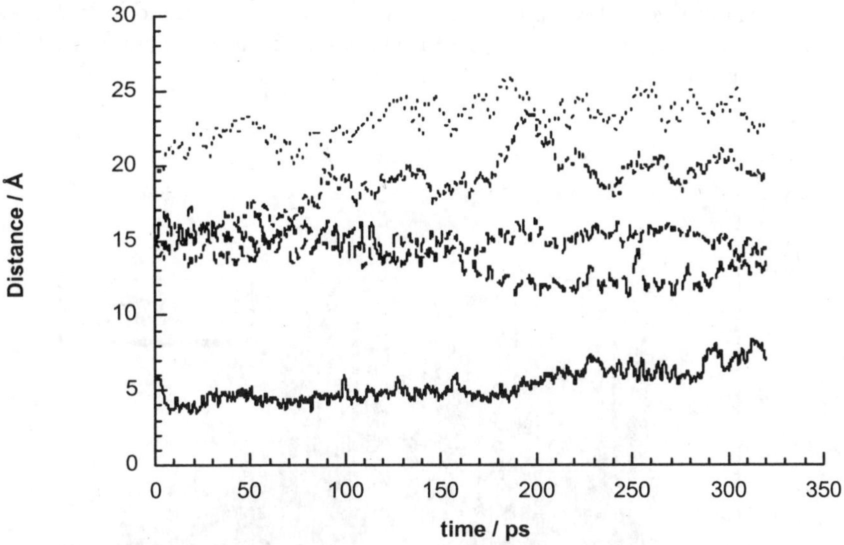

Figure 5. Distances of the eosins centres of mass with the centre of mass of 4D-POPAM.

Despite the simplicity of the explanation for the two lifetimes of eosin@4D-POPAM, one may try to make it more quantitative and exploit eq. 3 directly. To determine the k_i values, we start by ascribing the decay mechanism of the eosins to the interaction between the dendrimer and the guest molecules. Each decay channel therefore arises from their interaction with the nitrogens of the guest molecules (interaction with the CH_2 groups of 4D-POPAM should not modify the photophysics of eosin). For simplicity, to describe the fall off of the quenching probability of the excited eosin, we use the exponential of a distance, r_i, multiplied by a constant, $\exp(-\alpha r_i)$. Of course, functions with more parameters could also be used, but would complicate the picture unnecessarily. Depending on the nature of the deactivation process, the final point of interaction inside eosin can be taken as either the oxygen of the carbonyl, or the hydrogen atoms of the COOH or of the OH groups, or in a generic way, as its center of mass.

Figure 6 shows the interactions that occur dynamically during the MD. The curves are the sum of the exponentials of the distances between the eosins and the amine nitrogens. In very short times, the values vary strongly and prove how greatly such interactions vary and the high mobility of the guests inside 4D-POPAM. For illustrative purposes, one can compare the plots for the distances of each eosin from a) the oxygen atom of the carbonyl group, and b) the hydrogen atom of the more acidic carboxyl moiety. Regardless of what triggers the decay, the larger the value of the individual sums the greater the probability of relaxation to the ground state of the photoexcited eosins. Qualitatively, the curves show that different probabilities of deactivation exist for the individual guest molecules.

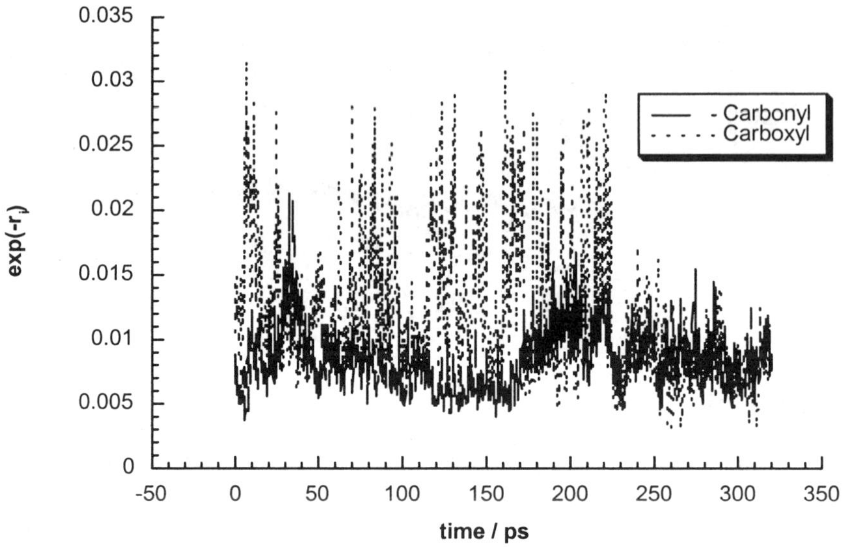

Figure 6. Sum of the exponentials of the distances between the oxygen atom of the carbonyl and the nitrogen atoms of the amines (solid line) and the hydrogen atom of the carboxyl and the nitrogen atoms of the amines (dashed line).

The probability of decay in the Δt interval between step n and step n+1 of the MD is proportional to the interaction that is represented by the areas of the curves of figure 6 between two steps. For simplicity, we average the sum of such areas. A multiplicative factor, α, for the distance is required, both formally, to make the argument of the exponential dimensionless, and in practice, to achieve a good fit with the experimental decay. Such proportionality factor is therefore optimized by calculating a numerical grid to reproduce the experimental decay. In the curve describing the experimental decay, the two lifetimes contribute equally. Figure 7 shows the comparison between experimental and calculated decays for the distance between the nitrogen of the atoms with a) the oxygen of the carbonyl, 2) the hydrogen of the carboxyl, 3) the hydrogen of the hydroxyl, and 4) the center of mass of the eosins. The fitted α values are 1.445 $Å^{-1}$ for the carbonyl, 1.675 $Å^{-1}$ for the carboxyl, 1.745 $Å^{-1}$ for they hydroxyl, and 1.970 $Å^{-1}$ for the center of mass. It is apparent that only the first two closely follow the experimental trend, while the third and the fourth produce a more complicated behavior. Experimentally, the oscillating behavior would result in a multi-exponential decay.

In this more physical picture, it is not the distance from the center of the dendrimer that differentiates the eosins, rather it is their environment, which is not static and changes dynamically, that leads directly to a double-exponential decay.

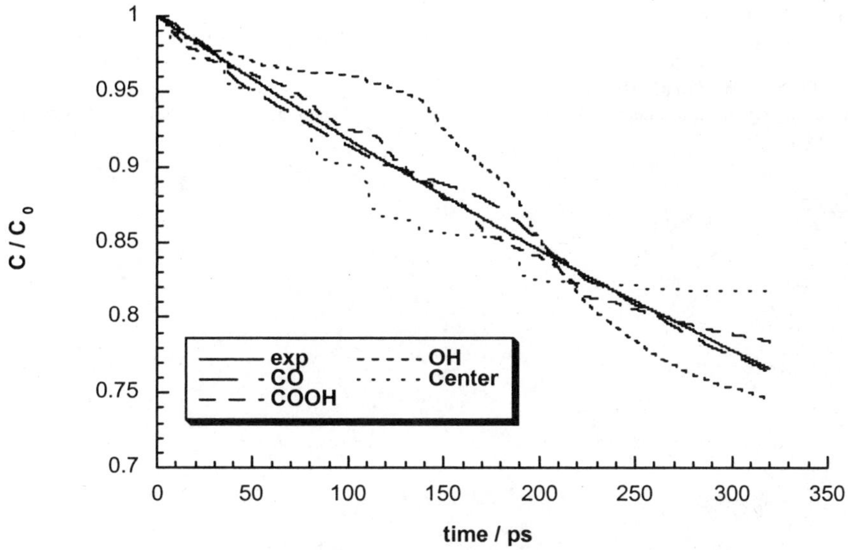

Figure 7. Decay of the excited eosins for several types of interactions.

4. Conclusion

The MD finds that eosin molecules have a good mobility inside 4D-POPAM. Simulations with an excess of guest molecules show that the extra eosin departs. The 6 dyes that remain inside 4D-POPAM spread radially from the center of the dendrimer. The analysis of the dynamics, coupled with a finite difference description of the electronic decay, can reproduce the bi-exponential decay and shows that the environment of the eosins is subject to large dynamical variations. The decay is best reproduced when the interactions between the carbonyl, or the carboxyl, with the amine groups are considered, and not when the same interactions are with either the center of mass of the eosins or with the hydroxyl group.

References

1. Newkome, G.R., Charles, M.N., Vogtle, F. (2001) *Dendrimers and Dendrons*, Wiley-VCH,.
2. Tomalia, D. A., Hedstrand, D. M., Ferritto, M. S. (1991) *Macromol.* **24**, 1435.
3. Balzani, V., Ceroni, P., Gestermann, S., Gorka, M., Kauffmann, C., Maestri, M., Vogtle, F. (2000) *CHEMPHYSCHEM* **1**, 224 – 227.
4. Balzani, V., Ceroni, P., Gestermann, S., Gorka, M., Kauffmann, C., Vogtle, F. (2002) *Tetrahedron* **58,** 629-637.
5. Miklis, P., Cagin, T., Goddard III, W.A., (1997) *J. Am. Chem. Soc.* **119**, 7458-7462.
6. Allinger, N. L., Yuh, Y. H., Lii, J.-H. (1989) *J. Am. Chem. Soc.* **111**, 8551-8566.

7. Ponder, J. W., Richards, F. (1987) *J. Comput. Chem.* **8**, 1016-1024.
8. Berendsen, H.J.C., Postma, J.P.M., van Gusteren, W.F., Di Nola, A., Haak, J.R., (1984) *J. Chem. Phys.* **81**, 3684.
9. Bermudez, V., Capron, N., Gase, T., Gatti, F. G., Kajzar, F., Leigh, D. A., Zerbetto, F., Zhang, S. W. (2000) *Nature*, **406**, 608-611.
10. Fustin, C.-A., Leigh, D. A., Rudolf, P., Timpel, D., Zerbetto, F. (2000) *CHEMPHYSCHEM*, **1**, 97-100.
11. Cavallini, M, Lazzaroni, R., Zamboni, R., Biscarini, F., Timpel, D., Zerbetto, F., Clarkson, G. J., Leigh, D. A. (2001) *J. Phys. Chem. B* **105**, 10826-10830.
12. Biscarini, F., Cavallini, M, Leigh, D. A., Léon, S., Teat, S. J., Wong, J. K. Y., Zerbetto, F. (2002) *J. Am. Chem. Soc.* **124**, 225-233.

PROTEIN NANOSTRUCTURES LIGHT CONTROL FOR DATA STORAGE

P. MEYRUEIS - I. EL-HAFIDI - R. GRZYMALA
Photonics Systems Laboratory
Louis Pasteur University
API Center - Illkirch Innovation Park
Bld. Sebastien Brant - 67400 ILLKIRCH - France
Tel. (33) 3 902 44 614 - Fax (33) 3 902 44 619
e-mail: meyrueis@sphot.u-strasbg.fr

1. Introduction

Biophotonics can be considered as the crossing of photonics physics, and biotechnology. A few years ago by a fundamental research collaboration between our laboratory : « the Photonics Systems Laboratory », and the « laboratory of the Strasbourg School of Biotechnology », we have identified a method and process for producing specific proteins molecules adapted to some expected applied photonic uses by local light modulation at a nanometric scale inside a protein molecule distribution organized as a layer.

Our first specific protein production was achieved by biotechnology by changing the genetic code of bacterias[1]. Then we use chemical methods for producing desired molecules out of selected animals biological materials.

Photonics crystals (PC) properties are very promising for Optical Signal processing, data storage, and photonic wavefronts processing[2]. However, effective PC production is very difficult up to now. To create a significant photonic effect, in the visible spectrum for instance, it is necessary to generate a PC including tridimensional (3D) material modulations. These modulation sizes are not very far away from the average organic molecular scale. A photonic crystal is a 3D periodic structure formed by subwavelength sized features[3].

In a semiconductor crystal, we have conduction properties modulated by the atomic distribution. The result is band gaps existing for specific energies in this crystal.

In a P.C, the same approach yields photonic band gaps related to light frequencies. If the photonic crystal is perfect (without defects), it will behave like a perfect mirror for a given wavelength under a given set of angles.

If the P.C has defects in its structure, the light propagation, in and through, the PC modulated material will follow complex laws. We use these laws for protein PC data storing used as a validation.

F. Charra et al. (eds.), Organic Nanophotonics, 217–230.

2. Protein Photonics Crystal Basis

2.1 INTRODUCTION

Compared to a crystal material, a fibrous protein, that is different for globular proteins which has a periodically structured organisation, is close to liquid crystals by its tendency to form spirals as well as other more complex structures[4,5].

One of the main specificity of protein is that their morphology can change while being chemically stable. This change of shape can occur through many stimulations. One of these stimulations can be light illumination which would trigger morphological changes in some doped protein molecules. Prior to stimulation, these proteins are doped with a mineral material (chromium for instance).

The PC material modulating input wavefront processing operates through light induced quasi molecular protein aggregate. Every aggregate is usually called a tecton. This tecton is then acting locally on photons, following Maxwell equations. Tectons operate collectively inside a protein PC. So PC tectons will process an input photonics wavefront into another one, that can be programmed. The dimensions of the PC tectons, to be effective in wavefront processing, are typically smaller than one half of the wavelength. Therefore, the PC modelling approach can be used to describe the behavior of light propagating in a spatial organisation of conveniently selected and processed protein molecules.

The protein once prepared as a powder are solved in water. With a temperature annealing process, a gelation phenomena is triggered that produces wrapped chain molecules[6]. Then the solution is layered and doped (with a metal as chromium). Illuminating of molecules with a selected light frequency will cause local rearrangements of the protein molecules shapes and aggregates due to chromium cross linking[7,8,9]. The input PC structuring light is partially modulated through an interference patterns coming from interferences between several appropriate wavefronts. We will obtain a resulting light amplitude modulation inside the material. The creation of stabilised cross links modulates the material in 3D tectons desired distribution and reshaping. A stabilisation process will freeze the PC tecton modulation. In this way the PC can be manufactured and stabilised. The protein tecton spatial organisation gives rise to the protein PC modulated band gap functionalities.

2.2. PROTEIN SPECIFICATIONS FOR ELABORATING OPTIMISED PC WITH IMPLEMENTED FUNCTIONS:

Proteins are composed of amino acids along a back bone. Amino acids can include carboxy group (COOH) amino group (NH2), and a residue that can contain sulphur and benzene.

Different proteins have different sequences of amino acids. A protein can include in 3 dimensions, several polypeptide chains. Some segments of the polypeptide chain are folded in a 3D structure. The secondary structure may form an helix[10].

We have selected protein for our experimental validation out of selected collagen which is formatted in a triple strands helix about 3 nm long and 0,2 nm wide. The collagen is hydrolysed and broken into polypeptide protein chains. These protein chains molecular weight ranging from 10 to 1000 KiloDaltons. We have selected the proteins having a molecular weight around 95 KD for practical operational reasons. Some potentials of protein photonic crystal (PPC) technology that can be considered for protein PC future development are listed below:

- PCs could provide highly integrated photonic chip functionalities for optical telecom
- PCs could provide new type of optical fibers
- PCs can constitute very good frequency filters that can be used in DWDM systems
- PCs can constitute excellent selective mirror
- PCs can support very dense 3D interconnection system
- PCs can be also used in optical data processing as an adapted filter
- PCs can constitute high gain laser cavities

Before implementing the physic PC model on protein we have validated this it with plastic balls tectons sized to fit microwaves application (fig.1).

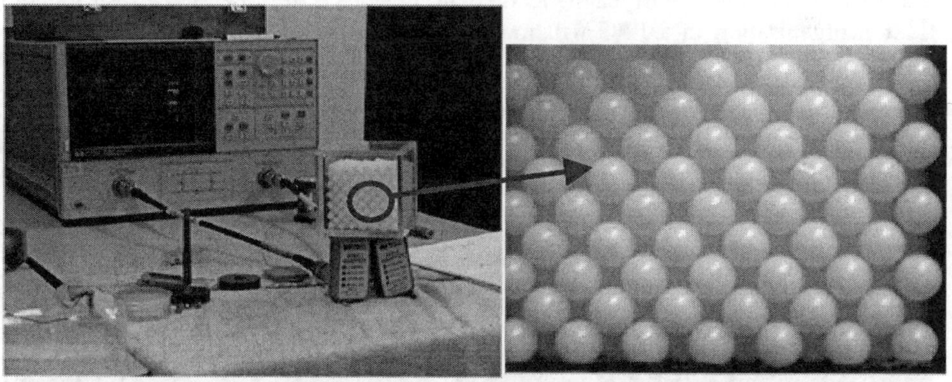

Figure 1 Preliminary test using microwave generator and plastic balls

These tests were performed to check the modelling by using microwaves easier to manipulate than visible light waves. We have created a regular tecton organisation then we easily perturb it to check if this perturbation fit the modelling. We have tested band gap mirror effect by changing tectons organisation.

Proteins are bio molecules produced in living cells out of DNA encoding. Protein have biological properties[11,12] but also they have specific physical photonics properties that we considered for opening the way to innovative applications.

Some of these physical properties are concerning interaction of protein once doped with light. They are the ones that we have tried to identify and to control for signal storing and processing, by investigating protein PC layers organised by wavefront modulation effects and a specific processing.

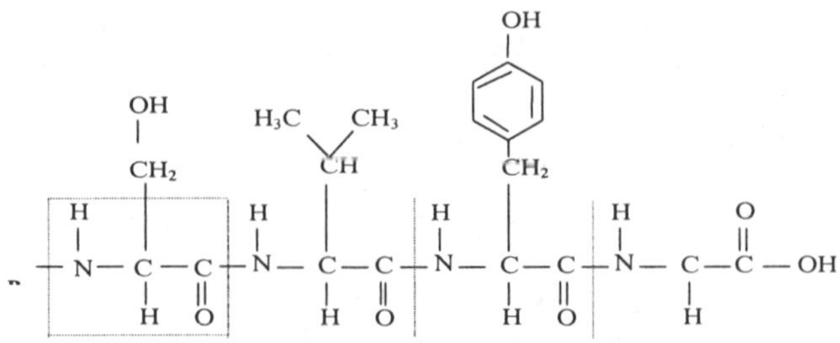

Figure 2 Typical chemical configuration of the family of protein molecules which used for PC implementing

Depending on the PPC defects protein tecton organisation the PPC will have different effects on an input photonics wavefront. The photons are going to propagate through PPC and modulated by the programmed tecton distribution. at its surface and/or inside following complex paths depending on their wavelength.

The prediction of the behavior of photons in a PPC can be modeled, but not yet completely for several frequencies[13]. In this way it is theoretically possible to determine approximatively the electromagnetic field in any place of a PC and its surrounding. Tecton organisation in a PPC will modulate the PC band gap related to frequencies effects and beam directions.

a **b**

Figure 3 a: Protein powder before gel coating, b: One of the coating machine used

2.3. PROTEIN PC REALISATION TECHNOLOGY:

Once the proteins are identified, the production method that we have first selected is a biotechnology based method. In our patented method we change the genetic code of bacterias to produce the appropriate protein molecule[14]. This biotechnology method is very expensive, that is why we have later developed a method which can select by both physical and chemical processes the desired molecules out of protein produced by animals[15,16].

Once the protein molecules are produced, with an appropriate method, we coat a layer of these molecules on a glass substrate. After several baths, we build the PPC inside the protein layer by illuminating this layer with a modulated light followed by a stabilisation processing[17]. We obtain the modulated light pattern by interference between several wavefronts. The thickness of the PC protein layer is 30 microns. Depending of the recording wavelength and subsequent phase modulation, different PPC structures can be built. We can thus program the recording wavefront to implement different functions within a PC. We have generated PPC mirrors and transmissive direction routers to check the band gap effect by selecting the interfering wavefronts (without defects) and the wavelength. A good angular selectivity was achieved.

a

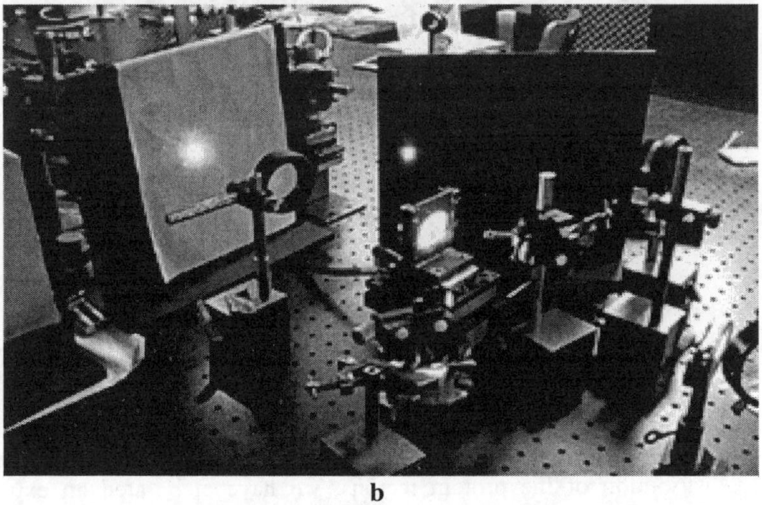

b

Figure 4 a: Set up for testing the PPC used as a reflective wavefront processor,
b: Set up for testing the PPC used as a transmissive wavefront processor

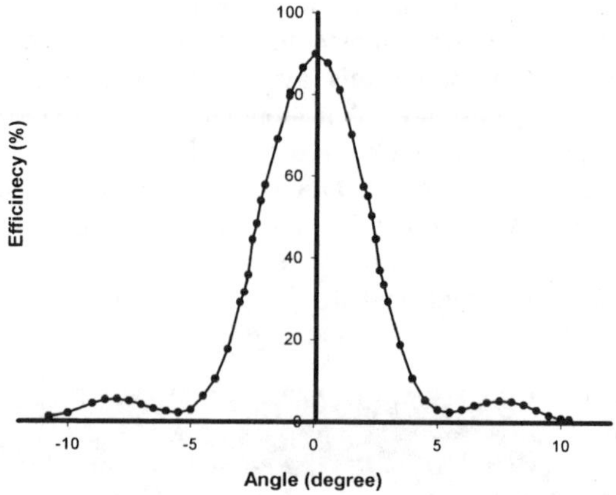

Figure 5 band gap response of a validating PC

We have tested a PPC band gap implementation (fig. 4 a, b) operating as a selective mirror with 90% efficiency (fig. 5). We have noticed that it is necessary to protect the protein PC layer by a coating to avoid scratches and other damages mainly from moisture. This cover glass can be frame cured in order to seal the layer. Once the layer is sealed, it is very resistant to surrounding constraints, moisture, temperature. We have identified no significant defects from temperature modification ranging from – 20°C to + 100°C (fig.6).

Figure 6 Example of layered photonic crystal operating as band gap mirror protected by a glass cover.

This mirror is implemented several times on this plate with different reflecting angles. Since it is theoretically possible to record an information in a PPC by perturbing the regular distribution of the protein tectons, we have elaborated an experimental methodology through which we can create a modulation by interfering several perturbed wavefronts. The wavefront data encoded in the "defects" of the interfering beams create 3D interference patterns carrying out images of these defects, which are then transferred to the PPC as material modulation (modulation of the tectons distribution) through tectons light sensitivity (fig.7)

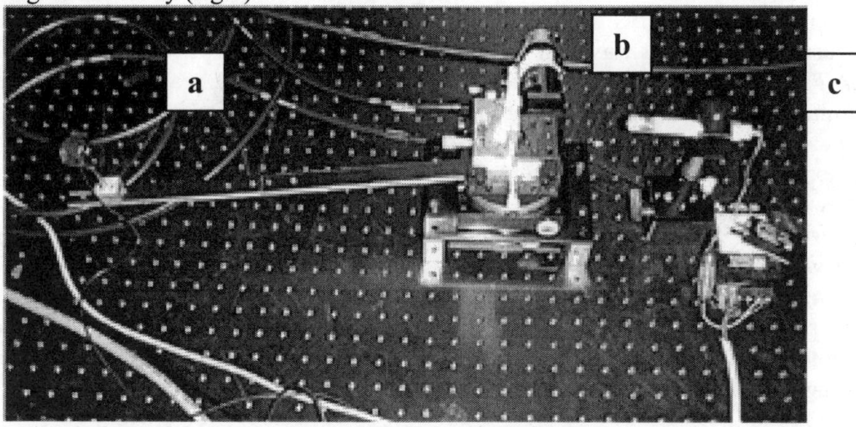

Figure 7 Systems for measuring the angular energy produced by transmissive excitation of the PC From left to right : rotating sensor (a), photonic crystal mounted on a rotating table (b), laser(c).

Data loaded interferences modulate the PPCs. Once the modulated PPC is modulated and stabilised, by illuminating it with a wavefront, having a wavelength as close as possible to the one that was used to create the structuring interferences, we will obtain, as a light output, a programmed wavefront loaded with functionalities. In this way, we have checked that if a PPC for this purpose is illuminated with the appropriate wavefront (angle, phase and polarisation), data can be recorded and retrieved with a low level of noise allowing an effective data storage use. Noise is coming mostly from errors in the wavefront generation which is performing the desired modulation in the PPC tectons structure. We have read packets of recorded data with simple reading device, including a CCD camera. Data packets are physically dispersed in the PPC volume because of the interferometric encoding method used. The storage capacity is so very important (fig.8).

Figure 8 PPPC used as data storage : simplified set up for reading the data from left to right : CCD, lens, PC, laser.

3. PPC Data Storage Technologies

3.1. INTRODUCTION:

Volume patterns inside the protein material layer are the result of interference between coherent light wavefronts beams having appropriate wavelength (example 514 nm). The data packet to be recorded is carried by one object beam where it is loaded. An other beam is incident to the protein PC material layer with a specific angle, it will produce 3D interferences with the data loaded beam[18].

The experimental set up is presented in figure 9. The laser beam is splited into two beams. One the these beams is processed to be a large and collimated one to illuminate the surface of the spatial light modulation device (SLM). After passing through this SLM which receive the data information, for instance from a computer, the laser beam carrying the data is sized about 1 square centimeter. The protein layer is mounted on a glass plate two translation stages operating on X and Y to provide a spatially multiplexing of the recording. The second beam is reflected by a mirror mounted on a rotating support for scanning a set of small mirrors mounted according a sequence of angles.

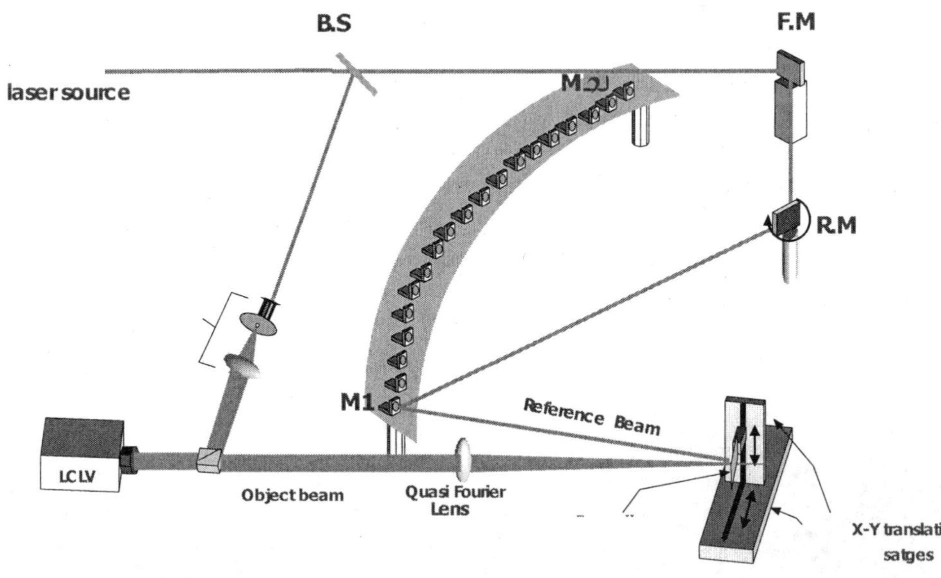

F. M: fixed mirror
R. M: rotating mirror
B. S: beam splitter

Figure 9 Recording set up design for PPC multiplexing

The number of mirrors in our experiment is twenty five, corresponding to the number of selected multiplexing angles. So, twenty five data packets are stored in same Voxel. The planar projected surface of this voxel is about 1 mm^2. After recording the data packets on one first voxel using angular multiplexing, a translation of the recording protein plate is done to change the recording area location on this PC plate. We realise in this way a matrix of data loaded multiplexed (fig.10).

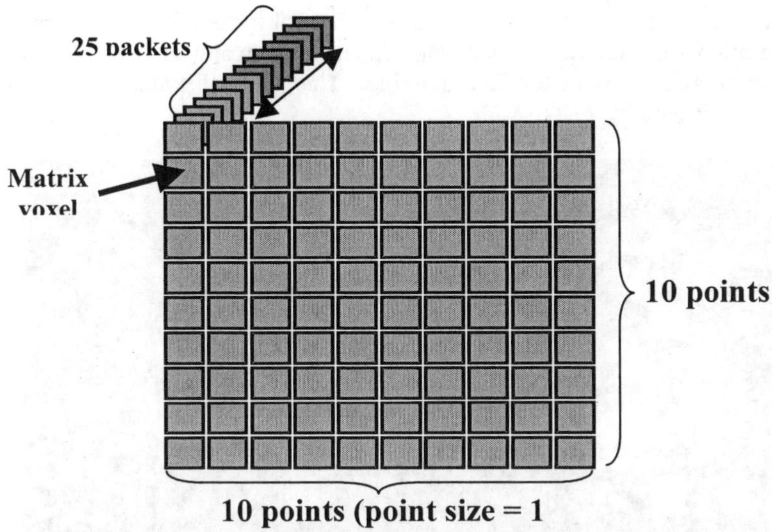

Figure 10 Geometrical strategy for data loading on a PPC, organised in matrix of Voxel implementing multiplexed packets.

So we can say that dimensions spatial and angular multiplexing are jointly used for photonic data storage on PPC layer (fig.11). For recording material modulation inside the protein layer we use wavefronts interferences. One of these wavefronts will be loaded with data to be recorded. So these data will be transformed into perturbations inside the protein layer. By wavefronts angular multiplexing it is possible to increase the number of efficient protein material quasi-molecular perturbations as tecton distribution and so to increase the data volume loaded.

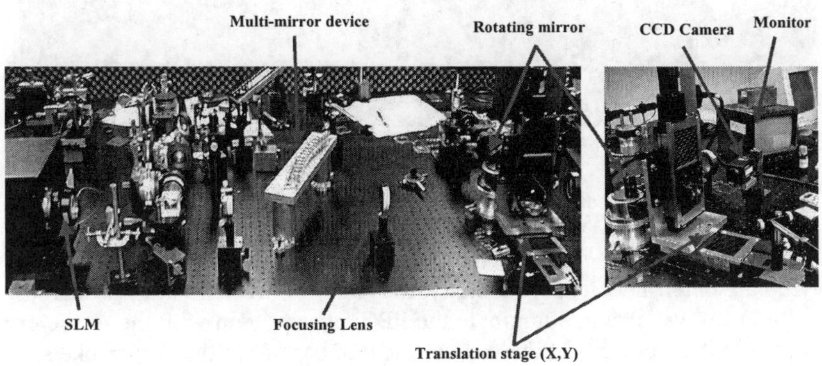

Figure 11 data storage PC recording set up overview

3.2 CROSS-TALK AND RADIOMETRIC CONTROL:

Figure 12 represent images output from our experimental reading system. In figure 12.a we have a cross-talk effect between image 1 and 2 related to packets recorded with neighbouring angle. In figure 12.b, this cross-talk has been controlled and the two neighbouring images are correctly separated, this control is done by finding an optimal angular separation between the two adjacent reference beam angles. This is done for all reference beam angles used for multiplexing. The optimal values of angles are determined experimentally.

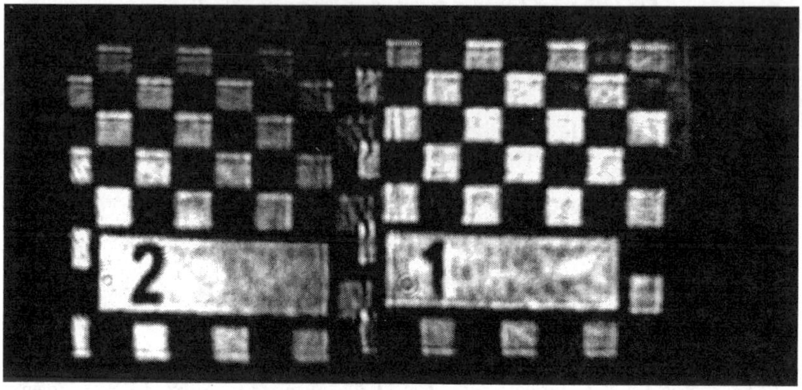

a

b

Figure 12 a: data reading With cross talk, b: without cross-talk

the values angles used in angular multiplexing to avoid any cross talk effect are summarized are determined experimentally, they vary between 10 and 75 degrees.
If the multiplexed wavefronts illuminate the PPC layer protein with the same exposure energy, the output efficiency decreases with the number of multiplexed packets.
This is represented below for 25 multiplexed patterns.

We experimentally studied the optimal radiometric values for recording every multiplexed packets to determine the optimal beam modulation allowing an uniform and stable data loaded output. In this way a data recording and reading system and process can provide good quality output. As it is shown in figures 13 and 14.

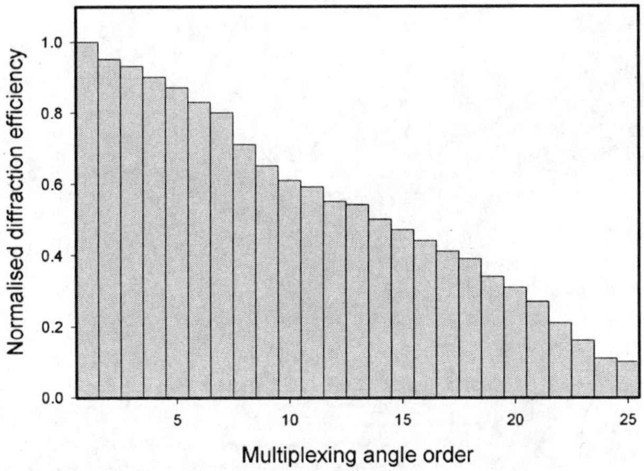

Figure 13 Histogram of diffraction efficiency without exposure correction

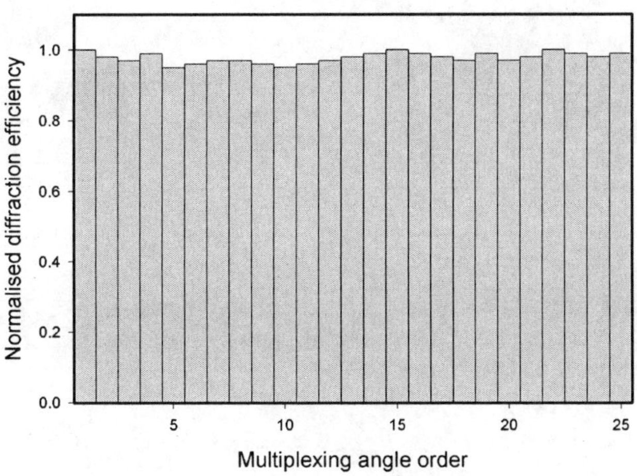

Figure 14 Histogram of diffraction efficiency with exposure correction

3.3 PPC READING SYSTEM:

The reading system will allow to address a reading beam at an appropriate XY location and appropriate angle on the PPC. We have elaborated a hardware design allowing to implement this function (figure 15).

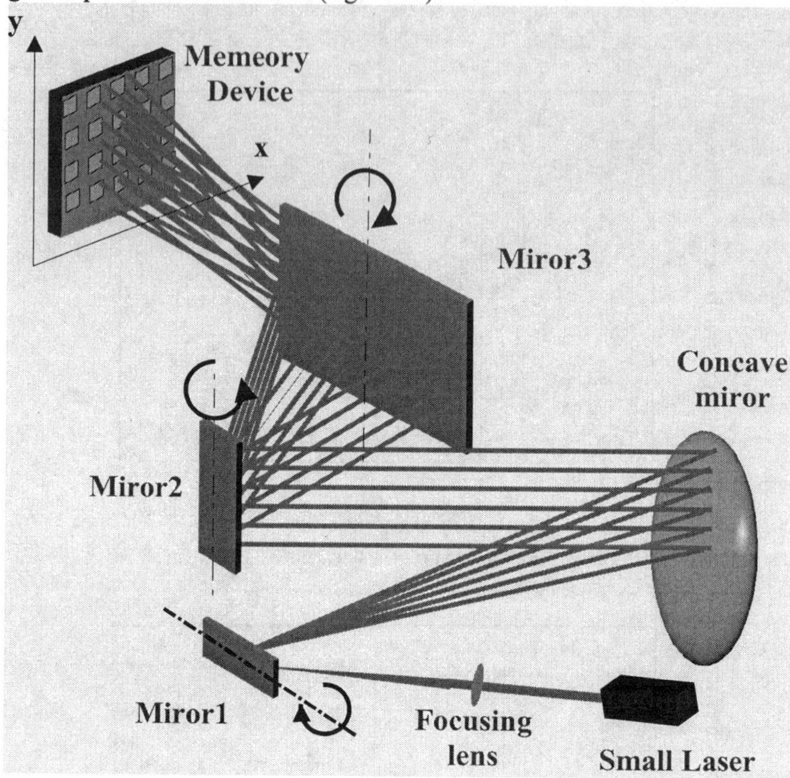

Figure 15 diagram of the experimental PC system for reading multiplexed recording

In our first prototype we use micro galvanometric mirror. Basic hardware implementation for recorded PC data reading is shown on figure 16. In other prototypes actuators are MEOMS and acousto-optic.

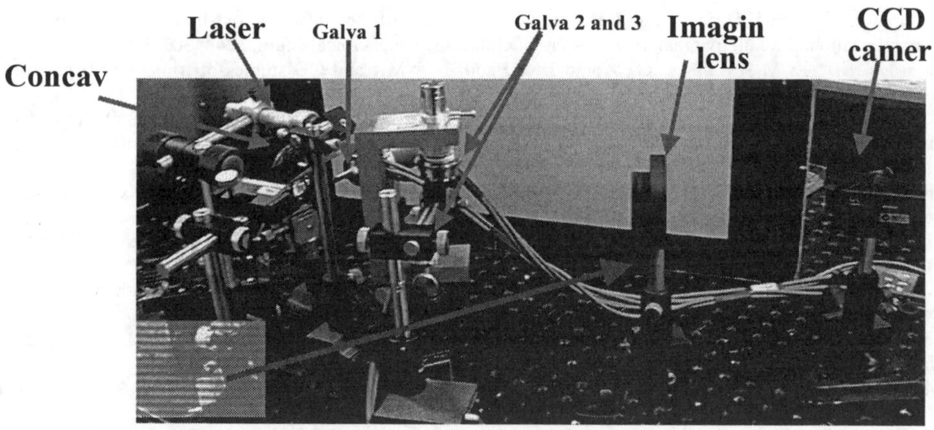

Figure 16 Overview of a PC reading system prototype

This Reading system design is based on an optimised nodal spatial architecture, optical element and actuators (Mirror1, Mirror2, Mirror3).

4. Conclusion

The protein material that we used for PPC applications was validated for data storage as an example. This protein material has a high level of performance including flexibility. Protein material was efficiently associated with the addressing technology that we have developed. Micro laser beam scanning techniques initially elaborated for the optical telecom provide satisfying reading addressing for PPC used in data storage. Many other applications are possible for producing micro optical component on PPC with a high level of performance. It will be soon possible to realise micro laser, display, router, switch, multiplexer, detector etc, on which we are working and that will be introduced in papers to come.

The authors thank DVA-Pioneer company for their support to the biophotonic research program of the photonics system laboratory[19].

References:

1. Thesis Gauthier OBRECH Louis Pasteur University Strasbourg`France, "Bacteria production of triple helix polypeptide perspective for photonics uses", June 1993.
2. B. Kress and P. Meyrueis, " digital diffractive optics", Willey.
3. J.K. Gimzewski, C. Joachim, " Nanoscale Science of Single Molecules Using Local Probes", Science, vol.283, p.1683-1688, 12 March 1999.
4. Bansal M., Ramakrishnan C. & Ramachandran "A triple-Helical Model for (Gly-Pro-Hyp)n with cis Peptides Units". Biopolymers 14, 2457-2466, 1975.
5. Germann H.P. & Heidenmann E. "A synthetic model of collagen : an experimental investigation of the triple-helix stability". Biopolymers 27, 157-163, 1988.
6. Richardson J. "Advances in protein", Chemistry 34, 174-175, 1981.

7. Barnard K., Light N.D, Sims T.J. & Bailey A.J. "Chemistry of the collagen cross-links : Origin and partial characterization of a putative mature cross-link of collagen". Biochemical Journal 244, 303-309, 1987.
8. Brodsky B., Li M-H, Long C.G., Apigo J. & Baum J. "NMR and CD studies of triple-helical peptides" Biopolymers 32, 447-451, 1992.
9. Yamauchi M., London R.E., Guenat C., Hashimoto F., Mechanic G.L. "Structure and formation of a stable histidine-based trifunctional cross-link in skin collagen". The Journal of Biological Chemistry 262, 11428-11434, 1987.
10. M.G. Moharam and T.K. Gaylord, "Chain-matrix analysis of arbitrary-thickness dielectric reflection gratings", J. Opt. Soc. Am. 72, 187-190, 1982.
11. T. Strick, J-F Allemand, V. Croquette, D. Bensimon, " The manipulation of single biomolecules ",Physics Today, p.46-51, October 2001.
12. A.D. Mehta, M.Rief, J.A. Spudich, D.A. Smith, R.M. Simmons, " Single-Molecule Biomechanics with Optical Methods ", Science, vol.283, p.1689-1695, 12 March 1999.
13. Thesis, I. El-Hafidi, " interférométrie holographique multiplexée en longueur d'onde ", Louis Pasteur University, France, 26 Nov. 1999.
14. Johnson Jr. W.C. "Secondary structure of proteins through circular dichroism spectroscopy". Annual revue of biophysic & biophys. Chemistry 17, 145-166 , 1988.
15. Blout E.R. "Fundamental aspects and recent developments in optical rotator dispersion and circular dichroism – polypeptides and proteins". Heyden & Son Ltd, Ciardelly & Salvadori éd., p 352-372, 1973.
16. Capello J. "The biological production of protein polymers and their use". Tibtech 8, 309-311, 1990.
17. Lang K., Schmid F.X. & Fischer G. "Catalysis of protein folding by propyl isomerasse". Nature 329, 268-270
18. D. Psaltis, "Parallel optical memories", Byte 17(9), 179-182, 1992.
19. ULP-DVA-PIONEER patent international application : PCT/FR01/02386, PCT/EP01/15420, PCT/EP02/05445, PCT/EP02/05454, PCT/EP02/05451, PCT/EP02/05457

HIGHLY EFFICIENT MULTIPHOTON ABSORPTION IN A NEW QUADRUPOLAR HETEROCYCLIC DYE

R. SIGNORINI, C. FERRANTE, D. PEDRON, M. SLAVIERO, R. BOZIO
Department of Physical Chemistry, University of Padova
Via Loredan 2, I-35131 Padova, Italy

L. BEVERINA, A. ABBOTTO, G. A. PAGANI
Department of Materials Science, University of Milano-Bicocca
Milano, Italy

Abstract

The multiphoton absorption properties of an organic push-molecule are investigate with nano- (ns) and femto-second (fs) laser sources. The active molecule is a new quadrupolar dye, named PEPEP, where heterocyclic units act as strong donor-acceptor moieties. Z-scan and pump-probe experiments (both single-color and with white light pulses) allow us to measure the value and the wavelength dispersion of the cross section for instantaneous two photon absorption (TPA). When Z-scan experiments are performed with nanosecond pulses, much larger effective cross sections are measured than with femtosecond pulses and with remarkably different wavelength dispersion. This is interpreted as due to two-photon absorption followed by one-photon absorption from the excited state (TPA-ESA). The TPA (or TPA-ESA) spectrum is measured, through pump-probe and z-scan measurements in a large visible range, in order to find the peak wavelength for multiphoton absorption.

Keywords: Push-pull heterocyclic dye, two photon absorption, multiphoton absorption, optical limiting

1. Introduction

During the recent years interest in new materials possessing high nonlinear (NL) optical response for advanced photonic applications has grown tremendously. Examples include materials for up-conversion lasing[1] for new laser sources, optical power limiting (OL),[2] modulation and stabilisation of laser sources,[3] optical switches and logic gates[4] and three-dimensional optical data storage and imaging.[5] OL is a NL phenomenon useful for protection against high intensity laser beams. In fact the absorption of an OL

F. Charra et al. (eds.), Organic Nanophotonics, 231–240.

material is a function of the intensity of the input beam: the NL absorption increases as the intensity increases. In this way it is possible to use OL materials to protect optical sensors and the human eye, against damages from lasers.

OL is obtained by different mechanisms, like NL absorption, NL refraction and NL scattering. Among the NL absorption mechanisms the two photon absorption (TPA) and the excited state absorption (ESA) are widely investigated. With TPA two photons are simultaneously absorbed and the absorption increases with increasing input beam intensity, while with ESA single photons are absorbed sequentially within the pulse duration, the first absorption leads to excited state population and allows a subsequent absorption. The reverse saturable absorption (RSA) is an example. In order to discriminate between pure TPA and ESA processes the choice of the laser pulse duration is a crucial point. With fast pulses (femtosecond, fs), the sequential absorption does not contribute to the measured signal, since most nonradiative processes occur on the picosecond (ps) or nanosecond (ns) timescale. When ns pulses are used, instead, two photon process and multiphoton sequential absorption may contribute to the total absorption.

Two- and multi-photon absorption of organic push-pull molecules, with heteroaromatic rings acting as efficient acceptor (A) and donor (D) moieties, have been investigated for second- and third-order NL optical activity. They allowed the preparation of highly efficient NLO-phores for frequency-upconverted lasing and OL applications. The NL optical properties of push-pull molecules have been characterized in order to investigate the relation between the structure and the resulting NL response. In fact, heteroaromatic groups offer the opportunity to vary widely the donor or acceptor strength as well as to add functionalities that favor the processing of these dyes into useful photonic materials.

In order to be useful for OL applications, the dye molecules should possess large two-photon absorption (TPA) cross-sections (σ_{TPA}). Up to a few years ago, the number of systems that showed relatively good σ_{TPA} values was very scanty. It is only very recently that a number of research results have been reported on the design and development of very efficient TPA dyes.[6] In particular, it is now apparent that a successful design strategy involves symmetrically substituted systems having a general structure of the type D-π-D and D-π-A-π-D or the complementary structures A-π-A and A-π-D-π-A, where A is an acceptor group, D a donating group, and π a conjugated moiety.

The vast majority of efficient TPA dyes now available are based on substituted conjugated benzene rings and rely on the electronic properties of primary organic simple functionalities such as amino or alkoxy groups. The lack of a systematic use of heteroaromatic compounds is very surprising if one considers that π-deficient and π-excessive heterocycles may act as very efficient acceptor and donor moieties, respectively.[7] Furthermore, the intrinsic chemical versatility of the heteroaromatic rings makes these systems particularly appropriate in order to finely tune electronic and optical properties. The precious electronic properties of heteroaromatic rings have already been successfully employed for the design of efficient push-pull derivatives for second- and third-order non-linear optical (NLO) activity.[8] We have recently reported the design of new heterocycle-based push-pull dyes[9] and prepared new highly efficient

NLO-phores with large second-order NLO activity[10] or for frequency-upconverted lasing applications.[11]

The synthesis, linear optical absorption and fluorescence, and some preliminary data on the TPA-based optical limiting activity of the novel heterocycle-based two-photon dye N-methyl-2,5-bis[1-(N-methylpyrid-4-yl)eten-2-yl]-pyrrolo triflate belonging to the general structure A-π-D-π-A, hereafter denoted as PEPEP, (Scheme 1) have already been reported.[12]

We herein provide a more detailed report of the TPA and multiphoton absorption properties for both femtosecond and nanosecond laser pulses.

2. Two-photon and multi-photon absorption properties in solution

The TPA cross section of PEPEP was been measured at 790 nm by means of open aperture Z-Scan experiments performed with a femtoseconds (fs) laser source. Open

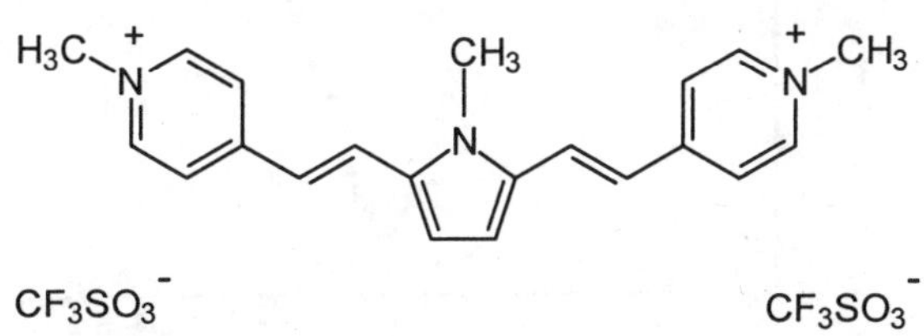

Scheme 1. Chemical structure of the TPA dye. PEPEP

aperture Z-scan[13] allows one to measure the NL absorption of the samples by recording the transmittance as a function of the input intensity. To this end, the sample position is moved along the propagation direction (z-axis) of a focused beam, whose energy is fixed, while the intensity increases by approaching the focal plane. Far from the focus position only the linear transmittance contributes to the signal, while in the vicinity of the focus the intensity increases very rapidly and the non linear absorption process generates a dip in the transmittance. The dip is symmetric with respect to the focal plane position.

This technique performed with fast (fs) laser pulses is capable of discriminating between coherent two photon processes and multiphoton sequential absorption implying excited states populated by non radiative processes after the initial TPA. Sequential absorption does not contribute to the measured signal when fast (~100 fs) pulses are employed since most nonradiative processes occur an a picosecond (ps) or nanosecond (ns) timescale. Experiments performed with nanosecond pulses, instead, may be affected by both processes.[14, 15]

The laser source was an amplified Ti-Sapphire fs laser system (Spectra Physics). It delivers ~130 fs pulses, with 0.7 mJ pulse energy, at 790 nm. By controlling the Pockels cells in the regenerative amplifier, the repetition rate was reduced to 100 Hz. The input beam was suitably attenuated by a set of neutral density filters. It was focused with a 160 mm focal length lens onto a 1 mm thick optical cell, mounted on a motorized translation stage (Newport M-UTM100P1HL), with 5 μm resolution. The transmitted beam was focused on a large area photodiode, whose signal was sampled by a 500 GHz oscilloscope (Tektronik TDS 520B). A personal computer stored the signal from the oscilloscope as a function of the sample position, and controlled the motion of the translation stage. The temporal and spatial profiles of the input beam were measured independently through autocorrelation experiments and by using a CCD camera (Pulnix TM-7CN), placed on the beam focus, respectively. The pulse energy was sampled with a pyroelectric detector (Molectron J3-05) .

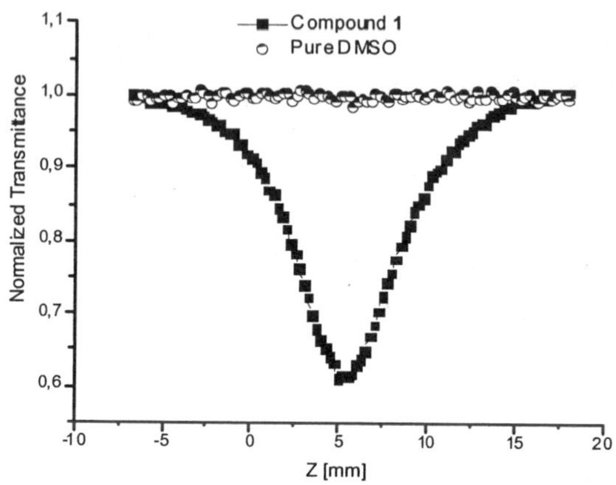

Figure 1. Z-Scan experimental data of PEPEP in DMSO (squares), with concentration of $3{\cdot}10^{-2}$ M and pure DMSO (circles), in a 1 mm cell, at the input intensity laser beam of 0.35 μJ.

PEPEP was dissolved in DMSO at $3x10^{-2}$ M concentration. At the wavelength (790 nm) of the fs pulses used for NL absorption experiments PEPEP does not absorb linearly. Open aperture Z-Scan measurements (Figure 1) were performed with input pulse energies in the range 0.05÷0.35 μJ. The upper limit is chosen so that pure DMSO does not show multiphoton absorption.

In Figure 1, the Z-Scan data of PEPEP in DMSO (squares) and of pure DMSO (circles) are reported as normalized transmittance (i.e. $I(z)/I(\infty)$, with $I(\infty)$ the linearly transmitted intensity far from the focal plane) as a function of the sample position (z).The normalized transmittance of pure DMSO is flat, indicating that NL absorption from the solvent does not occur at 0.35 μJ, which was the upper limit of pulse energy employed. PEPEP, instead, shows a deep dip typical of NL absorption.

Figure 2 reports Z-Scan data of PEPEP in DMSO, at two different input laser energies: 0.05 μJ (circles) and 0.16 μJ (squares). As expected, the depth of the dip is clearly a function of the input energy.

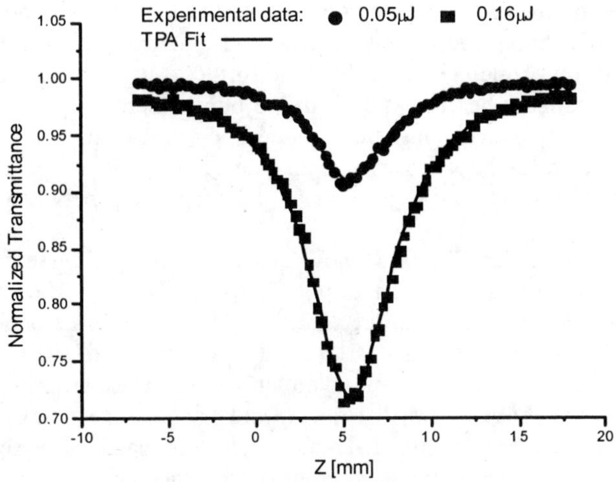

Figure 2. Z-Scan experimental data (points) and TPA fit (lines) of PEPEP in DMSO, in a 1 mm cell, with concentration of $3 \cdot 10^{-2}$ M. Input pulse energy was 0.05 μJ (circles) and 0.16 μJ (squares).

The model used to fit the Z-Scan data assumes a coherent two photon absorption process and was already thoroughly discussed in Ref.13. The function describing the TPA processes and used to fit the Z-Scan data is

$$T(z) = \sum_{m=0}^{\infty} \frac{-q_0(0)}{(m+1)^{3/2}} \frac{1}{1+(z/z_0)^m} \qquad (1)$$

where $q_0(0) = \beta I_0 L_{eff}$ is a parameter related to the TPA coefficient β, the intensity of the input beam I_0, and the effective length of the sample L_{eff}; z_0 is the Rayleigh range. The molecular two photon absorption coefficient σ, is related to β through

$$\sigma = \frac{\hbar \omega \beta}{N_{AV} C} \qquad (2)$$

where N_{AV} is the Avogadro's number and C the concentration.

The full lines in Figure 2 are examples of the fit to the Z-scan data obtained by including up to the sixth order term in the summation of Eq. (1). Using a TPA model, we obtain a good fit at both input energies. The parameters obtained from the fit of th Z-Scan data at 0.05 μJ and at 0.16 μJ coincide within experimental error: $\beta = 0.08 \pm 0.01$ cm^4/GW; $\sigma_{TPA} = (112 \pm 10) \times 10^{-50}$ cm$^4 \cdot$s/(photon·molecule). This, as well as the good fit to the shape of the Z-scan data, is consistent with the assumption that the basic mechanism for NL absorption is an instantaneous (coherent) TPA one.

To confirm this conclusion, we have performed a two-beam, time resolved TPA experiment. The fs laser beam was split into a pump beam and a probe one that were then made to spatially overlap onto the sample. The probe beam was kept at an intensity well below the threshold for observing any detectable TPA. The attenuation of the probe beam was measured as a function of the time delay between pump and probe pulses.

The data reported in Figure 3 show that absorption of photons from the probe beam is remarkable only when the two pulses overlap in time. The width of the dip in the data of Figure 3 corresponds to the autocorrelation time of the fs pulses measured in a separate SHG experiment using a BBO crystal. The shape of the transmittance dip is in fact slightly asymmetric. Such an asymmetry might arise from a weak one photon absorption from the probe beam that occurs only at positive delays. The one photon transition should start from an excited state of PEPEP that is accessed after the initial TPA process in which one photon is taken from the probe and one from the pump beam. The overall process would be a three-photon absorption, i.e. a TPA-induced ESA. Comparing the shape of the wings at negative and positive delays, one could argue that such weak one photon absorption is in fact slowly increasing with a comparatively long characteristic time.

In order to measure the wavelength dispersion of the TPA cross section, we have carried out a two-beam TPA experiment in which white light pulses, generated by using a sapphire window, were used as the probe beam. The values of the wavelength dependent TPA cross section are reported in Figure 4.

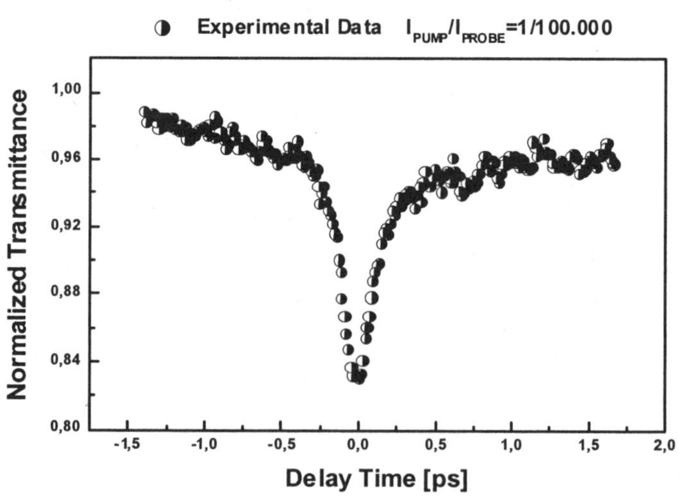

Figure 3. Two-beam, time resolved TPA data measured for PEPEP in DMSO solution.

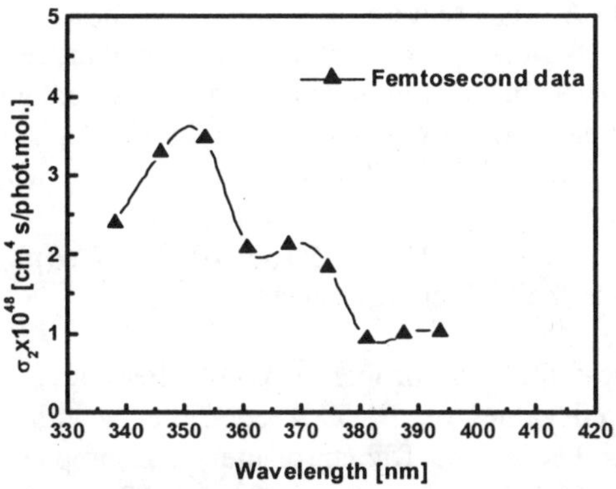

Figure 4. Wavelength dispersion of the TPA cross section measured by a two-beam experiment using white light fs pulses.

Note that the wavelength reported in the abscissa scale corresponds to the total energy of the two absorbed photons. The data show that the peak value of σ_{TPA} occurs around 350 nm, corresponding to a pump wavelength around 700 nm, that is at energies considerably higher than that of the lowest one-photon transition. More peaks could be found, and are indeed expected, by extending the TPA Spectrum towards longer wavelengths.

In view of OL applications, we have investigated the behavior of the PEPEP dye when nanosecond pulses are used instead of femtosecond ones. In fact, while the data of Figure 3 demonstrate that the main contribution to the absorption of fs pulses at 790 nm arises from an instantaneous TPA process, they also suggest that there might be contributions, growing at increasing delays after the pump pulse, due to sequential TPA ESA processes. Effective values of σ_{TPA} (which actually turn out to be intensity dependent as expected for a sequential three- or multi-photon absorption) have been measured by the Z-scan technique using 12 ns pulses delivered by an excimer-pumped dye laser with wavelength ranging from 670 to 840 nm. The wavelength dispersion of the σ_{TPA}^{eff}, measured both at low and at high pulse energies, is shown in Figure 5.

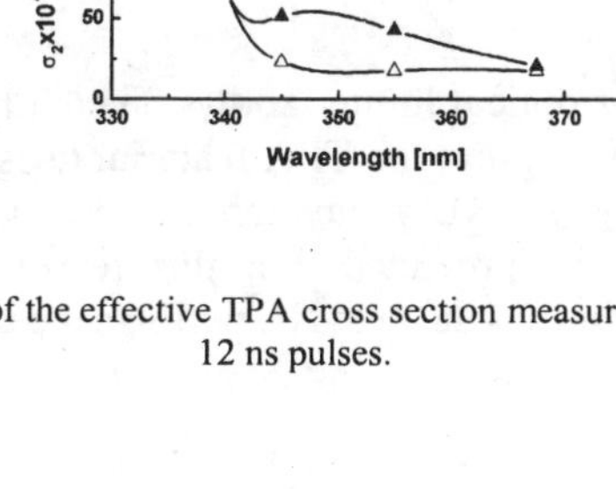

Figure 5. Wavelength dispersion of the effective TPA cross section measured by the Z-scan technique using 12 ns pulses.

The effective TPA cross sections are almost two orders of magnitude greater than those measured with fs pulses and attributed to instantaneous TPA. The shape of the spectrum is also remarkably different from that measured with white light fs pulses (Figure 4). A huge peak has grown at wavelengths shorter than the lower limit of our measurements.[16] The peak around 350 nm appears stronger when high energy ns pulses are used. For instance, considering a ns laser wavelength at 690 nm, the effective cross section increases from $2{,}320\times10^{-50}$ to $5{,}160\times10^{-50}$ $cm^4\cdot s/(photon\cdot molecule)$ in going from the low energy to the high energy spectrum.

In order to assess the TPA properties of our heterocyclic dye it is convenient to compare the measured values of the TPA coefficients with some available literature data. Table I reports the TPA coefficient of literature samples measured with fs pulses. AF-50[15] (*N,N*-Diphenyl-7-[2-(4-pyridinyl)ethenyl]-9,9-di-*n*-decyl-9*H*-fluoren-2-amine) was considered one of the best TPA sample, especially in the nanosecond regime[17]. PEPEP performs better than AF-50 and AF-250 (7-(7-benzothiazol-2-yl-9,9-diethylfluoren-2-yl)-9,9,diethylfluoren-2-yl) diphenylamina) [15] in the femtosecond regime and has similar or greater peak values of σ_{TPA}^{eff} when measure with nanosecond pulses.

TABLE I. Comparison of TPA cross sections of PEPEP in DMSO with literature data.

Femtosecond data			
	Laser pulse		σ_{TPA}
Compound	**λ (nm)**	**Duration (fs)**	**(cm^4 s/photon·molecule)×10^{50}**
PEPEP	790	130	112
	707	130	347
AF50	796	150	22
AF250	796	150	30
Stilbene 5	750	N.A.	1270
Stilbene 6	810	N.A.	3670
Nanosecond data			
	Laser pulse		σ_{TPA}
Compound	**λ (nm)**	**Duration (ns)**	**(cm^4 s/photon·molecule)×10^{50}**
PEPEP	670	12	22,750
	690	12	2,320
AF50	800	8	19,400
Stilbene 5	775	5	1,250
Stilbene 6	835	5	1,940

3. Conclusions.

The use of a TPA material as optical limiter allows one to obtain protection against laser sources in wavelength regions where the TPA material does not absorb linearly. This is one of the requirements for an NL protection device, since it does not restrict the visibility of the sensors to be protected. For this reason TPA materials have been intensively studied in order to find the best NL performances.

Within this work, a properly designed and synthesized heterocyclic push-pull molecule belonging to the general structure A-π-D-π-A, has been shown to be a very efficient multiphoton absorber. The NL optical properties have been studied with fs and ns laser pulses, through Z-scan and pump and probe measurements.

A remarkable difference between the TPA coefficients measured with fs pulses and with ns pulses has been observed.

The cross sections obtained with fs pulses, are independent of the input intensity, as appropriate for an instantaneous TPA process. A clear evidence of this is also provided by the time resolved pump-probe data. The peak value of σ_{TPA} is observed around 350 nm.

For nanosecond pulses the multiphoton absorption shows the largest effective cross section, more than one order of magnitude larger than the fs value. This fact can be explained by considering the different pulse duration. The cross sections obtained with fs pulses take into account only instantaneous, coherent TPA contributions, whereas with ns pulses sequential absorption following the TPA (TPA-ESA) contributes to the total absorption. A complex dependence on intensity is observed with ns pulses. The understanding of this behaviour requires a deeper knowledge of the excited state dynamics which is presently pursued in our laboratory. A peak at 350 nm, similar to that observed with short pulses, is observed only at higher pulse energies. The largest value is obtained at short wavelengths, where a possible contribution from one photon absorption and RSA may arise.

The comparison with TPA materials reported in literature shows, finally, that PEPEP turns out to be one of the best two-photon absorbers in the visible range.

4. Acknowledgements

This work was supported by the Italian National Research Council (C.N.R.) through the program "Materiali Innovativi (legge 95/95)", by the Ministry of Education and Research (MIUR - PRIN 2001) and by the University of Padova.

5. References

1. (a) Mukherjee, A., (1993) Two-Photon pumped upconverted lasing in dye doped polymer waveguides, *Appl. Phys. Lett.*,**62** (26), 3423-3425. (b) Gvishi, R.;He, G.S.; Prasad, P.N.; Narang, U.; Li, M.; Bright, F.V., Reinhardt, B.A.; Bhatt, J.C.; Dillard, A.G. (1995) Spectroscopic Studies if New Blue Laser Dyes in Tetrahydrofuran Solution and Composite Glass, *Appl. Spectr.*, **49** (6), 834-839.
2. (a) Harter, D.J.; Shand, M.L.; Band, Y.B. (1984) Power/Energy limiter using reverse saturable absorption, *J. Appl. Phys.*, **56** (3), 865-868. (b) Tutt, L.W.; McCahon, S.W. (1990) Reverse saturable absorption in metal cluster compounds, *Optics Letters*, **15** (12), 700-702.
3. He, G.S.; Weder, C.; Smith, P.; Prasad, P.N (1998) Optical Power Limiting and Stabilization Based on a Novel Polymer Compound, *IEEE J. of Quant. Electr.*, **34** (12), 2279-2285.
4. Li, C.; Zhang, L.; Wang, R.; Song, Y., Wang, Y. (1994) Dynamics of reverse saturable absorption and all-optical switching in C_{60}, *J. Opt. Soc. Am. B*, **11** (8), 1356-1360.
5. (a) Sonnleitner, M.; Schütz, G.J.; Schmidt, Th. (1999) Imaging individual molecules by two-photon absorption, *Chem. Phys. Lett.*, **300**, 221-226. (b) Cumpston, B.H.; Ananthavel, S.P.; Barlow, S.; Dyer, S.; Ehrlich, J. E.; Erskine, L.L.; Heikal, A.A.; Kuebler, S.M.; Sandy Lee, I.Y.; McCord-Maughon, D.; Qin,

J.; Röckel, H.; Rumi, M.; Wu, X.-L.; Marder, S. R.; Perry, J. W., (1999) Two-photon polymerization initiators for three-dimensional optical data storage and microfabrication, *Nature*, **398**, 51-54.

6. (a) Albota, M.; Beljonne, D.; Bredas, J. L.; Ehrlich, J. E.; Fu, J. Y.; Heikal, A. A.; Hess, S. E.; Kogej, T.; Levin, M. D.; Marder, S. R.; McCord-Maughon, D.; Perry, J. W.; Röckel, H.; Rumi, M.; Subramaniam, C., Webb, W. W.; Wu, I. L.; Xu, C. (1998) Design of Organic Molecules with Large Two-Photon Absorption Cross Sections, Science, **281**, 1653-1656. (b) Reinhardt, B. A.; Brott, L. L.; Clarson, S. J.; Dillard, A. G.; Bhatt, J. C.; Kannan, R.; Yuan, L. X.; He, G. S.; Prasad, P. N. (1998) Highly Active Two-Photon Dyes: Design, Synthesis, and Characterization toward Application, *Chem. Mat.*, **10**, 1863-1874. (c) Barzoukas, M.; Blanchard-Desce, M. J. (2000) Molecular engineering of push–pull dipolar and quadrupolar molecules for two-photon absorption: A multivalence-bond states approach, *Chem. Phys.*, **113**, 3951-3959.
7. Katritzky, A. R. (1983) Handbook of Heterocyclic Chemistry; Pergamon Press: Oxford.
8. Characterization Techniques and Tabulations for Organic Nonlinear Optical Materials; (1998) Kuzyk, M. G., Dirk, C. W., Eds.; Marcel Dekker: New York,.
9. (a) Abbotto, A.; Bradamante, S.; Pagani, G. A., (2001) Pyridoneimines and pyridonemethides: Substituent- and solvent-tunable intramolecular charge transfer and geometric isomerism, *J. Org. Chem.* **67**, 8883-8892. (b) Bradamante, S.; Facchetti, A.; Pagani, G. A., (1997) Heterocycles as donor and acceptor units in push-pull conjugated molecules, *J. Phys. Org. Chem.*, **10**, 514-524.
10. (a) Abbotto, A.; Bradamante, S.; Facchetti, A.; Pagani, G. A., (1997) Facile, regioselective synthesis of highly solvatochromic thiophene-spaced N-alkylpyridinium dicyanomethanides for second-harmonic generation, *J. Org. Chem.* **62**, 5755-5765. (b) Abbotto, A.; Facchetti, A.; Pagani, G. A.; Yuan, L. X.; Prasad, P. N., (1997) Enhanced second-order optical non-linearity of thiophene-spaced N-alkylpyridinium malononitrile zwitterions, *Gazz. Chim. Ital,* **127**, 165-166. (c) Abbotto, A.; Bradamante, S.; Facchetti, A.; Pagani, G. A.; Ledoux, I.; Zyss, J., (1998) *Mat. Res. Soc. Symp. Proceed.* **488**, 819.
11. (a) Abbotto, A.; Beverina, L.; Bozio, R.; Bradamante, S.; Pagani, G. A.; Signorini, R. (2001) Heterocycle-based materials for frequency-upconverted lasing, *Adv. Mater. Synth. Met.* **121**, 1755-1756. (b) Abbotto, A.; Beverina, L.; Bozio, R.; Bradamante, S.; Ferrante, C.; Pagani, G. A.; Signorini, R., (2000) Push-pull organic chromophores for frequency-upconverted lasing, *Adv. Mater.* **12** (24), 1963-1967.
12. A. Abbotto, L. Beverina, R. Bozio, A. Facchetti, C. Ferrante, G.A. Pagani, D. Pedron, and R. Signorini, (2002) Novel heterocycle-based two-photon absorbing dyes, *Org. Lett.* **4**, 1495-1498.
13. M.Sheik-Bahae, A.A.Said, T.-H.Wei, D.J.Hagan, E.W.Van Stryland, (1990) Sensitive Measurements of Optical Nonlinearities Using a Single Beam, *IEEE J. Quantum El.* **26**, 760-769; T.H.Wei, D.J.Hagan, M.J.Sense, E.W.Van Stryland, J.W.Perry, D.R.Coulter (1992) Direct measurement of nonlinear absorption and refraction in solutions of phthalocyanine, *Appl.Phys.B* **54**, 46-51.
14. R.A.Negres,E.W.Van Stryland,D.J.Hagan,K.D.Belfield,K.J.Schafer,O.V.Przhonska,B.A.Reinhardt, (1999) Nonlinear spectrometer for characterization of organic and polymeric molecules, *SPIE* **3796**, 88-97.
15. J.Swiatkiewicz, P.N.Prasad, B.A.Reinhardt, (1998) Probing two-photon excitation dynamics using ultrafast laser pulses, *Opt.Comm.* **157**, 135-138.
16. We cannot rule out the possibility that different mechanisms contribute to such an enhancement. For wavelengths shorter than ca. 680 nm the laser enters the far tail of the one-photon ground state absorption. Under these circumstances a conventional RSA mechanism may become operative adding to the one-photon enhanced instantaneous TPA and to the TPA-ESA.
17. B.A.Reinhardt, L.L.Brott, S.J.Clarson, R.Kannan, A.G.Dillard, (1997) Optical Power Limiting in Solution Via Two-Photon Absorption: New Aromatic Heterocyclic Dyes with Greatly Improved Performance, *SPIE* **3146**, 2-11; B.A.Reinhardt, L.L.Brott, S.J.Clarson, R.Kannan, A.G.Dillard, (1997) The Design and Synthesis of new organic molecules with large two-photon absorption cross-sections for optical limiting applications, *Mat. Res. Soc. Symp. Proc.* **479**, 3-8; M.K.Casstevens, R.Burzynski, J.F.Weibel, C.Spangler, G.He, P.N.Prasad, (1997) A New Classe of Materials for Optical Power Limiting, *SPIE* **3146**, 152-159; B.A.Reinhardt, L.L.Brott, S.J.Clarson, A.G.Dillard, J.C.Bhatt, R.Kannan, L.Yuan, G.He, P.N.Prasad, (1998) Highly active two-photon dyes: Design, synthesis, and characterization toward application, *Chem.Mater.* **10**, 1863-1874.

NANOPRECISE SELF-ASSEMBLY OF ELECTRO-OPTIC AND ELECTROLUMINESCENT MOLECULAR ARRAYS

QINGLAN HUANG, HU KANG, JONATHAN VEINOT, HE YAN, PEIWANG ZHU, AND TOBIN J. MARKS
Department of Chemistry and the Materials Research Center
Northwestern University, Evanston IL 60208-3113 USA

1. Introduction

Although the current state of molecular synthesis has reached an impressive level of sophistication in the past decades, the ability of scientists to organize array of engineered molecular building blocks to achieve certain specific materials functions is currently in its infancy. Of particular attraction would be pathways which led via spontaneous self-organization processes to robust arrays of molecules with designed electronic and/or opto-electronic characteristics and spatial relationships. Ideally, the structural control of the assembly process would be at the sub-nm level and the structures would "click" into place after initial self-assembly. In the present contribution, we briefly discuss two complementary efforts to achieve such capabilities. In one effort, we focus on self-assembly routes to fabricate nanoscopically acentric arrays of high-hyperpolarizability building blocks for electro-optic function and in the other on the nanoscopic interfacial phenomena that tune charge injection and radiative recombination efficiencies in electroluminescent molecule-based organic heterostructures. Each approach makes use of well-characterized siloxane condensation chemistry, which can be used in successions of self-limiting chemisorptive processes, to deposit conformal, adherent, and virtually pin hole-free layers of molecular entities with sub-nm precision. An important consideration here is to apply a battery of physical characterization techniques to precisely define nanostructure and to correlate this with photonic response. In this account, we review recent progress in: 1) self-assembled superlattices (SASs) with very large electro-optic response, 2) organic light-emitting diode (OLED) structures with self-assembled charge injection/adhesion layers, 3) emissive organic nanodiodes.

F. Charra et al. (eds.), Organic Nanophotonics, 241–263.

2. Electro-Optic Materials Based on Nanoscale Self-Assembled Superlattices

Molecule-based photonic materials represent a promising direction in the scientific quest to develop novel electro-optic (EO) modulators for greatly increased rates of information transmission by enhancing optical network speed, capacity, and bandwidth.[1,2] Non-centrosymmetry is one of the basic requirements of such materials. Currently, three principal approaches are being used to achieve molecular orientation: electric-field poling of glassy polymers[3], Langmuir-Blodgett film (LB) transfer[4], and layer-by-layer self-assembly (SA).[5-8] In the first, high-hyperopolarizabilty chromophores are either doped into or covalently bonded to a polymer to fabricate films. An external electric field is then applied while the films are heated near the glass transition temperature (Tg) to align the chromophore dipoles in the direction of the field. It is a straightforward procedure to fabricate thick poled films. However, the drawbacks are: 1) the modest orientation achieved by poling is intrinsically unstable in the absence of the field; 2) strong dipole-dipole interactions among chromophore molecules limit chromophore concentrations; 3) micro-domains formed during poling can increase the optical loss in waveguiding devices. For the LB films, only limited chromophores with long alkyl groups are applicable. Since weak van der Waals interactions are the main structural driving force, polar orientation becomes progressively randomized as films become thicker. Other drawbacks include low NLO response and poor mechanical strength. In contrast, layer-by-layer self-assembly approaches [5-8] are particularly attractive because: i) self-assembled superlattices (SASs) offer much higher EO coefficients and lower dielectric constants than those established inorganic materials (e.g., $LiNbO_3$);[9] ii) the chromophore molecules are covalently linked to the substrates, and the films are closely packed and robust; iii) molecular orientation is intrinsically acentric, and post-deposition processing steps such as poling are not necessary; iv) chromophore orientation is locked into place with strong covalent crosslinks, hence the microstructural orientation is very stable; v) precise control of film refractive index can be achieved at the sub-wavelength molecular level; vi) large area films can be easily prepared; vii) chromophore arrays can be fabricated on silicon or a variety of related substrates, allowing ready device integration and therefore significantly reduced device design complexity.[7,10]

Previous studies[6,7,11] developed a three-step procedure to fabricate robust, acentric multilayer superlattices composed of stilbazolium and related chromophores, and exhibiting very large NLO / EO responses ($\chi^{(2)}$ =150-200 pm/V). Devices such as ultrafast frequency-selective switching devices,[12] TE mode EO-phase modulators,[7a-c]and frequency doubling devices[7d] have been fabricated with such SAS materials. These results stimulated our interest in elaborating and refining layer-by-layer deposition methodologies for EO film growth. Since the thickness required for optimum single mode waveguiding EO film devices should be ~ 0.5 μm or more, at least 150 SAS trilayers (450 growth steps) are required for this three-step procedure. Therefore, to be practical, the procedure must be automated and should equal or surpass other well-established layer-by-layer growth techniques (e.g., LB methods) in terms of film microstructural regularity, growth efficiency, and flexibility. In the three-step "first-generation" procedure,[11] a coupling layer is first self-assembled onto substrates. Secondly, the chromophore is covalently linked to the coupling layer through a quaternization reaction. Thirdly, a planarizing capping layer is deposited. Since the quaternization reaction is very slow, spin-coating followed by vacuum oven treatment at 110 °C is necessary to obtain full chromophore coverage.[11c,d] This step is a serious impediment to automating the growth process.

A greatly improved all-"wet-chemical" approach to the expeditious self-assembly [13] of these large-response optically/-electronically functional siloxane-based SAS films has recently been developed using reactive $-SiCl_2I$ derivatives of a protected azobenzene chromophore and a chlorosiloxane deprotecting/capping reagent (Scheme I). The deposition technique reported here employs the iterative combination of *only two steps*: (i) Self-limiting polar chemisorption of a protected high-β chromophore monolayer (*t* ~15 min.), and (ii) *in situ* trialkylsilyl group removal and self-limiting capping of each chromophore layer using octachlorotrisiloxane (*t* ~25 min.; Scheme I). This latter step deposits a robust polysiloxane layer which stabilizes the polar microstructure via interchromophore crosslinking and regenerates a reactive hydrophilic surface.[11,14] This two-step, layer-by-layer approach affords thermally robust[13] superlattices exhibiting large NLO/EO responses ($\chi^{(2)}$ ~180 pm/V, r_{33} ~65 pm/V at 1064 nm) and can be efficiently carried out in a single reaction vessel using standard cannula techniques or by consecutive vertical dipping of the substrates in a simple nitrogen-filled glovebag. Since only "wet-chemical" processes are involved, it is easy to automate. The film assembly time is now comparable to LB and related layer-by-layer deposition techniques,[4,8] thus offering an efficient route to thicker self-assembled EO films.

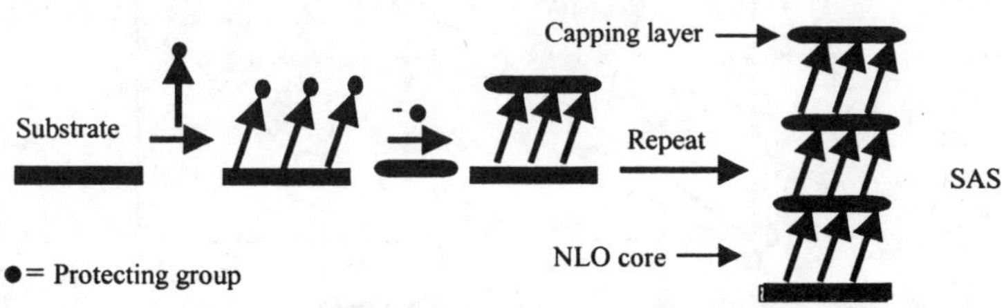

Scheme I. Schematic representation of the two-step layer-by-layer self-assembly of self-assembled superlattices.

2.1. SAS MICROSTRUCTURAL CHARACTERIZATION. X-RAY REFLECTIVITY.

Specular x-ray reflectivity was used to probe microstructural evolution and details (e.g., film thickness, density, and interfacial roughness etc.) of the SAS films. Data for a series of 1-5 bilayers were obtained by fitting the multilayer reflectivity data to a Gaussian-step model.[11c-e,15] The SAS x-ray reflectivity data exhibit pronounced minima, "Kiessig fringes" (Figures 1; inset),[11c-e,16] corresponding to destructive interference of reflections from the film multilayer substructure and indicating a narrow distribution of layer thicknesses. Variation in layer thicknesses would smear out the first-order fringe [16], which is not observed. It can be seen that the film thickness (D_{film}) increases linearly as a function of the number of bilayers (Figure 1), underscoring the high structural regularity and efficiency of this synthetic approach. An average interlayer spacing of 32.6 Å can be deduced from the slope of the line obtained by linear regression. The "Bragg" peak in the reflectivity due to scattering by identical layers also yields an interlayer spacing of about 32 Å (Figure 1; inset), meaning that these films are extremely well-ordered in the surface normal direction, although there are small density variations between the interlayers.

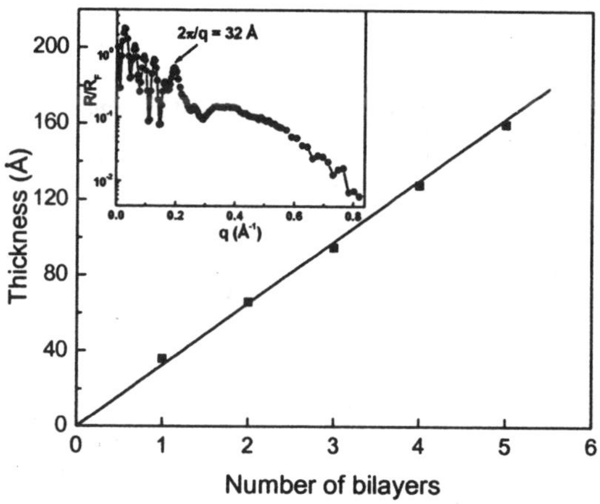

Figure 1. Specular x-ray reflectivity (XRR)-derived film thickness (Å) data as a function of the number of chromophore-based bilayers prepared via the procedure of Scheme I. The solid line is the fit by linear regression indicating an average interlayer spacing of 32.6 Å. Inset: XRR data for a 5-bilayer sample. The"Bragg" feature independently yields an interlayer spacing of ~ 32 Å.

"Bragg" peaks are observed in highly ordered LB films and covalently bonded self-assembled multilayers.[11c-e,16] It is known that relatively small structural variations in layered structures significantly broaden or eliminate the Bragg features. The present measurements further reveal that the films have well-defined interfaces, consistent with good structural ordering as also indicated by the UV-vis and SHG data (Figures 2, 3). Based on the electron density profiles obtained, we have calculated the number of electrons per unit of substrate area for one bilayer, $N_{film,exp}$, according to: $N_{film,\exp} = \int \rho(z)dz$, where the integration is taken over the whole film. The molecular footprint is calculated as $N_{mol}/N_{film,exp}$, where N_{mol} is the calculated number of electrons in one molecular unit, using the chemical formula of the chromophore and the capping layer. In the case of two or more bilayers, $N_{film,exp}$ is calculated for the entire film and N_{mol} is the quantity of electrons in one molecular unit (= chromophore + capping layer) multiplied by the number of bilayers. Importantly, these estimations show that the molecular 'footprint' of ~40 Å/chromophore remains constant upon multilayer formation.

2.2. SAS MICROSTRUCTURAL CHARACTERIZATION. OPTICAL AND SHG SPECTROSCOPY

Homogeneous substrate coverage and deepening of the purple color of the SAS films assembled on transparent surfaces upon repeating the two deposition steps can be clearly observed by eye. UV-*vis* measurements unambiguously evidence a linear dependence of the HOMO-LUMO CT chromophore optical absorbance at $\lambda = 575$ nm on the number of bilayers, demonstrating that essentially equal quantities of uniformly aligned chromophore units are incorporated in each sublayer up to 80 bilayers (Figure 2). No

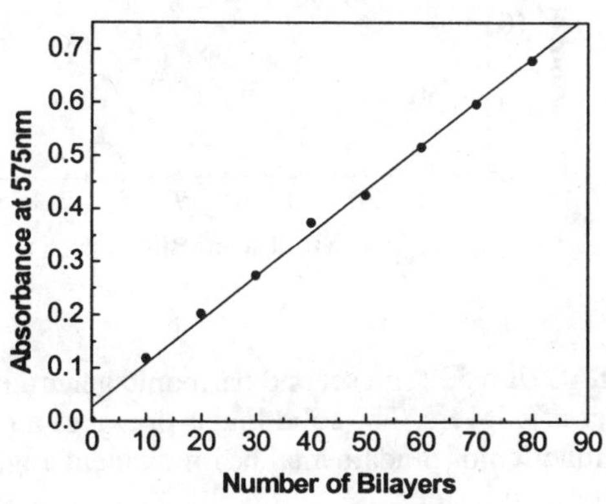

Figure 2. Optical absorption of SAS films at 575 nm as a function of the number bilayers. The solid line is the fit by linear regression.

other bands or shifts in the optical absorbance maxima are observed, arguing against significant chromophore aggregation or decomposition.

Polarized transmission SHG measurements on the present SAS structures at λ_o = 1064 nm were carried out on samples placed on a computer-controlled rotation stage, enabling the incidence angle of the input radiation to the sample surface normal to be varied. Angle-dependent SHG interference patterns for glass substrates coated on both sides with SAS films demonstrate that the present deposition process affords identical film quality and uniformity on both sides of the substrate (Figure 3; inset). A quadratic dependence of the 532 nm output intensity ($I^{2\omega}$) on the number of bilayers is observed for thin samples (< 40 bilayers; Figure 3). For thicker samples, SHG response intensities exhibit a saturation effect which has been observed in other SA materials [17] and presumably arises from self-absorption of the SHG light. For a regular, thin polar multilayer the light output intensity ($I^{2\omega}$) should scale quadratically with the number of layers.[18] The multilayer SAS films exhibit large second-order macroscopic EO responses, $\chi^{(2)}$ ~180 pm/V, which doubtless reflects non-negligible resonant enhancement

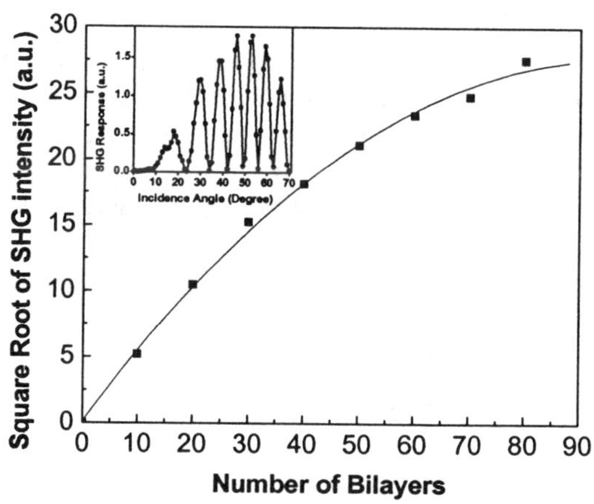

Figure 3. Square-root of SAS film 532 nm second harmonic generation (SHG) response as a function of the number of bilayers. The solid line is drawn as a guide to the eye. Inset: SHG response as a function of fundamental beam incident angle from a float glass slide having a SAS bilayer on both sides.

as established previously.[11,19] Since the present films are too thin for measurements of the electro-optic coefficient by conventional reflectometry,[20] the EO coefficient r_{33} was estimated from the SHG response to be ~ 65 pm/V at 1064 nm by calibration vs quartz and known EO-films using the established relationship $r_{33} = -2\chi^{(2)}_{zzz} / (n_z)^4$.[21,22] The chromophore number density is estimated using $N_s = (l \times \chi^{(2)}_{zzz})/(\beta_{zzz} \times \cos^3 \Psi)$.[11c] This relationship of course assumes a narrow distribution of molecular tilt angles. Note however, that the SHG-derived tilt angle, $\Psi \sim 40°$, is close to the so-called magic angle.[22] An average chromophore surface density, N_s ~2.5 × 10^{11} molecules/cm^2, is estimated for each bilayer using the ZINDO-derived chromophore molecular hyperpolarizability β_{zzz} at 1064 nm,[11c] the experimental $\chi^{(2)}_{zzz}$, and the average interlayer spacing (l = 32.6 × n Å; n = number of bilayers). The derived N_s corresponds to an average "footprint" of ~ 40 Å^2/chromophore, which agrees with the XRR-derived "footprint" (vide supra) and those reported for SA and LB films containing structurally similar chromophore building blocks.[11,23]

2.3. SAS MICROSTRUCTURAL CHARACTERIZATION. ATOMIC FORCE MICROSCOPY

Contact mode AFM measurements on 10 and 80 bilayer SAS samples reveal a slightly grainy texture and an rms surface roughness of ~3.3 and ~6.5 nm, repectively, for 1 × 1 μm^2 scan areas (Figure 4). Grain formation is a common,[11] but still poorly understood phenomenon in films fabricated using layer-by-layer deposition processes. Scanning force microscopy studies of related organic-inorganic hybrid films prepared with group 13 oxide interlayers and group IV metal-coordination based SASs reveal somewhat similar grain-type topographies.[11]

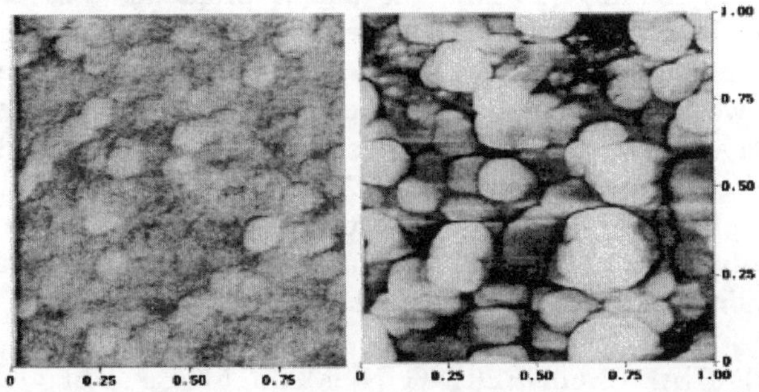

Figure 4. AFM image at 1 × 1 µm scan area of an SAS structure composed of 10 (left) and 80 bilayers (right).

2.4. SAS SUMMARY

We have demonstrated the expeditious deposition of structurally regular, polar films consisting of up to 80 alternating chromophore and capping layers. The efficiency of growth is ~ 10x that previously achieved. The present film quality, assembly time, chromophore density and microstructural polarity, and experimental simplicity are comparable to well-known layer-by-layer techniques, which are based on relatively weak interactions between chromophore layers (i.e., hydrogen bonding, van der Waals interactions, or electrostatic forces). The large numbers of strong, covalent Si–O bonds renders the present films chemically and durable thermally. Indeed, for such a quality of film microstructure, robustness, and morphology, this growth rate rivals or exceeds that of other typical solution phase layer-by-layer growth process.[4,8] No post-deposition steps such as crosslinking of specially designed building-blocks to enhance thermal stability are required. Moreover, film deformation due to evaporation of included solvents is not observed. The potentially automatable process demonstrated here with a model azobenzene chromophore should be applicable to a wide range of electro-optic and other chromophores, and provides an unprecedented level of accessibility to intrinsically polar molecular EO materials for future photonic devices.

3. Nanoscale Modification of Oled Electrode-Organic Interfaces. Structure and Function.

In the intense pursuit of next-generation electroluminescent display technologies, nanoscale interfacial tuning represents a promising and intriguing area to improve organic light-emitting diode (OLED) performance in terms of brightness, quantum efficiency, and stability [24-26]. The unique properties associated with interfaces in OLEDs have been revealed from the electronic, chemical, and morphological points of view. These properties have been shown to be crucial to OLED response characteristics, stimulating worldwide studies to gain control over them, including insertion of Å-scale insulating layers (i.e., LiF [27, 28], CsF [29], alkali metal acetates[30]) between the cathode and electron transport layers (ETLs), thin layers of various organic acids [31, 32], copper phthalocyanine [33], dielectrics [34], metals [35], or thicker (300-1000 Å) layers of polyaniline [36] or polythiophene (PEDOT) [37] between the anode (usually tin-doped indium oxide, ITO) and hole transport layers (HTLs).

3.1. INTERFACIAL ENGINEERING IN SMALL MOLECULE OLEDS

We find that layer-by-layer self-limiting chemisorptive organosiloxane-based self-assembly enables the growth of triarylamine-based charge transport layers with sub-nm control over total film thickness, as exemplified by using a silyl-functionalized triarylamine as the precursor (Figure 5). These self-assembled hole transport thin films were characterized by atomic force microscopy, synchrotron x-ray reflectivity, cyclic voltammetry, ultraviolet photoelectron spectroscopy, and spectroelectrochemical analysis, and reveal the formation of smooth, adherent, essentially pinhole-free films with electrical properties comparable to those of similar vapor-deposited hole transport layers [38, 39]. OLEDs fabricated with these films as ITO- HTL interfacial layers exhibit considerably higher luminous intensities and thermal stabilities than bare ITO-based devices, supporting the validity of the interface tuning concept. Interestingly, OLEDs with the commonly used ITO buffer layer, copper phthalocyanine (Cu(Pc)), pale by comparison to the aforementioned devices in terms of thermal robustness, luminance, and external quantum efficiency (Figures 6,7). In devices having the structure ITO/TAA- or $TPDSi_2$- /TPD/Alq/Al, a 10x increase in light output, as well as a ~3-4 × external quantum efficiency enhancement is observed compared to devices having the structure ITO/Cu(Pc) /TPD/Alq/Al. Thermal stress tests were performed by heating the devices at 90 °C for 0.5 h in vacuum. A catastrophic degradation at the Cu(Pc)-TPD interface occurs for Cu(Pc)-based devices, revealed by optical microscopy [40]. In contrast, this nanoscale decohesion is completely suppressed by TAA or $TPDSi_2$ interfacial modifications, which stabilize the HTL-TAA or HTL-$TPDSi_2$ interface (Figure 8).

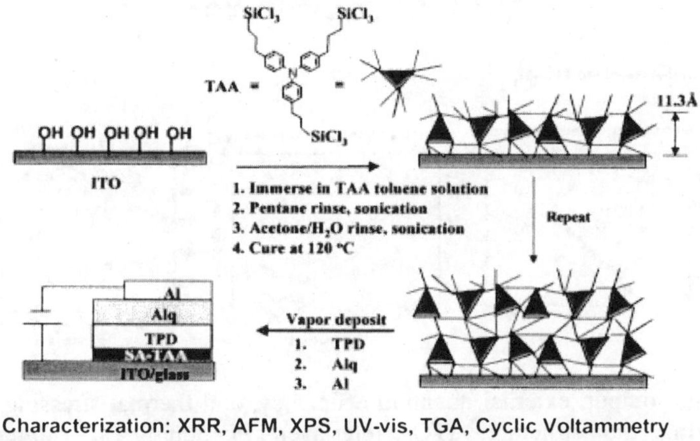

Figure 5. Self-assembly process for the layer-by-layer deposition of a crosslinked triarylamine hole transport layer (exemplified by TAA) for use as an OLED charge injection/adhesion layer.

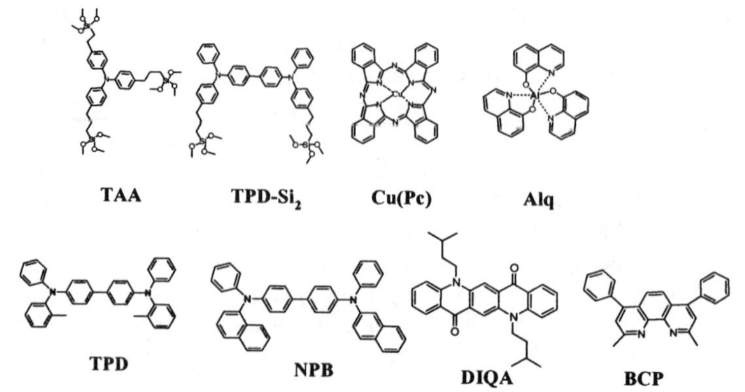

Figure 6. Organic molecular structures used in small molecule OLEDs.

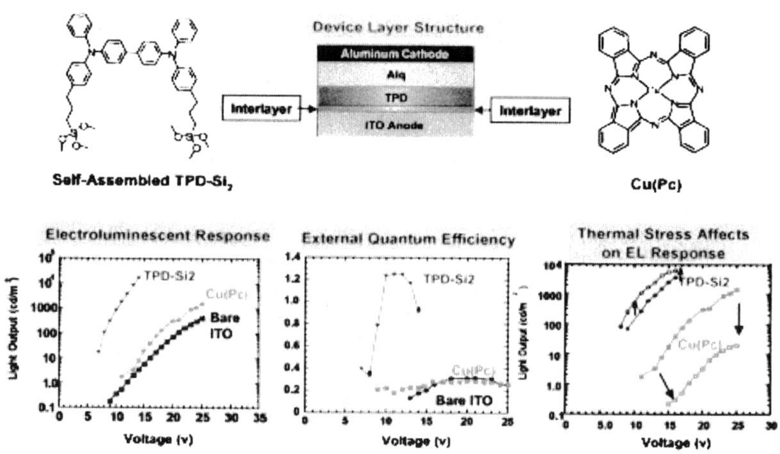

Figure 7. Light output, external quantum efficiency, and thermal stressing response for OLEDs having the structure: ITO/ interlayer/TPD(50nm)/Alq (60nm)/Al, where injection/adhesion interlayer = Cu(Pc) or TPDSi$_2$. The arrows in the thermal stressing data indicate change of OLED light output vs. voltage characteristics after heating at 90 °C under vacuum for 1 h.

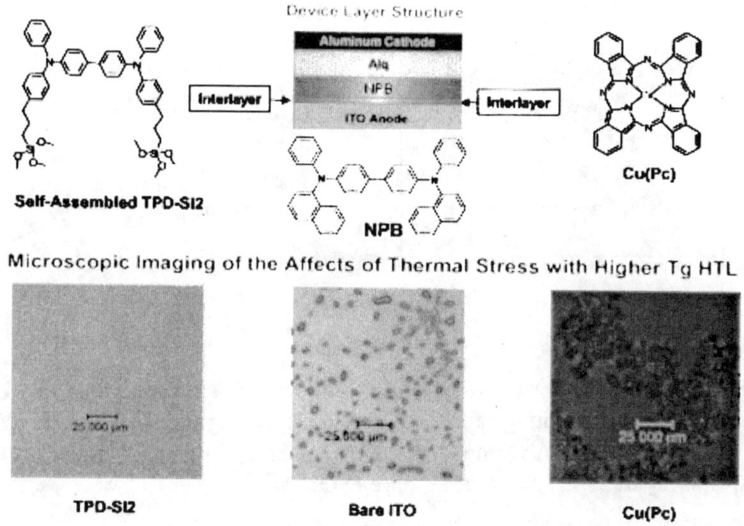

Figure 8. Optical images of the effects of a self-assembled interfacial layer on the stability of the OLED anode-HTL interface vs. that of bare ITO and the commonly used buffer layer, copper phthalocyanine.

To explore the extent of nanoscale tuning on ITO-HTL interfacial properties, four ITO substrates with 1, 2, 4, and 10 self-assembled TAA layers were used for device fabrication. The deposition of organic molecules and metal cathodes on these substrates was performed in a single run under identical experimental conditions. In devices with structure of ITO/TAA(n layers)/NPB/Alq:DIQA/BCP/Li/Al, introduction of $1 \rightarrow 2 \rightarrow 4 \rightarrow 10$ SA-TAA layers at the ITO-HTL interface results in progressive maximum luminance increasing to up to 80,000 cd/m^2. The external quantum efficiencies are altered by different number of TAA layers also, with 10-layer TAA-based devices having as high as ~4% efficiency (Figure 9). These results prove a strong correlation between interface nanostructures and OLED performances, which is very sensitive to the nanoscale thickness and structure in the interfacial region. Another technique to study the

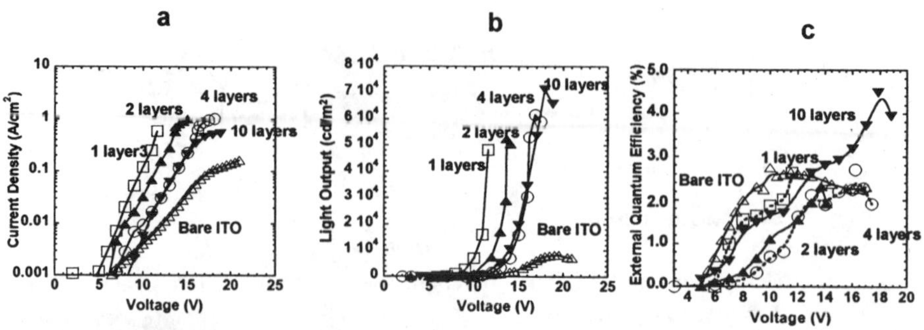

Figure 9 (a-c). Current density, light output and external quantum efficiency as a function of operating voltage for OLED devices having the structure: ITO/TAA (n layer)/ NPB (20nm)/Alq :DIQA (60nm)/BCP/Li/Al, where n = 1, 2, 4, and 10. □, 1 layer; ▲, 2 layers; ○, 4 layers; ▼, 10 layers; Δ, bare ITO.

effects of nanoscale OLEDanode-organic interface modification is to study the degree to which hole injection characteristics are modulated using "hole-only" devices which block electron injection and only allow holes to be injected. As shown in Figure 10, the effects of TAA and TPD-Si2 modification of the ITO-TPD interface are dramatic in terms of

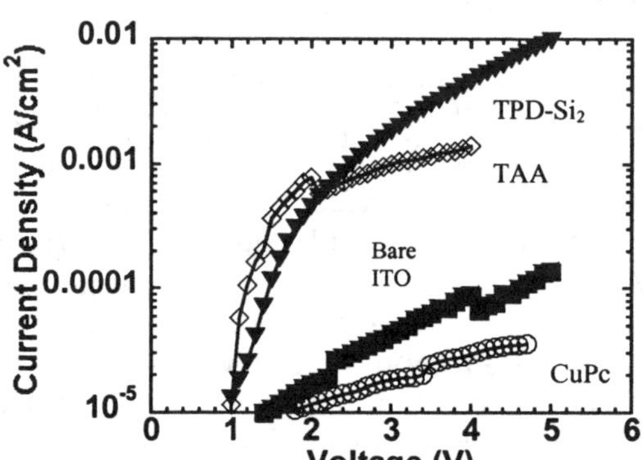

Figure 10. Hole-only device data showing greatly enhanced hole injection in OLED structures with self-assembled triarylamine charge injection/adhesion layers at the ITO-TPD interface.

facilitating hole injection, while Cu(Pc) is significantly less effective. Conventional wisdom has been that Alq-based OLEDs were electron-starved and that buffers that attenuated electron injection led to the highest quantum efficiencies and device stabilities by preventing excess holes from reaching the ETL or cathode. Recent results such as those shown here argue that enhanced hole injection can in fact create an additional field within the device active regions which in effect enhances electron injection.

3.2. INTERFACIAL ENGINEERING IN POLYMER-BASED OLEDS

The present nanometer-precise self-assembly techniques are also applicable to organic LEDs based on conjugated polymers (polymer light-emitting diodes; PLEDs). Thus, modifying the ITO anode with a thin TPD-Si_2 self-assembled monolayer, dramatically improves the performance of poly(9,9-dioctylfluorene) (PFO)-based blue PLED devices. Hole injection has been generally accepted to be the principal limiting factor in blue PLED device performance due to the high ionization potentials of most blue-emitting polymers[41,42]. For PFO[43], with a HOMO level of 5.9eV, the hole injection barrier is 1.2eV for an ITO anode. Conventionally, a hole transport layer (HTL) is necessary in blue PLED device structures to achieve high performance. The commonly used HTL materials are doped conductive polymers such as poly(3,4-ethylene dioxythiophene) – poly(styrene sulfonate) (PEDOT-PSS)[44]. In our work, instead of using a conventional doped conductive polymer as the HTL, we modified the ITO surface with a self-assembled monolayer (SAM) of TPD-Si_2 and used the SAM-modified ITO as the blue PLED anode. Devices of the structure ITO/TPD-Si_2 SAM/PFO/Ca/Al were then characterized. For comparison, PLEDs having bare ITO and ITO/PEDOT-PSS as anodes were also fabricated. The device characteristics of the present PLED devices are compared in Figure 11 in current-voltage (I-V), luminance-voltage (L-V), and current efficiency-voltage response. As seen from Figure 11, the SAM-modified device turns on at ~5.5 V and reaches maximum luminance of about 7,000 cd/m^2 at ~15 V and with a current efficiency of 1.1 cd/A, while the bare ITO-based device turns on ~12 V and reaches maximum luminance of ~100 cd/m^2 near 21 V, and with a current efficiency of 0.03 cd/A. With the ITO anode modified by this ultra-thin TPD-Si_2 SAM, the maximum brightness and current efficiency for the PFO-based blue PLED device are increased by one to two orders of magnitude, and the turn-on voltage is reduced by 6.5 V. To the best of our knowledge, this is the first report of a high-performance blue PLED device fabricated without a conventional HTL, which usually has a thickness on the order of tens of nanometers. Compared with a double-layer PLED control device having PEDOT-PSS as the HTL, the SAM-modified device exhibits comparable current efficiency and turn-on voltage, but three times higher maximum light output.

We further studied the hole injection efficiency into PFO of the SAM modified ITO anode by examining the current-voltage characteristics of hole-only devices[45] having the structure ITO/TPD-Si_2 SAM/PFO/Au/Al, in which electron injection at the cathode is blocked by

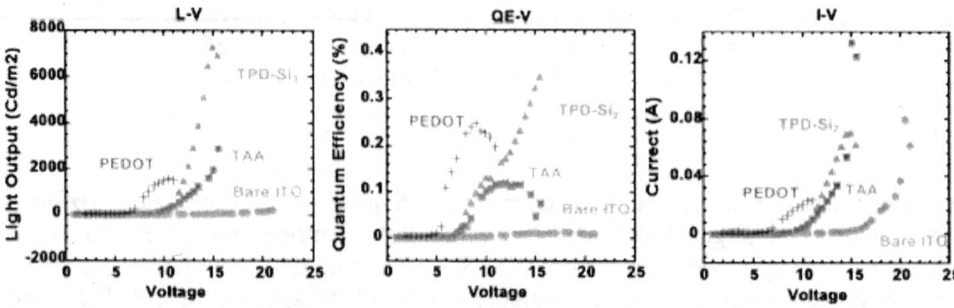

Figure 11. Comparative device response characteristics of PLEDs fabricated with a self-assembled TPD-Si_2 (or TAA) adhesion/hole injection monolayer, a conventional doped polymer buffer layer, and bare ITO.

a gold layer. As shown in the hole-only data of Figure 12, compared to bare ITO, the SAM modified ITO anode greatly enhances the hole injection current by two orders of magnitude and to the level comparable to that of the PEDOT-PSS modified ITO anode. This dramatic increase in hole injection can be attributed to the higher interfacial affinity

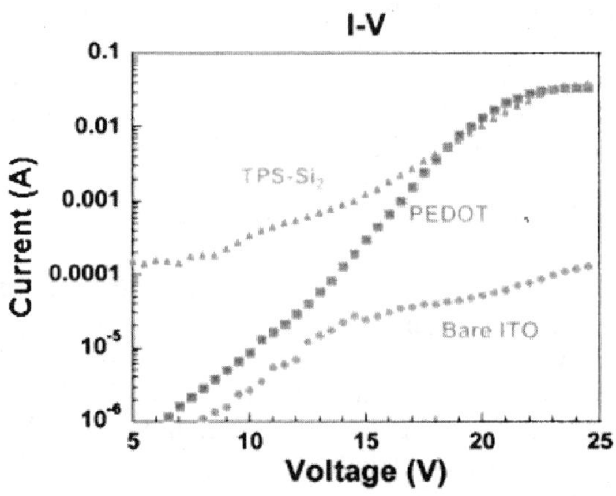

Figure 12. Hole-only device data for PFO-based PLEDs having TPD-Si_2-functionalized, PEDOT-PSS-functionalized, and bare ITO anodes.

between SAM-modified ITO and PFO. The large surface energy mismatch and consequent poor electronic contact between the hydrophilic ITO and the hydrophobic emissive polymer is likely one of the causes of the inefficient hole injection observed here. This surface energy mismatch can be minimized by modifying the ITO anode with the TPD-Si2 triarylamine SAM, as evidenced by aqueous contact angle measurements. The advancing aqueous contact angles of ITO, SAM-modified ITO, and PFO are 20°C, 80°C, and 90°C, respectively.

4. Organic Nanodiodes

The ever-increasing drive toward miniaturization of optoelectronic components to the nanosize regime offers significant challenges for conventional inorganic semiconductors as revealed by extensive experimental and theoretical studies of sub-100nm structures and their confinement-dependent electronic and optical properties[46,47]. The pronounced and band structure- connected size sensitivity of quantum confinement[48a], particle shape uniformity[48b], lattice strain[48b], surface defect[48c], and compositional nonuniformity[48d] phenomena profoundly influence photonic response in inorganic light-emitting diode (LED) materials such as GaAs[48a], doped GaAs[48b,c,e], InAs[48d], InP[49], and porous Si[50]. For these reasons, fabrication of efficient nanoscale inorganic light-emitting devices, such as LEDs has presented daunting challenges.

As macroscale OLEDs advance in sophistication and durability, their remarkable properties (high quantum efficiencies, ease of fabrication, color tunability) raise interesting possibilities at nanoscale dimensions. In contrast to band structure-dominated inorganic diodes, the transport characteristics of disordered organic heterostructures are largely dominated by carrier hopping[51]. This presents the intriguing situation that: i) organic heterostructures should be largely immune to quantum confinement effects, ii) a host of self-assembly and soft lithographic techniques should be applicable for fabrication and patterning[26,52,53], iii) optoelectronic properties should be tunable at the molecular architecture level. To date, available fabrication techniques (using either electrode blocking[53,54] or direct object formation[55-57] approaches) have not proven capable of producing discrete, ordered, structurally uniform heterostructures in the size regime (<100nm) needed to investigate nanoscale diode response and confinement effects. In principle, the availability of emitting and detecting diodes of such dimensions could enable a variety of new, high-resolution sensing, imaging, and storage technologies. With these issues in mind, we have pursued an adaptation of nanosphere lithography[58,59], combined with the nm-scale self-assembly processes discussed above[26], to achieve fabrication of highly ordered, structurally regular OLED heterostructure arrays at length scales one-two orders of magnitude smaller than heretofore possible. We also describe initial data on the confinement properties of such organic "nanodiode" arrays[60].

Nanosphere lithography offers a convenient approach to well-ordered sub-μm patterns via self-organization of hexagonally close-packed μ-bead arrays on a variety of substrates. Using a variant of a procedure reported to afford uniform coverage over relatively large areas (~1 cm^2)[61], monolayers of 160 and 400 nm diameter carboxy-terminated polystyrene beads were assembled from suspensions of suitable concentration (1.5% solids) and solvent composition (3:1 ethanol:water) on the surface of ultra-smooth ITO/glass substrates (5-10Å rms roughness) coated with a self-assembled triarylamine monolayer (TAA, Figure 13) to facilitate hole injection and cohesion of the subsequently applied hole transport layer[26,40]. Prior to use, the substrates were dried in a stream of dry nitrogen. They were then mounted at a 9° tilt angle from horizontal in a dust- and vibration-free enclosure containing a saturated atmosphere of the bead latex solution vapor, and were drop-coated with the bead latex. Slow solvent evaporation yielded uniform, self-assembled monolayers of the polystyrene beads. The coated substrates were then dried for 4 days, after which they were transferred to a glove box equipped with two interfaced thermal evaporators for diode fabrication. Prior to diode fabrication, the structural regularity of the bead arrays was verified by ultra-high resolution SEM.

Using the aforementioned close-packed μ-bead arrays as shadow masks, multilayer OLED heterostructures, NPB (Figure 6; 50nm at 0.3nm/s)/ Alq:quinacridone

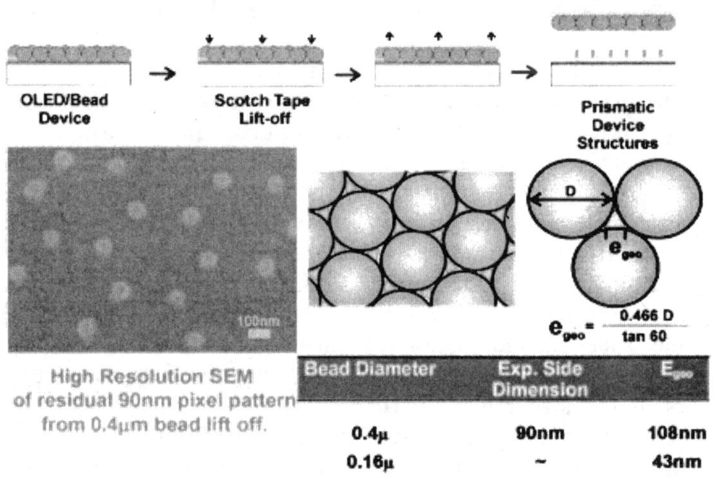

Figure 13. Self-assembled nanosphere route to nanodiodes, showing the shadow mask fabrication strategy and images of the completed structures after lift-off.

(40nm at 0.3nm/s)/ Al (300nm at 0.5nm/s) (Figure 13) were fabricated by vapor phase deposition as described above for macroscale OLEDs. Devices produced with 400nm bead masks are readily imaged via high-resolution SEM following tape lift-off of the beads, revealing regular patterns of trigonal prismatic posts with 90nm sides (Figure 13). Pixel dimensionality (43nm/side) from the 160nm bead arrays can be readily estimated from geometrical considerations and is in excellent agreement with published results for analogous metal nanoparticle depositions[58d]. OLED I-V characteristics were evaluated using the techniques described above. Importantly, negligible evidence for quantization effects in device I-V response are found upon comparison of nanoscopic and macroscopic diodes having the same layer dimensions and compositions (Figure 14). Furthermore, no

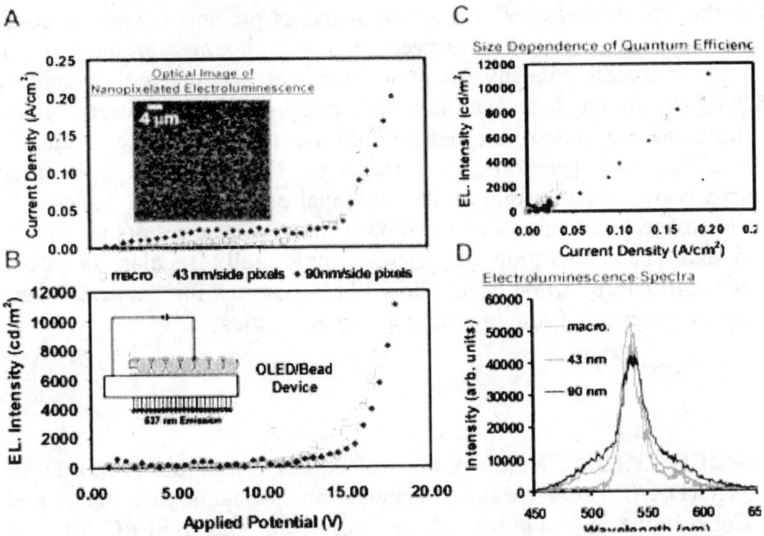

Figure 14. Photophysical properties of 90 and 43 nm nanodiodes vs. a macroscopic analogue. A. I-V response of nanodiodes vs. a macroscopic OLED of the same basic structure. Inset: optical image of an emitting field of 90 nm nanodiodes. B. Electroluminescent intensity vs. applied potential of nanodiodes vs. an OLED. Inset: schematic diagram of nanodiode configuration. C. Electroluminescent intensity vs. current density of nanodiodes vs. an OLED of the same structure. D. Electroluminescence spectra of nanodiodes vs. a comparable OLED of the same structure.

significant shift in the Alq:quinacridone EL maximum (λ_{EL}= 537nm) arising from the nanoscale dimensions is manifested. Similarly, no shift in photoluminescence maximum (λ_{em}= 532nm; $\lambda_{ex.}$= 256nm) vs. the macroscopic diodes is observed following bead lift-off for either 90nm or 43nm nanopixel arrays. Minimal effects of device compression are also evident in electroluminesence intensity vs. current density (Figure 14) – the forward external quantum efficiency (~2%) corrected for fill factor is essentially independent of device dimensions. Clearly minimal influence of excitonic or carrier confinement is imposed by length scale compression of Alq-based OLEDs to the nanoscopic length scale.

The decreased emissive fill factor in the nanodiode experiments manifests itself in expected decrease of total collected electroluminescence intensity. Geometrical analysis of the 400 nm and 160 nm SA-bead masks predicts decreased emissive region areas in good agreement with the observed attenuation in emission intensity. Despite this decreased EL emission intensity, there is sufficient brightness to image the spatially homogeneous EL arising from the nanodiode array (Fig. 14 A, inset), underscoring the minimal influence of patterning defects on the present device evaluation and the uniformity of the nanopixelation. Not surprisingly, the diffraction limits of conventional optical microscopy hinder resolution of individual pixels (Fig. 14 A, inset). Efforts are underway to image these structures via other means. These results demonstrate the first fabrication and characterization of discrete, structurally regular, nanoscale emissive organic heterostructure arrays and show that they exhibit minimal size-dependent compromise of either rectifying or emissive characteristics.

ACKNOWLEDGMENT. We thank the NSF-MRSEC program (Grant DMR 0076077), DARPA/ARO(DAAD19-00-1-0368), Nanovation Technologies Inc., and the U.S. Display Consortium for support of this research, and NSERC of Canada for a postdoctoral fellowship to JV. We thank M. Troost of the Northwestern EPIC Microscopy Facility for assistance with SEM imaging and Drs. M. van der Boom, G. Hutchison, and Y. Koide for helpful discussions.

References

1. For some reviews: (a) Prasad, P. N. and Williams, D. J. (1991) *Introduction to Nonlinear Optical Effects in Molecules and Polymers*, John Wiley & Sons, New York. (b) Zyss, J. (ed.), (1994) *Molecular Nonlinear Optics – Materials, Physics and Devices*, Academic Press, San Diego, CA . (c) Burland, D. M. (ed.), (1994) Optical nonlinearities in chemistry – introduction. *Chem. Rev.* special issue on "*Optical Nonlinearities in Chemistry*" **94**, 1-278.

2. (a) Steier, W. H., Chen, A., Lee, S.-S., Garner, S. , Zhang, H. , Chuyanov, V. , Dalton, L.R. , Wang, F. , Ren, A. S. , Zhang, C. , Tordorova, G. , Harper, A. , Fetterman, H. R. , Chen, D. , Udupa, A. , Bhattacharya, D. , Tsap, B. (1999) Polymer electro-optic devices for integrated optics *Chem. Phys.* **245**, 487-506. (b) Marder, S. R., Kippelen, B., Jen, A. K. Y., Peyghambarian, N. (1997) Design and synthesis of chromophores and polymers for electro- optic and photorefractive applications. *Nature* **388**, 845-851. (c) Marks, T. J., Ratner, M. (1995) Design, Synthesis, and Properties of Molecule-Based Assemblies with Large 2nd-Order Optical Nonlinearities. *Angew. Chem. Int. Ed. Engl.* **34**, 155-173. (d) Shi, Y., Zhang, C., Zhang, H., Bechtel, J. H., Dalton, L. R., Robinson, B. H., Steier, W. H. (2000) Low (sub-1-volt) halfwave voltage polymeric electro-optic modulators achieved by controlling chromophore shape. *Science* **288**, 199-122.
3. (a) Ye, C., Marks, T. J., Yang, J., Wong, G. K. (1987) Synthesis of Molecular Arrays with Nonlinear Optical-Properties - 2nd-Harmonic Generation by Covalently Functionalized Glassy- Polymers. *Macromolecules* **20**, 2322-2324. (b) Ma, H., Chen, B., Sassa, T., Dalton, L. R., Jen, A. K.-Y. (2001) Highly efficient and thermally stable nonlinear optical dendrimer for electrooptics. *J. Am. Chem. Soc.* **123**, 986-987. (c) Yitzchaik, S., Di Bella, S., Lundquist, P. M., Wong, G. K., Marks, T. J. (1997) Anomalous second-order nonlinear optical response of in-plane poled glassy polymers. Spectroscopic and theoretical support for the importance of charged chromophore aggregates. *J. Am. Chem. Soc.* **119**, 2995-3002.
4. (a) Schwartz, H., Mazor, R., Khodorkovsky, V., Shapiro, L., Klug, J. T., Kovalev, E., Meshulam, G., Berkovic, G., Kotler, Z., Efrima, S. (2001) Langmuir and Langmuir-Blodgett films of NLO active 2-(p-N-alkyl-N-methylamino)benzylidene-1,3-indandione-pi/A curves, UV-vis spectra, and SHG behavior. *J. Phys. Chem. B.* **105**, 5914-5921. (b) Ricceri, R., Neto, C., Abbotto, A., Facchetti, A., Pagani, G. A. (1999) Morphological characterization of H aggregates in Langmuir-Blodgett films of pyridinium-dicyanomethanide dyes. *Langmuir* **15**, 2149-2151. (c) Wijekoon, W. M. K. P., Wijayu, S. K., Bhawalkar, J. D., Prasad, P. N., Penner, T. L., Armstrong, N. J., Ezenyilimba, M. C., Williams, D. J. (1996) Second harmonic generation in multilayer Langmuir-Blodgett films of blue transparent organic polymers. *J. Am. Chem. Soc.* **118**, 4480-4483.
5. (a) Sagiv, J. (1980) Organized monolayers by adsorption.1. Formation and structure of oleophobic mixed monolayers on solid-surfaces. *J. Am. Chem. Soc.* **102**, 92-98. (b) Netzer, L., Sagiv, J. (1983) A new approach to construction of artificial monolayer assemblies. *J. Am. Chem. Soc.* **105**, 674-676. (c) Bain, C. D., Whitesides, G. M. (1988) Molecular-level control over surface order in self-assembled monolayer films of thiols on gold. *Science* **240**, 62-63. (d) Troughton, E. B., Bain, C. D., Whitesides, G. M., Nuzzo, R. G., Allara, D. L., Porter, M. D. (1988) Monolayer films prepared by the spontaneous self-assembly of symmetrical and unsymmetrical dialkyl sulfides from solution onto gold substrates - structure, properties, and reactivity of constituent functional-groups. *Langmuir* **4**, 365-385.
6. For the application in EO materials: (a) Yitzchaik, S., Roscoe, S. B., Kakkar, A. K., Allan, D. S., Marks, T. J., Xu, Z., Zhang, T., Lin, W., Wong, G. K. (1993) Chromophoric Self-Assembled NLO Multilayer Materials - Real- Time Observation of Monolayer Growth and Microstructural Evolution by in-Situ 2nd-Harmonic Generation Techniques. *J. Phys. Chem.* **97**, 6958-6960. (b) van der Boom, M. E., Evmenenko, G., Dutta, P., Marks, T. J. (2001) Nanoscale refractive index tuning of siloxane-based self-assembled electro-optic superlattices. *Adv. Funct. Mat.* **11**, 393-397.
7. For SAS based electro optical modulators: (a) Zhao, Y.-G, Wu, A., Lu, H.-L, Chang, S., Lu, W.-K., Ho, S.-T., van der Boom, M. E., Malinsky, J. E., Marks, T. J. (2001) Traveling wave electro-optic phase modulators based on intrinsically polar self-assembled chromophoric superlattices. *Appl. Phys. Lett.* **79**, 587-589. (b) van der Boom, M. E., Malinsky, J. E., Zhao, Y.-G., Chang, S., Lu, W. K., Ho, S. T., Marks, T. J. (2001) High-response electro-optic phase modulator based on an intrinsically acentric layer-by-layer self-assembled molecular superlattice. *Polym. Prepr.* **42**, 550-551. (c) Zhao, Y.-G., Chang, S., Wu, A., Lu, H.-L., Ho, S.-T., van der Boom, M.E., Marks, T.J., submitted for publication. (d) Lundquist, P. M., Lin, W., Zhou, H., Hahn, D. N., Yitzchaik, S., Marks, T. J., Wong, G. K. (1997) Frequency doubling in two-component self-assembled chromophoric waveguide structures. *Appl. Phys. Lett.* **70**, 1941-1943.

8. For other examples of materials construction via layer-by-layer self-assembly see: (a) Doron-Mor, I., Hatzor, A., Vaskevich, A., van der Boom-Moav, T., Shanzer, A., Rubinstein I., Cohen, H. (2000) Controlled surface charging as a depth-profiling probe for mesoscopic layers. *Nature* **406**, 382-385. (b) Neff, G. A., Helfrich, M. R., Clifton, M. C., Page, C. J. (2000) Layer-by-layer growth of acentric multilayers of Zr and azobenzene bis(phosphonate): Structure, composition, and second-order nonlinear optical properties. *Chem. Mater.* **12**, 2363-2371. (c) Frutos, A. G., Brockman, J. M., Corn, R. M. (2000) Reversible protection and reactive patterning of amine- and hydroxyl-terminated self-assembled monolayers on gold surfaces for the fabrication of biopolymer arrays. *Langmuir* **16**, 2192-2197. (d) Flory, W. C., Mehrens, S. M., Blanchard, G. J. (2000) Structural contributions to second-order optical nonlinearities in oriented interfacial multilayers. *J. Am. Chem. Soc.* **122**, 7976-7985. (e) Fang, M., Kaschak, D. M., Sutorik, A. C., Mallouk, T. E. (1997) A "mix and match" ionic-covalent strategy for self-assembly of inorganic multilayer films. *J. Am. Chem. Soc.* **119**, 12184-12191. (f) Maoz, R., Matlis, S., DiMasi, E., Ocko, B. M., Sagiv, J. (1996) Self-replicating amphiphilic monolayers. *Nature* **384**, 150-153.
9. Wooten, E. L., Kissa, K. M., Yi-Yan, A., Murphy, E. J., Lafaw, D. A., Hallemeier, P. F., Maack, D., Attanasio, D. V., Fritz, D. J., McBrien, G. J., Bossi, D. E. (2000) A review of lithium niobate modulators for fiber-optic communications systems. *IEEE, J. Sel. Topics Quantum Electron.* **6**, 69-82, and references therein.
10. For poled polymer devices, see: a) Zhang, H., Oh, M.-C., Szep, A., Steier, W. H., Zhang, C., Dalton, L. R., Erlig, H., Chang, Y., Chang, D. H., (2001) Push-pull electro-optic polymer modulators with low half-wave voltage and low loss at both 1310 and 1550 nm. *Appl. Phys. Lett.* **78**, 3136-3138. b) Dalton, L. R. (2000) Polymeric electro-optic materials: optimization of electro-optic activity, minimization of optical loss, and fine-tuning of device performance *Opt. Eng.* **39**, 589-595.
11. (a) Li, D. Q.; Ratner, M. A.; Marks, T. J.; Zhang, C. H.; Yang, J.; Wong, G. K. (1990) Chromophoric Self-Assembled Multilayers - Organic Superlattice Approaches to Thin-Film Nonlinear Optical-Materials. *J. Am. Chem. Soc.* **112**, 7389-7390. (b) Roscoe, S. B.; Yitzchaik, S.; Kakkar, A. K.; Marks, T. J. (1994) Ion-Exchange Process and Environmental-Effects in Chromophoric Self-Assembled Superlattices - Manipulation of Microstructure and Large Enhancements in Nonlinear-Optical Response. *Langmuir.* **10**, 1337-1339. (c) Lin, W., Lin, W., Wong, G. K., Marks, T. J. (1996) Supramolecular approaches to second-order nonlinear optical materials. Self-assembly and microstructural characterization of intrinsically acentric [(aminophenyl)azo]pyridinium superlattices *J. Am. Chem. Soc.* **118**, 8034-8042. (d) Yitzchaik, S., Marks, T. J. (1996) Chromophoric self-assembled superlattices. *Acc. Chem. Res.* **29**, 197-202. (e) Malik, A., Lin, W., Durbin, M. K., Marks, T. J., Dutta, P. (1997) Specular x-ray reflectivity studies of microstructure and ordering in self-assembled multilayers. *J. Chem. Phys.* **107**, 645-651. (f) Lin, W., Lee, T.-L., Lyman, P. F., Lee, J., Bedzyk, M. J., Marks, T. J. (1997) Atomic resolution X-ray standing wave microstructural characterization of NLO-active self-assembled chromophoric superlattices. *J. Am. Chem. Soc.* **119**, 2205-2211
12. Wang, G., Zhu, P., Marks, T.J., Ketterson, J.B. (2002) Ultra-fast frequency-selective optical switching based on thin self-assembled organic chromophoric films with a large second-order nonlinear response *Appl. Phys. Lett.*, in press.
13. (a) Zhu, P., van der Boom, M. E., Evmenenko, G., Dutta, P., Marks, T. J. (2001) Efficient consecutive assembly of large-response thin-film molecular electro-optic materials. *Polym. Prepr.* **42(2)**, 579-580. (b) Zhu, P., van der Boom, M. E., Kang, H., Evmenenko, G., Dutta, P., Tobin, T. J. (2002) Realization of Expeditious Layer-by-Layer Siloxane-Based Self-Assembly as an Efficient Route to Structurally Regular Acentric Superlattices with Large Electro-Optic Responses. *Chem. Mater.* in press. (c) van der Boom, M. E., Zhu, P., Evmenenko, G., Malinsky, J. E., Lin, W., Dutta, P., Marks, T. J. (2002) Nanoscale consecutive self-assembly of thin-film molecular materials for electrooptic switching. Chemical streamlining and ultrahigh response chromophores. *Langmuir* **18**, 3704-3707. (d) van der Boom, M. E., Richter, A. G., Malinsky, J. E., Lee, P. A., Armstrong, N. R., Dutta, P., Marks, T. J. (2001) Single reactor route to polar superlattices. Layer-by-layer self-assembly of large-response molecular electrooptic materials by protection-deprotection. *Chem. Mater.* **13**, 15-17.

14. Malinsky, J. E., Jabbour, G. E., Shaheen, S. E., Anderson, J. D., Richter, A. G., Marks, T. J., Armstrong, N. R., Kippelen, B., Dutta, P., Peyghambarian, N. (1999) Self-assembly processes for organic LED electrode passivation and charge injection balance. *Adv. Mater.* **11**, 227-231.
15. (a) Tidswell, I. M., Ocko, B. M., Pershan, P. S. (1990) X-ray specular reflection studies of silicon coated by organic monolayers (alkylsiloxanes). *Phys. Rev. B* **41**, 1111-1127. (b) Pomerantz, M., Segmuler, A., Netzer, L, Sagiv, J. (1985) Coverage of Si substrates by self-assembling monolayers and multilayers as measured by ir, wettability and x-ray-diffraction. *Thin Solid Films* **132**, 153-162. (c) Tidswell, I. M., Rabedeau, T. A., Pershan, P. S., Kosowsky, S. D., Folkers, G. P., Whitesides, G. M. (1991) X-ray grazing-incidence diffraction from alkylsiloxane monolayers on silicon-wafers. *J. Chem. Phys.* **95**, 2854.
16. (a) Richter, A. G., Durbin, M. K., Yu, C.-J., Dutta, P. (1998) In situ time-resolved X-ray reflectivity study of self-assembly from solution. *Langmuir* **14**, 5980-5983. (b) Malik, A., Durbin, M. K., Richter, A. G., Huang, K. G., Dutta, P. (1995) Structures of headgroup and tail-group monolayers in a Langmuir-Blodgett-film*Phys. Rev. B.* **52**, 11654-11657.
17. Roberts, M. J., Lindsay, G. A., Herman, W. N., Wynne, K. J. (1998) Thermally stable nonlinear optical films by alternating polyelectrolyte deposition on hydrophobic substrates. *J. Am. Chem. Soc.* **120**, 11202-11203.
18. Bloembergen, N., Pershan, P. S. (1962) Light waves at the boundary of nonlinear media. *Phys. Rev.* **128**, 606-622.
19. Lundquist, P. M., Yitzchaik, S., Zhang, T., Kanis, D.R., Ratner, M.A., Marks, T. J., Wong, G. K. (1994) Dispersion of 2nd-order optical nonlinearity in chromophoric self-assembled films by optical parametric amplification - experiment and theory. *Appl. Phys. Lett.* **64**, 2194-2196.
20. (a) Teng, C. C., Man, H. T. (1990) Simple reflection technique for measuring the electrooptic coefficient of poled polymers. *Appl. Phys. Lett.* **56**, 1734-1737. (b) Schildkraut, J. S. (1990) *Appl. Opt.* **29**, 2839-2847.
21. Shuto, Y., Amano, M. (1995) Reflection measurement technique of electrooptic coefficients in lithium-niobate crystals and poled polymer-films. *J. Appl. Phys.* **77**, 4632-4638.
22. Simpson, G. J., Rowlen, K. L. (1999) An SHG magic angle: Dependence of second harmonic generation orientation measurements on the width of the orientation distribution. *J. Am. Chem. Soc.* **121**, 2635-2636.
23. Ashwell, G. J., Hargreaves, R. C., Baldwin, C. E., Bahra, G. S., Brown, C. R. (1992) Improved 2nd-harmonic generation from langmuir-blodgett-films of hemicyanine dyes. *Nature* **57**, 393-394.
24. Baldo, M.A., Forrest S.R. (2001), Interface-limited injection in amorphous organic semiconductors. *Phys. Rev. B.* **64**(085201), 1-17.
25. Ishii, H., Sugiyama, K., Ito, E., Seki, K. (1999) Energy level alignment and interfacial electronic structures at organic/metal and organic/organic interfaces. *Adv. Mater.*,**11**(12), 605-625.
26 Cui, J., Huang, Q., Wang, Q., Marks, T. J. (2001) Nanoscale Covalent Self-Assembly Approach to Enhancing Anode/Hole-Transport Layer Interfacial Stability and Charge Injection Efficiency in Organic Light-Emitting Diodes. *Langmuir*, **17**(7), 2051-2054.
27. Jabbour, G.E., Kawabe, Y., Shaheen, S.E., Wang, J.F., Morrell, M.M., Kippelen, B., Peyghambarian, N. (1997) Highly efficient and bright organic electroluminescent devices with an aluminum cathode. *Appl. Phys. Lett.*, **71**(13), 1762-1764.
28. Mason, M.G., Tang, C.W., Hung, L.S., Raychaudhuri, P., Madathil, J., Giesen, D.J., Yan, L., Le, Q.T., Gao, Y., Lee, S.T., Liao, L.S., Cheng, L.F., Salaneck, W.R., Santos, D.A., Gredas, J.L. (2001) Interfacial chemistry of Alq(3) and LiF with reactive metals. *J. Appl. Phys.*, **89**(5), 2756-2765.
29. Piromreun, P., Oh, H.-S., Shen, Y., Malliaras, G.G., Scott, J.C., Brock, P.J. (2000) Role of CsF on electron injection into a conjugated polymer *Appl. Phys. Lett.*, **77**(15), 2403-2405.
30. Ganzorig, C., Suga, K., Fujihira, M. (2001) Alkali metal acetates as effective electron injection layers for organic electroluminescent devices. *Materials Science and Engineering B-Solid State Materials for Advanced Technology*,. **85**(2-3), 140-143.
31. Nüesch, F., Forsythe, E.W., Le, Q.T., Gao, Y., Rothberg, L.J. (2000) Importance of indium tin oxide surface basicity for charge injection into organic materials based light emitting diodes. *J. Appl. Phys.*, **87**(11), 7973-7980.

32. Appleyard, S.F.J., Day, S.R., Pickford, R.D., Willis, M.R. (2000) Organic electroluminescent devices: enhanced carrier injection using SAM derivatized indium tin oxide electrodes. *J. Mater. Chem.*, **10**(1), 169-173.
33. Slyke, S.A.V., Chen, C.H., Tang, C.W. (1996) Organic electroluminescent devices with improved stability. *Appl. Phys. Lett.*, **69**(15), 2160-2162.
34. Malinsky, J.E., Veinot, J. G. C., Jabbour, G. E., Shaheen, S. E., Anderson, J. D., Lee, P., Richter, A. G., Burin, A. L., Ratner, M. A., Marks, T. J., Armstrong, N. R., Kippelen, B., Dutta, P., Peyghambarian, N. (2002) Nanometer-scale dielectric self-assembly process for anode modification in organic light-emitting diodes. Consequences for charge injection and enhanced luminous efficiency. *Chem. Mater.*, **14**(7), 3054-3065.
35. Shen, Y., Jacobs, D.B., Malliaras, G.G., Koley, G., Spencer, M.G., Ioannidis, A. (2001) Modification of indium tin oxide for improved hole injection in organic light emitting diodes. *Adv. Mater.*, **3**(16), 1234-1238.
36. Carter, S.A., Angelopoulos, M., Karg, S., Brock, P.J., Scott, J.C. (1997) Polymeric anodes for improved polymer light-emitting diode performance. *Appl. Phys. Lett.*, **70**(16), 2067-2069.
37. Ho, P.K.H., Kim, J.-S., Burroughes, J.H., Becker, H., Li, S.F.Y., Brown, T.M., Cacialli, F., Friend, R.H. (2000) Molecular-scale interface engineering for polymer light-emitting diodes. *Nature*, **404**, 481-484.
38. Li, W., Wang, Q., Cui, J., Chou, H., Marks, T.J., Jabbour, G.E., Shaheen, S.E., Kippelen, B., Peyghambarian, N., Dutta, P., Richter, A.J., Anderson, J., Lee, P., Armstrong, N. (1999) Covalently interlinked organic LED Transport layers via Spin-coating/Siloxane condensation. *Adv. Mater.*, **11**(9), 730-734.
39. Li, W., Malinsky, J.E., Chou, H., Ma, W., Geng, L., Marks, T.J., Jabbour, G.E., Shaheen, S.E., Kippelen, B., Peyghambarian, N., Anderson, J., Lee, P., Armstrong, N. (1998) Molecular Self-assembly routes to optically functional thin films. *Mat. Res. Soc. Symp. Proc.*, **488**, 385-387.
40. Cui, J., Huang, Q., Veinot, J.G.C., Yan, H., Marks, T.J. (2002) Interfacial microstructure-function in organic light-emitting diodes: assembled tetraaryldiamine and copper phthalocyanine interlayers. *Adv. Mater.*, **14**(8), 565.
41. Grice, A. W. Bradley, D. D. C., Bernius, M. T., Inbasekaran, M., Wu, W., Woo, E. P. (1998) High brightness and efficiency blue light-emitting polymer diodes. *Appl. Phys. Lett.* **73**(5), 629-631.
42. Yu, W. Pei, J., Cao, Y., Huang, W. (2001) Hole-injection enhancement by copper phthalocyanine (CuPc) in blue polymer LED. *J. Appl. Phys.* **89**(4), 2343-2348.
43. Anietz, S. et al., (1998) Electrochemical determination of the ionization potential and electron affinity of poly(9,9-dioctylfluorene). *Appl. Phys. Lett.* **73**(17), 2453-2455.
44. Cao, Y., Yu, G., Zhang, C., Menon, R., Heeger, A. J. (1997) Polymer light-emitting diodes with polyethylene dioxythiophene- polystyrene sulfonate as the transparent anode. *Synth. Met.* **87**(2), 171-174.
45. Parker I. D. (1994) Carrier tunneling and device characteristics in polymer light-emitting diodes. *J. Appl. Phys.* **75**(2), 1656-1666.
46. (a) Lan, S., Akahane, K., Jang, K., Kawamura, T., Okada, Y., Kawabe, M., Nishimura, T., Wada, O. (1999) Two-dimensional In0.4Ga0.6As/GaAs quantum dot superlattices realized by self-organized epitaxial growth, *Jpn. J. Appl. Phys.* **38**, 2934-2943. (b) Alivisatos, A.P. (1996) Semiconductor clusters, nanocrystals, and quantum dots. *Science*, **271**, 933-937.
47. Ulbrich, R.G. (2000) Handbook of Semiconductor Technology-Electronic Structure and Properties of Semiconductors, Chapter 2.
48. (a) Fukui, T., Ando, S., Tokura, Y., Toriyama, T. (1991) GaAs tetrahedral quantum dot structures fabricated using selective area metalorganic chemical vapor-deposition. *Appl. Phys. Lett.* **58**, 2018-2020. (b) Zou, J., Liao, X.Z., Cockayne, D.J.H., Leon, R. (1999) Transmission electron microscopy study of InxGa1-xAs quantum dots on a GaAs(001) substrate. *Phys. Rev. B* **59**, 12279-12282. (c) Temkin, H., Dolan, G.J., Panish, M.B., Chu, S.N.G. (1987) Low-temperature photoluminescence from InGaAs/InP quantum wires and boxes. *Appl. Phys. Lett.* **50**, 413-415. (d) Joyce, P., Krzyzewski, T.J., Bell, G.R., Joyce, B.A., Jones, T.S. (1998) Composition of InAs quantum dots on GaAs(001): Direct evidence for (In,Ga)As alloying.

Phys. Rev. B. **58** 15981-15984. (e) Fasol, G. (1997) Semiconductor lasers - Fast, cheap, and very bright. *Science* **275**, 941-942.

49. Kobayashi, M., Kitamura, K., Umeya, H., Jia, W., Yoshikawa, A., Shimotomai, M., Takahashi, K. J. (2000) Growth of CdS self-organized quantum dots by molecular beam epitaxy and application to light emitting diode structures. J. *Vac. Sci. Technol. B* **18**, 1684-1687.
50. Malinin, A., Ovchinnikov, V., Novikov, S., Tuovinen, C., Hovinen, A. (2000) Fabrication of a silicon based electroluminescent device *Mat. Sci. Eng. B* **74**, 32-35.
51. Barth, S., Wolf, U., Bässler, H., Müller, P., Riel, H., Vestweber, H., Seidler, P.F., Rieß, W. (1999) Current injection from a metal to a disordered hopping system. III. Comparison between experiment and Monte Carlo simulation. *Phys. Rev. B* **60**, 8791-8797.
52. (a) Malinsky, J.E., Jabbour, G.E., Shaheen, S.E., Anderson, J.D., Richter, A.G., Armstrong, N.R., Kipplelen, B., Dutta, P., Peyghambarian, N., Marks, T.J. (1999) Self-assembly processes for organic LED electrode passivation and charge injection balance. *Adv. Mater.* **11**, 227-231. (b) Zuppiroli, L., Si-Ahmed,L., Kamars, K., Nüesch, F., Bussac, M.N., Ades, D., Siove, A., Moons, E., Grätzel, M. (1999) Self-assembled monolayers as interfaces for organic opto-electronic devices. *Europ. Phys. J. B* **11**, 505-512.
53. Koide, Y., Wang, Q.W., Cui, J., Benson D.D., Marks T.J. (2000) Patterned luminescence of organic light-emitting diodes by hot microcontact printing (HμCP) of self-assembled monolayers. *J. Amer. Chem. Soc.* **122** *(45)* 11266-11267.
54. Hulteen, J.C., Martin, C.R. (1997) A general template-based method for the preparation of nanomaterials. *J. Mater. Chem.* **7**, 1075-1087.
55. Duffy, D.C., Jackman, R.J., Vaeth, K.M., Jensen, K.F., Whitesides, G.M. (1999) Patterning electroluminescent materials with feature sizes as small as 5 μm using elastomeric membranes as masks for dry lift-off. *Adv. Mater.* **11**, 546-552.
56. Kim, C., Burrows, P.E., Forrest, S.R. (2000) Micropatterning of organic electronic devices by cold-welding. *Science* **288** , 831-833.
57. Roger, J.A., Bao, Z., Dhar, L. (1998) Fabrication of patterned electroluminescent polymers that emit in geometries with feature sizes into the submicron range. *Appl. Phys. Lett.* **73**, 294-296.
58. (a) Haynes C.L., VanDuyne R.P. (2001) Nanosphere Lithography: A Versatile Nanofabrication Tool for Studies of Size-Dependent Nanoparticle Optics. *J. Phys. Chem. B* **105**, 5599-5611. (b) Hulteen, J.C., Treichel, D.A., Smith, M.T., Duval, M.L., Jensen, T.R., Van Duyne, R.P. (1999) Nanosphere Lithography: Size-Tunable Silver Nanoparticle and Surface Cluster Arrays. *J.* Phys. *Chem. B,* **103**, 3854-3863. (c) Hulteen, J.C., Van Duyne, R.P. (1995) Nanosphere lithography: A materials general fabrication process for periodic particle array surfaces. *J.*; *Vac. Sci. Technol. A* **13**, 1553-1558.
59. (a) Schnitzer, I., Yablonovitch, E., Caneau, C., Gmitter, T.J., Scherer, A. (1993) 30% external quantum efficiency from surface textured, thin-film light-emitting diodes. *Appl. Phys. Lett.* **63**, 2174-2176. (b) Windisch R., Heremans, P., Dutta, B., Kuijik, M., Schoberth, S., Kiesel, P., Dohler G.H., Borghs, G. (1998) High-efficiency non-resonant cavity light-emitting diodes. *Electron. Lett.* **34**, 1153-1155.
60. Veinot, J. G. C., Yan, H., Smith, S.M., Cui, J., Huang, Q., Marks, T.J. (2002) Fabrication and Properties of Organic Light-Emitting "Nanodiode" Arrays. *Nano Letters* **2**(4), 333-335.
61. Micheletto, R., Fukuda, H., Ohtsu, M. (1995) A simple method for the production of a 2-dimensional, ordered array of small latex-particles; *Langmuir* **11**, 3333-3335.

MESOSCOPIC AGGREGATION CONTROL OF ORGANIC FLUOROPHORES IN DEWETTED THIN FILMS

O. KARTHAUS[1,2], R. CHIBA[1], T. IMAI[1], K. KAGA[1], S. KURIMURA[1], R. NAKAMURA[1], K. OKAMOTO[1], J. SATO[1]
1 Chitose Institute of Science and Technology
Bibi 758-65, Chitose 066-8655, Japan
2 PRESTO, JST, Japan

1. Introduction

Many electronics, optoelectronics and photonics devices contain organic compounds as the functional metarial. Photographic films, organic light emitting diodes (OLEDs), solar cells, rewritable compact disks (RW-CDs), or liquid crystal displays are just a few examples of the vast variety of devices that exist. The function of many of these devices depend on the aggregation state of the organic material. Even a simple device such as a photographic film contains cyanine dyes that aggregate onto silver halide microcrystals to form so-called J-aggregates. These J-aggregates show a narrow absorption band and thus are ideal materials for the absorption of a certain wavelength, and thus can be used as photosensitizers in color films [1]. In most OLEDs the emissive and/or conductive compounds have to be in a molecular dispersed state. Aggregation of the molecules would lead to microcrystals that form trap sites for electrons or excited states and thus lead to a decrease in device function, or even to device failure. In the case of LCDs, the orientation of the molecules in the liquid crystalline state can be controlled by an external electric field. The switching between different orientational states then leads to a change in the transmission of light and thus to a visible contrast.

The dependence of device function on molecular aggregation and molecular orientation is not restricted to artificial, man-made devices. A striking biological example for the necessity of controlled molecular orientation is the photosynthetic reaction center and the surrounding antenna proteins. Here, a defined number of photofunctional porphyrin molecules have to be held in a precise orientation towards each other to ensure rapid and efficient energy- and electron transfer [2,3].

The above examples illustrate that it is crucial to control the aggregation state of the photofunctional material as precise as possible on the nanometer to micrometer scale. Here, results will be presented that focus on two examples, the aggregation control of cyanine dyes in polymer "microdomes" and the controlled crystallization of non-linear optically active dyes on the submicrometer to micrometer scale.

F. Charra et al. (eds.), Organic Nanophotonics, 265–277.

2. J-Aggregation of Cyanine Dyes in Polymer Microdomes

2.1. FORMATION OF POLYMER MICRODOMES

It has been reported that the casting of a dilute polymer solution on a horizontal substrate leads to dewetting and the formation of micrometer-sized polymer droplets, or "domes" on the substrate [4]. The origin of the dewetting phenomenon lies in complex nonlinear dynamics and non-equilibrium processes during solvent evaporation. The mechanism of dome formation has been explained in a previous paper [5]. Even though the evaporation of the solvent is a complex and far-from-equilibrium phenomenon, well ordered mesoscopic structures can be obtained. In these structures the dome, which have a diameter of a few hundred nm to a few μm, can have a very narrow size distribution and a regular spacing on the micrometer scale to form large 2-D patterns which can cover several cm^2. This self-organization into ordered structures can be explained by so-called dissipative structures which develop as the result of temperature and surface tension differences within the evaporating solution. Since dissipative structures are a physical phenomenon which does not depend on the chemical structure of the solvent nor the solute, many different polymers can form dewetted dome and line structures [6-9]. For the same reason, many different substrates can be used, for example glass, mica, or silicon wafers.

Figure 1. Optical micrograph of a polystyrene pattern on a mica surface. The bar corresponds to 50μm.

The size of the domes can be controlled by the initial polymer concentration in the casting solution and by the evaporation speed. Higher concentration and slower speeds lead to larger domes. Patterns with domes having diameters between 300 nm and several ten micrometer can be realized [4,5]. The hight of a dome is typically in the range of 5% of its diameter. AFM investigations showed that the contact angle of polystyrene of those domes is between 5 and 8°. Asuming a truncated spherical shape of the dome, the volume can be calculated, which is 0.8 attoliter, or 8×10^{-19} l for a dome with a diameter of 1 μm.

2.2. INTRODUCTION TO J-AGGREGATES OF CYANINE DYES

As already mentionned previously, cyanine dyes tend to aggregate in solution or in polymer films, and to form so-called J-aggregates [10,11]. Due to strong intermolecular interactions in these aggregates the absorption spectra are red-shifted and very narrow, compared to the molecularly dispersed state [12]. In addition, the fluorescence spectra have only a small Stokes shift and also a narrow shape. J-aggregates are used as sensitizers in photographic films and the aggregation behavior of numerous dyes have been studied in detail. Recent advances in imaging on the sub-wavelength scale of light [13] and femtosecond spectroscopy [14] have revealed that the J-aggregates consist of fibrous aggregates with a hierarchic structure. The smallest units are the so-called "coherent aggregates" which determine the extend of the exciton, and thus the position and shape of the absorption and fluorescence bands. A coherent aggregate contains a few molecules only [15]. These coherent aggregates are thought to form larger "structural aggregates", which form the anisotropic fibrils. Even though the exciton is thought to be restricted within a coherent aggregate, interactions and annihilation of excitons has been observed, too.

Even though J-aggregates have been studied in detail in the past seven decades, they still pose some intersting and unresolved questions. Is is possible to form J-agregates with a defined size? How do the spectral properties relate to the size of the J-aggregegate? Is it possible to observe a single coherent aggregate?

Our approach of using a dewetted polymer film for the incorporation of cyanine dyes will hopefully lead to the answers of some of these questions.

2.3. CYANINE DYES IN POLYMER MICRODOMES

It is well known that cyanine dyes form J-aggregates in polyvinylalcohol films that were produced by spin coating of a mixed dye/PVA solution [13]. But since these films are spatially restricted in only one dimension, the z-direction, the size of the aggregates cannot be controlled. Diffusion in the x-y plane of the film during and after the spin coating will lead to a variety of aggregate sizes.

If the matrix would form small micrometer-sized domes (see chapter 2.1), the diffusion of the dye molecules would be restricted to one polymer dome. Since the number of dye molecules per unit volume of solution is constant, this will lead to a certain number of dye molecules in each dome after the solvent has evaporated. Thus there will be an upper limit for the number of dye molecules in the J-aggregate, assuming that a single J-aggregate is formed in each dome.

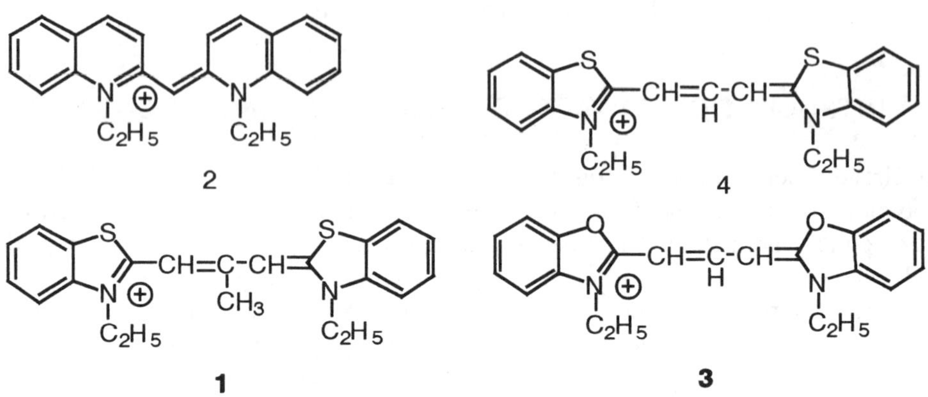

Figure 2. Chemical formulae of cyanine dyes that are known to form J-aggregates

2.3.1. *Optical Microscopy*

We chose polystyrene as the matrix polymer, since polystyrene dewets upon casting on hydrophilic substrates. By mixing aliquots of chloroform solutions of polystyrene (1mg/ml) and cyanine dye (0.1mg/ml) casting solutions were prepared which contain 5, 10, and 15 % (w/v) of cyanine dye with respect to polystyrene. Solutions were cast onto glass or mica surfaces and investigated by optical and fluorescence microscopy and fluorescence microspectrometry. The experimental conditions and the equipment have been described in detail in a previous publication [16]. Figure 3 shows an optical micrograph of a sample which contains 15 % (w/v) **1** in polystyrene. The dewetted structure of polystyrene domes can be clearly seen. In contrast to the pure polystyrene sample (Figure 1), here a phase separated structure within each dome suggests the formation of dye aggregates.

The fluorescence micrograph of a similar sample shows that, upon excitation with blue-violet light (440-480 nm), the polystyrene domes emit orange fluorescence (see inset of Figure 3). This is characteristic for aggregates of the cyanine dye and is a proof that the dye is located in the polymer domes. No background fluorescence on the substrate can be detected. Thus optical and fluorescence microscopy show that it is possible to form an environment in which a low molar mass material is restricted inside a micrometer-sized polymer dome. Taking the dome size and the concentration of the dye into account, each dome should contain 10^5 - 10^7 dye molecules.

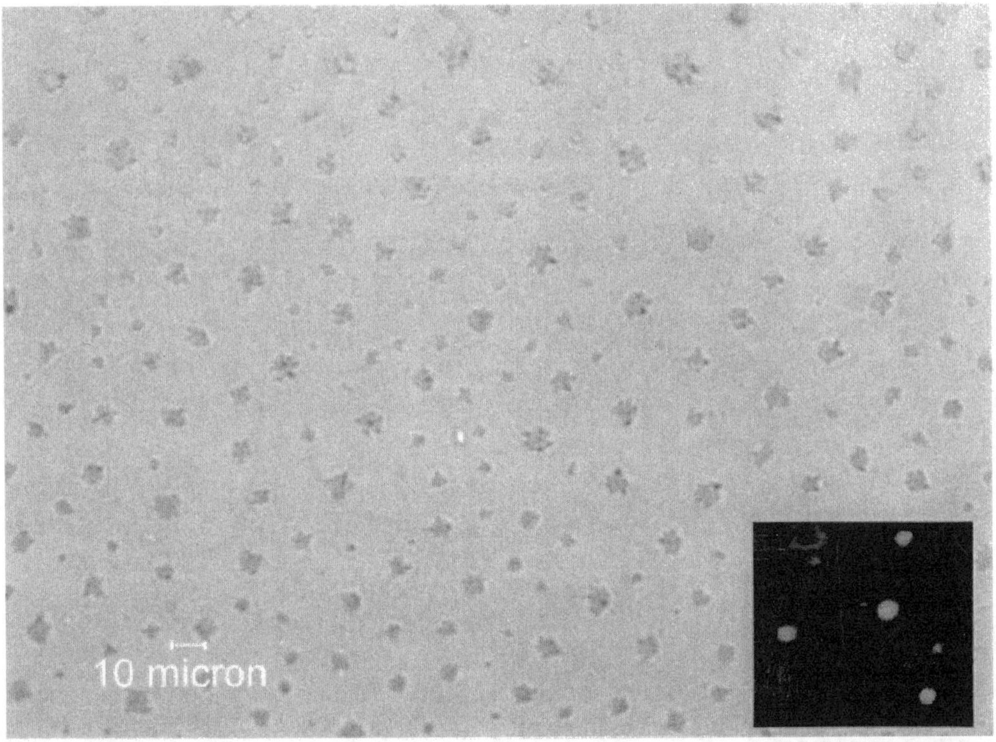

Figure 3. Optical micrograph of **1** in polystyrene. The inset is a close-up fluorescence micrograph of a similar sample.

2.3.2. *Microspectrometry*

In order to get a more detailed account on the aggregation state of the dye within the polymer domes, microspectroscopy was carried out. The lateral resolution of the microscope spectrometer was less than 3 μm, with a spectral resolution of 0.4 nm. Figure 4 shows the absorption and fluorescence spectra of single domes that contain 10 % (w/v) of **1**. The blue line is the absorption spectrum and the red line is the fluorescence spectrum. The width at half height of the absorption spectrum is 14 nm and that of the fluorescence spectrum is 23 nm. With an observed Stoke's shift of 9 nm this confirms the formation of J-aggregates of cyanine dyes within the polystyrene microdomes.

Since the microscope spectrometer has a lateral resolution in the micrometer range, it is possible to obtain the spectral characteristics of J-aggregates within individual single polymer microdomes. Thus the effect of the microenvironment of each polymer dome on the photophysics of J-aggregates can be evaluated.

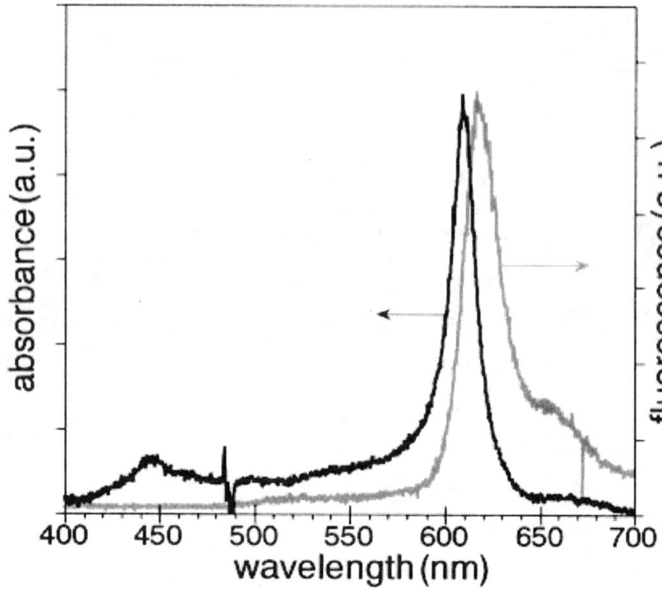

Figure 4. Absorption and fluorescence spectra of **1** in a single polystyrene dome (15 %(w/v)) (a).

The width at half height of the absorption spectrum is 14 nm and that of the fluorescence spectrum is 23 nm. With an observed Stoke's shift of 9 nm this confirms the formation of J-aggregates of cyanine dyes within the polystyrene microdomes.

Since the microscope spectrometer has a lateral resolution in the micrometer range, it is possible to obtain the spectral characteristics of J-aggregates within individual single polymer microdomes. Thus the effect of the microenvironment of each polymer dome on the photophysics of J-aggregates can be evaluated.

2.3.3. Mesoscopic Size Effect

The fluorescence spectra of 100 individual polymer domes were recorded for three dye concentrations, 5%, 10%, and 15%, respectively. All domes showed the characteristic sharp fluorescence peak at around 616 nm. For a detailed analysis of the spectral properties, the maxima of the fluorescence emissions were plotted against the dome diameter, as can be seen in Figure 5.

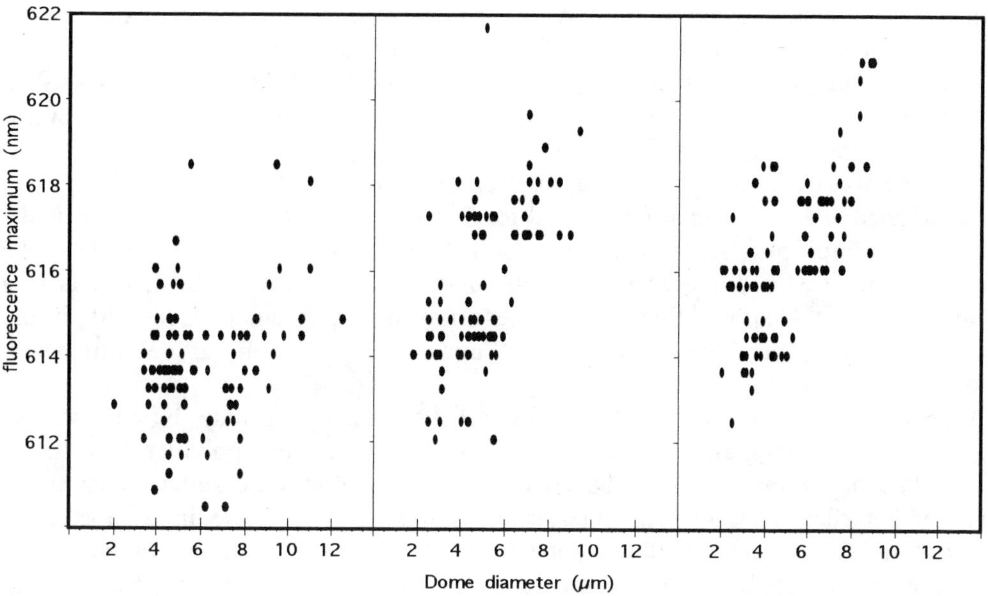

Figure 5. Plot of the fluorescence maxima of **1** against dome diameter for three different dye concentrations, 5% (left), 10% (middle) and 15% (right).

The sample with the lowest concentration (5%) showed fluorescence peaks between 610 nm and 619 nm. Most of the observed domes have the fluorescence maximum at around 614 nm, and no size effect was observed, that is the fluorescence maximum of the cyanine J-aggregate is independent on the dome diameter. On the otherhand, higher concentrated samples show a size effect. First of all it has to be noted that the fluorescence maxima shift towards slightly higher wavelengths, namely 612 nm to 622 nm. Second, the fluorescence maximum is dependent on the dome diameter. Larger domes show a more red-shifted fluorescence. In the case of the 10% sample, the fluorescence maxima are gouped into two areas. At around a diameter of 6 μm, the range of fluorescence maxima drastically changes. Below 6 μm the maxima range from 612 nm to 618, whereas for larger domes the range is from 617 nm to 620 nm. This quantized shift in fluorescence is also detected in the sample with 15% of **1**. Also there a threshold diameter of 6 μm can be seen. Below a diameter of 6 μm the fluorescence range is from 613 nm to 618 nm, above 6 μm the fluorescence is between 616 nm and 618 nm. Even though the data points for larger domes are few, a second threshold diameter at 9 μm can be seen, above which the fluorescence maximum reaches up to 621 nm. It has to be emphasized that these data not necessarily imply that the fluorescence of a single J-aggregate has been measured. Even though the domes are on the micrometer scale, we cannot rule out that multiple nucleation of J-aggregates occur simultaneous in a single dome, which then leads to several J-aggregates per dome. By using far field microscopy we have no information how many J-aggregates (which might have a diameter of a few nanometer) exist in each microdome, and there might be more than one J-aggregate per dome.

Still, our data is, to the best of our knowledge, the first time a quantized shift in fluorescence spectra of J-aggregates has been observed. We can formulate several models for the origin of this phenomenon.

First, the size of the J-aggregates are indeed different in domes of different size. As mentionned before, there might be more than one J-aggregate per dome. But the growth of multiple J-aggregates may still depend on the dome size. Thus, the average size of the J-aggregate might be different in a large dome than it is in a small dome. It is known that J-aggregates may show a size-dependent shift in the absorption spectrum, thus a shift in the fluorescence spectrum can be anticipated, too.

A second reason for the shift in fluorescence maximum may be the sample geometry. The J-aggregates are not embedded in a continuous polymer film, but in lens-shaped polymer domes. The coherent laser light may be reflected from the curved interface and form an interfering light field, not unlike in a shpere. By incorporating a light emitting material inside of such a sphere a so called whispering-gallery laser can be made. Even though we do not observe a lasing phenomenon in our samples, a strong light field from the laser may influence the excited state of the J-aggregates and lead to a shift in the fluorescence peak.

A third explanation might be the readsorption of the fluorescent light. The Stokes shift of J-aggregates is small and thus the fluorescence occurs at the long wavelenth sholder of the absorption. The re-absorption of fluorescence light increases both with higher concentration and with larger film thickness. Since larger domes are also thicker in the direction perpendicular to the substrate, a re-absorption seems plausible. But this effect should be linear with dome size and should not be quantized.

3. Microcrystallization of Low Molar Mass Dyes in Dewetted Films

In the previous chapter the dewetting of a polymer solution was described, which can be used for the control of dye aggregation. In the following the concept of using dewetted structures as a controlled microenvironment will be extended to completely low molar mass systems, that is without the need for a polymer to form dewetted microdomes. The extension to low molar mass compounds is not trivial, since low molecular compounds tend to crystallize during solvent evaporation. Crystallization leads to inhomogeneieties in the solution and the dewetting process will be disrupted and no micronsized domes can be formed. Instead the sample consist of randomly oriented and spaced microcrystals or even larger crystals.

Here we report the successful preparation of tw-dimensional arrays of dye microcrystals. In order to circumvent the undesired immediate crystallization, the compound has to form a metastable amorphous state. When these compounds are cast from dilute solution, dewetting takes place, as in the polymer case, and micronsized domes of the compound are formed. Subsequent annealing then leads to the crystallization of the compounds and the formation of microcrystals. In the present study a series of electron acceptor/donor compounds were investigated, but the method of dewetting of a dilute solution is not limited to this class of compounds, but is a general method for the preparation of microcrystals on surfaces.

Table 1 : Summary of the crystallization behavior of thienyl dyes **5-10**

	compound	m.p. (°C)	crystallization as cast	crystallization after annealing	crystallization solvent vapor
5	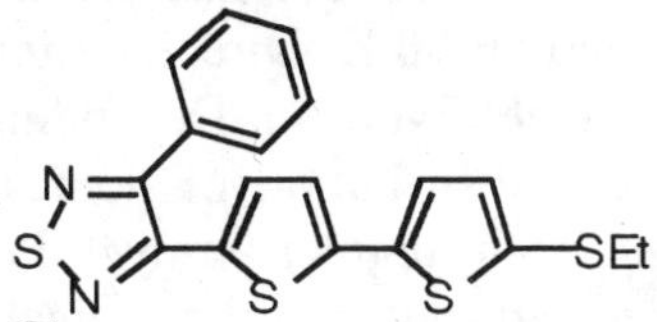	54	x	x	x
6	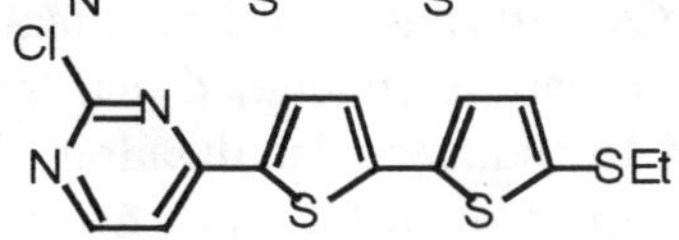	74	x	0	x
7	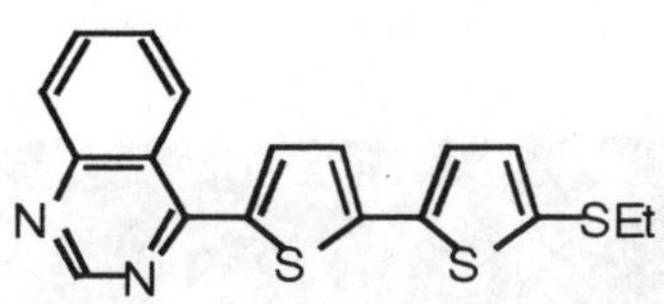	95	x	0	n.e.
8	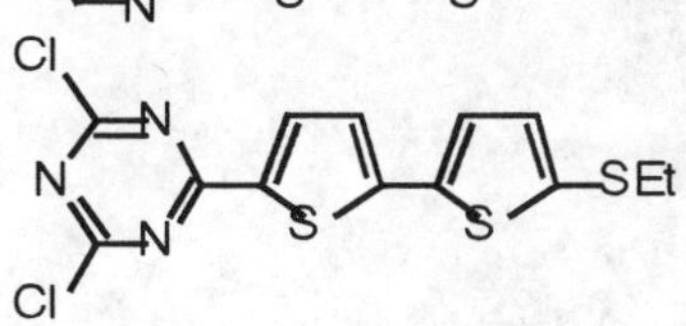	130	x	0	0
9	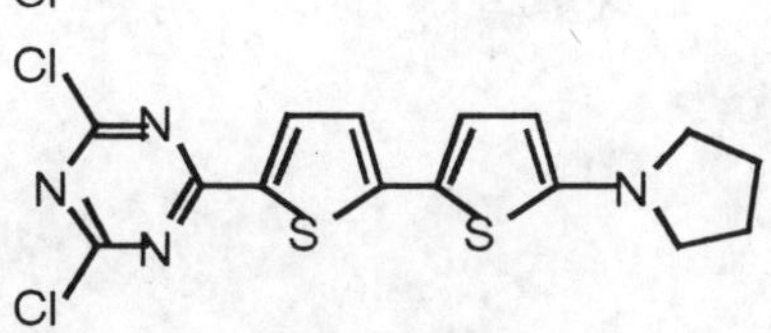	214	0	0	n.e.
10	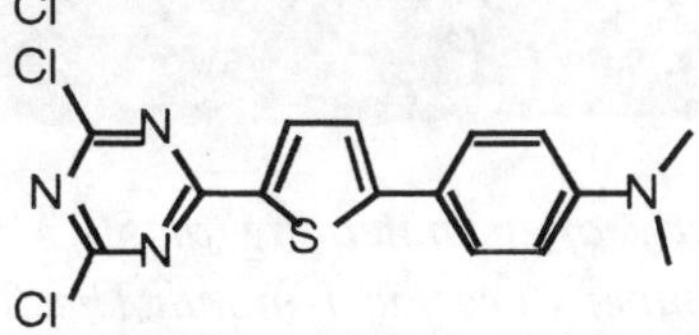	218	0	0	n.e.

x : no crystallization ; o : crystallization ; n.e. : not examined

As can be seen from Table 1, the bulk melting point determines the crystallization tendency and thus the possibility to form metastable amorphous dewetted structures. Compounds with a melting point above 200°C (**9** and **10**)

crystallize during the casting and thus are not available for patterning. Compound **5** with a melting point below 60°C, on the other hand, forms dewetted dome structures, but fails completely to crystallize. Compounds with an intermediate melting point form metastable amorphous domes which then crystallize, depending on the annealing conditions.

In the following, compounds **6** and **8**, which are very similar in structure, but have very different crystallization behavior will be compared.

3.1. MONOCHLORO-PYRIMIDINE DERIVATIVE 6

The monochloro compound **6** has a melting point of 75°C. Thus the metastable amorphous state at room temperature is undercooled by 50 K. Directly after casting the sample resembles the one of pure polystyrene (see Figure 1). Depending on the preparation condition, samples with a narrow size distribution of domes can be formed, even though the domes do not form such regular arrays. Polarizing microscopy, for which a polarizer and an analyzer are put into the optical path of the microscope, shows no birefringence so that it can be concluded that the domes are amorphouys. The sample is then annealed to induce crystallization. Figure 6 shows the dewetted sample after keeping the sample at 50°C for two months and then at room temperature for one month .

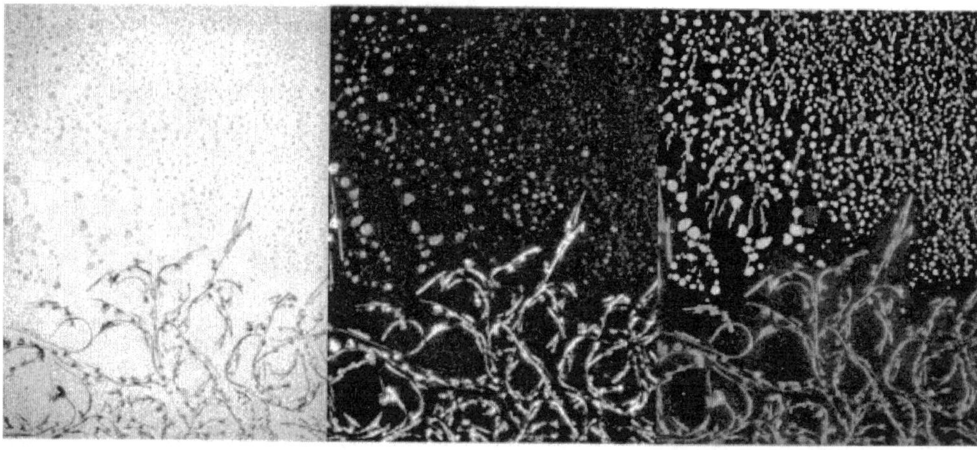

Figure 6 : optical micrograph of **6** *directly after annealing at 50 ° C for 2 months. Then the sample was kept at room temperature for 1 month.The fiber-like crystals develoedp during annealing at 50°C, while the round crystals formed after that and at room temperature. Transmission micrograph (left), polarizing micrograph (center) and fluorescence micrograph (right) of the same area. The scale of each micrograph is 1200 μm x 800 μm.*

Interestingly, two crystal morphologies with different fluorescence behavior were formed. After annealing at 50°C the formation of fiber-like crystals was observed (see that bottom parts of the pictures in Figure 6). The process of crystallization was followed by in situ microscopy and can be described as follows. Typically the fiber starts at a defect (scratch, dust particles, etc) or at the edge of the substrate. The formation of a nucleus leads to the crystallization of the first amorphous dome. During crystallization the shape of the initially spherical dome changes and becomes an elongated crystal fiber. Due to the lateral vicinity of the neighboring domes, this fiber may attach to the next dome and induce crystallization there. This process then repeates and leads to the formation of interconnected fibers. Polarizing microscopy (Figure 6, center) shows that the birefringence of the fibers is not homegeneous and it can be concluded that they are not single crystalline, even though the fluorescence micrograph shows homogeneous green fluorescence when excited with violet light (Figure 6, right). This crystallization process stops when either the sample is cooled to room temperature, or when a crystal fiber cannot elongate anymore due to the lack of neighboring domes. Thus the sample then contains regions with interconnected needles and regions with amorphous domes. When the sample is then subsequently kept at room temperature, the remaining amorphous domes crystallize. Interestingly their crystal morphology is different from those obtained at higher temperature. First, it has to be noted that the crystal shape is round (see the top part of Figure 6). Polarizing microscopy suggests that these domes form single crystals of micron size, since the birefringence is homogeneous within each dome. Unexpectedly, the excitation with violet light shows that these round crystals emit orange-red fluorescence. Fluorescence microscopy of a solution grown single crystal shows a needle-like crystal with greenfluorescing edges and orange fluorescing tips. Thus the different morphology and fluorescence behavior can be explained by a different orientation of the crystal axis, and is not the result of polymorphism. Further investigation into the control of crystal morphology and the formation of samples with one selected crystal orientation is present under way.

3.2. DICHLORO-TRIAZINE DERIVATIVE **8**

The dichloro derivative **8** has a higher molecular symmetry (due to the symmetric substitution of the triazole ring with two chloro atoms) than **6** and has a stronger dipole moment. These two factors lead to a higher melting point and a stronger tendency to crystallize. Thus the typical annealing times are 10-14 hours, compared with several weeks for **6** . Also here the annealing temperature is important, as can be seen in Fig. 7. Keeping the sample at room temperature leads to the formation of polycrystalline domes, which show inhomegeneous birefringence in the polarizing micrographs. On the other hand, keeping the sample at 80°C leads to the formation of small and isolated microcrystals which show homogeneous birefringence (Fig. 7) and fluorescence (data not shown) in each microcrystal.

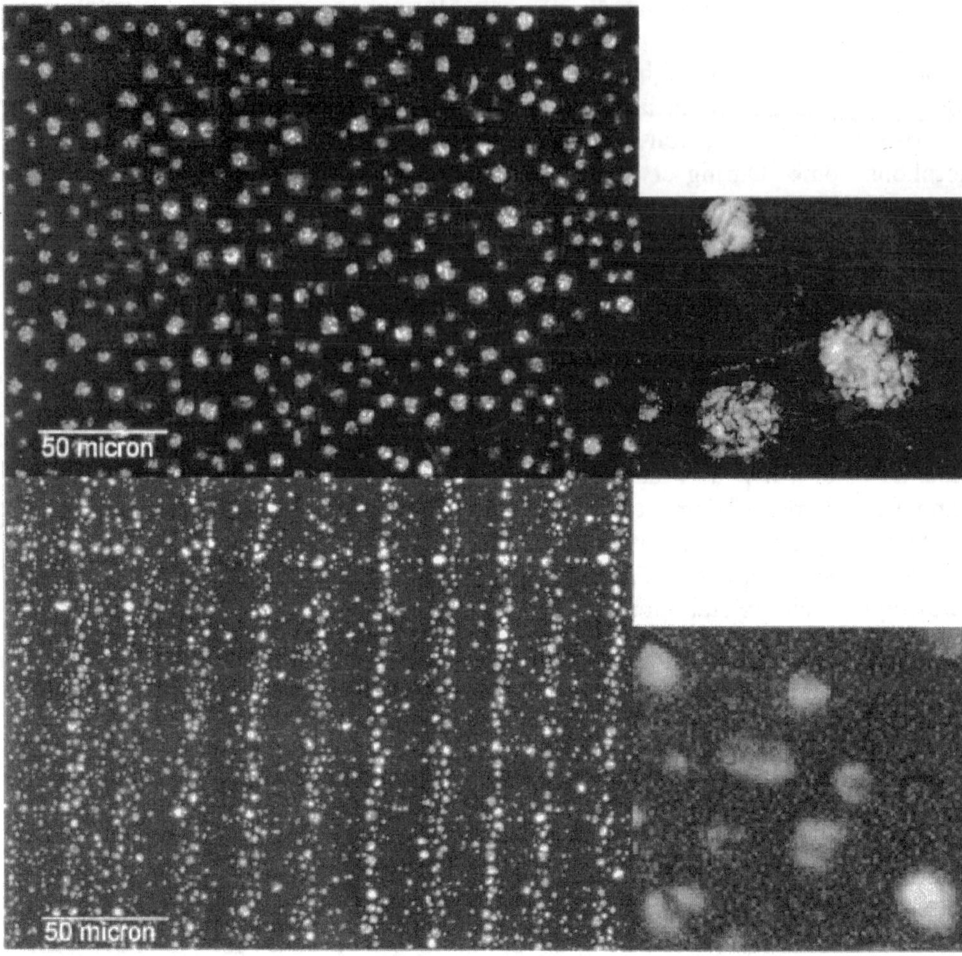

Figure 7: polarizing micrographs of crystallized microdomes of 8 after annealing for 1 day. Roomtemperature (top)annealing leads to polycrystalline sample domes, whereas annealing at 80°C (bottom) leads to micronsize single crystals.The small pictures on the righ are zoomed areas of the pictures on the left.

The preparation of micronsize crystals of dyes and pigments is important for many applications, e.g. organic photo conductors, in solar cells, in medicine, and so on. Thus our method may be usefull for the controlled preparation of a wide variety of dyes.

Acknowledgments

This work was partially funded by the Precursory Research for Embrionic Science and Technology "Organization and Function" of the JST and by the Scientific Research of Priority Area (A) "Dynamic Control of Strongly Correlated

Soft Materials" by the Japanese Ministry of Education, Culture, Sports, Science and Technology.

4. **Refererences**

1. W.West,P.B.Gilman:The Theory of the Photographic Process ,ed.by T.H.James (Macmillan,New York 1977)
2. G.McDermott,S.M.Prince,A.A.Freer,A.M.Hawthorn hwaite-Lawless, M.Z.Papiz,R.J.Cogdell,N.W.Issacs:Nature 374 ,517 (1995)
3. J.Koepke,X.Hu,C.Muenke,K.Schulten,H.Michel:Structure 4 ,58 (1996)
4.O.Karthaus,K.Ijiro,M.Shimomura:Chem.Lett.821 (1996)
5.O.Karthaus,L.Grasjo,N.Maruyama,M.Shimomura:Chaos 9 ,308 (1999)
6.J.Hellmann,M.Hamano,O.Karthaus,K.Ijiro,M.Shimomura,M.Irie:Jpn. J.Appl.Phys.37 ,L816 (1998)
7.O.Karthaus,T.Koito,M.Shimomura:Mater.Sci.Eng.C 8 –9 ,523 (1999)
8.O.Karthaus,H.Yabu,K.Akagi,M.Shimomura:Mol.Cryst.Liq.Cryst.364 , 395 (2001)
9. O. Karthaus, N. Maruyama, H. Yabu, T. Koito, K. Akagi, M. Shimomura Macromol. Symp., 160, 137-142 (2001)
10E.E.Jelly:Nature 138 ,1009 (1936)
11.G.Scheibe:Angew.Chem.49 ,563 (1936)
12. A.S.Davidov:Theory of Molecular Excitons ,(Plenum,1971)
13. D.A.Higgins,J.Kerimo,D.A.Vanden Bout,P.F.Barbara:J.Am.Chem.Soc. 118 ,4049 (1996)
14A.A.Muenter,D.V.Brumbaugh,J.Apolito,L.A.Horn,F.C.Spano, S.Mukamel:J.Phys.Chem.96 ,2783 (1992)
15. K.Minoshima,M.Taiji,K.Misawa,T.Kobayashi:Chem.Phys.Lett.218 ,67 (1994)

NONLINEAR-OPTICAL PROPERTIES OF SELF-ASSEMBLED MOLECULAR J-AGGREGATES OF PSEUDOISOCYANINE IN NANOMETER FILMS

R.V. MARKOV, A.I. PLEKHANOV,
Institute of Automation and Electrometry SB RAS.
Prosp. Acad. Koptyug 1, 630090 Novosibirsk - 90, Russia
Z.M. IVANOVA, V.V. SHELKOVNIKOV,
Novosibirsk Institute of Organic Chemistry SB RAS.
Prosp. Acad. Lavrentjev 9, 630090 Novosibirsk - 90, Russia
J.KNOESTER
Institute for Theoretical Physics and Material Science Center,
University of Groningen.
Nijenborgh 4, 9747 AG Groningen, The Netherlands.

1. Introduction

J-aggregates, first discovered by Jelley and Scheibe in 1936 [1,2] for cyanine dyes, have been studied for many years [3]. They are well-known for their unique optical properties, such as a strong absorption, narrow line widths, and high optical non-linearities. Their role as photographic sensitizers can hardly be overestimated [4-6]. Recent studies have shown that a number of biological light-harvesting systems, such as the LH1 and LH2 complexes of purple photosynthetic bacteria [7] and the chlorosome of green photosynthetic bacteria [8], consist of aggregated bacterio-chlorophylls. Thus, aggregates of dye molecules may be used to mimic light harvesting arrays and to prepare artificial photosynthetic systems. Another new development is the efficient electroluminescence revealed in single-layer light-emitting diodes based on electron-hole conducting polymers containing nano-crystalline phases of J-aggregates of cyanine dyes [9]. Finally, the property of J-aggregates that still holds most promise for the near future, is their high nonlinear (cubic) optical susceptibility, $\chi^{(3)} \sim 10^{-6} - 10^{-5}$ esu, with a fast response time (less than 1 ps) at the J-peak resonance [10-12]. As shown in [13], films of J-aggregates with bistable behavior may be the basis for two-dimensional optical switches, controllable by light. The nonlinear-optical properties of organic dye J-aggregates have been intensively studied for application in future optical telecommunication and signal processing systems with ultrahigh bit rates (Tbit/s) [14].

At present, methods are known to form J-aggregates in water solution, in frozen glasses, and in polymer matrices [3]. However, these methods do not give stable aggregates, which hampers their application as nonlinear optical materials. Besides, a thin-film geometry is preferable for applications. Therefore, the preparation of

F. Charra et al. (eds.), Organic Nanophotonics, 279–290.

large-area thin films of J-aggregates with thermal and photochemical stability and a high optical quality is vital for practical application and a matter of much current interest.

In this paper, we describe the results of detailed studies of the influence of adding appropriate substituents to cyanine dyes or different organic cations to the initial organic dye solution, as well as varying preparation techniques, with the goal to control the process of aggregation, and the structure and ordering of the aggregates and their nonlinear optical properties in thin films. Such control is of much scientific interest, as it allows for detailed studies of aggregation and the dynamics of excitons on the prepared aggregates. In particular, we have shown that by spin-coating from dye solutions, nanometer J-aggregate films may be prepared that are stable at room temperature and have a good optical quality over areas of 30 x 30 mm. We will also report on the first observation of the effect of size-enhancement of the nonlinear optical properties in one-dimensional molecular aggregates.

2. Materials and methods

2.1. PREPARATION OF J-AGGREGATE FILMS

2.1.1. *N-alkyl-substituted J-aggregates*

We first prepared J-aggregate films using the dye 1,1'-diethyl-2,2'-cyanine iodide [pseudo-isocyanine (PIC)]. The production of films without a stabilizing polymer can lead to an increased concentration of J-aggregates in the layer, resulting in a larger optical nonlinearity of the films. Such films, first obtained in [15], have a high optical nonlinearity $\chi^{(3)} \sim 10^{-5}$ esu. However, these J-aggregates proved to be unstable and were destroyed due to crystalisation in 1-2 days. To solve this problem, efficient routes to synthesize long-alkyl substituted PIC dyes were developed, leading to eight different dyes [16]. Some of these dyes were used to

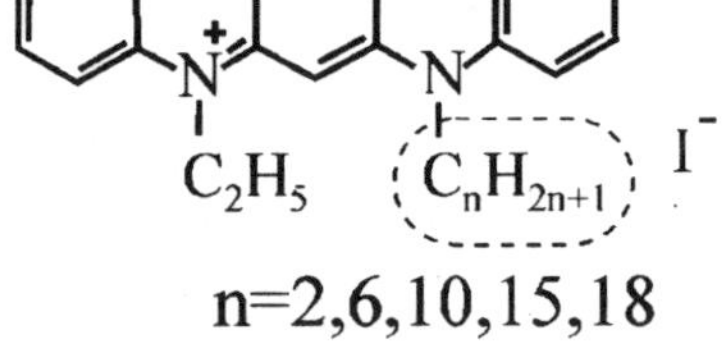

Figure 1. Molecular structure of PIC-I forming J-aggregates.

form new types of J-aggregates without stabilizing the aggregates by polymers. In particular, PIC derivatives were used to create solid films, with a thickness of 15-30 nm, by spin coating (at 1500 – 3000 rpm) from the organic solvents acetonitrile and dichloroethylene (at molar concentration $c = 5 \cdot 10^{-3}$ M) on a glass substrate (30 x 30 mm). It was established that the PIC dyes with non-symmetric substitution formed films with J-aggregate nature, characterized by a narrow absorption band (J-peak) red-shifted relative to the single-molecule absorption spectrum. We have also found that the symmetrically substituted dyes have a poor ability to form films on a glass substrate, as a result of the bad wetting of the glass surface by the fat tails. This problem was solved by silicone organic compound hydrofobization of the substrate surface. The thus obtained films of the symmetrical dyes do not form J-aggregates, but rather tend to form H-aggregates, which are characterized by an exciton absorption band that is blue-shifted compared to the monomer absorption.

At room temperature the formation of stable (few months) J-aggregates with a narrow J-peak (200 cm^{-1}, FWHM) at 583 nm is observed for the asymmetric PIC dyes (see Fig.1), containing a C_2H_5-group at one nitrogen atom and C_nH_{2n+1} (n=6, 10, 15, 18) group at the other one [17] (Fig 2a, curve 1). The humidity plays and important and positive role in the growth of the J-peak. We found that the growth is noticeable when the alkyl tail is longer. This shows the important role of

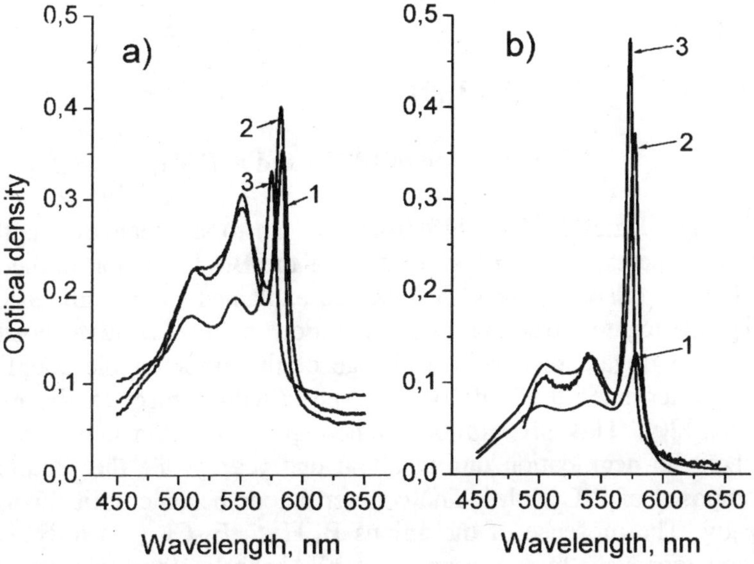

Figure 2. Absorption spectra of PIC J-aggregates in thin films: (a1) PIC iodide with alkyl tails C_2H_5-$C_{18}H_{37}$ at 290 K; (a2) PIC iodide with alkyl tails C_2H_5-$C_{18}H_{37}$ at 80 K; (a3) PIC with anion $B_{10}H_{10}^{2-}$ at 290 K. (b) The width of the J-peak is narrowed due to addition of tetrabuthylammonium iodide and decreasing the temperature. Curve b1 shows the spectrum for pure PIC_2^+ - $B_{10}H_{10}^{2-}$; curve b2 shows the spectrum for PIC_2^+ - $B_{10}H_{10}^{2-}$ / tetrabuthylammonium iodide at molar ratio 1:2 at 300 K; curve b3 is the same as curve b2 but at 77 K.

hydrophobic interactions in the J-aggregation process. Decreasing the temperature to 80 K gives a further narrowing of the J-peak (down to 75 cm-1) [18] (see Fig 2a, curves 1 and 2). Thus, temperature is an important tool to control the width of the J-peak [19].

2.1.2. *J-aggregates with borate as anion*

We have also found that the formation of stable films strongly depends on the nature of the counter-anion. For PIC with the symmetric substituents C_2H_5-C_2H_5, the anion decahydro-*closo*-decaborate ($B_{10}H_{10}^{2-}$) leads to efficient J-aggregate formation on a glass surface. It is sufficient to have only one $B_{10}H_{10}^{2-}$ anion per 20 dye molecules for the formation of J- aggregates to occur. Figure 3 illustrates schematically the reaction of (PIC)I and $K_2B_{10}H_{10}$. The absorption peaks of

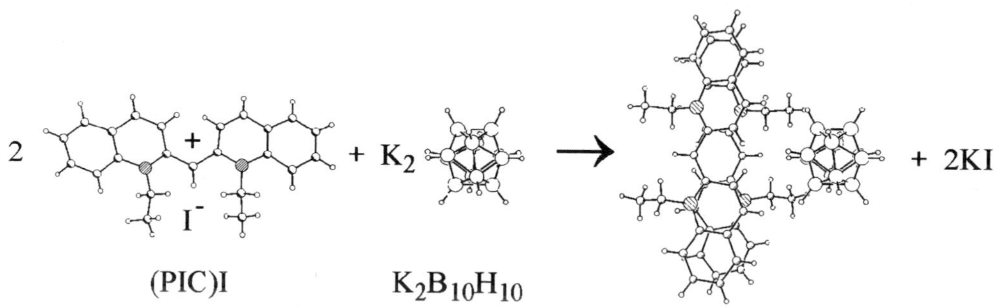

Figure 3. The reaction of (PIC)I and $K_2B_{10}H_{10}$.

the films with J-aggregates tend to shift from 583 nm to shorter wavelengths over about 10 nm upon increasing the concentration of the $B_{10}H_{10}^{2-}$ anion in the organic solution (see Fig 2a, curve 3). This shift can be explained by the influence of the anions on the aggregate structure. An alteration of the relative position of molecules in the aggregate results in a change of the dipole-dipole coupling, the value of which determines the shift of the J-peak with respect to the monomer absorption band [20]. This preparation method gives thin films of high optical quality that show no degradation during about one year, while the samples of J-aggregates composed of PIC with iodide counter anions become unfit for use after 2-3 days already. The presence of the anions $B_{12}H_{12}^{2-}$, $B_{10}Cl_{10}^{2-}$ in a dye solution destroys the spontaneously formed J-aggregates. Absorption spectroscopy and X-ray analysis show that the anion $B_{10}H_{10}^{2-}$ gives the complex $(PIC)_2B_{10}H_{10}^{2-}$, which forms J-aggregates more effectively than PIC iodide. The organizing role of the $B_{10}H_{10}^{2-}$ anion is due to close B-H···H-C contacts in the complex $(PIC)_2B_{10}H_{10}^{2-}$ [21].

Recently, we have found the same best stability for PIC J-aggregate thin films doped by the cluster borane derivative $[Ni(B_9C_2H_{11})_2]^0$ [22].

2.1.3. *J-aggregates with different organic cations*

We have experimentally found for the first time that adding organic cations (methyl quinolinium iodide, tetrabuthyl ammonium iodide, thiocholine iodide, dodecylpypidinium bromide) to pseudo-isocyanine *closo*-decaborate leads to narrowing of the J-band from (HWHM) 250 cm^{-1} to 140 cm^{-1} and allows to prepare J-aggregated films with a controlled width of the J-peak. Typical absorption spectra of the films are shown in Fig 2b. As we shall see later, this narrowing of the J-peak leads in turn to an increase of the third-order optical susceptibility of J-aggregates in thin solid films. As the narrowing of the J-peak is related to the ordering of the J-aggregates, our results stress the important role of the ordering of the aggregates to improve the nonlinear optical properties of solid films of organic dyes.

2.1.4. *J-aggregates in polymer matrices*

In spite of the fact that long-alkyl substituted PIC dyes allow to form effectively aggregates without stabilizing polymers, with optical and nonlinear optical properties that are stable during several months, it is important for applications to have even more stable films that are well-protected against environment. To this end, we also studied J-aggregates stabilized by polymers. In this work thin solid films ($\approx$ 100 nm) of PIC J-aggregates were obtained by spin coating of the proper choice of an organic-organic or a water-organic mixture of the solvents and the optimal dye/polymer weight ratio. We have used five polymers of different types, such as hydrophobic polymethylmetacrylate (PMMA), polyvinilacetate (PVA), amphiphylic polyvinylpyrrolidone (PVP), hydrophilic poly-(sodium-p-styrene-sulfonate) (PSS) and poly-(anetholesulfonate) (PAS). It was established that PIC J-aggregates are effectively formed in PMMA, PVP, PAS and PSS matrices and are not formed in a PVA matrix at the weight concentration ratios of dye:polymer from 1:2 up to 1:8. The ratio 1:4 was found to be optimal, giving the maximum conversion from the monomeric to the aggregated form of the dye. We have used PIC with the symmetric substituents C_2H_5-C_2H_5 and PIC with non-symmetric substituents C_2H_5-$C_{18}H_{37}$ at concentration $\sim 10^{-2}$ M and in both cases found the same efficiency of J-aggregate formation in polymer matrices. In that case, the optical density at the J-peak maximum was about one and the width of the J-peak was about 160-200 cm^{-1} (HWHM). The J-aggregates prepared in PAS, PSS, and PMMA matrices were stable at room temperature and could be stored without decomposition for more than six months. These J-aggregates were selected for studying their nonlinear optical properties.

2.2. OPTICAL AND NONLINEAR OPTICAL METHODS FOR CHARACTERIZING THIN J-AGGREGATE FILMS.

Absorption spectra were recorded on Specord M-40, Shimadzu-3000UV, and HP-s8453 (Hewlett Packard) spectrophotometers. The thermal stability of the thin films was studied by measuring their optical spectra during heating at a rate of 1.5 K per minute in a thermostat optical cell. The absorption spectrum of the film was recorded during 1-2 seconds on the HP-s8453 spectrophotometer, so that the change of the sample temperature during each measurement could be neglected.

The film thickness and the frequency dispersion of the refractive index (n_o) and the absorption coefficient (k) were determined by spectral ellipsometry. We have used a spectral ellipsometer developed at the Institute of Semiconductor Physics, Siberian Branch of RAS. The optical constants of the thin films were measured with an accuracy of ±0.01 for n_o and k and ±0.5 nm for the thickness. We have also used the Nanoscope III Multimode SFM (Digital Instrument Inc., Santa Barbara, CA) in the Max Planck Institute for Colloid and Interface Research (Potsdam) to obtain an image of the dye film surface on a glass plate and to determine the film thicknes.

We have used the Z-scan method [23] and pump-probe experiments to study the third-order nonlinear optical susceptibility in the region of the exciton resonance of the J-aggregates. The real and imaginary parts of the cubic susceptibility of the prepared films (both with iodide counter-ions and doped with $B_{10}H_{10}^{2-}$ anions) were measured using the Z-scan method at nanosecond pulse duration. A pulsed dye (Rodamin 6G) laser pumped by the second harmonic of an LTI-411 pulsed Nd^{3+}YAG laser was used in the experiments (we used 5 nanoseconds pulses). The generation linewidth of the dye laser was 0.3 cm^{-1}. In these experiments, the sample moved along the propagation path (z) of a focussed Gaussian beam. The radiation intensity at the focal point was 1 MW/cm^2 and about 500 W/cm^2 far from the focal point. We have measured the transmittance of the nonlinear medium by a photodiode in the far field as a function of the sample position z measured with respect to the focal plane. These measurements make it possible to calculate the nonlinear optical absorption and refraction coefficients, which are related to the imaginary and real parts of cubic susceptibility. Due to the low laser intensity far from the focal point, we could also measure the linear transmittance in the studied point of the film. Using Kramers-Kronig relations, the dispersion of the real part of the nonlinear susceptibility was also calculated starting from the measured imaginary part.

The pump-probe method was used to perform measurements at low temperatures in a cryostat. In these measurements, we used two pulsed dye lasers. The intensity of the pump radiation was 0.5 – 1 MW/cm^2, and the probe beam intensity was three orders of magnitude lower. The pump and probe pulses acted on the sample simultaneously. The duration of both pulses was 5 ns. The pump radiation frequency in each measurement run was tuned to the absorption maximum of the J-aggregates, while the probe frequency was scanned in the region of the exciton absorption band. Measuring the absorption of the probe beam in the presence and in the absence of a strong pump field allows for the determination of the nonlinear absorption coefficient or the imaginary part of the nonlinear optical susceptibility. Note that in the Z-scan method the self-action of a high-intensity laser beam in a nonlinear medium is measured, whereas in the pump-probe technique, the absorption coefficient of weak radiation on a nonlinear medium is measured in the presence of high-intensity laser radiation.

3. Results and discussion

3.1. DISPERSION OF OPTICAL CONSTANTS OF J-AGGREGATE FILMS

The frequency dispersion of the absorption coefficient k and the refractive index n_o of the J-aggregate films were determined in the wavelength region from 500 to 650 nm. As an illustration the resulting spectra for two films are shown in Fig.4. The spectra for other films were similar. One observes normal dispersion for n_o within the long-wave spectral region of film transparency and anomalous dispersion for n_o within the region of dye absorption (500-570 nm), with an n_o value 2.3 – 2.7 at the

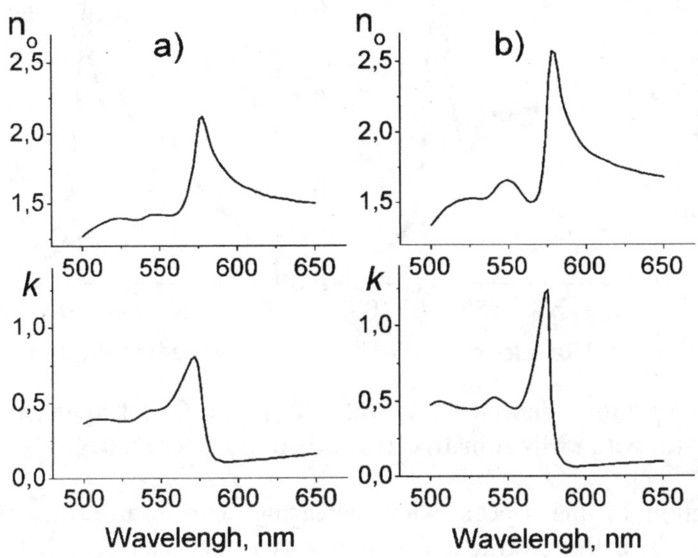

Figure 4. Refractive index n_o and absorption coefficient k of the film: (a) PIC borate on a glass substrate (the film thickness is 15 nm); (b) PIC borate with addition of tetrabuthyl ammonium iodide at the weight concentration ratio dye/organic salt = 1:2 (the film thickness is 23 nm).

maximum of the J-peak. The absorption coefficient k at the J-peak maximum corresponds to $\alpha = 4\pi k/\lambda = 2.4\cdot10^5$ cm^{-1}. The comparison of Figs. 3a and 3b shows that upon adding organic salt, n_o and k increase near the J-peak. From a practical standpoint it is interesting to note that the refractive index of the films in this spectral region greatly exceeds those of fused silica and optical glass.

3.2. THERMAL STABILITY OF J-AGGREGATES IN THIN FILMS

We have studied the thermal stability of the J-aggregates prepared from the PIC dyes with long-N-alkyl non-symmetric substituents [24] and of the J-aggregates in a polymeric matrix.

At room temperature the J-aggregated films of non-symmetrical long alkyl-substituted PIC are stable. Above room temperature, however, the thermal stability is poor.

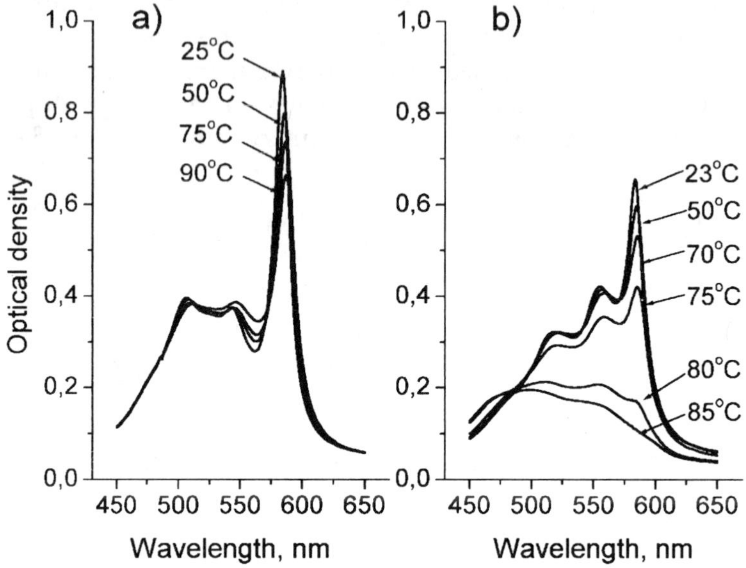

Figure 5. Temperature dependence of the PIC $\{C_2H_5\text{-}C_2H_5\}$ aggregate absorption in thin films: (a) with PMMA matrix; (b) without polymer matrix.

The destruction of the J-peak with increasing temperature was studied. The temperature where the destruction started was established as follows: PIC with tails $\{C_2H_5\text{-}C_2H_5\}$ (73°C) > PIC $\{C_2H_5\text{-}C_6H_{13}\}$ (67°C) > PIC $\{C_2H_5\text{-}C_{18}H_{37}\}$ (55°C) > PIC $\{C_2H_5\text{-}C_{10}H_{21}\}$ (47°C) > PIC $\{C_2H_5\text{-}C_{15}H_{31}\}$ (37°C). The dye PIC with tails $\{C_2H_5\text{-}C_2H_5\}$ tends to decay spontaneously by re-crystallization during storage at room and the J-peak disappears completely and irreversibly, in spite of the rather high temperature of the thermal decay. On the contrary, the J-aggregates of the dyes with long-alkyl tails has a tendency to grow during storage and the J-peak is recovered (up to 90% of their initial optical density) after thermal decay. The J-aggregate recovering speed depends on the humidity and length of the alkyl tail. The recovering speed is faster when the alkyl tail is longer. This result once more shows the important role of the hydrophobic interaction in the formation of J-aggregates.

The investigations of thermal stability of PIC J-aggregates in polymeric matrices have shown that the J-aggregates were more stable than thin film J-aggregates without polymer. For comparison the absorption spectra of thin films of PIC iodide J-aggregates in polymethylmetacrylate matrix and without polymer matrix are shown in Fig. 5. Furthermore, the thin solid films with PIC J-aggregates in polymer matrices (PMMA, PAS, PSS) almost completely recover their initial optical density upon cooling down after heating.

3.3. NONLINEAR OPTICAL PROPERTIES OF J-AGGREGATE THIN FILMS AND OBSERVATION OF SIZE ENHANCEMENT

The magnitude and frequency dispersion of the imaginary part of the third-order nonlinear optical susceptibility $\chi^{(3)}$ for the samples with J-aggregates and polymer was measured using the pump-probe method. The maximal value of $\mathrm{Im}\chi^{(3)} \sim -10^{-5}$ esu was observed for the J-aggregates in PMMA and PAS matrices. The threshold of the radiation destruction of J-aggregates in polymer films lies at the laser intensity of 1 $\mathrm{MW/cm^2}$.

The magnitude and frequency dispersion of $\mathrm{Im}\chi^{(3)}$ for the samples with controlled exciton absorption line width was measured both by the Z-scan and the pump-probe techniques. Here we have used thin films of molecular J-aggregates of PIC with $B_{10}H_{10}^{2-}$ as counter ion. The absorption line width was varied by two

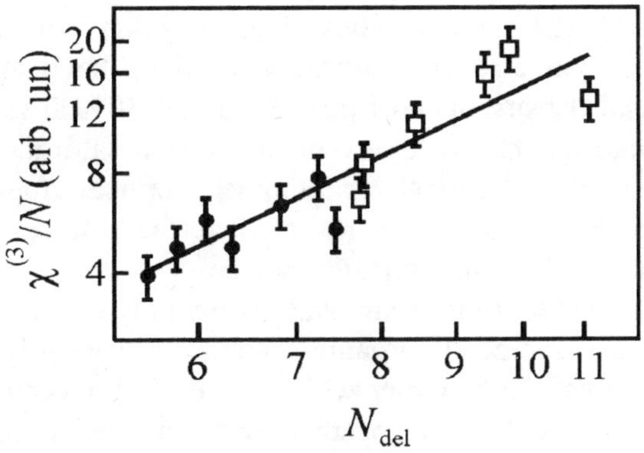

Figure 6. Third-order optical susceptibility normalized to the concentration of molecules in the aggregate phase plotted versus the delocalization length in the log-log coordinates. The data obtained by pump-probe technique at room (solid dots) and nitrogen (open squares) temperature for the samples with addition of tetrabuthyl ammonium iodide salt.

techniques. First: by addition of organic salts to the J-aggregate sample, second: by decreasing the sample temperature. The absolute values of the nonlinear susceptibility were about $\chi^{(3)} \sim 10^{-5}$ esu at room temperature and $\chi^{(3)} \sim 10^{-4}$ esu at liquid nitrogen temperature. A power scaling of the nonlinear susceptibility as a function of the absorption line width $\chi^{(3)} \sim W^{\delta}$ was observed with $\delta = -(1.36 \div 1.58)$. This scaling was found not to depend on the method used to vary exciton absorption line width. Under the assumption that the line width is governed by the exciton delocalization length according to the scaling relation published in Ref. [25], the above scaling can be translated into a relation between the nonlinear susceptibility and the delocalization length. We then obtain $\chi^{(3)} \sim N_{del}^{\gamma}$, with $\gamma = 2.2 \pm 0.3$ (see Fig.6). This result agrees very well with the theoretical prediction $\gamma = 2.36$ for the effect of enhancement of the nonlinear optical properties of molecular aggregates with growing exciton delocalization length [26].

4. Conclusion

We have shown that long-alkyl substituted pseudo-isocyanine (PIC) dyes form new types of J-aggregates. In particular, the PIC dyes with non-symmetric substitution were used to create solid films, with a thickness of 15-30 nm, by spin coating from the organic solvents on a glass substrate. Such films were prepared without stabilizing the aggregates by polymers. It was found that the substituent length, the nature of the counter-anion, temperature, and humidity have a strong influence on the film stability and the J-band. The resulting thin films are of high optical quality and show no degradation for more than six months. The obtained films have a high optical quality on an area of more than 30x30 mm and high photochemical stability: they withstand laser intensities of up to 2 MW/cm^2 at nanosecond pulse duration. They show an effective saturation of the J-band at its maximum. The change of the optical density of the J-peak by about 30% at radiation intensities of about 1MW/cm^2 and the fast recovery of absorption saturation (faster than 1 ps) make these films very attractive for nonlinear optical applications. The large absorption coefficient of more than 10^5 cm^{-1} at the resonant wavelength is also important for constructing actual thin-film devices.

Nanometer J-aggregate films with stabilizing polymer were obtained by spin coating of the proper choice of organic-organic or water-organic mixture of the solvents and the optimal dye/polymer weight ratio. The maximal value of the cubic susceptibility $\mathrm{Im}\chi^{(3)} \sim -10^{-5}$ esu together with the technological advantages of polymer protected J-aggregated films make them attractive for the application in optical devices. A useful property of the polymer is also that it prevents the coalescence of unstable assemblies such as J-aggregates as well as metal clusters or semiconductor nanostructures, which can be embedded into a matrix for amplification of the nonlinear optical properties or other functionalities.

We have also found that the addition of different organic cations in the initial dye solution leads to the formation of an intense and narrow J-band. By using these different influences, the aggregation and the optical properties of the aggregate may be controlled. In particular, the width of the absorption band and the value of the nonlinear susceptibility are controllable. This has allowed us to observe for the first time the predicted effect of size enhancement of the nonlinear optical properties for molecular aggregates.

5. Acknowledgments

Financial support by the NWO 047.009.019, INTAS 97-10434, and RFBI 02-03-33336 grants is gratefully acknowledged

9. References

1. Jelley E.E. (1936) Spectral absorption and fluorescence of dyes in the molecular state, *Nature* **138**, 1009-1010.
2. Scheibe G. (1936) Variability of the absorption spectra of some sensitizing dyes and its cause, *Angew. Chem.* **49**, 563.
3. Kobayashi T., ed. (1996) *J-aggregates,* World Scientific Publish. Co Pte. Ltd., Singapore.
4. Tani T. (1996) J-aggregates in spectral sensitization of photographic materials, in T. Kobayashi (eds.), *J-aggregates,* World Scientific Publish. Co Pte. Ltd., Singapore, pp. 209-228.
5. Trosken B., Willig F., Spitler M. (1995) The primary steps in photography: excited J-aggregates on AgBr microcrystals, *Advanced Materials* **7,** 448-450.
6. Shapiro B.I. (1994) Aggregates of cyanine dyes: photographic problems, *Russian chemical reviews* **63,** 231-241.
7. McDermott G., Prince S.M., Freer A.A. et al. (1995) Crystal structure of an integral membrane light-harvesting complex from photosynthetic bacteria, *Nature* **374**, 517-521.
8. Blankenship R.E., Olson J.M., Miller M. (1995) Antenna complexes from green photosynthetic bacteria, in R.E. Blankenship, M.T. Madigan, and C.E. Bauer (eds.), *Anoxygenic photosynthetic bacteria.* Kluwer Academic Publish., Dordrecht, pp. 399-435.
9. Mal'tsev E.I., Lypenko D.A., Shapiro B.I., and Brusentseva M.A. (1999) Electroluminescence of polymer/J-aggregate composites, *Appl. Phys. Lett.* **75**, 1896-1898.
10. Wang Y. (1991) Resonant third-order optical nonlinearity of molecular aggregates with low-dimensional excitons, *Journal of the Optical Society of America B* **8,** 981-990.
11. Bogdanov V.L., Viktorova E N., Kulya S V., and Spiro A.S. (1991) Nonlinear cubic susceptibility and dephasing of exciton transitions in molecular aggregates, *JETP Letters* **53**, 105-108.
12. Zhuravlev F.A., Orlova N.A., Plekhanov A.I., Safonov V.P., Rautian S.G., Shelkovnikov V.V. (1992) Giant nonlinear susceptibility of thin-films with (molecular J-aggregate)-(metal cluster) complexes, *JETF Lett* **56,** 260-263.
13. Glaeske H., Malyshev V.A., Feller K.-H. (2001) Mirrorless optical bistability of an ultrathin glassy film built up of oriented J-aggregates: Effects of two-exciton states and exciton-exciton annihilation, *J. Chem. Phys.* **114,** 1966-1969.
14. Furuki M., Tian M., Sato Y., Pu L.S. (2000) Terahertz demultiplexing by a

single short time-to-space convertion using a film of squaryliun dye J-aggregates, *Appl. Phys. Lett.* **77**, 472-474.

15. Shelkovnikov V.V. Zhuravlev F.A., Orlova N.A., Plekhanov A.I., Safonov V.P. (1995) Polymer films of J-aggregated cyanine dyes and metal clusters for non-linear optical applications, *J. Mater. Chem.* **5**, 1331-1334.
16. Shelkovnikov V.V., Orlova N.A., Kolchina E.F., Zhuravlev F.A., et al. Synthesis 2,2'-quinocyanines with long-chain N-alkyl groups, (2002) accepted in *Journal of Heterocyclic Compounds*.
17. Markov R.V., Chubakov P.A., Plekhanov A.I. et al. (2000) Optical and nonlinear optical properties of low-dimensional aggregates of amphyphilic cyanine dyes, *Nonlinear Optics* **25,** 365-371.
18. Shelkovnikov V.V., Markov R.V., Plekhanov A.I., et al. (2002) Optical and nonlinear-optical properties of pseudoisocyanine J-aggregates with the controlled exciton absorption linewidth, *High Energy Chemistry* **36**, 260-264.
19. Renge I., Wild U.P. (1997) Solvent, temperature, and excitonic effects in the optical spectra of Pseudoisocyanine monomer and J-aggregates, *J. Phys. Chem. A* **101,** 7977-7988.
20. Knapp E.W. (1984) Line shapes of molecular aggregates exchange narrowing and intersite correlation, *Chem.Phys.* **85,** 73-82.
21. Gerasimova T.N., Orlova N.A., Shelkovnikov V.V. et al. (2000) The structure of pseudoisocyanine decahydro-closo-decaborate and its nonlinear optical properties in thin films, *Chemistry for Sustainable Development* **8,** 109-114.
22. Shelkovnikov V.V., Ivanova Z.M., Orlova N.A. et al. (2002) Thin solid films of the pseudoisocyanine doped by cluster borane derivatives for nonlinear optics, *Abstracts of XI international conference on boron chemistry*, July 28- August 2, Moscow, Russian Federation, p.144.
23. Sheik-Bahae M., Said A.A., and Van Stryland E.W. (1989) High sensitive single beam n_2 measurement, *Opt. Lett.* **14,** 955-957 .
24. Shelkovnikov V.V., Ivanova Z.M., Orlova N.A. et al. (2002) Preparation and properties of J-aggregates of the pseudoisocyanines with long alkyl substituents in thin films. Spectral properties and thermal stability, *Optics and Spectroscopy* **92,** 958-966.
25. Bakalis L.D., Knoester J. (2000) Linear absorption as a tool to measure the exciton delocalisation length in molecular assemblies, *J.Lumin.* **87-89,** 66-70.
26. Knoester J. (1993) Third-order optical response of molecular aggregates. Disorder and the breakdown of size-enhancement, *Chem.Phys.Lett.* **203,** 371-377.

CAVITY POLARITONS IN ORGANIC MATERIALS

V.M.AGRANOVICH
Institute for Spectroscopy, Russian Academy of Sciences
Troitsk Moscow obl., 142190 Russia

M. LITINSKAIA
Institute for Spectroscopy, Russian Academy of Sciences,
Troitsk Moscow obl., 142190 Russia

G.C. LA ROCCA
Scuola Normale Superiore
Piazza dei Cavalieri 7, 56126 Pisa, Italy

D.G.LIDZEY
Department of Physics and Astronomy, University of Sheffield,
Sheffield S3 7RH, United Kingdom

Abstract.

We analyze theoretically the spectrum of excitations in a microcavity, which utilizes organic materials with broad and dispersionless electronic resonances as the optically active semiconductor. The excitations in such non-cavity materials are localized, and for these excitations the wave vector is not a "good" quantum number. We show that as a result of strong light-matter coupling two polariton branches appear which are analogous to the cavity-polariton branches observed in inorganic semiconductor structures. However, the polariton branches only exist in a certain restricted intervals of wave vectors. We show that in such materials the majority of the electronic excited states do not strongly couple to the cavity photon, and these states can be regarded as essentially incoherent.

Applying this physical picture for the microcavities containing disordered cyanine dye J-aggregates, we examine the decay of the upper cavity polaritons accompanied by the emission of an intramolecular phonon. We show that the main contribution to the upper polariton nonradiative decay rate arises from the transition to incoherent states, and the transition is very fast (of the order of 50 fs). Finally, we show that the exciton-exciton annihilation in organic microcavities is strongly suppressed due to the polaritonic effects.

F. Charra et al. (eds.), Organic Nanophotonics, 291–315.

1. Introduction

In the last decade planar semiconductor microcavities have attracted much attention, as they provide a method to enhance and control the interaction between light and electronic excitations [1]. The character of the light-matter interaction depends on the competition between the processes of exciton-photon coupling and damping (both optical damping and exciton dephasing). If the damping prevails, the system is in the weak coupling regime, and the effect of light is just to modify the radiative decay rate and the emission angular pattern of the cavity mode. In contrast, if the damping processes are small in comparison with the interaction, the system is said to be in the strong coupling regime. Then the true eigenstates of the system are the doublet of the cavity polaritons which are separated by the Rabi splitting energy [2], which is proportional to the square root of the oscillator strength of the transition.

The strong coupling regime was observed for the first time in inorganic semiconductors [3], and since then it has been investigated extensively both experimentally and theoretically (see, e.g., [2] for the review). The dynamics of microcavity polaritons in inorganic structures are now reasonably well understood [4]. Commonly, the structures associated with the strong coupling regime are inorganic crystalline materials with a small width of the electronic resonance, where the large radii Wannier-Mott excitonic states form. The hybridization of the excitonic states with the cavity photon results in the appearance of the cavity polaritons. If the structural disorder is moderate, for all the polaritonic states the wave vector is a "good" quantum number. The typical value of the Rabi splitting in these structures ranges between 5 and 10 meV, i.e. it is moderate.

In this connection, the use of organic structures as optically active material in microcavities is attractive. Their advantage is that they possess Frenkel excitons, which have an oscillator strength much larger than that in inorganic structures. Thus, it was noted [5] that the strength of their interaction with light (and, consequently, the resulting Rabi splitting) may be much larger than in the inorganic analogues. However, for many of organic materials the relation between the inhomogeneous broadening of the excitonic resonance and its oscillator strength is such that the strong coupling between the exciton and the cavity photon is washed out. That is why organic materials are usually associated with the weak coupling regime.

Quite recently, however, the strong coupling was observed in some organic materials. First, porphyrin dyes with a single optical absorption line of the width 90 meV were used as a semiconductor material [6]. A giant Rabi splitting (160 meV at room temperature) was observed. Soon after this, strong coupling was observed using another type of a semiconduc-

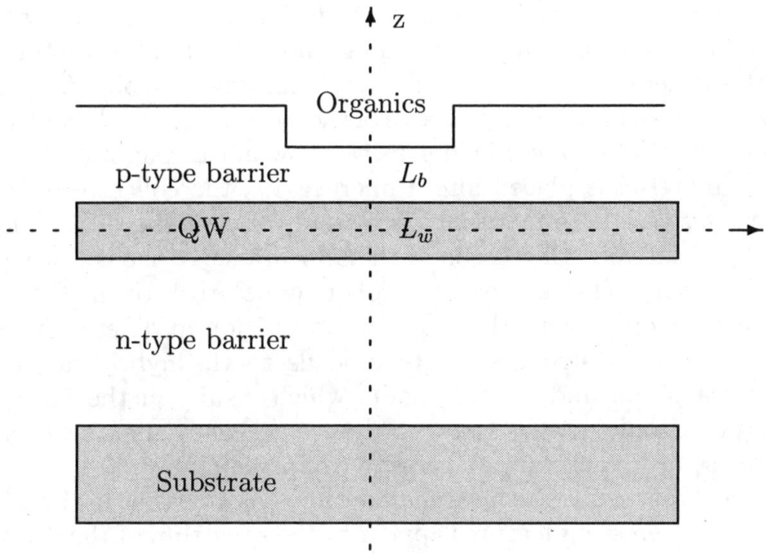

Figure 1. The experimental points and the fitting curves showing the anticrossing behavior (taken from Ref.[8]) for cyanine dyes based microcavity.

tor material, namely J-aggregates of a cyanine dye [7, 8], which have an absorption linewidth about 40 meV. In the experiments [7, 8, 9] two polaritonic branches were observed, separated by a giant Rabi splitting ranging between 80 meV and 300 meV at room temperature. The experimental points and the dispersion curves obtained by fitting to a simple two-level cavity-exciton model is shown in Fig.1 [8].

In this paper we examine theoretically the strong coupling regime in a microcavity, which utilizes an organic material as optically active semiconductor. The organic structures possess a number of features, which may strongly influence the picture of light-matter coupling in these structures. Here we take into account the following aspects. First, as it was noticed in [5], the energy of Rabi splitting may be large and thus comparable with the energy of an intramolecular phonon. Thus, in contrast to inorganic structures, intramolecular phonons might be involved in the dynamics of the microcavity polaritons. Second, many of suitable organic materials are far from being crystalline. For instance, the materials used in [7, 8, 9] are essentially disordered, and thus an adequate description of cavity polaritons in these structures is needed. Third, in many of organic materials resonance intermolecular interaction is weak.

Let us focus on a material (either crystalline or disordered) with weak resonance intermolecular interaction and relatively broad electronic transition. In this general case, for the non-cavity material the collective excitonic states do not form, and the eigenstates of the system are *incoherent* localized molecular excitations. Nevertheless, as we show in this paper (see also [10]), when such material is placed into a microcavity, the collective excitations can appear also in these materials as a result of coupling of the molecules by light. Qualitatively, the reason of the appearance of the collective states is that the cavity photon creates a coherent polarization in the medium and interacts strongly with the electronic transition in different molecules. If the interaction is strong enough, it leads to the hybridization of the electronic transition and photon mode, which results in the formation of *coherent* (polaritonic) states. Between the scattering events, this excitations propagate as wave packets with certain group velocity.

An important result which we shall outline below is that in these materials some new interesting features appear in the spectrum of the excitations. Namely, the coherent (polaritonic) states exist in certain restricted wave vector intervals. In particular, lower branch polaritonic states exist for relatively small wave vectors only. The rest of the excited states are incoherent (localized) molecular excitations, so that the coherent (polaritonic) states and incoherent molecular states coexist in these materials. Moreover, the number of the lower branch polaritonic states is much smaller than the number of the incoherent states. Qualitatively, such a situation can be understood if we take into account the fact that cavity photons with large wave vectors are not resonant with the molecular excitations and a strong dephasing of the molecular excitations destroys any short wavelength coherent polarization. As a result, the coherent states do not form for large wave vectors. Note that in the case of crystalline semiconductors, the interaction of small wavelength cavity photons with exciton states is also negligible; however in this case the coherence of short wavelength excitons occurs due to resonance Coulomb intermolecular (or interatomic) interactions and not as a result of exciton-photon interactions.

Another challenging question is the relaxation dynamics in organic systems. Besides of the fact that it is enriched by the possibility of the upper polariton nonradiative decay to the lower energy states with the emission of an intramolecular phonon, in the materials with broad dispersionless electronic transition it may have a series of peculiarities due to the traits of the spectrum of excitations. While in crystalline organic materials with strong resonance intermolecular interaction the only possible final states are lower branch polaritonic states, in the materials with broad dispersionless resonance a huge amount of incoherent states may serve as very effective channel for the nonradiative decay of the upper branch polaritonic states.

Finally, in organic non-cavity materials the process of exciton-exciton annihilation, which is sometimes refereed to as the bimolecular quenching process, is rather effective for high concentrations of excitons [11, 12]. A question arises about its role in the cavity polariton dynamics when organic material is placed into a microcavity. We shall show by the example of a crystalline material that the polaritonic effect results in strong suppression of the exciton-exciton annihilation process.

The paper is organized as follows. In Section 2 we consider a general case of a microcavity filled by an organic material with weak resonance intermolecular interaction and broad electronic transition. We obtain the spectrum of excitations from the phenomenological Maxwell equations and discuss its peculiarities. This approach allows one to see clearly the origin of the transition from coherent to incoherent excited states. The results of this consideration are equally applicable both for the cases of disordered and crystalline medium as far as the above mentioned conditions (strong damping and weak resonance intermolecular interaction) are satisfied.

In Section 3 we apply this physical picture for the particular case of the J-aggregates based microcavities, which is one of a few organic systems where the strong coupling was observed. A microscopic (quantum) description will be given for the case of the disordered organic material. The relation between the weight amplitudes which determine the structure of the polaritonic states in the disordered medium will be found.

Following this, in Section 4 we discuss some peculiarities of the dynamics of cavity polaritons in organic structures. For cyanine dyes based microcavities, we calculate the transition rate from the upper branch polariton states to the incoherent excited states due to the emission of an intramolecular phonon and discuss the role of this transition in the cavity polariton dynamics (Section 4.1). The suppression of exciton-exciton annihilation in a microcavity is illustrated in Section 4.2 for the case of a crystalline material with strong resonance intermolecular interaction.

2. Spectrum of excitations for a material with broad dispersionless resonance

In this Section we discuss the spectrum of excitations in an organic material with weak resonance intermolecular interaction and broad electronic transition (or, in other words, with broad dispersionless resonance). The most natural and most general approach to describe the properties of optical modes in general and cavity polaritons in particular is to find their dispersion curves as the solutions of the phenomenological Maxwell equations. These equations account also for the effects of medium polarization, the interaction between optical and electronic resonances, and the processes of damping. We use the simplest expression for the dielectric constant, which accounts for the total (homogeneous plus inhomogeneous) broadening of the electronic resonance. We show that the energy uncertainty of polaritonic states is always moderate, while the wave vector uncertainty strongly increases at certain values of the wave vector. As a result, polaritonic states form in certain restricted intervals of the wave vector only (see also [10]).

2.1. MACROSCOPIC APPROACH

As usual, we treat a microcavity with an optically active material as a slab of width L_c bounded by two perfect mirrors. Let the axis z be directed normal to the cavity plane, with $z = 0$ corresponding to the middle of the microcavity. The eigenmodes of the empty microcavity are characterized by a two dimensional (in-pane) wave vector q, and the values of the wave vector in the z direction are quantized, with $k_z = \pi/L_c$ for the lowest energy eigenmode. The mode is polarized in the (x, y) plane, and is normal to the photon wave vector $\mathbf{q}$. Its dispersion is given by:

$$E_{cav}(q) = \frac{\hbar c}{\sqrt{\varepsilon_b}}\sqrt{k_z^2 + q^2} = \frac{\hbar c}{\sqrt{\varepsilon_b}}\sqrt{\frac{\pi^2}{L_c^2} + q^2}, \tag{1}$$

where ε_b is the background dielectric constant.

Let the microcavity be filled with a medium with a single dispersionless resonance at an energy E_0. Then the Maxwell equations for transverse wave read:

$$\frac{\hbar^2(q^2 + k_z^2)c^2}{E^2} = \varepsilon_b + \frac{A}{E_0^2 - E^2 - 2iE\gamma_0}, \tag{2}$$

where γ_0 is the total broadening of the excitonic resonance, and A is proportional to the oscillator strength of the transition. Then for $E \approx E_{cav}(q) \approx E_0$:

$$[E - E_{cav}(q)][E - E_0 + i\gamma_0] = \Delta^2, \tag{3}$$

where $\Delta^2 = (A/4\varepsilon_b)$. Eq.(3) is just a well-known dispersion relation for microcavity polaritons which accounts for the broadening of the excitonic (dispersionless, in our case) resonance. Its well-known solutions for the case of strong light-matter coupling (i.e. $\gamma_0 \ll \Delta$) can be written in the form:

$$E_{U,L}(q) = E_{U,L}^{(0)}(q) - i\,\delta E_{U,L}(q); \qquad \delta E_{U,L}(q) = \gamma_0 |c_{ex}^{(U,L)}(q)|^2, \tag{4}$$

where

$$E^{(0)}_{U,L}(q) = \frac{E_{cav}(q) + E_0}{2} \pm \sqrt{\Delta^2 + \frac{(E_{cav}(q) - E_0)^2}{4}} \tag{5}$$

are the polariton dispersion curves in the absence of the exciton broadening (i.e. with $\gamma_0 = 0$ in Eq.(3)), and $\delta E_{U,L}(q)$ is the polariton *energy broadening*, which is proportional to the weights of the exciton in the polaritonic state [4]

$$|c^{(U,L)}_{ex}|^2 = \frac{\Delta^2}{\Delta^2 + (E_0 - E^{(0)}_{U,L})^2}. \tag{6}$$

Eq.(4) shows that the energy broadening of the lower polariton state moderately grows with the increase of the wave vector, while the upper polariton broadening decrease with the increase of the wave vector (see the wave vector dependencies of $|c^{(U,L)}_{ex}(q)|^2$ shown in the inset of Fig.3). As $E^{(0)}_{U,L}$ is of the order of 2 eV, and γ_0 is of the order of 20 meV, we see that the energy is well-defined, since the uncertainty of the energy $\delta E_{U,L}(q) \ll E^{(0)}_{U,L}(q)$.

2.2. THE END-POINTS OF THE POLARITONIC SPECTRUM

Now let us consider the *wave vector broadening* of the polaritonic states. The uncertainties of the wave vector and of the energy of a quasiparticle are related through its group velocity [13], which for the cavity polaritons is:

$$v^{(U,L)}_g = \frac{\hbar c^2 q}{\varepsilon_b E_{cav}(q)} \frac{(E_0 - E^{(0)}_{U,L}(q))^2}{\Delta^2 + (E_0 - E^{(0)}_{U,L}(q))^2}, \tag{7}$$

and then (see Eqs.(4) and (6))

$$\delta q_{(U,L)} = \frac{\delta E^{(U,L)}(q)}{\hbar v_g} = \frac{\gamma_0 \varepsilon_b E_{cav}(q)}{c^2 \hbar^2 q} \frac{(E_{cav}(q) - E^{(0)}_{U,L}(q))^2}{\Delta^2}. \tag{8}$$

The polaritonic state is well defined as far as $\delta q \ll q$. However, from Eq.(8) it is seen that this condition is violated (i) for $q \to 0$ for both branches and (ii) for large q for the lower polariton. In both these cases $v_g \to 0$, and thus δq diverges. For $q \to 0$ it is clear that at some point the uncertainty of the wave vector becomes large in comparison with the wave vector itself. In turn, from Eqs.(1), (5) and (8) it follows that $\delta q_{(L)} \propto q^2$ for large q, and thus at some point the condition $\delta q \ll q$ again fails.

From this macroscopic consideration it is seen that there exist certain limiting values of the wave vector ($q^{(L)}_{min}$, $q^{(L)}_{max}$ and $q^{(U)}_{min}$), which can be treated as the end-points of the coherent (polaritonic) spectrum. The positions of these end-points can be estimated from the condition $q \sim \delta q$. These positions depend on the parameters of the given material (the broadening γ_0, the Rabi splitting 2Δ and the detuning $u = E_{cav}(0) - E_0$).

In Fig.2a the obtained limiting values of the wave vector are shown as functions of the detuning for fixed broadening $\gamma_0 = 20$ meV and Rabi splitting $2\Delta = 80$ meV. With the increase of the detuning (from negative to positive values) the lower branch becomes more excitonic-like. Thus, when u increase $|c^{(L)}_{ex}|^2$ increase, while $v^{(L)}_g$ decrease. As a result, $\delta q_{(L)}$ grows with the increase of the detuning (see Eqs.(8) and (4)), and the interval where $\delta q_{(L)} < q$ diminishes. In contrast, the upper branch polaritons becomes more photon-like with the increase of detuning, so that $|c^{(U)}_{ex}|^2$ diminishes, while $v^{(U)}_g$ grows. Thus, the value of $q^{(U)}_{min}$ decrease when the detuning grows from negative to positive values.

The region of the strong coupling is not affected by the existence of the end-points as far as $q^{(L)}_{min}, q^{(U)}_{min} < q_{res} < q^{(L)}_{max}$, where q_{res} is the wave vector at which the resonance between the cavity photon and the electronic transition occurs: $E_{cav}(q_{res}) = E_0$. The dependence of q_{res} on the detuning is shown in Fig.2a by dashed line. It is seen that this condition is satisfied for negative values of detuning ($u < -10$ meV) only. In turn, for large positive values of detuning ($u > 18$ meV) $\delta q_{(L)}$ always is large in comparison with q, and the lower branch polaritonic states do not form at all.

In Fig.2b the obtained limiting values of the wave vector are shown as functions of the Rabi splitting for fixed broadening $\gamma_0 = 20$ meV and detuning $u = -130$ meV. In accordance with Eq.(8), the larger is Δ, the smaller is δq, and thus the wider is the coherent part of the spectrum of the lower polariton branch. An example of the spectrum of the excitations in a material with broad dispersionless resonance is shown in Fig.3. The used parameters are those of the cyanine dyes based microcavities of Refs.[7, 8] ($\gamma_0 = 20$ meV, $2\Delta = 80$ meV, $u = -130$ meV). The corresponding values of $q^{(L)}_{min} = 4.2 \cdot 10^3$ cm^{-1}, $q^{(L)}_{max} = 9.3 \cdot 10^4$cm^{-1} and $q^{(U)}_{min} = 3.5 \cdot 10^4$ cm^{-1} were found from Figs.2. The dashed lines in Fig.3 show the uncoupled exciton and photon, and the solid lines show the coherent part of the spectrum. The crosses show the end-points of the polariton spectrum.

To summarize, the existence of the end-points of the polaritonic spectrum is a direct consequence of the trend of the polaritonic dispersion curves, since they appear when v_g vanishes. In the following discussion we consider the case of large negative detuning.

For vanishing wave vectors both polariton branches form for $q > q^{(U,L)}_{min}$

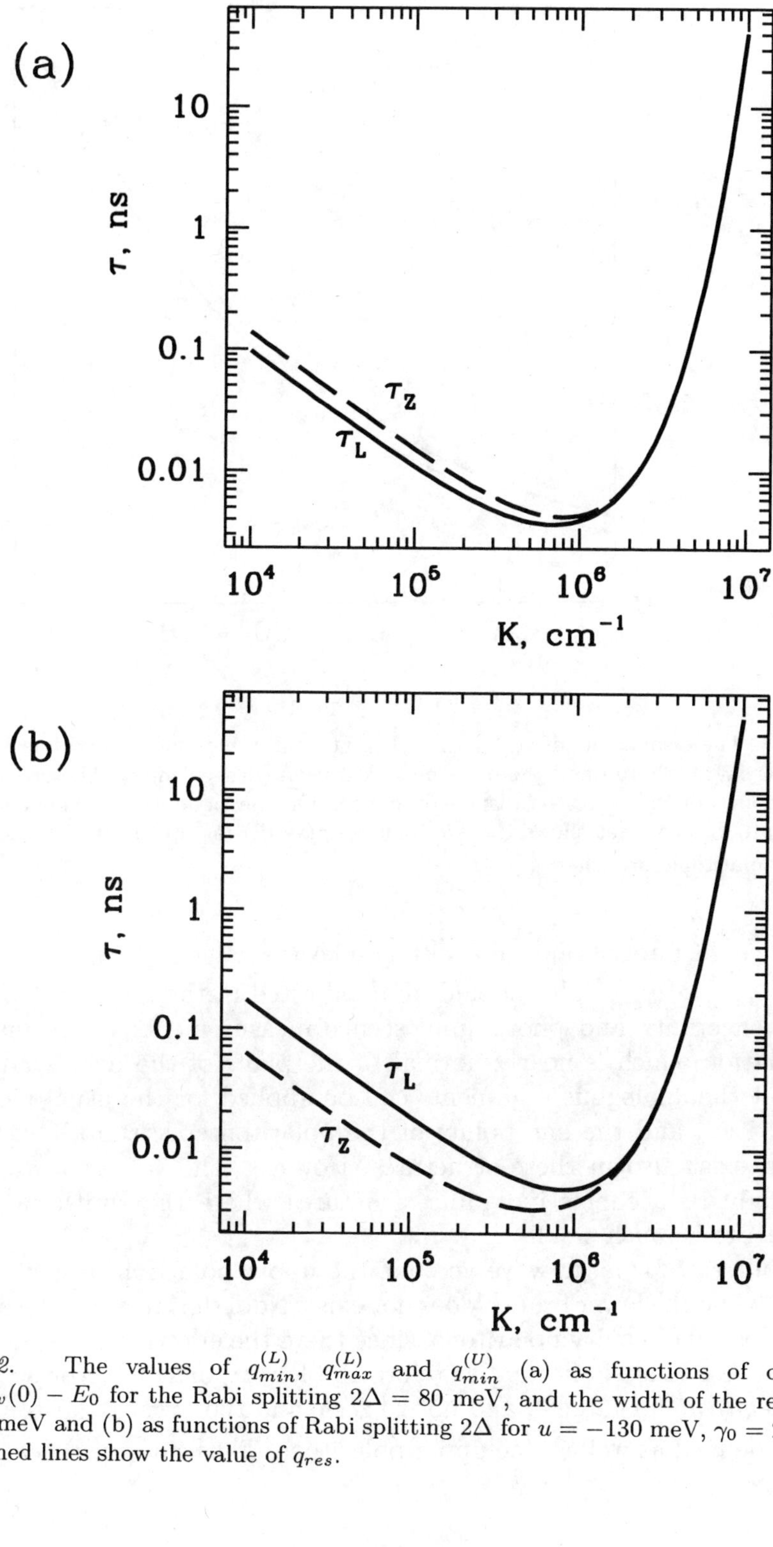

Figure 2. The values of $q_{min}^{(L)}$, $q_{max}^{(L)}$ and $q_{min}^{(U)}$ (a) as functions of detuning $u = E_{cav}(0) - E_0$ for the Rabi splitting $2\Delta = 80$ meV, and the width of the resonance $\gamma_0 = 20$ meV and (b) as functions of Rabi splitting 2Δ for $u = -130$ meV, $\gamma_0 = 20$ meV. The dashed lines show the value of q_{res}.

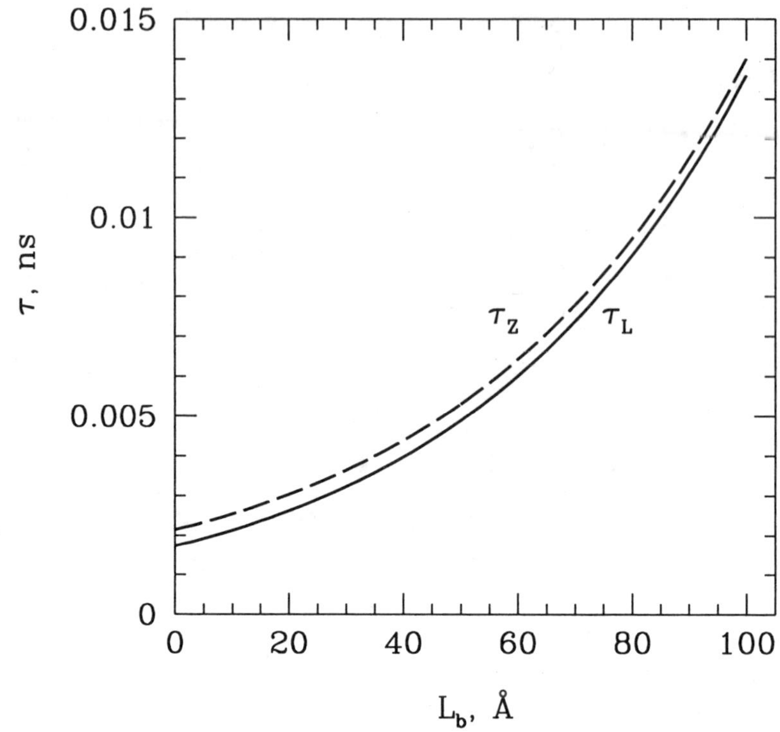

Figure 3. The dispersion curves for the coherent polaritonic states (solid lines) and for uncoupled cavity photon and the molecular excitation (dashed lines). The crosses show the end-points of the polariton dispersion curves. On the right, the broadened exciton line is drafted. The inset shows the excitonic weights (6) for upper ($|c_{ex}^{(U)}|^2$) and lower ($|c_{ex}^{(L)}|^2$) polaritonic branches.

only. Since the upper branch is excitonic-like at small q, its dispersion is very flat and $q_{min}^{(U)} > q_{min}^{(L)}$. This should be manifested in the angular dependencies of reflectivity and photoluminescence measurements when analyzing the radiation which is nearly normal to the plane of the microcavity. We note that the analogous arguments can be applied for the usual inorganic microcavities, and the end-points of the polariton spectrum at vanishing q should exist also in these structures. However, the values of q_{min} may be strongly reduced for inorganic structures where the width γ_0 at low temperature is much smaller.

In contrast, for large wave vectors, the upper polariton branch is well-defined, while the lower branch does not exist. Note that it is not the case for usual (inorganic) cavity polaritons, since there the effective mass m_* of the excitonic resonance is finite. Therefore the lower polariton group velocity tends to $q\hbar/m_*$ for large q, and not to zero, and the lower polariton states are well-defined as well as the upper polaritons. Thus, the existence of $q_{min}^{(L)}$

is the characteristic feature of the organic materials with weak resonance intermolecular interaction.

Finally, let us estimate the relative number of the lower polariton states. If we treat the molecules of the medium as two-level systems, then the total number of the excited states is equal just to the number N of the molecules. In turn, it is clear that the number of new coherent states is of the order of $(\bar{R}q_{max}^{(L)})^2 N$, where $\bar{R}$ is the mean distance between the molecules (the absence of the coherent states in the vicinity of $q \simeq 0$ in this estimation can be neglected). As $q_{max}^{(L)} \ll 1/\bar{R}$, the number of the coherent states is:

$$N_{coherent}^{(L)} \simeq (\bar{R}q_{max}^{(L)})^2 N \ll N. \tag{9}$$

Since the total number of the states is a constant equal to the total number of the molecules N, it means that the rest of the states are the incoherent excitations of individual molecules. These weakly coupled to light states are drafted by a broadened line centered at the energy E_0 in Fig.3. The important thing is that quite generally they form the majority of the electronic excited states of the system. For example, for the microcavities with cyanine dye J-aggregates investigated experimentally in Refs.[7, 8], $\bar{R} \simeq 200$ Å, and thus $N_{coherent}^{(L)} \simeq 0.035N$.

In all these estimations we did not take into account the vicinities of the end-points, which correspond to the regions of transition from coherent to incoherent states. This approximation does not change our main conclusions on the coexistence of coherent and incoherent states and on the predominance of the number of incoherent states. However, the investigation of the states in the intermediate regions is also interesting and has to be carried out. In these transition regions the excited states are also the wave packets formed by the hybridized cavity photon mode and electronic excitations of the optically active medium, but the expression for the group velocity (7) used in our estimations already is inapplicable, since $\delta q \sim q$ in these regions. For such wave packets one has to apply the approach used by Loudon [14] in the calculation of the velocity v_E of the energy flow in a medium. The approach was based on the damping oscillator microscopic model of dielectric. This velocity was found to be the Pointing vector divided by the energy density W stored in the monochromatic wave.

3. Disordered organic semiconductors

The consideration of Section 2 is applicable, in particular, to a disordered system with a broad dispersionless electronic transition as far as the typical length scales of the system (the mean distance between the molecules, the length of the polymers or J-aggregate chains, etc.) are much smaller than the cavity photon wave length. Then the photon "feels" just the averaged characteristics of the system, which in this situation behaves as a homogeneous medium. In this section we consider a particular case of such material, namely, the cyanine dyes based microcavities examined experimentally in Refs.[6, 7, 8, 9], since it is one of a few organic materials where strong light-matter coupling was observed.

A theoretical description of microcavity polaritons in strongly disordered systems as far as we know had not been addressed before. In this Section we present a simplest microscopic model (see also [10]), which allows us to find out the structure of the polaritonic states in the simplest approximation, in which the scattering of the cavity photon by the inhomogeneities of the medium is neglected. The obtained results allow us to discuss some peculiarities of the dynamics of microcavity polaritons in disordered systems (Section 4.1).

3.1. CYANINE DYES BASED MICROCAVITIES

To start with, several words about the structures under consideration should be said. As it was discussed in the Introduction, the strong coupling regime requires that the width of the electronic transition is smaller than the Rabi splitting energy. Though the light-matter interaction in organic materials may be strong, the broadening of the electronic transition not in all cases is small enough, and this relation is not satisfied for many materials. In this connection, so-called J-aggregates may be useful, since in these structures the reduction of the width of the electronic transition in comparison with the width of the electronic transition in a monomer occurs.

J-aggregates are one-dimensional chains self-organized from monomers, which carry a net charge. The interaction between the monomers within a single chain results in the appearance of collective excitonic states. In such systems the total oscillator strength is concentrated (up to 85 %) in the lowest excitonic state [15]. The interaction between this "bright" state with other excitonic states can be neglected, and thus this "bright" state can be treated as a single *dispersionless* resonance. The total oscillator strength of the aggregate is enhanced compared to the oscillator strength of a monomer in proportion to the number n of coupled monomers in the aggregate [15]. In addition, the inhomogeneous broadening of a J-aggregate line scales as $n^{-1/2}$ compared to the inhomogeneous broadening of a single monomer due to the process of motional narrowing [16]. As a result, the excitonic transition in J-aggregates comprising of many molecules has both a relatively narrow linewidth and a large oscillator strength, allowing the strong-coupling regime to be reached. In the following, we shall speak about J-aggregates formed from cyanine dyes.

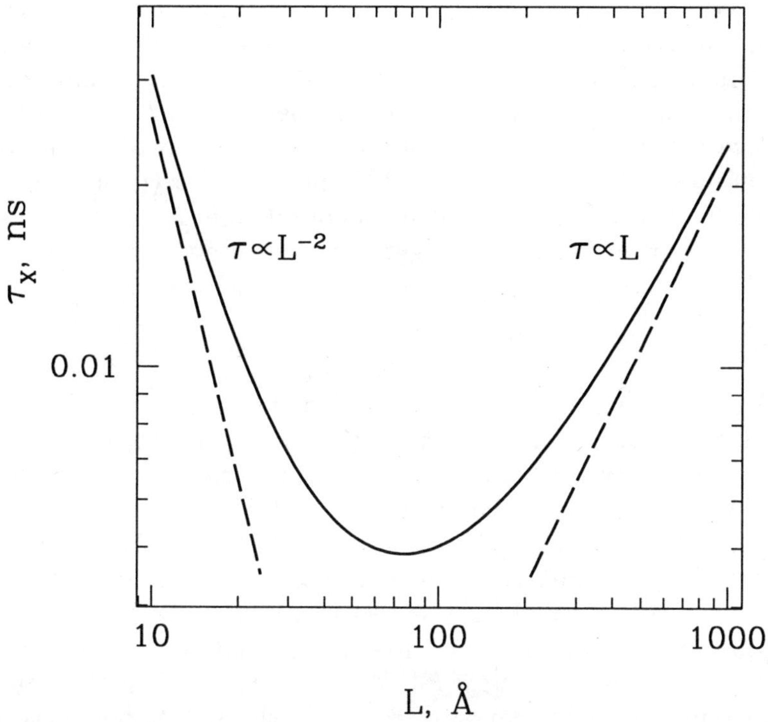

Figure 4. A draft of a microcavity with J-aggregates formed from a cyanine dye suspended in a transparent polymer matrix. The dipole moments of the J aggregates are indicated for two particular chains.

The mean distance between the J-aggregate chains is about 200 Å(this quantity may be estimated from the concentration η of J-aggregates: $\bar{R} \simeq \eta^{-1/3}$, and $\eta = 1.3 \cdot 10^{17}$ cm^{-3}). Thus, the J-aggregate chains are well separated, so that the resonance Coulomb interaction between them is very weak. Therefore this material belongs to the class of materials with weak resonance intermolecular interaction discussed in Section 2.

In contrast to microcavities containing inorganic semiconductors, in which a thin quantum well with an excitonic resonance usually is located in the antinode of the electric field (in the middle of a half-wavelength cavity), the optically active material in the samples of Refs.[7, 8] filled the whole microcavity. A typical cyanine dye based microcavity consists of a thin (2000 Å) film of J-aggregates in a transparent polymer matrix placed between two high-reflectivity mirrors (see Fig.4). In our calculations, we assume that all the J-aggregates are elongated in the microcavity plane ((x, y)-plane in Fig.4) and fill the whole volume between the mirrors. For the sake of simplicity we ignore the possibility of the formation of substructure and assume that the centres of the J-aggregates are homogeneously distributed within the cavity. The orientation of the aggregates chains (and thus, the orientation of their dipole moments, which are directed along each single chain, as indicated in Fig.4 for two chains) is random within the (x, y)-plane. Therefore in contrast to inorganic semiconductor nanostructures which have a true in-plane translation invariance, two different types of disorder are present in J-aggregate based microcavities: (i) the orientational (in-pane) disorder of the dipole moments of different J-aggregates, and (ii) the random distribution of the J-aggregates themselves.

3.2. MICROSCOPIC MODEL

Here we develop a microscopic quantum model which accounts for positional and orientational disorder in the system of J-aggregates in the intervals where the wave vector of the cavity polaritons is a good quantum number. Again, we treat microcavity as a slab of the thickness L_c and dielectric constant ε_b, bounded by two perfect mirrors. However, the slab now contains J-aggregates chains having both position and orientation distributed chaotically throughout the microcavity volume (see Fig.4). Again, we suppose that each J-aggregate has a single intense and dispersionless optical transition at an energy E_0. We are interested in the eigenstates in the system, which appear as a result of the interaction between the cavity photons and the excitations supported by the J-aggregates. As the length of each J-aggregate chain is assumed to be much shorter than the typical photon wavelength, we do not account for the finite extension of the J-aggregates. Instead we model each J-aggregate as a single "molecule", which possesses a point dipole moment, again distributed homogeneously and randomly throughout the cavity volume.

Let $a^+_{\mathbf{q}}$ and $a_{\mathbf{q}}$ be the creation and annihilation operators of cavity photons, and B^+_i and B_i be the creation and annihilation operators of an excitation on the i-th "molecule". We do not account for the dipole-dipole interactions between different "molecules", as we consider the coupling by light to be the principal coupling mechanism. The Hamiltonians of non-interacting cavity photons and molecular excitations are

$$H_c = \sum_{\mathbf{q}} E_{cav}(q) a^+_{\mathbf{q}} a_{\mathbf{q}}; \qquad H_J = \sum_i E_0 B^+_i B_i. \tag{10}$$

Following the usual procedure [17], we write the bosonic Hamiltonian of the exciton-photon interaction near the anticrossing region in the form:

$$H_{int} = \sum_i \sum_{\mathbf{q}} \Big[T_i(\mathbf{q}) a_{\mathbf{q}} B^+_i + T^*_i(\mathbf{q}) a^+_{\mathbf{q}} B_i \Big], \tag{11}$$

$$T_i(\mathbf{q}) = i\sqrt{\frac{4\pi}{SL_c\varepsilon_b E_{cav}(q)}} p_{0f} E_0 \, (\mathbf{n_q s}_i) \, \cos\left(\frac{\pi z_i}{L_c}\right) \, e^{i\mathbf{q}\mathbf{r}_\parallel^{(i)}}, \tag{12}$$

where $\langle 0|\mathbf{P}_i|f\rangle \equiv \mathbf{s}_i p_{0f}$ ($\mathbf{s}_i$ is a unit vector directed along i-th "molecule"), $\mathbf{n_q} = [\mathbf{e}_z \times \mathbf{q}]/q$ ($\mathbf{e}_z$ is the unit vector in the direction of the z-axis), and the coordinates $(\mathbf{r}_\parallel^{(\mathbf{i})}, z_i)$ determine the position of the i-th "molecule".

To find the eigenstates of the full Hamiltonian, which is the sum of the Hamiltonians (10) and (11), we introduce new Bose operators ξ and ξ^+ for the mixed exciton-photon states as follows:

$$\xi^+ = \sum_{\mathbf{q}} \alpha(\mathbf{q}) a_{\mathbf{q}}^+ + \sum_i \beta_i B_i^+; \qquad \xi = \sum_{\mathbf{q}} \alpha^*(\mathbf{q}) a_{\mathbf{q}} + \sum_i \beta_i^* B_i. \tag{13}$$

The coefficients $\alpha(\mathbf{q})$ and β_i determine the fraction of light (with the wave vector $\mathbf{q}$) and the fraction of the excitation on the i-th "molecule" in the polaritonic states respectively. From the Hamiltonian eigenvalue problem we find that the coefficients $\alpha(\mathbf{q})$ and β_i obey the following system of equations:

$$\begin{aligned} \alpha(\mathbf{q})[E - E_{cav}(q)] &= \sum_i T_i^*(q)\beta_i; \\ \beta_i[E - E_0] &= \sum_{\mathbf{q}} T_i(\mathbf{q})\alpha(\mathbf{q}). \end{aligned} \tag{14}$$

Substituting β_i from the second of Eqs.(14) into the first equation, we find that the relation which determines the energies and amplitudes of cavity polaritons is:

$$\alpha(\mathbf{q})[E - E_{cav}(q)] = \sum_i T_i^*(\mathbf{q}) \sum_{\mathbf{q}_1} T_i(\mathbf{q}_1)\alpha(\mathbf{q}_1)/[E - E_0]. \tag{15}$$

The amplitude of cavity photon $\alpha(\mathbf{q})$ enters in the right-hand side of Eq.(15) with all possible values of its argument $\mathbf{q}_1$ due to the fact that the medium does not possess true translation symmetry. However, those values of the argument, which differ from $\mathbf{q}$ appearing in the left-hand side of Eq.(15), just describe the scattering of light by the inhomogeneities in the cavity medium. If this scattering is neglected, then only the term with $\mathbf{q}_1 = \mathbf{q}$ survives in the sum over $\mathbf{q}_1$, and the zero-order relation which determines the dispersion equation for the cavity polaritons is:

$$[E - E_{cav}(q)] \; [E - E_0] = \sum_i |T_i(\mathbf{q})|^2. \tag{16}$$

In this approximation the in-plane positions of the "molecules" $\mathbf{r}_{\parallel}^{(i)}$ do not enter into the solution. The final form for the dispersion equation of cavity polaritons repeats Eq.(3) obtained within the framework of the classical approach with $\gamma_0 = 0$ and

$$\Delta^2 = \sum_i |T_i(\mathbf{q})|^2 \simeq \frac{4\pi p_{0f}^2 E_0^2}{SL_c\varepsilon_b E_c} \sum_i (\mathbf{n_q s}_i)^2 \cos^2\left(\frac{\pi z_i}{L_c}\right), \tag{17}$$

where $E_c = E_{cav}(0)$. In the particular case when the dipole moments of all the "molecules" are completely randomly distributed with the centers of the "molecules" being distributed homogeneously in the z direction, the sum over i in Eq.(17) can be calculated using

$$\sum_i \rightarrow \frac{N}{2\pi L_c} \int_0^{2\pi} d\phi \int_{-L_c/2}^{L_c/2} dz, \tag{18}$$

where N is the total number of the "molecules", and ϕ is the angle between $\mathbf{q}$ and $\mathbf{s}_i$. Then

$$\Delta^2 = \frac{\pi p_{0f}^2 E_0^2}{\varepsilon_b E_c} \eta, \tag{19}$$

where $\eta = (N/SL_c)$ is the concentration of the "molecules". The dispersion relations of the cavity polaritons are again given by Eq.(5), and the expressions for the weight coefficients $\alpha(\mathbf{q})$ and β_i can be found for each branch from Eqs.(14) together with the normalization condition. If we neglect the scattering of the polaritons by the inhomogeneities of the medium, then $\beta_i^{(U,L)} = \beta_i^{(U,L)}(\mathbf{q})$, and the normalization condition for each polariton mode reads:

$$|\alpha(\mathbf{q})|^2 + \sum_i |\beta_i(\mathbf{q})|^2 = 1. \tag{20}$$

Then, from Eqs.(14), (20) we find:

$$|\alpha^{(U,L)}(\mathbf{q})|^2 = \frac{(E_{U,L} - E_0)^2}{(E_{U,L} - E_0)^2 + \Delta^2};$$
$$|\beta_i^{(U,L)}(\mathbf{q})|^2 = \frac{\Delta^2 (4/N)(\mathbf{n_q s}_i)^2 \cos^2\left(\frac{\pi z_i}{L_c}\right)}{(E_{U,L} - E_0)^2 + \Delta^2}. \tag{21}$$

In this Section we completely ignored the damping of the molecular and the cavity photon states. In other words, the cavity photon wave vector was treated as a good quantum number. Therefore, based on the results the Section 2, we conclude that the relations we have obtained are only applicable for the wave vectors $q_{min}^{(L)} < q < q_{max}^{(L)}$ for lower branch polaritons, and $q > q_{min}^{(U)}$ for upper branch polaritons.

4. On the cavity polariton dynamics

The study of the dynamics of cavity polaritons requires the consideration of a variety of scattering processes of elementary excitations. In organic microcavities with strong light-matter coupling the coexistence of coherent and incoherent states essentially enriches the physical picture. In this Section we consider only two processes which are typical for organic structures. In Section 4.1 the nonradiative decay of the upper branch polaritons into incoherent states is considered for the case of organic material with broad dispersionless resonance (see also [10]). In Section 4.2 we examine the polaritonic effect in the process of exciton-exciton annihilation in strongly coupled microcavity.

4.1. UPPER POLARITON DECAY TO THE INCOHERENT STATES

Here we calculate the rate of the resonance inelastic transition from the upper polariton states to the incoherent states for a J-aggregate based microcavities discussed in Section 3. It was found [8] that the cyanine dye material studied possessed a strong Raman-active mode having an energy $E_{vib} = 74$ meV, which was close to the Rabi splitting energy $2\Delta = 80$ meV. We consider the upper branch polariton nonradiative decay accompanied by the emission of the intramolecular optical phonon. We assume strong vibration-electronic coupling, so that as final incoherent states we consider the excited states in which the electronic and vibrational quanta are localized on the same "molecule" of the medium (i.e. just the next vibronically excited states of the "molecules", which are resonant with the upper polariton branch). We also assume that these excited molecular states weakly couple to light, so that all these states are incoherent (no second Rabi splitting was observed for these structures).

We introduce creation and annihilation operators of an intramolecular phonon on i-th "molecule" b_i^+ and b_i. The Hamiltonian of the free phonons and the Hamiltonian of the interaction between the phonons and the molecular excitations are then given by

$$H_{phon} = E_{vib} \sum_i b_i^+ b_i; \qquad H_{ex-phon} = \sum_i H_{ex-phon}^{(i)} = gE_{vib} \sum_i B_i^+ B_i (b_i^+ + b_i) \tag{22}$$

respectively, where g is the coupling constant. Let $|0\rangle \equiv |0_{ex}, 0_{phot}, 0_{phon}\rangle$ denote the ground state of the system. If we then consider the exciton-phonon coupling as a weak perturbation, we write the wave functions of zero approximation for (i) the upper polariton with wave vector $\mathbf{q}$ and (ii) for the j-th incoherent excited electronic state with one intramolecular phonon on the same "molecule" in the form:

$$\Psi_{\mathbf{q}}^{(U)} = \alpha^{(U)}(\mathbf{q})a_{\mathbf{q}}^{+}|0\rangle + \sum_i \beta_i^{(U)}(\mathbf{q})B_i^{+}|0\rangle; \qquad \Psi_j^{(incoh)} = B_j^{+}b_j^{+}|0\rangle. \quad (23)$$

It is apparent that these wave functions are orthogonal and normalized. We assume that the excitonic resonances are distributed with the following distribution function:

$$\rho(E_0^{'}) = \frac{1}{\sqrt{\pi}\gamma_0}\, e^{-[E_0 - E_0^{'}]^2/\gamma_0^2}, \quad (24)$$

and we suppose that the main contribution to the level broadening comes from the disorder in chains.

The transition rate can be computed using Fermi's golden rule. The matrix elements of the Hamiltonian $H_{ex-phon}$ (22) for the wave functions (23) are:

$$\langle \Psi_j^{(incoh)} | H_{ex-phon}^{(j)} | \Psi_{\mathbf{q}}^{(U)} \rangle = g E_{vib} \beta_j^{(U)}(\mathbf{q}). \quad (25)$$

The transition rate is therefore given by:

$$W^{(incoh)}(\mathbf{q}) = \frac{2\pi}{\hbar} g^2 E_{vib}^2 \frac{1}{\sqrt{\pi}\gamma_0}\, e^{-[E_0 - (E_U(q) - E_{vib})]^2/\gamma_0^2} \sum_i |\beta_i^{(U)}(\mathbf{q})|^2. \quad (26)$$

The sum over i in the right-hand side of Eq.(26) gives (see Eqs.(20), (21), and compare with Eq.(6)):

$$\sum_i |\beta_i^{(U)}(\mathbf{q})|^2 = 1 - |\alpha^{(U)}(\mathbf{q})|^2 = \frac{\Delta^2}{(E_0 - E_U(q))^2 + \Delta^2}, \quad (27)$$

and thus

$$W^{(incoh)}(\mathbf{q}) = \frac{2\pi g^2 E_{vib}^2}{\hbar\sqrt{\pi}\gamma_0}\, e^{-[E_0 - (E_U(q) - E_{vib})]^2/\gamma_0^2}\, \frac{\Delta^2}{(E_0 - E_U(q))^2 + \Delta^2}. \quad (28)$$

The upper polariton lifetime $\tau^{(incoh)}(q) = 1/W^{(incoh)}(q)$ is shown in Fig.5 for those values of the wave vector, where the upper polariton branch

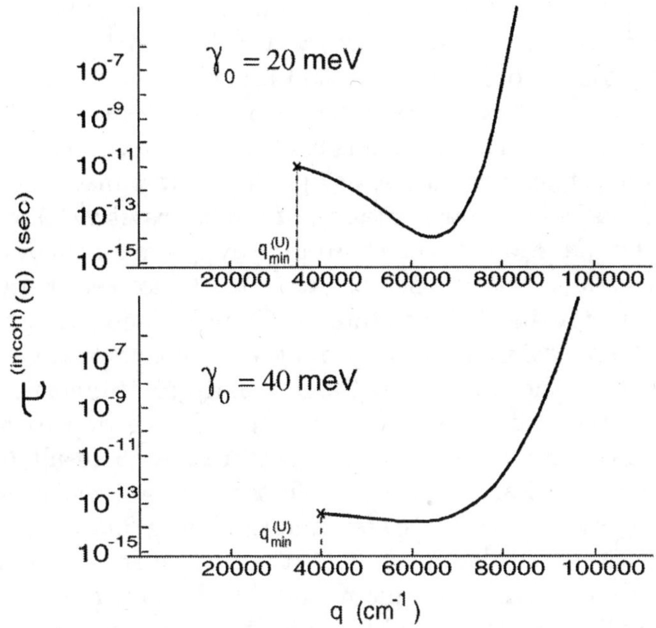

Figure 5. The transition time $\tau^{(incoh)}(q) = 1/W^{(incoh)}(q)$ as function of the upper polariton wave vector for two values of the broadening of the excitonic resonance: $\gamma_0 = 20$ meV and $\gamma_0 = 40$ meV.

exists. The two values of γ_0 were used: $\gamma_0 = 20$ meV and $\gamma_0 = 40$ meV. The following values of the parameters were taken from Refs.[7, 8]: $2\Delta = 80$ meV, $E_{vib} = 74$ meV, $E_0 = 1.85$ eV, $\varepsilon_b = 2.6$, $L_c = 2260$ Å, so that $E_c = 1.72$ eV. The value of g has to be determined from optical measurements. For our estimation we use the typical value $g = 0.5$. The typical upper polariton lifetime related to its decay to the incoherent states thus ranges between 20 and 70 fs. This very short lifetime is in good agreement with the rough estimation of Ref.[18].

Another situation occurs in a microcavity where all the excited states are coherent. In particular, in Ref.[5] the polaritons in a microcavity containing crystalline organic material with strong resonance intermolecular interaction were considered. It was assumed that the excited states in the crystalline material were two-dimensional coherent Frenkel excitons with large oscillator strength. Besides, it was assumed that the energy of the Rabi splitting was close to the energy of the intramolecular vibration, and the rate of the transition from the upper to the lower polaritonic branch was calculated. Since all the excited states in the microcavity were coherent, the number of possible final states was relatively small, and the rate of such transition was found to be in picosecond range.

Based on this result, we propose the following scenario for the excitation dynamics in a microcavity containing an organic semiconductor with broad dispersionless electronic transition as optically active material. Under non-resonant high energy optical excitation of a microcavity, the overwhelming majority of the radiation incident into the cavity creates high energy incoherent states. Then, due to fast relaxation to the lowest energy electronic excitation state the incoherent states (having an energy E_0) are mainly populated. The photoluminescence from these incoherent states pumps the coherent states of the upper and lower polaritonic branches. The upper branch polaritonic states have a very short lifetime due to nonradiative decay to the large number of incoherent states. As a result of this rapid decay, the cw population of upper branch states is significantly smaller than the population of lower branch states, and thus the experimentally detected emission intensity from the upper polariton branch is very weak [18].

We note that, though the main part of the excited states is in the regime of weak light-matter coupling, the relaxation dynamics qualitatively differ from that in the case of usual weak coupling regime, when the coherent states do not form at all. In this last case, the photoluminescence from the excited electronic states simply pumps the *cavity photons* states, which eventually escape from the cavity, and no Rabi splitting appears. In contrast, in microcavities containing organic materials with broad dispersionless electronic resonances, the photoluminescence from incoherent states pumps the *cavity polariton* states which are seen in experiments.

4.2. EXCITON-EXCITON ANNIHILATION IN MICROCAVITIES

Here we examine the role of the polaritonic effect in the bimolecular quenching of excitons. The process of bimolecular quenching [11] is a result of intermolecular electronic excitation energy transfer. As a first step, the energy of two molecular excitations becomes concentrated on one of the two molecules in the state with the energy equal to $E_2 = 2E_1$ (E_1 is the lower molecular excitation energy). Then the amount of energy approximately equal to E_1 relax irreversibly turning into the vibrational energy of the medium. As a result of this fast (in picosecond range) relaxation, only one excitation with the energy E_1 remains instead of two initial excitations (see Fig.6). In non-cavity organic materials at high pumping the process of bimolecular quenching drastically decrease the concentration of the excitations. We show below that in microcavity this process may be strongly suppressed.

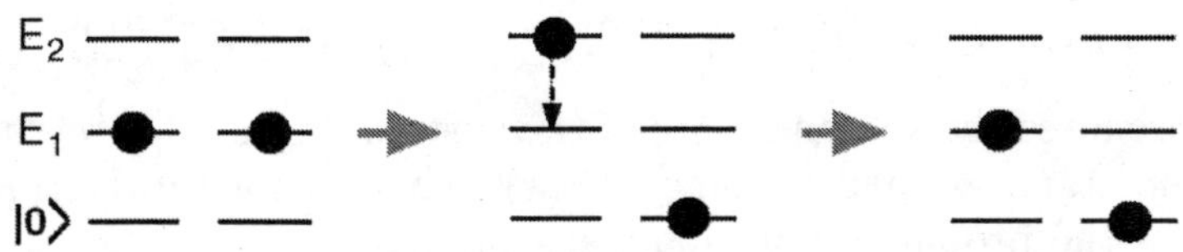

Figure 6. An illustration to the process of bimolecular quenching.

For the sake of simplicity and to demonstrate the reason of this suppression in the most simple way, in what follows we consider the case of an ordered two-dimensional crystalline layer with strong resonance intermolecular interaction. We assume that two-dimensional Frenkel excitons form in the layer, and the wave vector is a "good" quantum number for all these states.

To model the process of bimolecular quenching we assume that the crystalline organic material has two excitonic bands with the energies $E_1(q)$ and $E_2(q)$, such that $2E_1 \approx E_2$. We assume also that the second excitonic state weakly couples to light and has very small lifetime associated with its irreversible transition to the lower energy state ($E_2(q) \rightarrow E_1(q)$). In this situation the rate of the bimolecular quenching is equal to the rate of the fusing of two excitons with the energies $E_1(q_1)$ and $E_1(q_2)$ with the formation of one exciton with the energy $E_2(q)$.

Let $B^{+(1)}_{\mathbf{q}}$ and $B^{(1)}_{\mathbf{q}}$ be the creation and annihilation operators for the lower energy exciton, and $B^{+(2)}_{\mathbf{q}}$ and $B^{(2)}_{\mathbf{q}}$ be the creation and annihilation operators for the higher energy exciton. Then the bosonic Hamiltonian for the non-cavity material is the sum of two Hamiltonians of noninteracting excitons

$$H^{(1)}_{exc} = \sum_{\mathbf{q}} E_1(q) B^{+(1)}_{\mathbf{q}} B^{(1)}_{\mathbf{q}}, \qquad H^{(2)}_{exc} = \sum_{\mathbf{q}} E_2(q) B^{+(2)}_{\mathbf{q}} B^{(2)}_{\mathbf{q}} \tag{29}$$

and the Hamiltonian of interaction between the excitons

$$H_A = \frac{1}{2\sqrt{N}} \Big\{ \sum_{\mathbf{q}_2,\mathbf{q}_2} \Big[W(\mathbf{q}_1) + W(\mathbf{q}_2) \Big] B^{+(2)}_{\mathbf{q}_1+\mathbf{q}_2} B^{(1)}_{\mathbf{q}_1} B^{(1)}_{\mathbf{q}_2} + h.c. \Big\}, \tag{30}$$

where $W(\mathbf{q}) = \sum_n W_{nm} e^{i\mathbf{q}(\mathbf{r}_n - \mathbf{r}_m)}$ is the Fourier transform of the interaction potential, $W_{nm} \sim p^{01}p^{02}/(\mathbf{r}_n - \mathbf{r}_m)^2$, $\mathbf{r}_i$ is the two-dimensional position vector of i-th molecule, and p^{01} and p^{02} are the matrix elements of the dipole moment operator for the first and second electronic transitions, respectively. The initial and final states for the transition under consideration are:

$$\Psi_i(\mathbf{q}_1,\mathbf{q}_2) = B^{+(1)}_{\mathbf{q}_1} B^{+(1)}_{\mathbf{q}_2}|0\rangle; \qquad \Psi_f(\mathbf{q}) = B^{+(2)}_{\mathbf{q}}|0\rangle, \tag{31}$$

where $|0\rangle$ denotes the vacuum state of the system. Let $\rho^{(2)}(q)$ be the density of excitonic states with the energy $E_2(q)$. Then using the Fermi's golden rule we find the probability of this transition as

$$W_{exc-exc}(\mathbf{q};\mathbf{q}_1,\mathbf{q}_2) = \frac{2\pi}{N\hbar}|\mathrm{Re}\{W(\mathbf{q}_1) + W(\mathbf{q}_2)\}|^2\ \rho^{(2)}(E_2(q)), \tag{32}$$

where $E_2(q) = E_1(q_1) + E_1(q_2)$.

Now let us examine what happens if this material is placed into a microcavity. The total Hamiltonian will then be the sum of Hamiltonians (29) – (30), the Hamiltonian of free cavity photons H_c (10), and the Hamiltonian which describes the interaction between the low-energy excitons and cavity photons:

$$H^{(ex_1-cav)}_{int} = \sum_{\mathbf{q}} \left[T(\mathbf{q}) a_{\mathbf{q}} B^{+(1)}_{\mathbf{q}} + h.c.\right], \tag{33}$$

where $T(\mathbf{q})$ is a coupling constant. The sum of the Hamiltonians (33), H_c and $H^{(1)}_{ex}$ can be diagonalized [17] by the transformation to the new Bose operators for the polariton states $\xi^+_{\mathbf{q}} = \alpha(\mathbf{q})a^+_{\mathbf{q}} + \beta_i(\mathbf{q})B^{+(1)}_{\mathbf{q}}, \xi_{\mathbf{q}} = \alpha^*(\mathbf{q})a_{\mathbf{q}} + \beta^*(\mathbf{q})B^{(1)}_{\mathbf{q}}$. Then the total Hamiltonian can be written as

$$H = \sum_{\mathbf{q},\rho} \mathcal{E}_\rho(q)\xi^+_{\mathbf{q},\rho}\ \xi_{\mathbf{q},\rho} + \sum_{\mathbf{q}} E_2(q)B^{+(2)}_{\mathbf{q}}B^{(2)}_{\mathbf{q}} + H_A, \tag{34}$$

where the index $\rho = (U, L)$ corresponds to two polaritonic branches with the energies which satisfy the usual dispersion equation (2Δ is again the Rabi splitting energy):

$$[\mathcal{E}_\rho(q) - E_{cav}(q)][\mathcal{E}_\rho(q) - E_1(q)] = \Delta^2. \tag{35}$$

Let us consider the process in which due to the exciton-exciton interaction two lower polaritons transform to an exciton with the energy $E_2(q)$. The initial and final states now are:

$$\Psi_i(\mathbf{q}_1,\rho;\mathbf{q}_2,\rho') = \xi^+_{\mathbf{q}_1,\rho}\ \xi^+_{\mathbf{q}_2,\rho'}|0\rangle; \qquad \Psi_f(\mathbf{q}) = B^{+(2)}_{\mathbf{q}}|0\rangle. \tag{36}$$

Again, using the Fermi's golden rule we find the probability of this transition as

$$W_{pol-pol}(\mathbf{q};\mathbf{q}_1,\rho;\mathbf{q}_2,\rho') = |C^{(\rho)}_{ex}(\mathbf{q}_1)|^2|C^{(\rho')}_{ex}(\mathbf{q}_2)|^2 W_{exc-exc}(\mathbf{q};\mathbf{q}_1,\mathbf{q}_2), \tag{37}$$

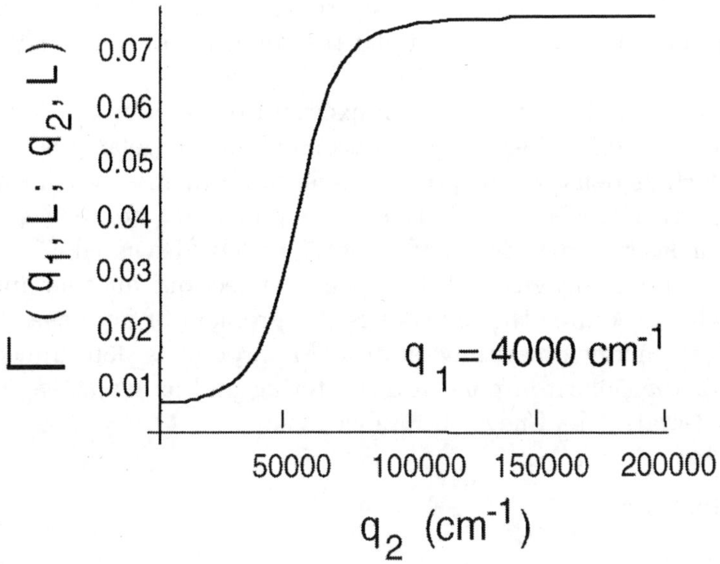

Figure 7. The factor $\Gamma(\mathbf{q}_1\rho; \mathbf{q}_2, \rho')$ for $\rho = \rho' = L$, $q_1 = 4000\text{cm}^{-1}$ as function of the wave vector q_2.

where $|C_{ex}^{(\rho)}(\mathbf{q})|^2$ are the excitonic weights (6) with E_0 replaced by $E_1(q)$, and $E_{U,L}(q)$ replaced by $\mathcal{E}_\rho(q)$. Then the total probability of the bimolecular quenching for a polariton with a wave vector $\mathbf{q}_1$ is:

$$W(\mathbf{q}_1, \rho) = \sum_{\mathbf{q}_2, \rho'} W_{pol-pol}(\mathbf{q}_1 + \mathbf{q}_2; \mathbf{q}_1, \rho; \mathbf{q}_2, \rho')n(\mathbf{q}_2, \rho'), \tag{38}$$

where $n(\mathbf{q}_2, \rho')$ is the number of polaritonic states of the branch ρ' with the wave vector $\mathbf{q}_2$.

The important thing is that in a microcavity the bimolecular quenching $W_{pol-pol}$ (37) is strongly suppressed in comparison with the non-cavity material due to the factor $\Gamma(\mathbf{q}_1, \rho; \mathbf{q}_2, \rho') = |C_{ex}^{(\rho)}(\mathbf{q}_1)|^2 |C_{ex}^{(\rho')}(\mathbf{q}_2)|^2$. This factor is of the order of unity when both polaritons are excitonic-like, and is much smaller than unity otherwise. For the illustration, the dependence of $\Gamma(\mathbf{q}_1, \rho = L; \mathbf{q}_2, \rho' = L)$ on the wave vector $\mathbf{q}_2$ for fixed value of $\mathbf{q}_1 = 4000\text{cm}^{-1}$ is shown in Fig.7. Other parameters again are: $2\Delta = 80$ meV, $u = -130$ meV. It is seen that the quenching of lower branch polaritons with small wave vector is suppressed in more than ten times in comparison with the non-cavity material. Note that the similar suppression of bimolecular quenching may be expected in other cases of organic materials (for instance, for disordered materials or materials with weak intermolecular interaction) under the condition of strong light-matter coupling.

It is clear that the suppression of exciton-exciton annihilation is essential for the creation of high concentrations of cavity polaritons in organic structures. High densities of polaritons became of interest, for instance, in relation with the speculations on the possibility to realize the idea of polariton laser. This idea was recently widely discussed for inorganic semiconductor microcavities [19]. In this connection, another interesting question which requires the analysis is the problem of nonlinear pumping of the lowest energy polaritonic state. This process is determined by the efficiency of the polariton-polariton scattering and as far as we know, for organic materials it had never been discussed.

5. Conclusions

We have discussed the strong light-matter coupling regime in the microcavities which contain organic semiconductor as an optically active material. This field of investigation is relatively new, and though a vast store of knowledge is accumulated about the strong coupling regime in inorganic microcavities, a few of them can be directly applied to the organic microcavities. Besides, the light-matter coupling in organic microcavities can possess many features qualitatively new with respect to that in inorganic materials.

In this paper we have demonstrated the coexistence of two types of the excitations (coherent and incoherent) for an organic material with weak resonance intermolecular interaction. We have shown that the polaritonic states form in certain restricted intervals of the wave vector only. We have estimated the positions of the end-points of the polaritonic spectrum and demonstrated that the number of incoherent states strongly exceeds the number of lower branch polaritonic states. We then have studied the nonradiative decay of the upper branch polaritonic states accompanied by the emission of an intramolecular optical phonon for the case of strong exciton-phonon coupling. This transition was found to be very fast (20 – 70 ps) due to the fact that the number of final (incoherent) states was very large. Finally, the bimolecular quenching of excitons in a microcavity has been considered. It is known that in non-cavity material this quenching is very important at high concentration of excitons. We have shown that in a microcavity, where the excitons become cavity polaritons, this process is essentially suppressed. We expect more and more intriguing problems to be found in this field as both the experimental and theoretical study of these systems goes on.

6. Acknowledgments

The support of this research both by the UK EPSRC and the Russian Ministry of Science and Technology and Russian Foundation of Basic Research is acknowledged. We also acknowledge financial support from UNESCO UVO-ROSTE grant 875650.2(20) and from MIUR PRIN grant 2001028432.

References

1. See, e.g., "Confined Electrons and Photons: New Physics and Applications" ed. by E. Burstein and C. Weisbuch (Plenum, New York, 1995).
2. "Microcavities and Photonic Bandgaps: Physics and Applications", ed. by C. Weisbuch and J.G. Rarity; Kluwer, Dordrecht (1996);
   M.S. Skolnick, T.A. Fisher and D.M. Whittaker; Semicond. Sci. Technol., **13**, 645 (1998).
   G. Khitrova, H.M. Gibbs, F. Jahnke, M. Kira, and S.W. Koch; Rev. Mod. Phys, **71**, 1591 (1999).
3. C. Weisbuch, M. Nishioka, A. Ishikawa, Y. Arakawa; Phys. Rev. Lett., **69**, 3314 (1992).
4. F. Tassone, C. Piermarocchi, V. Savona, A. Quattropani, P. Schwendimann; Phys. Rev. B, **56**, 7554 (1997).
5. V. Agranovich, H. Benisty, C. Weisbuch; Sol. St. Comm., **102**, 631 (1997).
6. D.G. Lidzey, D.D.C. Bradley, M.S. Skolnik, T. Virgili, S. Walker, D.M. Whittaker; Nature, **395**, 53 (1998);
7. D.G. Lidzey, D.D.C. Bradley, T. Virgili, A. Armitage, M.S. Skolnik; Phys. Rev. Lett., **82**, 3316 (1999).
8. A.I. Tartakovskii, M. Emam-Ismail, D.G. Lidzey, M.S. Skolnik, D.D.C. Bradley, S. Walker, V.M. Agranovich; Phys. Rev. B, **63**, 121302 (2001).
9. P.A. Hobson, W.L. Barnes, D.G. Lidzey, G.A. Gehring, D.M. Whittaker, M.S. Skolnick, S. Walker; Appl. Phys. Lett. (submitted for publication).
10. V.M. Agranovich, M. Litinskaia, G.G. Lidzey; Phys. Rev. B (2003, in press).
11. V.M. Agranovich, M.D. Galanin; Electronic Excitation Energy Transfer in Condensed Matter, (North-Holland, Amsterdam, 1982), Chapter 5;
    M.D. Fayer, in: "Spectroscopy and Excitation Dynamics of Condensed Molecular Systems", (Eds.: V.M.Agranovich, R.M.Hochstrasser, North-Holland, Amsterdam, 1983), p.185.
12. M. Pope, C.E. Swenberg; "Electronic Properties in Organic Crystals and Polymers", Oxford University Press, New York (1999).
13. See, e.g., R.E. Peierls; "Quantum Theory of Solids", Chapter 2; Oxford, Clarendon Press (1955).
14. R. Loudon; J. Phys. A, **3**, 233 (1970).
15. R.M. Hochstrasser, J.D. Whiteman; J. Chem. Phys. **56**, 5945 (1972);
    H. Fidder, J. Knoester, D.A. Wiersma; J. Chem. Phys., **95**, 7880 (1991).
16. E.W. Knapp; Chem. Phys., **85**, 73 (1984).
17. See, e.g., V.M. Agranovich, O.A. Dubovsky; JETP Lett., **3**, 223 (1966).
18. D.G. Lidzey, A.M. Fox, M.D. Rahn, M.S. Skolnik, V.M. Agranovich, S. Walker; Phys. Rev. B, **65**, 195312 (2002).
19. See, e.g., J.J. Baumberg; Physics World, **15**, 31 (2002).

PHOTOINDUCED LIGHT ASSISTED PATTERNING OF AZOPOLYMER FILMS: TOWARDS A NEW PROCESS FOR NANOSTRUCTURATION

C. HUBERT[1], C. FIORINI-DEBUISSCHERT[1], P. RAIMOND[1], J. M. NUNZI[2]
[1] *DRT-LIST/DECS/SE2M – Lab. Composants Organiques, Bât. 451, CEA Saclay, F-91191 Gif-sur-Yvette (France)*
[2] *Lab. POMA, UMR-CNRS 6136, Université d'Angers, 2 bvd. Lavoisier, F49045 Angers cedex (France)*

1. Introduction

Optical induction of quasi permanent birefringence and dichroïsm in azo-dye polymers has been the subject of many studies. It is now well established that azobenzene molecules, when excited with a polarized light beam whose wavelength lies in their absorption band, undergo a reversible trans-cis-trans isomerization process leading to a molecular quasi-permanent reorientation in a direction perpendicular to the light polarization direction. More recently, it has been shown that irradiation of an azo-polymer thin film with an interference pattern between polarized laser beams could lead to a direct and reversible topographic modification of the polymer film surface resulting in the induction of surface relief gratings in conjunction with the light interference pattern. [1-3] The position of the maximum heights correspond to the light intensity minima. It has been demonstrated that gratings dynamics and modulation amplitude depend on several parameters among which the polymer film thickness, the grating pitch, the power and time of irradiation, or the writing beams polarization: p-: p-polarizations leading to diffraction efficiencies ten times larger than s-: s- polarizations, modulation amplitudes being larger as the grating pitch is higher. Maximum amplitudes of modulation can be of the order of the initial polymer film thickness. Saturation was shown to depend only on the total energy accumulated although saturation is reached faster as the irradiation power increases.

Several models have been proposed to explain such surface relief grating formation and tend to prove that a mass transport phenomenon is at the origin of this structuration process.[4-6]

Surprisingly, we show in this paper that uniform and coherent irradiation of an azo-dye polymer film at normal incidence results in a spontaneous structuration process leading to the creation of an hexagonal surface relief grating.[7] The influence of different experimental parameters on these photo-induced structures are studied and a review of the possible origins of such process is given with discussion.

F. Charra et al. (eds.), Organic Nanophotonics, 317–325.

2. Light assisted hexagonal patterning

Samples are polymer films of disperse red one (DR1) azo-dye molecules grafted to a polymethyl-methacrylate (PMMA) skeleton (DR1MMA). The λ=514nm laser line of an argon laser is used to excite the azo-chromophores in their absorption band. The incoming light intensity together with its polarization direction can be varied using polarizers. The sample is set perpendicularly to the incident laser beam. Figure 1 shows the typical structures obtained after atomic force microscopy (AFM) characterization.
We can see that the structures appear as circular peaks, regularly spaced out. If we consider six circles, we can define one elongated hexagon whose principal axis is along the light polarization direction. On the transverse section, we can observe that the period of these structures is closed to the irradiation wavelength. The structures height appears quite uniform and can be as large as 80nm.

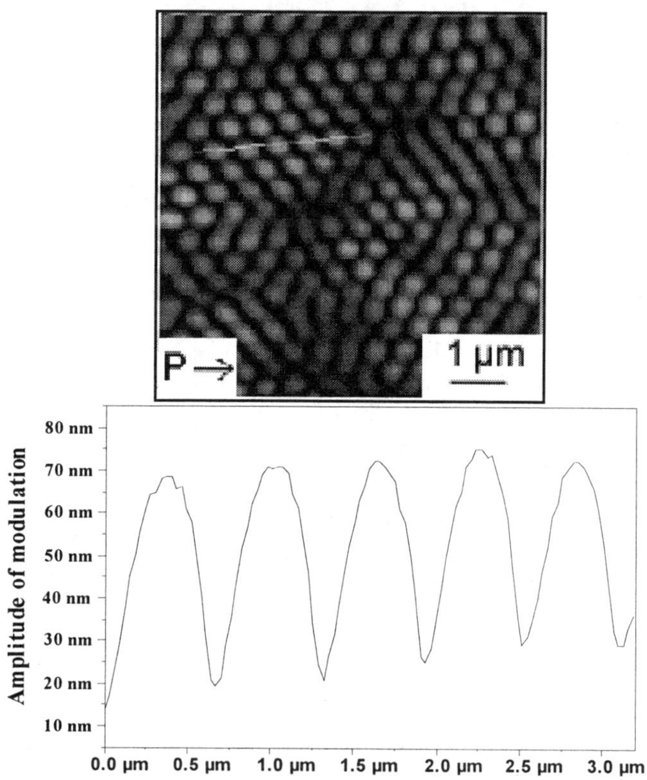

Figure 1. A.F.M. scan (7µm x 7µm) and cross section of the typical hexagonal structures obtained following uniform irradiation of a DR1MMA copolymer film. Incident laser beam polarization is set horizontal. Thickness of the sample is 220 nm, irradiation time and wavelength are respectively 100 minutes and 514 nm.

2.1 INFLUENCE OF THE MAIN EXPERIMENTAL PARAMETERS

The hexagonal structures orientation control was shown to depend on the incident light polarization direction. Indeed, in the case of a vertical or horizontal polarization, the principal axis of each elongated hexagon is oriented along this direction, whereas circular polarization does not lead to any preferential orientation of the structures.[7] Experiments have shown that for a comparable polymer film thickness, intensity laser beam and irradiation time, whatever the light polarization direction, the same modulation amplitudes are obtained. [7]

Working in the same experimental conditions than in the case of laser irradiation, i.e same energy absorbed by the polymer film, we have performed experiments with white polarized light. No structures can be observed in these conditions which prove that light coherence is a necessary condition and that coherent effects must lie at the origin of these structures.

Influence of the incident laser beam intensity together with irradiation time was also investigated. Whatever the laser intensities, the general dynamic growth of these photo-induced structures is the same, with a first phase of rapid growth followed by a second phase of saturation, around 70nm depth. Although saturation is reached faster as the laser intensity increases, the structures modulation amplitude only depend on the total energy accumulated by the polymer film. However, a threshold exists and if the intensity is lower than 200mW/cm^2, no structures can be observed. [7]

We have also studied the influence of the laser beam incidence angle on the structures modulation amplitude, using different light polarization directions (namely s and p). Results are presented on figure 2.

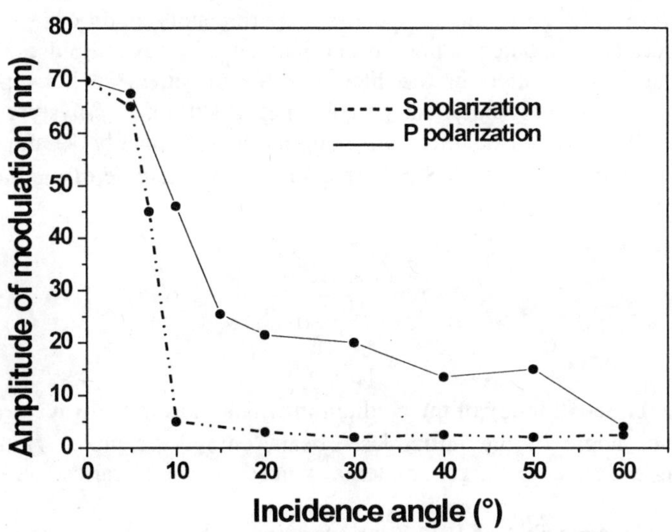

Figure 2. Dependence of the modulation amplitudes with the incidence angle. Irradiation time, wavelength, laser intensity and thickness of the polymer film were respectively 90 minutes, 514nm, 450mW/cm^2 and 445nm.

We can observe that, whatever the light polarization direction, s or p, the modulation amplitudes strongly decrease as the incidence angle is higher than 5° and in the case of s polarization, there exist a maximal incidence angle equal to 10° above which practically no structures are observed.
At the moment, the physical origin of these photo-induced structures is not fully understood and complementary experiments need to be performed to determine the relevant key parameters governing such hexagonal spontaneous light-assisted patterning. However, some quite similar features appear, either with previously developed numerical models or previously observed experimental results: this is discussed in the following section.

2.2 THEORITICAL MODELS AND POSSIBLE ORIGINS

Spatially periodic structures have been the subject of many studies. Hexagons and other regular patterns are predicted in a wide number of systems such as e.g. nonlinear optical systems, fluids... . In this latter case, convection due to thermal effects give rise to Rayleigh-Benard or Benard-Marangoni instabilities and was shown to lead to different structures like rolls or hexagons.[8] However, in our case, the fact that the hexagonal patterning process is quite dependent on the beam polarization and requires coherent excitation permits to rule out a possible purely thermal origin. In the following, we compare the results obtained in our experiments with different light-assisted patterning processes already proposed and whose physical origins was clearly established.
One possible explanation of this spontaneous patterning could originate from interferences between the incident beam and a surface scattered wave. Indeed, it has been recently demonstrated[9-11] that excimer laser irradiation can induce periodic surface structures (LIPSS), under specific experimental conditions, namely when polymer surfaces are irradiated in non normal incidence by several pulses. This is more precisely, what was reported in the literature for polymers such as poly(ethylene terephthalate) (PET), poly(butylene terephthalate) (PBT), polystyrene (PS), or polycarbonate (PC) with, irradiation wavelengths being typically 193nm and 248nm. The period Λ of the induced LIPSS patterns is given, in the case of s polarization, by the following formula :

$$\Lambda = \frac{\lambda}{n - \sin\theta} \tag{1}$$

where n is the refractive index of the medium in which the surface wave propagates and θ the incidence angle of the initial laser beam onto the sample. Thus, increasing incidence angle leads to larger patterns whose spacing varies with irradiation wavelength.
The mechanism responsible of LIPSS structuration is explained by the existence of a surface scattered wave. In fact, an interference pattern is created due to the interference between an incident wave and a scattered surface wave whose origin comes from the initial roughness of the polymer film. It results in spatial modulation of the fluence on the polymer giving rise to a periodic structuration of the surface. The periodicity of the

absorbed energy by the polymer film leads indeed to a periodic temperature elevation/melting of the surface or a periodic photolysis of the polymer.
In order to investigate the accuracy of such a model in the case of our experimental observations, we have studied the dependence of the hexagonal structures periodicity with the laser beam incidence using s-polarization. Indeed, according to the LIPSS model, structuration is only induced with s-polarization. Experimental results together with the theoretical dependence that would be expected in the case of LIPSS structures are shown in figure 3.

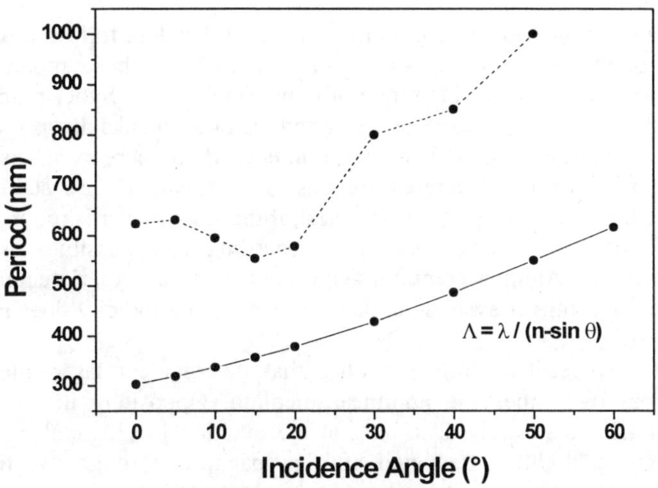

Figure 3. Influence of the incidence angle on the hexagonal structures periodicity (dashed line), and theoretical dependence expected from the LIPSS model (solid line) using n=1.7. Irradiation time and laser intensity were respectively 90 minutes and 450mW/cm^2 for a 445nm polymer film thickness. Irradiation wavelength was 514nm. Incident laser beam polarization was set to s.

Two phases can be distinguished. A first one where the period decreases, for incidence angles lower than 15°, and a second one where it increases, corresponding to higher values of the incidence angles. We can also notice that the variation of the incidence angle of the laser beam onto the polymer film surface does not lead to a modification of the patterns periodicity according to the expected theoretical formula of the LIPSS model and that the values of the period are much larger in our observations. AFM characterization reveals also a change in the structures type depending on the incidence angles: more precisely, the hexagonal structures, developed at normal incidence of the laser beam change to fringes or rolls as the incident angle is higher than 15°. Oppositely, only rolls structures can be observed in the LIPSS phenomena. Thus, due to these different observations, LIPSS can be ruled out to explain the observed phenomena in our azopolymer films.

Referring to literature, our experimental observations of light-assisted spontaneous patterning in azopolymer films show quite similar features with those observed in optical systems involving a feedback effect. Indeed, the fact that hexagons mainly develop at normal incidence is a good indication of a possible feedback mechanism and it has been predicted in some models that spontaneous structures can originate from interferences in a nonlinear medium (Kerr medium) with feedback element.[12-13] In these systems, also called mirror feedback systems, local perturbations of the nonlinear medium can induce a phase modulation of the input beam, which happens to be transformed into amplitude modulation due to diffraction or interferences, after propagation to the mirror and back, thus closing the feedback loop: this is the so-called "Talbot effect" .[14]

By increasing the pump power, concentric rings first develop, followed by bright spots in the center of the beam. Then, for higher pump values, these spots arrange in an hexagonal order, stable for a wide range of pump intensities. Higher pump power then leads to disorder, hexagons become unstable and the order breaks down.

Similar observations can be done in our experiments where there exists also an intensity threshold under which no structures are observed. Hexagons are stable for a wide intensity range but due to the polymer material, if the laser power is too high, ablation is observed thus preventing any conclusion regarding any possible disorder of the hexagonal structures. Another common feature between our experimental observations and single feedback mirror systems is that the patterns periodicity does not depend on the pump power.

By definition, any feedback effect implies that the incident laser intensity is high enough to cross two times the nonlinear medium. Oppositely in our experiments, azopolymer films are strongly absorbing at the exciting laser wavelength so that the incident energy is slightly attenuated after propagation through the film: typically optical density at λ=514 nm is about 1.5 for a DR1MMA 350nm azopolymer film.

However, due to its photoisomerization properties, photoinduced transparency results from DR1 excitation from its *Trans* form to its *Cis* form. The *Trans* and *Cis* form absorption spectra of DR1 were recently reported.[15] Based on these results, and considering λ=514 nm, the molecular extinction coefficient of the *Trans* form can be evaluated to be 2.3 times larger than for the *Cis* form. Such photoinduced transparency could thus enable a quite efficient feedback: roughly, the *cis* form absorption decreasing for wavelengths higher than 490nm, feedback could be more efficient as the laser excitation wavelength is higher than 490nm. Experimental results of the influence of the incident laser beam wavelength on the hexagonal structures modulation amplitude is presented in figure 4.

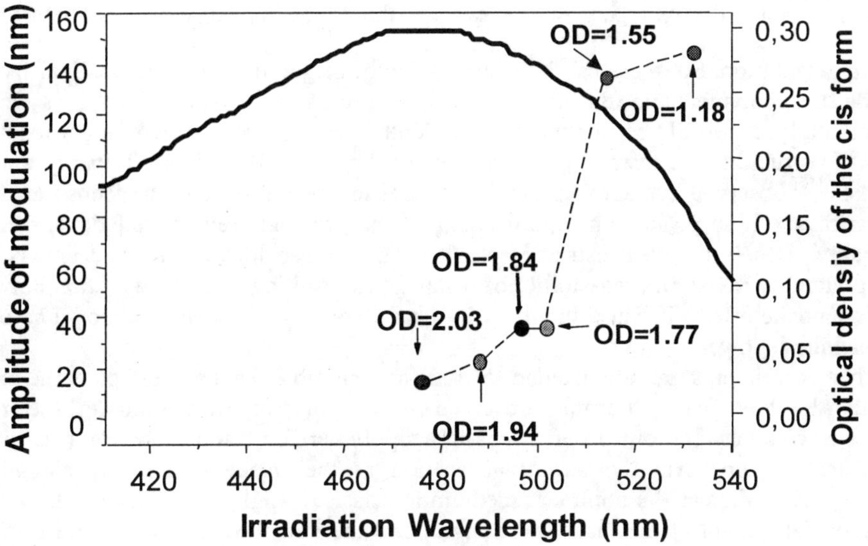

Figure 4. Modulation amplitudes measured at saturation in the case of different incident laser beam wavelengths for a 350 nm thick azopolymer film. Irradiation time and power were respectively 120 minutes and $300mW/cm^2$. The continuous line shows a zoom of the DR1 Cis form absorption spectrum. Optical densities values of the polymer film before irradiation are given for each wavelength.

The modulation amplitudes measured after saturation appear to be higher as the irradiation wavelength is higher thus meaning lower optical density values for both the *Trans* and *Cis* form of the azodye DR1 molecule. This tends to be in accordance with a possible feedback mechanism. Indeed, once they are excited and in their *Cis* form, azodye molecules are less absorbing to the irradiation wavelength and thus, due to this photo-induced transparency, more energy can be fedback from the back side of the sample leading to more efficient patterning.

However, quite surprisingly, experiments performed with different polymer film thicknesses show similar modulation amplitudes even for film thicknesses as high as 30 μm for which the photoinduced transparency cannot allow any significant feedback. Due to the high values of the optical densities of such thick film, practically all the incident energy is absorbed by the polymer film and we would have expected to observe a decrease of the modulation amplitudes with increasing the polymer film thickness. Parallely however, it can be noted that a quite important bleaching of our samples can be observed, the bleaching effect depending on the laser beam irradiation time and intensity. The origin of such bleaching is under investigation (FTIR spectroscopy) but, typically, the optical density change following hexagonal patterning can be as high as ΔO.D./O.D.=0.7.

After bleaching, the polymer is almost transparent. In order to size the parameters which can lead to feedback, the longitudinal interference pattern which may result from

reflexion along the exit size of the sample should be typically $\frac{\lambda}{2n} = 150nm$ for 514nm excitation. So, there can be a succession of bright and dark longitudinal fringes inside the polymer and the distance between dark and white fringes is 75nm. Surprisingly, this is almost the modulation depth. Besides models involving a feedback into the structure, an instability process could be invoked. Indeed, Tripathy and co workers[16] observed that above several watts per square centimeter, the deeps transform into peaks. These peaks can be attributed to the gradient field model developed by Tripathy, namely the electrostrictive effect which is used for optical tweezers. We do not pretend to invoke the possibility of instabilities resulting from competition between such opposite effects because the intensity used in our experiments is almost one order of magnitude lower.
Further experiments are now needed to confirm a possible feedback effect as the origin of the light-assisted patterning observed in azopolymer films: among the main discrepancies between our observations in azopolymer films and more classical single feedback systems are: the strong absorption at the incident laser wavelength in azopolymers whereas the nonlinear medium in classical single mirror feedback systems is fully transparent at the laser wavelength, and the structures sizes which are of the order of the laser wavelength in our experiments. In order to fully study the impact of such discrepancies, evolution of the classical single mirror feedback systems models is needed: more particularly, absorption, can no longer be neglected and small diffraction length together with a possible interferential feedback must be taken into consideration.

3 Conclusion

We have reported here experimental results of an original light-assisted spontaneous patterning process in azo-dye polymers films. A comparison with different already reported patterning processes or theoretical models leading to comparable photo-induced structures has been described and discussed.
It appears that such spontaneous hexagonal patterning process is still unclear although showing similar features with models or experiments already described in the literature. Actually, none of them can describe the totality of the observed phenomena. Among the different origin considered however, an optical feedback mechanism seems to be the most probable although requiring further complementary investigations. This should lead to the accurate determination of the relevant parameters at the origin of such spontaneous light-assisted patterning, further enabling a full modeling and control of the process. As demonstrated in more classical nonlinear systems such as liquid crystals, feedback systems enable easy generation of quite complex structures[13] which could open the way to new and easy ways of micro and nanopatterning in optically sensitive materials.

4. References

1. Rochon, P., Batalla, E., Natansohn, A.L. (1995) Optically induced surface gratings on azoaromatic polymer films, *Applied Physics Letters* **66**, 136-138.
2. Kim, D.Y., Tripathy, S.K., Li, L., Kumar, J. (1995) Laser-induced holographic surface relief gratings on nonlinear optical polymer films, *Applied Physics Letters* **66**, 1166-1168.
3. Kim, D.Y., Li, L., Jiang, X.L., Shivshankar, V., Kumar, J., Tripathy, S.K. (1995) Polarized laser induced holographic surface relief gratings on polymers films, *Macromolecules* **28**, 8835-8839.
4. Barrett, C. J., Natansohn, A. L., Rochon, P. L. (1996) Mechanism of optically inscribed high-efficiency diffraction gratings in azo polymer films, *Journal of Physical Chemistry* **100**, 8836-8842.
5. Lefin, P., Fiorini, C., Nunzi, J. M. (1998) Anisotropy of the photoinduced translation diffusion of azo-dyes, *Optical Materials* **9**, 323-328.
6. Kumar, J., Li, L., Jiang, X. L., Kim, D. Y., Lee, T. S., Tripathy, S. K. (1998) Gradient force: the mechanism for surface relief grating formation in azobenzene functionalized polymers, *Applied Physics Letters* **72**, 2096-2098.
7. Hubert, C., Fiorini-Debuisschert, C., Maurin, I., Nunzi, J.M., Raimond, P. (2002) Spontaneous patterning of hexagonal structures in an azo-polymer using light-controlled mass transport, *advanced Materials* **14**, **10**, 729-732.
8. Guyon, E., Hulin, J.P., Petit, L. (2001) *Hydrodynamique Physique*, Ed. CNRS, Paris.
9. Bolle, M., Lazare, S., (1993) Submicron periodic structures produced on polymer surfaces with polarized excimer laser ultraviolet radiation, *Applied Surface Science* **65/66**, 349-354.
10. Siegman, A. E., Fauchet, P. M., (1986) Stimulated wood's anomalies on laser-illuminated surfaces, *Journal of Quantum Electronics* **22**, 1384-1403.
11. Baüerle, D., (1995) *Laser Processing and Chemistry,* Springer Publisher, Berlin, pp. 187-330.
12. D'Allessandro, G., Firth, W. J., (1991) Spontaneous hexagon formation in a nonlinear optical medium with feedback mirror, *Physical Review Letters* **66**, 2597-2600.
13. Firth, W. J., (1995) *Self-Organization in optical Systems and applications to Information Technology*, in M. A. Vorontsov and W. B. Miller (eds.), Springer Berlin, pp. 69-96.
14. Mansuripur, M., (1997) The Talbot effect, *Optics and Photonics News*, pp. 42.
15. Loucif-Saïbi, R., Nakatani, K., Delaire, J. A., Dumont, M., Sekkat, Z., (1993) Photoisomerization and second harmonic generation in disperse red one doped and functionalized poly(methyl methacrylate) films, *Chemistry of materials* **5**, pp. 229-236.
16. Bian, S., Williams, J. M., Kim, D. Y., Li, L., Balasubramanian, S., Kumar, J., Tripathy, S. K., (1999) Photoinduced surface deformations on azopolymer films, *Journal of Applied Physics* 86, pp. 4498.

MAPPING THE DIPOLAR RADIATION OF COPPER NANOPARTICLES USING A PHOTOSENSITIVE FILM

P.A. CHOLLET,[1] P. ANDRÉ,[2,*] M.P. PILÉNI,[2] F. CHARRA[3]
[1]*Laboratoire des Composants Organiques,*
DECS/SE2M, CEA-DRT-LIST, 91191 GIF CEDEX (France).
[2]*Laboratoire des Matériaux Mésoscopiques et Nanométriques,*
Université PARIS VI, BP 52, 75231 PARIS CEDEX 05 (France).
[*]*Present address :NSF Science and Technology Center,*
University of North Carolina, Chapel Hill, NC 27599-3290 (USA).
[3]*Laboratoire Nanophotonique,*
SPCSI, CEA-DSM-DRECAM, 91191 GIF CEDEX (France).

1. Introduction

Enhancement of a static electric or an oscillating electromagnetic field close to a metallic tip is known since a very long time. Recently in the framework of SNOM (scanning near-field optical microscopy) and exalted fluorescence, several theoretical and experimental studies have been undertaken concerning enhancement of the electromagnetic field close to atomic force microscope (AFM) tips [1-4] and enhancement of two-photon excitation and fluorescence in the vicinity of metallic nanoparticles [5].
On the other hand, it has been shown that the surface of azo-polymer films undergoes structuration under laser beam intensity gradients [6,7]. The amplitude of the film relief can reach 200 nm for the *p*-polarization. The intensity gradients can be created through intensity gratings obtained by splitting the beam and producing interferences, or under the spot of a beam focused on the film by a microscope objective [8].

In this paper we describe film structuration generated by an intensity gradient arising from the interferences between the plane wave of a linearly polarized laser beam at normal incidence, and the dipolar radiation arising from copper nanoparticles deposited on the azo-polymer film. As described above, the light intensity gradient induces structuration of the film surface. Those structurations can be considered as a light intensity sensor.

F. Charra et al. (eds.), Organic Nanophotonics, 327–332.

2. Experimental techniques and sample preparation

2.1. APPARATUS

The film thickness was measured with a profilometer (Dektak 3ST). The film surface roughness and structuration was measured with an AFM (AutoProbe CP, Park Scientific Instruments) in the contact mode. The laser beam (average power: 120 mW) used to induce film structuration was obtained by frequency doubling the beam delivered by a Tsunami femtosecond Ti-Sapphire (Spectra Physics model n° 3980). The wavelength was $\lambda = 442$ nm, and the beam waist on the film 7.5 mm. This results in an average light intensity equal to 200 mW/cm^2.

2.2. MATERIALS

2.2.1. *Elaboration*

Copper nanoparticles were produced by chemical reduction of copper ions in water, isooctane and $Cu(AOT)_2$ (AOT is bis-(2-ethylhexyl) sulfosuccinate) reverse micellar procedure, as described previously [9]. A suspension of nanoparticles, coated with AOT, is made in toluene. A drop of this solution was deposited on a grid whose holes are bridged with a very thin carbon film, for examination by transmission electron microscopy (TEM), as shown in *Figure 1*.

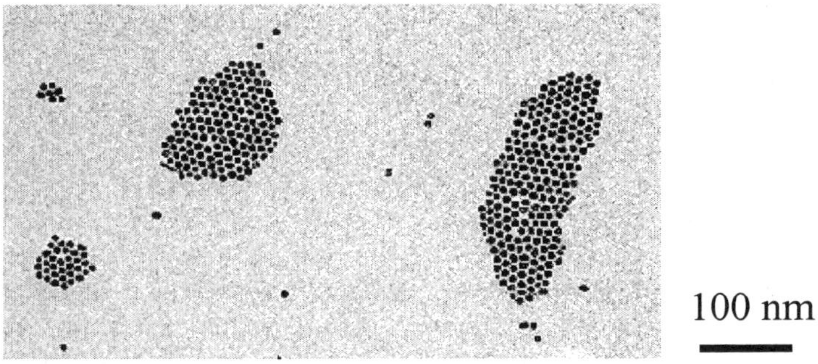

Figure 1. Transmission electron microscopy of copper nanoparticles.

On this TEM image it can be seen that the nanoparticles (diameter ~ 10 nm) have a strong tendency to cluster into islands

The azo-dye Disperse Red 1 (DR1: N-ethyl-N-(methacryloxyethyl)-4'-amino-4-nitroazobenzene) was attached to methylmethacrylate (MMA) in the molecular ratio 65:35 to form DR1-MA. Radical polymerization was used to achieve copolymerization of DR1-MA with MMA [10].

The copolymer film was dissolved in 1,1,2-trichloroethane (135 gL^{-1}) and deposited on a fused silica substrate by spin coating (1500 rounds.min^{-1} for one minute) and then baked at 80°C for 30 min to achieve complete elimination of the solvent. Its thickness was 875 ± 5 nm. Its roughness was found to be negligible.

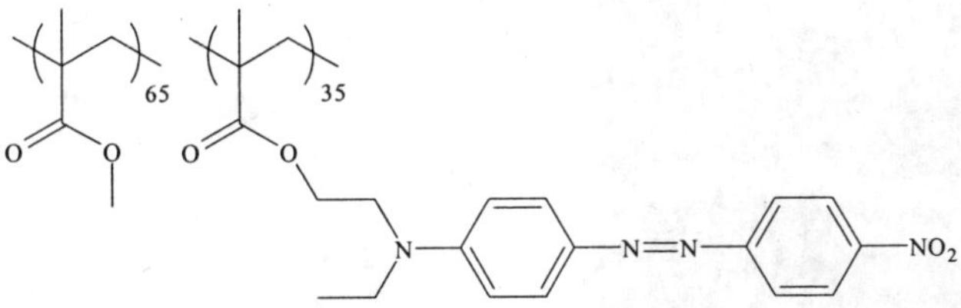

Figure 2. Chemical structure of the polymer film.

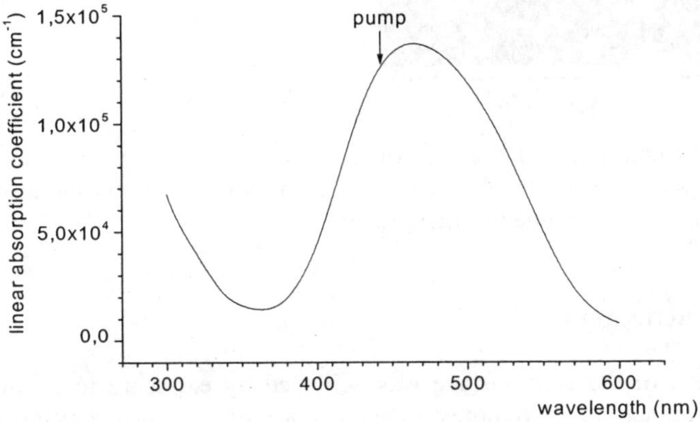

Figure 3. Linear absorption of the polymer film.
The arrow shows the laser wavelength.

The chemical structure of the copolymer is shown in Figure 2, and the linear absorption coefficient of the film in Figure 3.

2.2.2. *Deposition of nanoparticles on the polymeric film*

The AFM image of particles deposited on the film is shown in Figure 3. One can observe various sizes of island (the brightest are the thickest), due to cluster formation, similarly as in the TEM image on carbon films. Apart from these clusters, the film remains flat. In addition, it was checked that no surface modification of the film was induced either by deposing and letting evaporate a drop of pure toluene

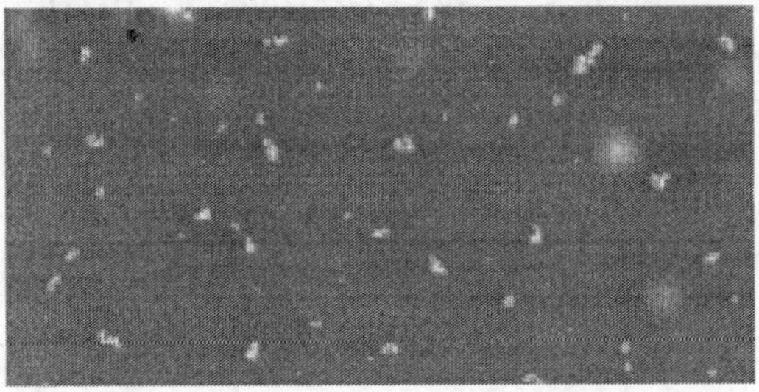

Figure 4. AFM image of the film, covered with copper nanoparticles. The horizontal width is 16 μm.

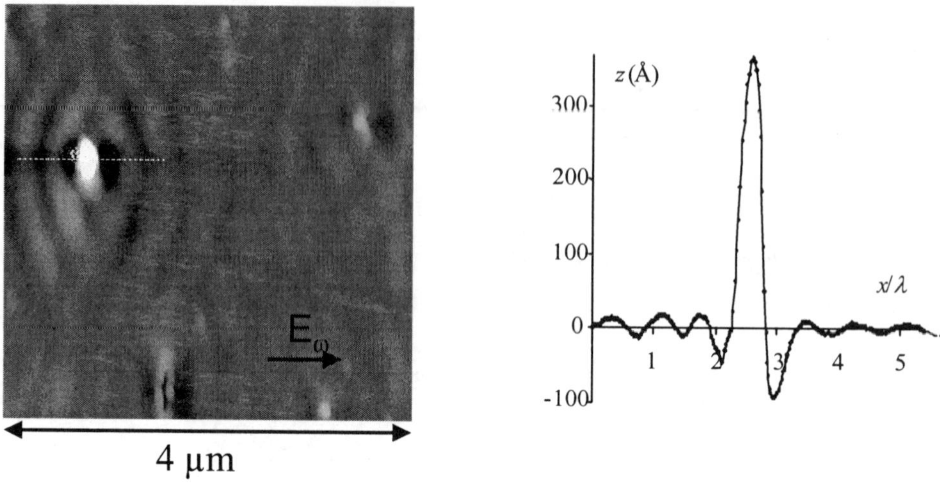

Figure 5. Structuration of the film surface around a nanoparticle cluster: AFM image on the left side (the laser polarization direction is indicated by the arrow), and section along the white dotted line on the right side.

3. Film structuration

Structuration of the film surface was achieved by exposure to a polarized beam, at normal incidence for 20 minutes. In such conditions, no structuration was observed in the absence of copper nanoparticles. The field gradients results from the inhomogeneous diffracted waves themselves and their interference with the homogeneous incident plane wave.

In Figure 5 is shown the surface film structuration in the neighborhood of a cluster of nanoparticles. One observes a strong bump at the location of the metal. Its height is larger than that measured over particle clusters before exposure. Thus, it doesn't correspond to the cluster thickness itself, but rather to a piling of the polymer under it, due to strong light intensity gradient. However, their is no plasmon resonance (see below) at the wavelength used (440 nm) where value of the copper complex dielectric constant is: $\tilde{\varepsilon}_{Cu} = -4{,}3 + 5{,}4i$ [Johnson 1972]. Nevertheless, since $|\tilde{\varepsilon}_{Cu}| >> 1$, in the vicinity of nanoparticles, inside the cluster, the field is locally enhanced by a factor close to 3, which results in a strong intensity gradient, whose magnitude order is $20\,\bar{I}/\lambda$, where $\bar{I}$ is the average intensity (200 mW/cm^2). As concerns the metallic cluster, if we assumed a spherical shape with radius a much smaller than the wavelength, its polarizability is (SI units):

$$\tilde{\alpha} = 4\pi\varepsilon_0\varepsilon_a a^3 \frac{\varepsilon_{Cu} - 1}{\varepsilon_{Cu} + 2\varepsilon_a} \qquad (1)$$

In this equation, ε_a is the dielectric constant of the ambient (unity). It results from Equation(1) the phase of $\tilde{\alpha}$ is equal to - 0,37 rd (- 21°). The oscillatory behavior observed in fig. 5 (left) is obviously the result of interference of the wave diffracted by the particle (i.e. the wave generated by the induced dipole) with the incident field. Hence, a careful analysis of this interference pattern permits to deduce the phase of the diffracted wave at its origin. This phase is found to be close to $\pi/2$ with respect to the. Since the polarizability $\tilde{\alpha}$ is nearly real, the phase of the diffracted wave observed here corresponds to a near-field term. This is consistent with the fast decay observed in fig. 5 and with the theoretical fact that the radiation pattern of a horizontal dipole on an interface vanishes in the direction along the interface.

4. Conclusions

We have shown that the surface of photosensitive film polymer can be structured under laser illumination when metal nanoparticles are deposited on it. This is due to the strong intensity gradient induced at the vicinity of those particles. The surface structuration map results from the interferences between the incident polarized laser beam (plane wave) and the dipolar radiation due to the nanoparticles. It appears clearly that the structuration vanishes when moving off from the particles. The calculation of this structuration is not straightforward since the radiated dipoles (naanoparticles) are located at the boarder between a perfect transparent medium (air, $n = 1$) and a strongly absorbing medium (polymer film, n = 1.8 + 0.45i). Even more complicated is the case of multiscattering by many nanoparticles. One of the author (FC) can perform the calculations. The other three authors are far too much mental deficient even to understand those calculations.

5. Acknowledgments:

This work arose from very fruitful discussions that we had with the regretted Sunkant K. TRIPATHY, two years ago. Thanks are due to Dr. P. Raimond for the synthesis of the DR1-copolymer.

6. References

[1] Novotny, L., Bian, R.X., and Xie, X.S. (1997), *Phys. Rev. Lett.* **79**, 645.
[2] Cory, H., Boccara, A.C., Rivoal, J.C., and Lahrech, A. (1998), *Microwave Opt. Technol. Lett.* **18**, 120.
[3] Zayats, A.V., *Opt. Commun.* (1999) **161**, 156.
[4] H'dhili, F. , Bachelot, R., Lerondel, G., Barchiesi, D., Royer, P. (2001), Appl. Phys. Lett. **79**, 4019
[5] Gryczynski, I., Malicka, J., Shen, Y., Gryczynski, Z., Lakowicz, J.R (2002)., J. Phys. Chem. B **106**, 2191-5.
[6] Kim, D.Y., Tripathy, S.K., Li, L., Kumar (1995), *J. Appl. Phys. Lett.* **66**, 1166.
[7] Rochon P., Battala E., Natansohn A. (1995), *Appl. Phys. Lett.* **66** 136.

[8] Shaoping Bian, S., Williams, J.M. , Kim, D.Y. ,Li, L. ,Balasubramanian, S. , Kumar, J., Tripathy, S. , 'Photoinduced surface deformations on azobenzene polymer films' (1999), *J. Appl. Phys*. **86**, 4498.
[9] Tanori, J., Piléni, M.P. (1997), *Langmuir* **13**, 632.
[10] Nahata, A., Shan, J., Yardley, J.T., Wu, C. (1993), *J. Opt. Soc. Am. B* **10**, 1553.
[11] Johnson, P. B. and Christy, R. W., *Phys. Rev. B* **6** (1972) 4370

COLLISION INDUCED LIGHT SCATTERING BY SF_6 - ISOTROPIC AND ANISOTROPIC SPECTRA

K. NOWICKA AND T. BANCEWICZ
Nonlinear Optics Division
Institute of Physics, Adam Mickiewicz University
60-780 Poznan, Grunwaldzka 6
Poland
Y. LE DUFF, J.-L. GODET AND F. RACHET
POMA Laboratory
University of Angers
2 boulevard Lavoisier, 49045 Angers
France

ABSTRACT. The two-body anisotropic and isotropic collision-induced light scattering spectra of the sulfur hexafluoride in the gas state at room temperature are presented on an absolute scale up to 210 cm^{-1}. Due to the octahedral symmetry of sulfur hexafluoride, the analysis of the high frequency wings is an excellent way for the determination of the dipole-octopole polarizability E. A value of E is proposed which is in good agreement with recent *ab initio* calculations.

Collision-induced light scattering has been the subject of many investigations over the past decade, and has been reviewed recently [1, 2, 3]. The scattered intensity depends on the polarizability change in a cluster of interacting molecules. The time-dependence of this change is a function of the intermolecular potential. The polarizability and the potential are electronic properties of the colliding molecules, and together they influence on scattering intensities [1]. The study of the collision-induced light scattering (CILS) intensities has been demonstrated to be an useful method to investigate multipolar polarizabilities in gaseous sample [2, 4, 3, 5].

During a collision the electronic distribution of the colliding molecules is disturbed by interatomic interactions producing an excess polarizability in the colliding complex [1]. The polarizability of pair of atoms or molecules A and B is no longer the simple sum of the internal polarizabilities. Due to dipole-induced dipole (DID) or multipolar mechanisms, to frame distortion or overlap and exchange effects, some additional polarizability appears, which is known as the collision-induced polarizability. The collision-induced polarizability is best manifested if it is the only rea-

F. Charra et al. (eds.), Organic Nanophotonics, 333–338.

son for the phenomenon observed. Globular molecules of SF_6 are optically isotropic and their depolarized (anisotropic) spectrum may be considered as purely of collision-induced nature.

It is possible to study two kinds of collision-induced spectra: the anisotropic and the isotropic ones. Generally the properties of the scattering spectra depend on the symmetry of the molecules involved. For the anisotropic CILS from gaseous SF_6, most of intensities come from translational effects including the DID and electron overlap interactions. The rototranslational contributions due to the dipole-octopole polarizability represent only a small part of the anisotropic intensities. In the case of the isotropic CILS spectrum of SF_6, the DID interaction contributes only at the second order and this gives a good opportunity to study multipolar contributions and the dipole-octopole polarizability tensor in particular. Several papers were published concerning the collision-induced light scattering of SF_6 [6, 7]. They concern the anisotropic intensities in the frequency shifts range of $0 - 70$ cm^{-1}. We present new experimental results of CILS intensities up to 210 cm^{-1}. The isotropic SF_6 spectrum is measured and analyzed for the first time.

Our experimental setup has been described previously [3]. We used a 15 watt argon-ion laser operating at 2 watt on the 514.5 nm line for excitation. The gas pressure was changed from 1 to about 21 bars. Densities were calculated from PVT data of Ref.[8]. The scattering spectra induced by binary interactions have been deduced from density behavior analyses of the spectral Stokes intensities [4]. Two types of binary scattering intensities, $I_{2,||}(\nu)$ and $I_{2,\perp}(\nu)$, were recorded with the laser polarization parallel and perpendicular to the scattering plane, respectively. These intensities were calibrated on an absolute scale, on the basis of the comparison with the intensity of $S_0(1)$ rotational line of hydrogen used as an external reference according to a procedure described previously [9]. However, some mixing between perpendicular and parallel components may occur because of the nonzero value of the collection angle of the light scattered by the sample and we apply a correction to take into account this effect [10]. Finally, we have:

$$I_{ani}(\nu) = I_{2,||}(\nu) \tag{1}$$

$$I_{iso}(\nu) = I_{2,\perp}(\nu) - \frac{7}{6} I_{2,||}(\nu). \tag{2}$$

In figures 1 and 2 we present the experimental binary isotropic spectrum $I_{iso}(\nu)$ and anisotropic one $I_{ani}(\nu)$ up to 210 cm^{-1} with the error bars. In order to deduce the value of the dipole-octopole polarizability E from our experimental data, we compared them with the theoretical spectra taking

Table 1: The isotropic and anisotropic Rayleigh light scattering correlation functions $F_{00}^{(m)}$ and $F_{22}^{(m)}$, respectively for the successive multipolar induction operators.

Mechanism	j_A	j_B	l_A	l_B	N	$F_{00}^{(m)}$
$\alpha\,\mathbf{T}_4\,\mathbf{E}$	0	4	1	3	4	$\frac{224}{9}\alpha^2 E^2 R_4(t) S_4(t)$
$\mathbf{E}\,\mathbf{T}_6\,\mathbf{E}$	4	4	3	3	6	$\frac{55}{3}E^4 (R_4(t))^2 S_6(t)$
Mechanism	j_A	j_B	l_A	l_B	N	$F_{22}^{(m)}$
$\alpha\,\mathbf{T}_4\,\mathbf{E}$	0	4	1	3	4	$\frac{220}{9}\alpha^2 E^2 R_4(t) S_4(t)$
$\mathbf{E}\,\mathbf{T}_6\,\mathbf{E}$	4	4	3	3	6	$\frac{11330}{21}E^4 (R_4(t))^2 S_6(t)$

into account both the DID and rototranslational contributions. To calculate these theoretical spectra, we chose the intermolecular potential of SF_6 proposed by Zarkova [11], because the theoretical zero moment obtained with this potential is the closest value to our experimental moment [5] (for others potentials see Refs.[4, 5]). These double differential cross sections are proportional to the Fourier transform of the autocorrelation function $F_{KK}(t)$ given in Table 1. The two-body anisotropic and isotropic collision-induced light scattered intensities are given by [3]:

$$I_{ani}^{(m)}(\nu) = k_s^4 \frac{1}{2\pi} \int \exp(-i\omega t) \frac{1}{5} F_{KK}^{(m)}(t) dt. \tag{3}$$

$$I_{iso}^{(m)}(\nu) = k_s^4 \frac{1}{2\pi} \int \exp(-i\omega t) \frac{1}{3} F_{KK}^{(m)}(t) dt. \tag{4}$$

were k_s is the light scattering vector. In Table 1 are reported the isotropic and anisotropic Rayleigh light scattering autocorrelation functions calculated for the two multipolar mechanisms: dipole-induced octopole ($\alpha\mathbf{T}_4\mathbf{E}$) and octopole-induced octopole ($\mathbf{E}\mathbf{T}_6\mathbf{E}$). The correlation function given in Table 1 is a product of translational:

$$S_N(t) = \left\langle D_{0\,0}^N(\delta\Omega_{AB}(t))\, R_{AB}(0)^{-(N+1)}\, R_{AB}(t)^{-(N+1)} \right\rangle \tag{5}$$

and orientational:

$$R_j(t) = \left\langle D_{n\,n}^j\,(\delta\Omega(t)) \right\rangle \tag{6}$$

correlation functions. In our equations (5) and (6) $\delta\Omega(t)$ and $\delta\Omega_{AB}(t)$ denote the reorientation angles of the molecule and intermolecular vector, respectively, at the time t.

The translational contribution due to the DID interaction was calculated from classical trajectories of colliding molecules. We investigated here the theoretical rototranslational contributions due to the dipole-octopole polarizability tensor **E** (($\alpha \mathbf{T}_4 \mathbf{E}$ and $\mathbf{E}\mathbf{T}_6\mathbf{E}$). Multipolar contributions to the spectrum were computed by the method described in details in Refs.[3] and [5]. We report in Fig.(1) the corresponding theoretical sets of curves for $\alpha = 4.549$ Å^3 and for value of E which allows the theoretical spectrum to fit experimental data in the 30 cm^{-1} – 85 cm^{-1} frequency range. This fitted value of E at $\lambda_L = 514.5$ nm is $E = 3.8 \pm 0.7$ Å^5. Moreover, we have checked that the values of E found in this way are compatible with the experimental anisotropic spectrum (Fig.(2)). The agreement of our fitted value with the one recently calculated by Maroulis ($E = 4.2$ Å^5) [12] as well as the good comparison between theoretical and experimental data point out that the $\alpha \mathbf{T}_4 \mathbf{E}$ mechanism may be considered as responsible for most of the spectral intensities in the 30 cm^{-1}-85 cm^{-1} frequency range. The discrepancy between theoretical intensities and experimental ones beyond 160 cm^{-1} may be explained by short range effects and/or the inaccuracy of the potential well shape [5].

In conclusion, we have shown that the anisotropic and isotropic spectra bring both informations on interactions between molecules. Anisotropic and isotropic collision-induced intensities are complementary to each other to obtain the experimental value of the dipole-octopole polarizability of the SF_6 molecule. However, the isotropic spectrum plays a leading role and improved significantly the accuracy of the dipole-octopole polarizability value deduced from experiment.

Figure 1: Two-body isotropic scattering spectra in absolute unit (cm^6) for SF_6 at 294.5 K. Our experimental data (full dots (•)) are given with error bars. Triangles (∇) represent estimated data. Partial and total theoretical spectra are computed for the Z potential [11], $\alpha = 4.549$ Å^3 and $E = 3.8$ Å^5: dipole-induced dipole, $\alpha \mathbf{T} \alpha$ (— - —); dipole-induced octopole, $\alpha \mathbf{T}\,\mathbf{E}$ (- - - -); octopole-induced octopole, $\mathbf{E}\,\mathbf{T}\,\mathbf{E}$ (— — —); total (——).

Figure 2: Two-body anisotropic scattering spectra in absolute units (cm^6) for SF_6 at 294.5 K. Our experimental data (circles (∘)) are given with error bars. Partial and total theoretical spectra are computed for the Z potential [11], $\alpha = 4.549$ Å^3 and $E = 3.8$ Å^5: dipole-induced dipole, $\alpha \mathbf{T} \alpha$ (— - —); dipole-induced octopole, $\alpha \mathbf{T}\,\mathbf{E}$ (- - - -); octopole-induced octopole, $\mathbf{E}\,\mathbf{T}\,\mathbf{E}$ (— — —); total (——). Total and $\alpha \mathbf{T} \alpha$ intensities are almost the same.

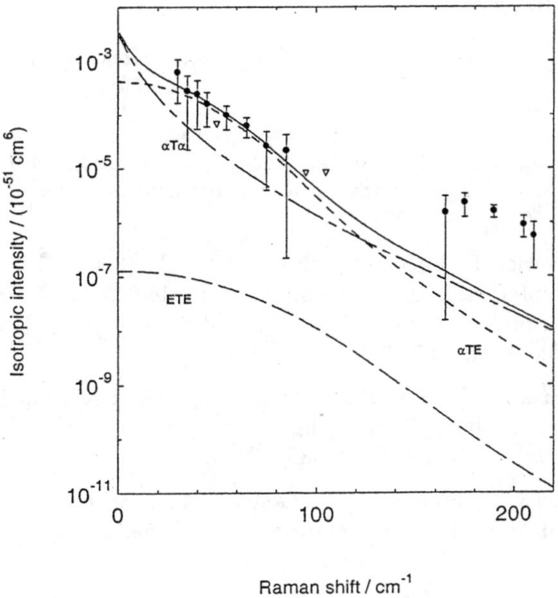
Isotropic intensity / (10^{-51} cm^6)
Raman shift / cm^{-1}
αTα
ETE
αTE
10^{-3}
10^{-5}
10^{-7}
10^{-9}
10^{-11}
0
100
200

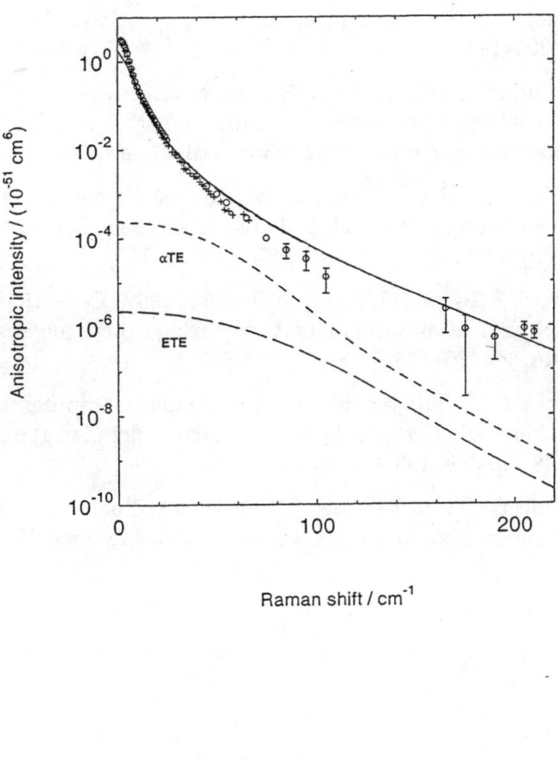
Anisotropic intensity / (10^{-51} cm^6)
Raman shift / cm^{-1}
αTE
ETE
10^{0}
10^{-2}
10^{-4}
10^{-6}
10^{-8}
10^{-10}
0
100
200

References

[1] Tabisz G.C. and Neuman M.N., 1995, *Collision–and Interaction–Induced Spectroscopy, NATO ASI Series C: Mathematical and Physical Sciences*, Vol. 452 (Dordrecht: Kluwer).

[2] Buckingham A. D. and Tabisz G.C. (1978) Collision-induced rotational Raman scattering by tetrahedral and octahedral molecules, *Molecular Physics*, **Vol. 36**, pp.583–596

[3] Bancewicz T., LeDuff Y., and Godet J. L. (2001) Multipolar polarizabilities from interaction-induced Raman scattering, *Modern Nonlinear Optics, Part 1, Advances in Chemical Physics*, **Vol.85**, edited by M. W. Evans (New York: J. Wiley), pp. 267–307.

[4] Godet J. L., Rachet F., Le Duff Y., Nowicka K., and Bancewicz T., (2002) Isotropic collision induced light spectra from gaseous SF_6, *Journal of Chemical Physics*, **Vol.116**, pp.5337–5340.

[5] Nowicka K., Bancewicz T., Godet J. L., Rachet F., Le Duff Y. (2002) Polarization Components of Rototranslational Light Scattering Spectra from gaseous SF_6, *to be published.*

[6] Shelton D. P. and Tabisz G. C. (1980) Binary collision-induced light scattering by isotropic molecular gases. II. Molecular spectra and induced rotational Raman scattering, *Molecular Physics*, **Vol.40**, pp.299–308.

[7] El-Sheikh S. M. and Tabisz G. C. (1989) Perturber dependence of the collision-induced light scattering by SF_6 and CH_4, *Molecular Physics*, **Vol.68**, pp. 1225–1238.

[8] Sigmund P. M., Silberberg I. H., and MacKetta J. K. (1972) Second and third virial coefficients for system tetrafluoromethane-sulfur hexafluoride, *Journal of Chemical and Enginering Data*, **Vol.17**, pp.168–174.

[9] Chapeau-Blondeau F., Teboul V., Berrué J., and Le Duff Y. (1993) An experimental and numerical study of high-frequency Raman scattering in argon gas, *Physical Letters A*, **Vol.173**, pp.153–159.

[10] Teboul V., Godet J. L., and Le Duff Y. (1992) Collection angle dependence of the depolarization ratio in light-scattering experiments, *Applied Spectroscopy*, **Vol.46**, pp.476–478.

[11] Zarkova L. (1996) An isotropic intermolecular potential with temperature dependent effective parameters for heavy globular gases, *Molecular Physics*, **Vol.88**, pp.489–495.

[12] Maroulis G. (1999) Hexadecapole moment, dipole and quadrupole polarizability of sulfur hexafluoride, *Chemical Physics Letters*, **Vol.312**, pp.255–261.

SUB-MICROSCOPIC PROBING OF INTRINSIC AND EXTRINSIC ENHANCEMENT OF SECOND HARMONIC GENERATION OF NANO-STRUCTURED GOLD SURFACES

S. BRASSELET[1], C. ANCEAU[1,2], P. GADENNE[2], J. ZYSS[1]

[1] *Laboratoire de Photonique Quantique et Moléculaire, Ecole Normale Supérieure de Cachan, 94235 Cachan, France*

[2] *Laboratoire de Magnétisme et d'Optique de Versailles, Université de Versailles Saint-Quentin en Yvelines, 78035 Versailles Cedex, France*

ABSTRACT

This paper describes the local enhancement of quadratic nonlinear optical properties of granular gold nano-structures and their effect on nonlinear molecules deposited on them. A two-photon excitation microscope has been employed in order to detect second harmonic generation with a sub-microscopic resolution. Surface-SHG from the gold granular structures exhibits intense signals in sub-micrometer sized regions randomly localized on the surface. The spatial distribution and polarization state of these signals provides significant information on the underlying physics of the local enhancement effect. We show furthermore that the signal from nonlinear chromophores deposited on these substrates is strongly enhanced with factors up to 2×10^3, before decaying within a few seconds. Quantitative measurements indicate a significant local electromagnetic contribution to the molecular enhancement. Molecular damage and possible metal-molecule interactions are analyzed in order to explain the magnitude and the fast decay of this effect.

1. INTRODUCTION

Due to their unique dielectric properties, metallic materials exhibit plasmon resonances which considerably enhance optical properties such as absorption or non-linear polarizations. Optical responses to an incoming field E^{ω} at the frequency ω arise from the interaction with the local field $E_l^{\omega} = L(\omega)E^{\omega}$, with $L(\omega)$ being the local field factor depending on the dielectric properties of the medium and especially on the local environment.[1] As a main consequence of this effect, local electric fields can significantly exceed the applied field amplitude when approaching plasmon resonances frequencies[2,3], thus enhancing the optical responses of molecules adsorbed or deposited

F. Charra et al. (eds.), Organic Nanophotonics, 339–353.

on metallic surfaces.[4-7] Local field enhancements, combined with a possible chemical interaction between the molecules and the metal surface, has lead to considerable developments in the fields of surface science and molecular spectroscopy whereby molecular responses which were too weak to measure were made visible using excitation at plasmon resonance frequency. In very specific regions, enhancement of Raman signals by Surface Enhanced Raman Scattering (SERS) with factors up to 10^{14} were observed on molecules adsorbed on metallic surfaces.[8] These large enhancements have allowed unprecedented detection levels down to the single molecule scale,[8,9] which have lead to significant advances in fields such as biomolecular spectroscopy. Despite more than 20 years of theoretical and experimental studies, a debate still remains regarding the relative importance of chemical contributions to SERS enhancements in resonant conditions. Indeed, charge and excitation transfer processes can occur between a molecule and the metal, thus mixing respective electronic states and therefore affecting the molecular optical properties.[4,10] The enhancement of molecular responses is believed to originate from the combination of chemical and electromagnetic effects, the latter being preponderant under non-resonant excitation. These two effects are however irresolvable in SERS and other related spectroscopic measurements such as infrared absorption, due to their inability to quantify the local field contributions to the optical process.

Despite the limitations discussed above, important information has emerged regarding the electromagnetic origin of optical signal enhancements. In particular, more than two decades of studies on the possible benefits of metallic surfaces for molecular spectroscopy have lead to the conclusion that Raman scattering or infrared absorption were further enhanced on rough surfaces or granular metal structures as opposed to planar surfaces.[4,11] This phenomenon was found to be due to a combination of two effects. Firstly, surface plasmon resonances which depend on the nature of the metal as well as the size and the morphology of the surface structures, and secondly the concentration of local fields on sharps features of metal grains (referred to as the "rod-light" effect)[4]. Enhancement processes are even more complex when a wide distribution of metal particles sizes and shapes comes into play. It has been shown theoretically that in metal structures close to percolation and having fractal geometries, the disorder of the structure provokes a strong confinement of local fields in nanometer to micrometer-sizes regions, randomly distributed on the surface.[12] Such localization occurs principally because the structure, in which metal clusters strongly interact with each other over small inter-distances, does not sustain propagating waves. This phenomenon has recently been observed experimentally by Scanning Near field Optical Microscopy (NSOM) measurements in the infrared regime, where local enhancement factors of up to 10-100 have been measured in comparison with the incident optical field.[13] Understanding these localization effects and the contribution of local geometric factors thus brought considerable progress in elucidating the local nature of electromagnetic enhancements. Such enhancements manifest themselves in the intense local molecular responses observed on granular substrates or colloids, which are less apparent when averaging over large spatial areas.

In this work, we propose to extend such studies to the context of nonlinear optics by focusing on a quadratic effect, second harmonic generation (SHG). Measurements of SHG are used to probe local optical enhancements on the surface of nano-structured metal films, as well as their effect on molecular layers which are deposited on these structures. A sub-microscopic optical resolution is provided by a SHG-microscope. SHG has been widely used for its high sensitivity to structural and electronic properties of surfaces and interfaces. The nature of this nonlinear optical process provides further advantages:

- Very high enhancement factors are expected in nonlinear optical spectroscopy,[6] as the signals depend on the local fields to the power 2n with n the order of the nonlinear effect. In the context of linear optics, measurements are sensitive to the square of the local field factor $|L(\omega)|^2$. SHG arising from a non-centrosymmetric medium is expected to be proportional to the factor $\left|L^2(\omega)L(2\omega)\right|^2$, where the local field factor L contributes to the effect at both frequencies ω and 2ω.

- In the SHG configuration, the electromagnetic nature of local field enhancements can be explored directly on metal surfaces through their intrinsic nonlinear surface second harmonic generation (S-SHG) signal. S-SHG from a metal surface mainly arises from the non-centrosymmetry breaking at the interface with another medium.[14] When using rough or granular surfaces, other contributions arise in the plane of the sample,[15,16] as will be discussed in this paper.

- The molecular responses of nonlinear chromophores are sensitive to the fourth power of the local field $L(\omega)$, and thus can be considerably enhanced on granular gold structures. Thus the wide variety of systems provided by molecular engineering with various optical, electronic and symmetry properties can be explored and enhanced.

Quadratic processes such as Second Harmonic Generation (SHG) and Sum Frequency Generation (SFG) of various molecular structures adsorbed on metallic surfaces[17,18] have previously been studied macroscopically. Although intense molecular nonlinear optical signals have been observed and analyzed, the macroscopic nature of these measurements prevented exploration of the contributions from local electromagnetic enhancements. Furthermore, difficulties in discriminating nonlinear signals from the molecules and those arising from the metal surface lead to complex studies involving various molecular concentrations or environment conditions.[18,19]

In this paper, we describe the properties of local second harmonic generation from molecular monolayers deposited on granular gold films, with independent studies on both components, metal and molecules, using a high-resolution SHG microscopy technique. The uniformity of the molecular distribution ensures homogeneous covering of possible randomly localized enhancement regions. The electromagnetic nature of the enhancement is demonstrated through studies performed on the metallic film itself, using samples of controlled roughness ranging from isolated gold grains to a continuous gold layer. The first part of this paper is devoted to the local nonlinear characterization of gold nano-structures deposited on glass substrates. Different parameters such as the spatial distribution and magnitude of S-SHG enhancements are discussed. Polarization studies show that the local S-SHG emission is locally polarization-dependent but depolarized when the signal is averaged over 1-10 micrometers sizes regions. The second part of this paper describes the first observation of local enhancement of nonlinear signals arising from molecular monolayers deposited on the gold nano-structures. Molecular nonlinear signals are analyzed separately from the gold surface contribution.

2. NON-LINEAR MICROSCOPY ON NANO-STRUCTURED GOLD SURFACES

In this section we describe S-SHG measurements at a sub-microscopic scale on granular and fractal gold structures deposited on silica substrates. S-SHG signals arising from the air-metal interface is generally attributed to electric-quadrupole and magnetic-dipole fields whose contribution has been historically modeled by Jha in 1965[20], based on the non-linear currents originating from induced oscillations in an isotropic free-electron gas. Although such a "Jellium" description is applicable to continuous plane metallic surfaces, this model has been shown to be insufficient for more complex surface morphology where local nonlinear anisotropy can be significant, such as on rough metal surfaces. In order to account for S-SHG observation on complex surfaces, a more refined model was later introduced by Rudnick and Stern,[21] which described surface-polarization density oscillating normal to the surface and within the surface, both involving electric-quadrupole contributions and accounting for the electron spatial density distribution.[22] Experimental observations generally show the surface contribution to be predominant and anisotropic,[15,23] which cannot be explained by the simple free electron model. In a metal nano-structured film, the overall SH polarization contains therefore surface and bulk contributions in the general form[24] $\vec{P}^{2\omega} = \vec{P}_s^{2\omega} + \vec{P}_b^{2\omega}$. The bulk contribution $\vec{P}_b^{2\omega}$, deduced from the Jellium-based model is traditionally written[25] $\vec{P}_b^{2\omega} = \alpha(\vec{\nabla}.\vec{E}^{\omega})\vec{E}^{\omega} + \beta(\vec{E}^{\omega}.\vec{\nabla})\vec{E}^{\omega}$, with the α and β being specific coefficients depending on the metal and the free electron density. The surface contribution $\vec{P}_s^{2\omega}$ lies in the plane of the metal film, and vanishes in the case of a centrosymmetric structure. Nonlinear properties of nano-structured metallic thin films and in particular structures with fractal geometries have been the object of theoretical investigations by Poliakov and Shalaev,[26] based on the Discrete Dipole approximation. They have shown that in this type of structure where confinement of electromagnetic fields occurs, the surface contribution is predominant in the local enhancement effect, and comes from oblate spheroids lying in the sample plane which strongly interact with each other.[12] Calculation of the spatial distribution of the local field intensity at the fundamental and harmonic wavelengths lead to the conclusion that this spatial distribution strongly fluctuates in time and space, over 1-10fs and 10-100nm scales respectively.[27] Theoretical estimations lead to enhancement factors (compared to the incoming fields intensity) of 10^5 and 10^2 for the in-plane and perpendicular contributions respectively.[26] Such factors are detectable in the near-field, where the calculations were performed. In order to access this local information in the near-field region where the electromagnetic fields are confined, SHG-NSOM experiments have been carried out recently. Enhancement factors as high as 10 were observed, with a spatial resolution of 150nm.[28] However, these experiments are difficult to implement technically and limited by their poor control of the polarization state at the end of the fiber tip.

Far-field configurations are easier to implement and are also capable of detecting intense localized plasmon modes via scattering coupling processes. Early studies of the enhancement of nonlinear optical properties in the far field regime explored such effects by comparing macroscopic signals from rough surfaces to continuous planar ones.[25] The reflected SHG signals have been shown to be nearly isotropic in angular distribution and depolarized.[29] An integration of this scattered signal in the 2π-sr emission solid angle resulted in enhancement factors up to 10^4. In order to address the local properties of such enhancements, we used in this work a 2-photon microscopy configuration in the reflection mode, in which S-SHG signals are detectable in the far field, with diffraction limited spatial resolution of a few hundred nanometers.

The excitation source used in the experimental setup is a Ti:Sa laser pumped by a doubled Nd:YAG laser, providing a tunable fundamental wavelength that ranges between 1μm and 1.05 μm, with a pulse duration of 120fs and a repetition rate of 86MHz. The fundamental beam is focused using a 60× oil immersion objective (N.A. 1.4) on the sample surface through an inverted microscope, which yields a spatial resolution of 400nm. The sample is mounted on a piezoelectric scanner providing a 7.6μm scanning range with a 10nm resolution in the (X,Y) directions defining the sample plane. An achromatic half-wave plate controls the incident beam linear polarization. The reflected SHG signal is then collected by the same objective and passes through the dichroic mirror that was used to steer the fundamental light. The SHG signal is then directed to either a set of avalanche photodiodes or a spectrometer coupled to a CCD camera. The samples are semicontinuous gold films consisting of a planar distribution of nanometer sized metal grains randomly distributed on the surface of the insulating silica substrate, which are prepared by evaporating gold onto the substrate at room temperature under ultra vacuum (10^{-9} Torr). Typically, the mass thickness (defined by the thickness that the structure would have if the gold layer was continuous) ranges between 20Å and 200Å. As the film grows progressively in the vacuum chamber, the metal filling factor increases and coalescence occurs, forming irregular shaped clusters on the surface. The mass thickness is continuously measured in situ by using a quartz microbalance of high sensitivity. Around 70Å, gold samples become conducting, which is characteristic of approaching the percolation threshold.

This results in self-affine films similar to fractal aggregates. Typical gold grain size determined by Transmission Electron Microscopy (TEM) measurements is about 10 to 30nm in diameter, depending on the deposition conditions. The roughness was determined by an atomic force microscope (AFM), and corresponds to approximately twice the mass thickness of film.

2.1. Spatial characteristics of the S-SHG response from the gold nano-structures

Optical excitation of the gold nano-structured films in the infrared regime required careful control of the incident peak intensities. At incident intensities above 150GW/cm2, we observed sample damage as well as the apparition of a wide spectral continuum, which has been observed and analyzed previously.[30,31] High intensities can also induce photo-modification of these samples.[32] By working below this intensity range, we observed pure S-SHG signals at the half wavelength, which depended quadratically on the incident intensity and were stable for hours.

The spatial distribution of S-SHG signals on the sample have been studied for several mass thicknesses, namely 159Å (a "continuous sample"), 60Å (close to the percolation threshold) and 20Å (below the percolation threshold). Typical SHG mapping of these structures are represented in Figure 1.

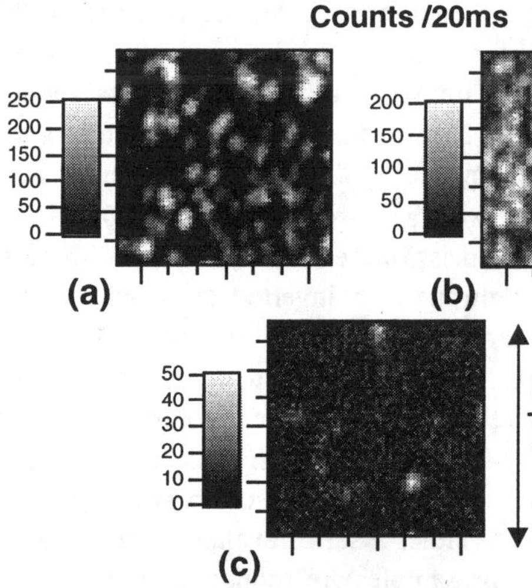

Figure 1. Spatial mapping of Surface Second Harmonic Generation signals of 15×15µm areas of different mass thickness samples and for different incident powers well below the damage threshold: (a) 20Å (107GW/cm^2), (b) 60Å (18GW/cm^2) and (c) 159Å (60GW/cm^2). The signal is integrated over 20ms for each pixel and the fundamental wavelength is 1026nm.

The 159Å sample exhibits a small continuous S-SHG signal, which is usually assigned to quadrupolar contributions to the non-linear induced dipole, as the surface is supposed to be centrosymmetric. For all granular samples, we observe clear localization of S-SHG signals in areas whose sizes approach our optical resolution. These experiments confirm that the disordered nature of the gold nano-structures provides a very high degree of localization of intense local field factors that is manifested in the S-SHG signals of the samples. When increasing the mass thickness from 20Å to 60Å, these areas become closer to each other and both background signal and peak intensities increase. The average signal obtained over a typical sample scan duration of about 80s is of $\sim 7.5\times 10^4$ counts for the 20Å sample, $\sim 4.6\times 10^6$ counts for the 60Å sample and $\sim 3.3\times 10^4$ for the 159Å sample. This tends to support earlier macroscopic observations that close to percolation, the average enhancement effect is higher on granular metal surfaces.[23,25] The visible localization of these enhancements in confined areas is in good agreement with theoretical and linear optical experimental investigations on the properties of these samples,[13] although the present experiment averages this effect over larger areas. Our experiments furthermore allowed us to quantify S-SHG enhancement factors in the different gold structures, defined as $\eta_{max}^{2\omega} = I_{max}^{2\omega} / \langle I_{cont}^{2\omega} \rangle$, where $\langle I_{cont}^{2\omega} \rangle$ is the average of the signal obtained in the SHG scanning of the continuous sample, and $I_{max}^{2\omega}$ is the maximum peak signal obtained on a typical scan for a granular sample. Normalizing the peak signals to the same incident intensity, we obtained local SHG factors of $\eta_{max}^{2\omega} \approx 6$ for the 20Å sample and $\eta_{max}^{2\omega} \approx 130$ for the 60Å sample. These values contain the contributions from the local field at both frequencies ω and 2ω, and thus cannot be directly compared with linear near field measurements. Moreover, the lower spatial scale resolution tends to underestimate the enhancement magnitude as compared to the near-field configuration. Considering however that the enhancement factor on a focal plane contains a proportion of 5-10% of enhancement peaks as estimated from near-field measurements,[13] these results lead to peak enhancement factors of $\left|L^2(\omega)L(2\omega)\right|^2 \approx 6$ in the case of the 20Å sample and $\left|L^2(\omega)L(2\omega)\right|^2 \approx 70$ in the case of the 60Å sample. These values show the potential of these gold nano-structures for the generation of local and large nonlinear signals, in particular when approaching the percolation threshold.

The localization of S-SHG signals was seen to be only slightly dependent on the incident wavelength λ. Figure 2 depicts this behavior in the case of a 70 Å sample, where λ has been varied between 1020nm and 1050nm. Slight spatial fluctuations of SHG intensity are visible in micrometer-size regions, which seems to correspond to an average of the stronger fluctuations observed in linear optical near-field measurements.[13] Our observation is consistent with the wavelength dependence of the local field factors at the scale of the elementary grain size, which varies spatially when the structures contains a wide distribution of morphology as well as interactions between gold clusters on the surface.[12]

Moreover, the wavelength dependence of the S-SHG response shows an increase of the average signal as λ increases, which is also consistent with a decrease of the damage threshold intensities of the structures.

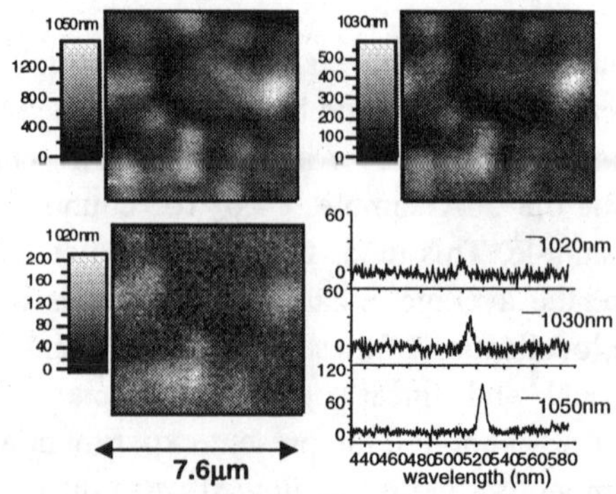

Figure 2. S-SHG mapping of a 7.6×7.6μm area of a 70Å sample for different incident wavelengths; 1020nm, 1030 and 1050nm. The S-SHG spectral response is given in each case for an integration time of 10s.

2.2. Local depolarization of S-SHG on gold nano-structures

One of the aims of this work is to provide more insight into the local nature of the S-SHG enhancement from granular gold structures, and especially the origin of the depolarization effect that has been observed at macroscopic scales.[25,29] In particular, we investigated whether this depolarization is a macroscopic propagation effect such as induced by topography-sensitive interferences at the reflection on the rough surface, or a local manifestation of the surface morphology. Figure 3 is a signature of the clear influence of the local surface properties on the spatial distribution of polarized S-SHG responses. Perpendicular excitation polarizations in the X and Y directions defining the sample plane show indeed very different S-SHG mapping images on a 20Å sample. Such an effect is also visible for higher mass thickness samples, indicating that the localization of the enhancement regions is strongly polarization dependent in these granular structures. This observation is consistent with the strongly depolarized reflected signals measured in far-field macroscopic S-SHG experiments, as signals are averaged over 100-10^4 μm^2 size areas where many local enhancements regions of different polarizations will contribute to the measurement.

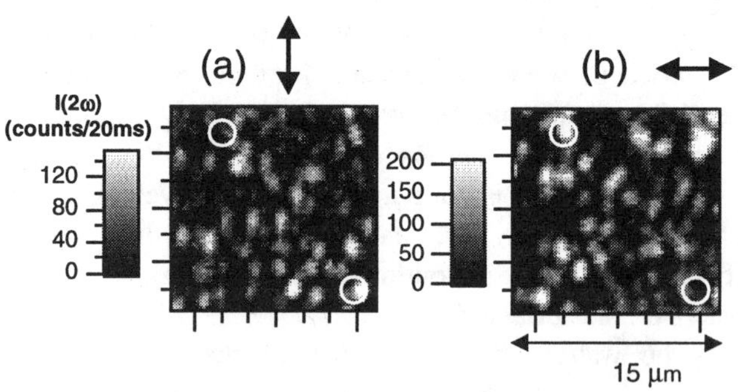

Figure 3. S-SHG mapping of a 15×15μm area of a 20Å sample for two polarization excitation directions, X (3a) and Y (3b), detected in the X analyzer direction. The incident intensity is $100GW/cm^2$.

Similar effects have been visible in linear optics measurements, which suggests that the anisotropy of the local environment properties affects the excitation fields. The different local S-SHG responses observed in Figure 3 are the signatures of very different local symmetries involved in the nonlinear optical interaction. The nonlinear polarization resulting from this interaction can be written using the phenomenological form $\vec{P}_S^{2\omega} = \chi_S^{(2)} \ (\vec{E}_l^{\omega}\vec{E}_l^{\omega}) = (L(2\omega)\chi_S^{(2)} : L(\omega)L(\omega)\vec{E}^{\omega}\vec{E}^{\omega})$, with the nonlinear S-SHG tensor $\chi_S^{(2)}$ being weighted by the tensorial local field factor $L = (L_{XX}, L_{YY}, L_{ZZ})$, with X and Y lying in the sample plane. In this optical process, the local field factor are integrated over the grain distribution in the focal area, and on a large time scale comparing to possible fluctuations affecting their position and their mutual interactions.[27,33] Although the focal areas statistically contains the same average number of gold grains from one region to another, the enhancement regions are less frequent, especially on 20Å samples. The local symmetry of the effective tensor $L(2\omega)\chi_S^{(2)} : L(\omega)L(\omega)$ in intense S-SHG regions is therefore strongly dependent on the local morphology of the gold grains or clusters that create the enhancement region. In particular, it is well known that geometric parameters strongly affect the anisotropy of the local field factors, as well as polarizability symmetries. The polarization dependence experiment carried out in Figure 3 is therefore a signature of a wide variety of local anisotropic plasmon modes, which is revealed in the enhancement regions.

These results clearly show that the depolarization effect of macroscopic S-SHG from gold granular structures is originating from a local phenomenon, which was the case for linear optical properties.[34] Such an effect is related to the local morphology of the sample and confirms that the S-SHG signals measured in this experiment are highly sensitive to the interface structure. This observation also supports earlier SHG-NSOM experiments that have been carried out under tilted excitation, which showed a strong correlation between the topography of the samples and the S excitation polarization responses, whereas the P polarization was more sensitive to the more traditional surface and bulk contributions. Further studies are under way in order to specify the undelying physics in the $\chi_S^{(2)}$ tensor, as well as the wide variety of nonlinear local anisotropies.

3. NONLINEAR MOLECULAR RESPONSE ENHANCEMENTS

The large local field factors produced by the gold nano-structures are potential sources of enhancement of molecular optical responses, provided that the molecules are deposited close to the gold surface. This effect has been widely used in Raman scattering or infrared absorption spectroscopy experiments, where molecules are either adsorbed on rough metal surfaces or homogeneously deposited. Second harmonic generation from gold surfaces covered with molecules is expected to originate from several contributions which can be written as $I^{2\omega} = I^{2\omega}_{S-SHG} + I^{2\omega}_{SHG} + I^{2\omega}_{int}$, with $I^{2\omega}_{S-SHG}$ being the S-SHG arising from the gold structure, $I^{2\omega}_{SHG}$ the SHG from the molecular layer itself, and $I^{2\omega}_{int}$ a correction accounting from the possible interactions between the metal and the molecules.[35] This correction term includes possible changes in the electronic properties of the molecules in contact with the metal, or possible modifications of the gold characteristics due to the presence of the molecules. Although this contribution has been widely evoked in Raman spectroscopy,[5,36] it has not been specifically addressed in the context of SHG processes.

The nonlinear efficiency of the molecular ensemble results from the non-centrosymmetric arrangement of nonlinear chromophores on the sample surface, which ensures the existence of a non-vanishing macroscopic molecular $\chi_m{}^{(2)}$ susceptibility tensor. The SHG contribution from a non-centrosymmetric molecular layer is then written $I^{2\omega}_{SHG} = \left| L(2\omega)\chi_m{}^{(2)} : L^2(\omega)E^{\omega}E^{\omega} \right|^2$, which is sensitive to the surrounding local field. The high local field factors provided by the metal structure are therefore expected to intensify the molecular nonlinear response with a factor $\left|L^2(\omega)L(2\omega)\right|^2$ similar to the one that has been introduced in the previous section. The $I^{2\omega}_{S-SHG}$ and $I^{2\omega}_{SHG}$ magnitudes are expected to depend on the morphology of the gold surface.

In order to take advantage of the high local field factors in the confined regions and achieve enhancement of nonlinear molecular responses, we coated the gold fractal structures with molecular homogeneous monolayers using the Langmuir-Blodgett technique. The nonlinear chromophore, (4-(4-dimethylamino)styryl)-1-docosylpyridium bromide) was deposited using a surface pressure of 30mN/m. Macroscopic SHG studies on such Langmuir-Blodgett layers on glass substrates have revealed that the molecules lies on the surface with an average tilt angle of ~70°.[37] The molecular surface density was 6.9×10^9 molecules/cm^2, as deduced from absorption measurements on a glass substrate, which corresponds to a single molecular layer. The maximum wavelength of absorption of this layer lies around 500nm, which lies in the resonant region for two-photon excitation using the 1026nm-excitation wavelength. After deposition on the gold surface, we confirmed the presence of the molecules by Atomic Force Microscopy measurements in tapping mode. In order to distinguish the molecular responses on and outside possible enhancements regions on the gold structures, we chose to study the molecular SHG signals on 20Å mass thickness gold samples, where these regions are spatially separated from each other as shown above.

SHG measured on the molecular layer deposited on glass was spatially homogeneous with an average signal of 40counts/20ms for an incident peak intensity of 100GW/cm^2. The SHG nature of this signal has been confirmed by spectral information and by its quadratic dependence with respect to the incident fundamental intensity.

Molecular damage occurred above an incident peak intensity of $500GW/cm^2$. Typical SHG mapping of two 20Å mass thickness gold samples, the first one without molecules and the second one covered by the Langmuir-Blodgett layer, are given in figure 4a and 4b, for a $100GW/cm^2$ incident intensity.

As seen in Figure 4a, the gold structure exhibits intense and localized S-SHG signals similar to those described in the first part of this paper. Figure 4b exhibits the same structure of localized and intense signals as for gold only, with somewhat higher signal levels. By performing this measurement on several locations on the sample, we observed that this peak intensification is on average higher than on gold only. Further distinctions between the behavior of the bare gold and the monolayer coating the gold structure were seen when by studying the temporal evolutions of local signals on different areas, outside the intense peaks (region 1 in figure 4b) and on an intense SHG peak (region 2 in figure 4b).

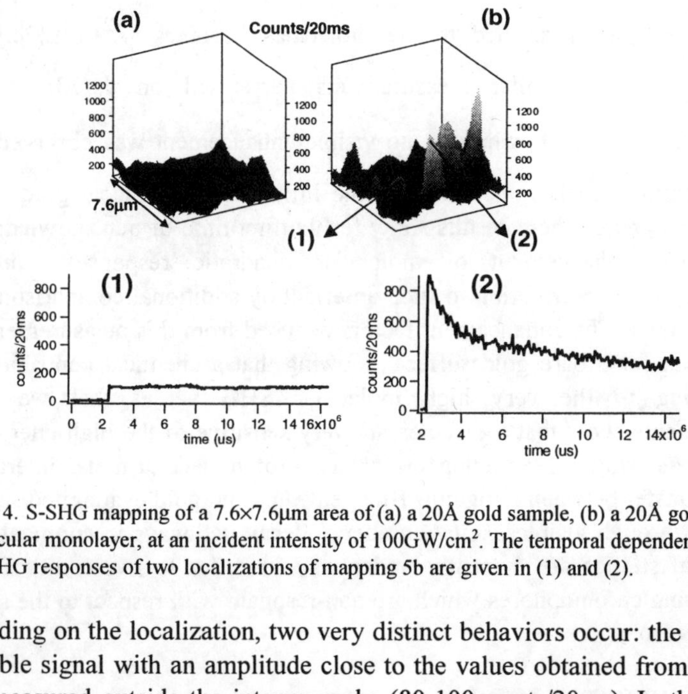

Figure 4. S-SHG mapping of a 7.6×7.6µm area of (a) a 20Å gold sample, (b) a 20Å gold sample covered by a molecular monolayer, at an incident intensity of $100GW/cm^2$. The temporal dependencies of the SHG responses of two localizations of mapping 5b are given in (1) and (2).

Depending on the localization, two very distinct behaviors occur: the first case (1) shows a stable signal with an amplitude close to the values obtained from the gold surface when measured outside the intense peaks (80-100counts/20ms). In the second case (2), the signal decays in several seconds to a stable value of 300 counts/20ms, which is the average level of signal of intense peaks on the gold 20Å structure, as seen from Figure 1. The intense signal observed in (2) before the fast decay is most likely a signature of molecular enhancement. The decay, observed in every intense SHG peak, is most probably a signature of photo-destruction. Indeed, we estimated that if the molecule is effectively sensitive to the electromagnetic enhancement provided by the gold nano-structures, the local concentration of energy that they experience approaches intensities close to the molecular damage threshold. Lowering the incident intensity to values that ensure stable molecular signals and therefore avoid this effect was however limited by the experimental signal to background ratio.

Before photo-destruction, the calculation of the enhancement factor corrected for the bare gold signal provides a value of $\left(I^{2\omega} - I^{2\omega}_{S-SHG}\right)/I^{2\omega}_{SHG,Glass} \approx 60$ in the case of a 20Å gold sample, and 200 in the case of a 70Å sample. Such values are representative of a non-negligible local SHG enhancement experienced by the molecules, which is furthermore amplified by the fact that the intense electromagnetic effect occurs in a limited number of confined regions that is averaged over the focus area. The low molecular signals outside of the enhancement regions come from molecules that are not in contact with gold, or from molecules that experience low enhancement factors from the gold. Considering that the enhancement regions occupy only 5-10% of the surface of the focal area for a 20Å gold sample, the local SHG enhancement factor applied to the molecules becomes $\left|L^2(\omega)L(2\omega)\right|^2 \approx 70$ before photo-destruction. Similar measurements performed on a 70Å sample lead to $\left|L^2(\omega)L(2\omega)\right|^2 \approx 2500$, whereas no visible enhancement was observed on continuous gold structures. Such values confirm the influence of the local gold structure on the molecular signals. These results show for the first time of our knowledge the evidence of a local enhancement of molecular quadratic responses, and confirm the electromagnetic contribution to such an effect by additional comparisons with the gold substrate itself. The enhancement factors deduced from this measurement are moreover higher than for a bare gold surface, showing that a chemical contribution is possibly contributing to the very high molecular SHG values that we observe. Such measurements show that molecules are very sensitive to the high energies involved in this process, without excluding the presence of molecular-metal interactions. Indeed, energy transfer between Langmuir Blodgett films and gold evaporated surfaces has been observed through visible excitation energy decay pathways in conjugated molecules.[38] Additional studies are currently under way in order to overcome photo-destruction effects using chromophores which are non-resonant with respect to the surface plasmon 2-photon resonances.

4. CONCLUSION

Two-photon microscopy and its implementation in SHG analysis has been shown to be very powerful in the local analysis of nonlinear processes such as local field enhancements. Furthermore this technique provides interesting information on the nature of nonlinear optical behavior in nano-structured media and SHG emission from complex environments involving localization of electromagnetic fields and strong interactions. Using this technique, we have observed and quantified for the first time localized enhancements in second harmonic generation using 2-photon microscopy, with local enhancement factors up to 150 comparing to continuous thin gold films. The features of the enhancement localization, density, and polarization response, illustrate the strong influence of the surface morphology and possible interactions between gold clusters on the substrate on the underlying optical process. Such measurements bring significant information that is not available in more traditional S-SHG far field measurements. Molecular enhancements by gold nano-structures has been shown with higher factors than on bare gold, although limited in time duration by the damage threshold of the molecules. Comparative studies showed that part of molecular enhancement lies in electromagnetic effects occurring on the gold structures, with a possible contribution from chemical interactions. Such structures are shown to provide a wide range of perspectives applications, such as nanostructured materials for photonic devices and supporting layers for molecular linear and non-linear spectroscopy.

ACKNOWLEDGEMENTS

We would like to acknowledge Bruno Bérini for the fabrication of gold samples, as well as André Clouqueur, Jean-François Roch, François Treussart and Dan Wright for technical assistance.

REFERENCES

1. Oldenburg, S.J., Averitt, R.D., Westcott, S.L. and Halas, N.J. (1998) Nanoengineering of optical resonances, *Chem. Phys. Lett.* **288**, 243-7.
2. Doremus, R.H. (1964) Optical properties of small gold nanoparticles, *Journ. Chem. Phys.* **40**, 2389-96.
3. Kreibig, U. and Vollmer, M. (1995). *Optical Properties of Metal Clusters* Springer Verlag
4. Chang, R.K. and Furtark, T.E. (1982). *Surface Enhanced Raman Scattering,* New York Plenum Press
5. Golab, J.T., Sprague, J.R., Carron, K.T., Schatz, G.C. and Van Duyne, R.P. (1988) A surface enhanced hyper-Raman scattering study of pyridine adsorbed onto silver: Experiment and theory, *J. Chem. Phys.* **88**, 7942.
6. Andrews, M., Kusyk, M.G. and Ghebremichael, F. (1993) Local field enhancement of the cubic optical nonlinearity in fractal silver nanosphere/poly(methylmetacrylate) composite, *Nonlinear Optics* **6**, 103-12.
7. Zhang, J., Zhao, J., He, H., Zhang, H., Li, H. and Liu, Z. (1998) Studies on the surface enhanced infrared spectroscopy of Langmuir-Blodgett monolayers of azobenzene carboxylic acid on silver island films, *Langmuir* **14**, 5521-5.
8. Kneipp, K., Wang, Y., Kneipp, H., Perelman, L.T., Itzkan, I., Dasari, R.R. and Feld, M.S. (1997) Single Molecule Detection using Surface-Enhanced Raman Scattering, *Phys. Rev. Lett.* **78**, 1667.
9. Nie, S. and Emory, S. (1997) *Science* **275**, 110-1106.
10. Delvin, J.P. and Consani, K. (1981) *J. Phys. Chem.* **85**, 2597.

11. Maxwell, D.J., Emory, S. and Nie, S. (2001) Nanostructured thin-film materials with surface-enhanced optical properties, *Chem. Mater.* **13**, 1082-8.
12. Shalaev, V.M. and Sarychev, A.K. (1998) Nonlinear optics of random metal-dielectric films, *Physical Review B* **57**, 13265-88.
13. Grésillon, S., Aigouy, L., Boccara, A.C., Rivoal, J.C., Quélin, X., Desmarest, C., Gadenne, P., Shubin, V.A., Sarychev, A.K. and Shalaev, V.M. (1999) Experimental Observation of Localized Optical Excitations in Random Metal-Dielectric Films, *Phys. Rev. Lett.* **82**, 4520-3.
14. Guyot-Sionnest, P., Chen, W. and Shen, Y.R. (1986) General considerations on optical second-harmonic from surfaces and interfaces, *Phys. Rev. B.* **33**, 8254-63.
15. Bloembergen, N., Chang, R.K., Jha, S.S. and Lee, C.H. (1968) Optical second-harmonic generation in reflection from media with inversion symmetry, *Phys. Rev.* **174**, 813.
16. Chang, C.S. and Lue, J.T. (1997) Optical second harmonic generation from thin silver films, *Surface Science* **393**, 232-9.
17. Pluchery, O., Zheng, W.Q., Marin, T. and Tadjeddine, A. (1999) Study of the absorption of 4-cyanopyridine on Au(111) using sum frequency generation nonlinear spectroscopy, *Physica Status Solidi A* **175**, 145-51.
18. Aktsipetrov, O.A., Keller, O., Pedersen, K., Nikulin, A.A., Novikova, N.N. and Fedyanin, A.A. (1993) Surface-enhanced second harmonic generation in C60-coated silver island films, *Phys. Lett. A* **179**, 149-53.
19. Manaka, T. and M., I. (2001) Enhancement of second harmonic generation from phtalocyanine Langmuir-Blodgett films on metal electrodes, *Thin Solid Films* **393**, 119-23.
20. Jha, S.S. (1965) *Phys. Rev. A* **140**, 2020.
21. Rudnick, J. and Stern, E.A. (1971) Second-harmonic radiation from metal surfaces, *Phys. Rev. B* **4**, 4274-90.
22. Weber, M. and Liebsch, A. (1987) Density-functional theory appraoch to second-harmonic generation at metal surfaces, *Phys. Rev. B* **35**, 7411-6.
23. Boyd, G.T., Rasing, T., Leite, J.R.R. and Shen, Y.R. (1984) Local-field enhancement on rough surfaces of metals, semimetals, and semiconductors with the use of optical second-harmonic generation, *Phys. Rev. B.* **30**, 519-26.
24. Van Driel, H.M. (1994) Second-harmonic generation from metal surfaces: beyond jellium, *Appl. Phys. A* **59**, 545-52.
25. Chen, C.K., Castro, A.R.B. and Shen, Y.R. (1981) Surface-Enhanced Second-Harmonic Generation, *Phys. Rev. Lett.* **46**, 145-8.
26. Poliakov, E.Y., Markel, V.A., Shalaev, V.M. and Botet, R. (1998) Nonlinear optical phenomena on rough surfaces of metal thin films, *Phys. Rev. B* **57**, 14901-13.
27. Podolskiy, V.A., Sarychev, A.K. and Shalaev, V.M. (2002) Temporal Dynamics of Local Optical Responses and Sub-fs Pulse Generation in Semicontinuous Metal Films, *Laser Physics* **12**, 292-9.
28. Zayats, A.V., Smolyaninov, I.I. and Davis, C.C. (1999) Observation of localized plasmonic excitations in thin metal films with near-field second harmonic microscopy, *Opt. Comm.* **169**, 93-6.
29. Breit, M., Podolsky, V.A., Grésillon, S., Von Plessen, G., Feldmann, J., Rivoal, J.C., Gadenne, P., Sarychev, A.K. and Shalaev, V.M. (2001) Experimental observation of

percolation-enhanced non-linear scattering from semi-continuous gold films, *Phys. Rev. B.* **64**, 106.
30. Ducourtieux, S., Grésillon, S., Boccara, A.C., Rivoal, J.C., Quélin, X., Gadenne, P., Drachev, V.P., Bragg, W.D., Safonov, V.P., Podolskiy, V.A., Ying, Z.C., Armstrong, R.L. and Shalaev, V.M. (2000) Percolation and Fractal composites: Optical Studies, *Jour.Nonlinear Opt. Phys. Mat.* **9**, 105.
31. Anceau, C., Brasselet, S., Gadenne, P. and Zyss, J. (2002) Local second harmonic generation enhancement on gold nanostructures probed by 2-photon microscopy, *submitted*
32. Safonov, V.P., Shalaev, V.M., Markel, V.A., Danilova, Y.E., Lepeshkin, N.N., Kim, W., Rautan, S.G. and Armstrong, R.L. (1998) Spectral dependance of selective photomodification in fractal aggregates of colloidal particules, *Phys. Rev. Lett.* **80**, 1102-5.
33. Stockman, M.I. (2000) Giant attosecond fluctuations of local optical fields in disordered nanostructured media, *Phys. Rev. B* **62**, 10494-7.
34. Ducourtieux, S., Podolskiy, V.A., Grésillon, S., Buil, S., Bérini, B. and Gadenne, P. (2001) Near-field optical studies of semi-continuous metal films, *Phys. Rev. B* **64**, 1554.
35. Shen, Y.R. (1994) Nonlinear optical studies of surfaces, *Appl. Physics A* **59**, 541-3.
36. Aroca, R.F. and Constantino, C.J.L. (2000) Surface-enhanced Raman scattering: imaging and mapping of Langmuir-Blodgett monolayers physically adsorbed onto siver island films, *Langmuir* **16**, 5425-9.
37. Simpson, G.J. and Rowlen, K.L. (2000) Measurement of Orientation in Organic Thin Films, *Acc. Chem. Res.* **33**,, 781-9,.
38. Zhang, Z.J., Verma, A.L., Tamai, N., Nakashima, K., Yoneyama, M., Iriyama, K. and Ozaki, Y. (1998) Excitation energy transfer in Langmuir-Blodgett films of 5-(4-N-octadecylpyridyl)-10,15,20-tri-tolyporphyrin on gold-evaporated surfaces studied by time-resolved fluorescence spectroscopy, *Thin Solid Films* **333**, 1-4.

STRUCTURAL AND DYNAMIC STUDIES OF THE POLAR ORIENTATION INDUCED BY CORONA POLING AND ALL-OPTICAL POLING IN CROSSLINKABLE POLYMER THIN FILMS

A. APOSTOLUK[1], V. BOUCHER[2], R. SEVENO[2], C. FIORINI[3], B. SAHRAOUI[1], J.-M. NUNZI[1]
[1]*Université d'Angers, Laboratoire POMA*
2 Boulevard Lavoisier, 49045 Angers Cedex 01, France
[2]*LPIO, Faculté des Sciences & Techniques*
2 rue de la Houssinière BP 92208, 44322 Nantes Cedex 3, France
[3]*DRT-LIST, DECS-SE2M, Groupe Composants Organiques, Bât. 451, CEA Saclay, 91191 Gif-sur-Yvette, France*

1. Introduction

The fast development of photonics and progress in laser technologies has created the necessity for materials possessing high nonlinear photoactivity. Polymers containing various dye molecules in grafted as well as doped systems have recently been in the limelight of many research groups because of their various potential applications in electro-optic devices and integrated optics, as they possess large second order nonlinear properties, low dielectric constants, the possibility of inducing large nonlinearities, low attenuation and ease of processing [1-3]. Yet, some important issues concerning temporal stability of induced order and optical loss are still a challenge to researchers [4-7]. A polymer matrix containing non-centrosymmetric molecules is a macroscopically centrosymmetric solution in which second order nonlinear processes are forbidden due to symmetry reasons. Yet, the system properties may be altered by a non-centrosymmetric external field. In order to create second order nonlinear properties in polymer materials, they have to be prepared with required noncentrosymmetry, which is usually accomplished by poling techniques. Some years ago it was demonstrated that the all optical poling technique permits purely optical orientation of molecules and thus induction of the large second order susceptibilities $\chi^{(2)}$ [1, 8]. The important advantage of the all-optical technique is that it leads to a periodic ordering of the molecules with the spatial period fulfilling automatically the phase matching condition for second harmonic generation (SHG) [8]. The experiment consists of two phases: the writing (so called "seeding") period and the readout one. In seeding phase two beams, the fundamental one (ω) and its second harmonic (SH, 2ω) illuminate the sample and encode in the material the second order $\chi^{(2)}$-susceptibility grating. In the readout period only the fundamental beam irradiates the sample and the second harmonic generation is observed at the sample output. The coherent superposition of two beams at ω and 2ω frequencies results in the presence of a polar field E(t) inside the material, which is able

F. Charra et al. (eds.), Organic Nanophotonics, 355–366.

to break the centrosymmetry of the medium. More accurately the temporal average cube $<E^3>_t$ of this field $E(t) = E_{\omega}(t) + E_{2\omega}(t)$, is non-zero, which can lead to an orientation-selective excitation of the molecules. The physical origin of the effect is attributed to an orientational hole burning in the initially isotropic distribution of molecules. Efficient all optical poling requires optimization of the seeding beam intensities and their relative phase difference. We report here on all-optical poling of a new material with improved stability of the induced order (when compared to Disperse Red 1) in a novel configuration of the experimental set-up where there is no need for taking into account the phase difference between seeding beams.

2. Sample preparation

The chemical structure of the electro-optic monomer (PIII) is given in the fig.1. Nitro- and tertiary amines were chosen as efficient acceptor and donor groups, respectively, and placed at the end of the p-extended molecule in order to obtain a large electro-optic coefficient. Moreover, the carboxylic group (cross-linking group) is placed on one of the aromatic groups in order to perform anchoring after the poling process. Because of its location and its electronic configuration, the anchoring function retains a high level of hyperpolarizability. This monomer can undergo free-radical copolymerisation with the methacrylic monomer bearing the second cross-linking function. This function is an epoxy one. The sample was 350 nm thick, coherence length $l_c = 2\pi/\Delta k = \lambda_{2\omega}/(n_{\omega} - n_{2\omega}) = \lambda_{2\omega}/\delta n$, where n_{ω} and $n_{2\omega}$ are the refractive indices for the fundamental and the second harmonic wave, respectively, is equal to 11,6 μm, optical density at 532 nm was equal to 1,57.
We also prepared the reticulated sample, of the similar - as close as it was possible - optical density at 532 nm as the non-reticulated sample, in which the reticulation reaction was performed via heating the sample to 150°C in an oven in ambient conditions (without applying the electric field). The sample thickness was 600 nm, optical density at 532 nm was equal to 1,35, the coherence length is 3,99 μm.

3. Poling and cross-linking procedure

We describe hereafter the poling and cross-linking processes to yield electro-optic copolymers with stable NLO properties. Preparation of thin films consists in dissolving the polymer powder in 1,2-dichlorethane. The solution is filtered through a Millipore Teflon filter (0,45 μm). The final polymer solution is then spin-coated onto glass substrates to form thin films. Corona poling involves heating of the NLO polymer to just above its glass transition temperature (T_g), applying a large electric field to orient the chromophore molecules, and cooling the NLO polymer while the field is still applied. As T_g of the non reticulated polymer is 65°C, the poling of the chromophores can be achieved at 70°C under an electric field (5,8 kV) applied by the Corona method for 30 minutes. Heating is performed in ambient conditions and the temperature is controlled with the thermocouple. The distance between the sample surface and the needle electrode is 20 mm and the poling current of about 1,5 μA (positive on the needle). The next step is the anchoring process, which has to be done at a temperature, T_c = 150°C, higher than the poling one. The two functions must not react at the poling temperature to ensure an optimum chromophore orientation in the electric field. On the other hand, the anchoring kinetic has been studied [9] to determine the most efficient temperature to achieve the cross-linking between the NLO poled molecules and the epoxy function of the copolymer. This study has shown that the reaction involving epoxy and carboxylic functions is thoroughly achieved after 30 minutes at 150°C. The reticulation process involves changing of the colour of the sample, from deep pink to red. The absorption spectra made before and after the reticulation procedure revealed the blue shift of 11 to 39 nm in the spectrum of the reticulated sample when compared to the spectrum of the non-reticulated one, the value of the blue shift depending on the sample (see fig.1).

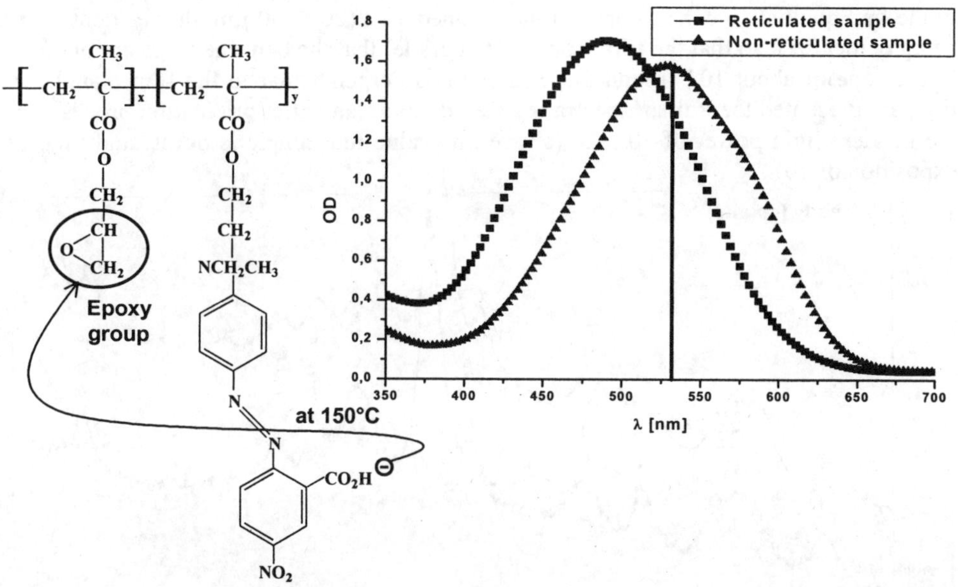

Fig.1. Structure of the cross-linkable PIII polymer synthesized for this study and absorption spectra of the non-reticulated sample (350 nm thick) and the reticulated one (600 nm of thickness). Optical densities (OD) at 532 nm were, respectively, 1,57 and 1,35. The straight line indicates the 532 nm wavelength. Blue shift between the two absorption spectra is equal to 37 nm.

4. Raman spectroscopy studies of the Corona poled samples

Raman spectroscopy permitted us to obtain information about the orientation of chromophores in a polymer film. The spectra were recorded in the backscattering geometry configuration with the use of a Jobin-Yvon T64000 spectrometer in conjunction with the microscope objective (×50). The preliminarily Corona poled film was illuminated by the linearly polarised in the plane of the sample beam from the continuous argon laser (514,5 nm). The radiation was collected by the microscope objective and its intensity measured and analysed by the spectrograph and the CCD camera. Different peaks observed in the spectrum characterise the vibrations of the dye molecules. It was demonstrated that the "breathing" mode intensity of para-substituted benzene rings (observed at 992 cm^{-1}) splits into several modes [10, 11]. Particularly, the intense peak observed at 1600 cm^{-1} is associated to the C-C stretching. This vibrational mode rings can be observed only if the incident field is polarized in the plane of the benzene rings. Therefore it is expected that it will disappear after the orientation profess is performed. Indeed, this peaks is no longer observed in the spectrum after orientation, when the chromophores are oriented perpendicularly to the sample surface. This leads to the first qualitative conclusion – in the scanned area (20 to 30 μm, the incident beam being defocalized so that the film would not degrade) that the benzene rings are oriented in the cone of about 10° around the 0*z* axis (axis perpendicular to the film plane). In fig.2 are presented the Raman spectra registered before and after orientation process, for the incident light power of 10 mW (above this value the sample is burnt) and time of exposition of 10 s.

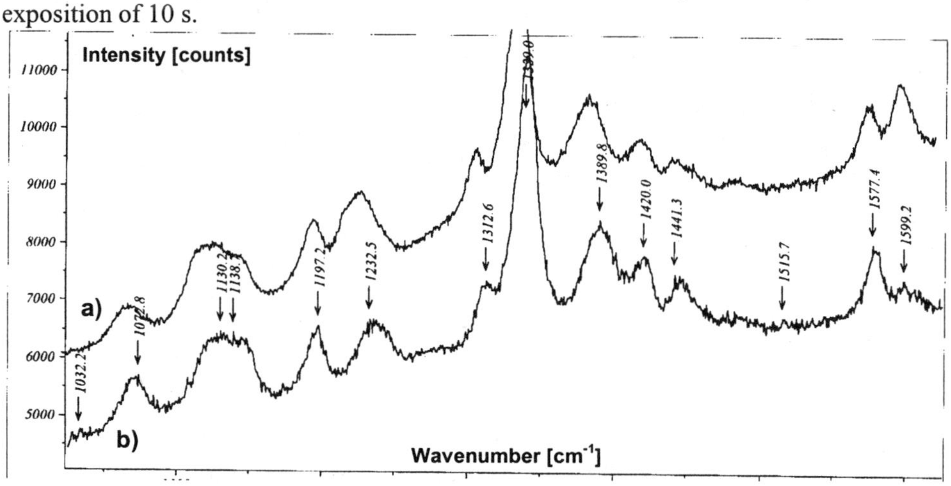

Fig.2. Raman spectra of the a) non-reticulated (upper curve) and b) reticulated and oriented via Corona poling method sample (lower curve).

5. Experimental set-up for all-optical poling

The scheme of the experimental set-up is presented in fig.3. The beam source was the actively Q-switched Nd:YAG laser (YG 580, Quantel) delivering 7 ns pulses at 1064 nm with a 10 Hz repetition rate. An external cavity potassium dihydrogen phosphate (KDP, KH_2PO_4) crystal (type II) permitted frequency doubling of the fundamental beam. At the crystal output the coherent superposition of copropagating generated SH beam and the residual fundamental one. The half-wave plate is used to turn the polarization of the fundamental beam. A polarizer assures linear and parallel polarization of the two beams. The pulse energies were 20 mJ and 17 μJ, respectively, for the fundamental and SH beam. The beams diameter was about 3 mm at the sample location. The experiment consists in two alternating phases: writing (seeding) and probing one. The sample was placed onto a 21,6° acute angle prism. In the writing periods, the two seeding beams ω and 2ω are incident onto the sample. The fundamental beam passing through the sample was blocked with the KG3 filter (heat absorber) inserted behind the prism and the sample and leaving only the SH beam pass. In the probing period the second order susceptibility was measured through the SH generation in the sample. The SH signal was detected by a photomultiplier tube (PMT), with the 2ω seeding beam blocked by a RG 630 filter (green cutting filter, 3 mm thick, infrared transmitting) and the ω beam left incident onto the sample. A set of calibrated filters was used to ensure correct scaling of the SH signal. The signal delivered by the PMT was measured and averaged by an oscilloscope. Moreover, an electronically controlled shutter was placed in front of the PMT in order to avoid the possible saturation due to the strong green seeding beam during seeding step. The fundamental beam served as a seeding and the probing beam, while the SH beam was used as a seeding beam only. The configuration with the prism and the beam arrangement at the sample output during the readout phase is shown in fig.4.a), wavevectors configuration – in fig.4.b). For the polymer system studied the intensity of the strong 2ω signal was registered.

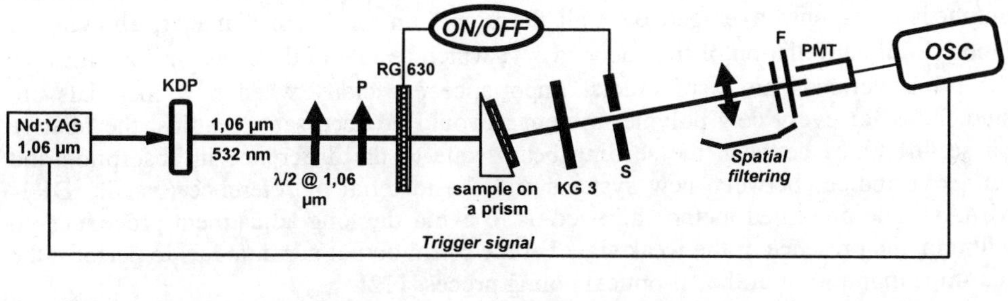

Fig.3. Experimental set-up for all-optical poling. P – polarizer, λ/2 for 1,06 μm – half wave plate at 1,06 μm, RG 630 – Schott green blocking filter, 3 mm thick, KG3 – Schott heat blocking filter; shutter (S) is synchronized with the insertion of the red filter (Schott RG 630) which cuts the 2ω beam during $\chi^{(2)}$ readout process; F - spatial filter for 532 nm. The signal is detected by the photomultiplier tube (PMT) and registered on the oscilloscope (OSC).

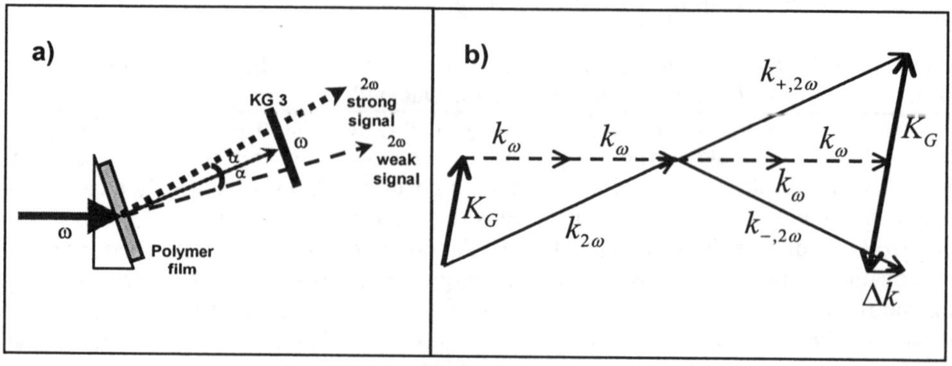

Fig.4. Part (a) shows schematically the beam configuration at the output of the prism; (b) scheme of the interacting fields' wavectors – k_ω , $k_{2\omega}$– wavevectors of the fundamental and second harmonic frequency fields, respectively, $k_{+,2\omega}$ – wavevector of the strong SH signal, $k_{-,2\omega}$ – wavevector of the weak SH signal, K_G – grating wavevector, Δk - wavevector mismatch.

6. Results and discussion

In order to test the real harmonic conversion potentialities in optimized conditions, it is necessary to determine the optimal phase difference between the seeding beams. Any other phase difference will lead to an "overpainting" of the polar interference pattern, and finally result in a limited (less effective) polarization of the sample. The configuration with the prism permits us to become independent of the influence of the relative phase difference ($\Delta\Phi$) between the seeding beams on the induced polar order [12]. In the case of a freestanding sample the intensity of the generated SH signal reveals a sinusoidal variation with $\Delta\Phi$ with a period of 180° [8]. In our experiments the value of the generated second harmonic signal remains constant, irrespective of the relative phase difference. The SH signal observed at the sample output is a response averaged over all the points on the beam diameter, allowing to ignore local cancellation of the encoded $\chi^{(2)}$, which results in the constant SH signal as the $\Delta\Phi$ is varied. This is of crucial importance especially when new materials are studied, as for every new polymer system it would be necessary to adjust the relative phase difference between the seeding beams due to the difference in absorption and refractive indices between new systems and the material of reference (usually DR1-MMA). The presented method allowed us to avoid the long adjustment procedure. In addition, the presence of the weak signal at the prism output permitted us to perform the real-time monitoring of the all-optical poling process [12].

The growth and dynamics of the seeding-type preparation of both samples is illustrated in fig.5. The preliminarily reticulated sample gave higher signal than the non-reticulated one, which was surprising. Both samples reveal a good stability of the photoinduced second order susceptibility $\chi^{(2)}$ under constant illumination with strong fundamental beam (the same one as in the seeding phase) over 5 hours. The decay dynamics of the normalized photoinduced $\chi^{(2)}$ is presented in fig.6 and fig.7.

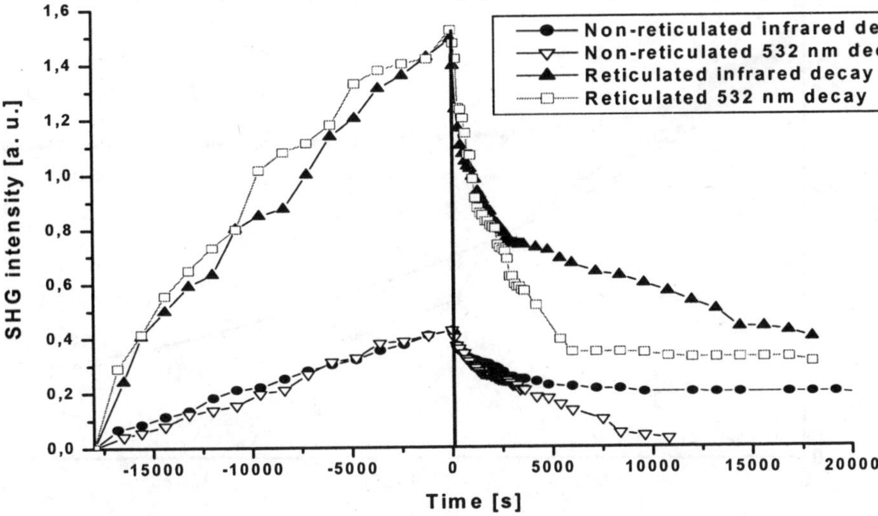

Fig.5. Time evolution of the SH generated signal (in arbitrary units) in the non-reticulated sample (circles – decay under infrared beam, down triangles – decay under 532 nm beam, sample 350 nm thick) and in the reticulated one (up triangles – decay under 1,06 μm beam, squares – decay under green beam, sample 600 nm of thick). Negative times correspond to the seeding preparation phase. At time zero the seeding process is stopped (straight vertical line). Positive times correspond to the temporal decay of the photoinduced $\chi^{(2)}$ susceptibility. The samples were placed on a prism.

When the material is irradiated by either of the two writing fields after the seeding process is stopped, it was observed that the dynamics of the relaxation of the induced susceptibility $\chi^{(2)}$ is much faster than for the dark decay [8]. However, for a mono-frequency 2ω irradiation, the induced birefringence persists [8]. It clearly indicates that the irradiation of a formerly seeded area with a beam at frequency ω or 2ω destroys only the photoinduced polar order [13, 14]. Under monochromatic irradiation, the molecules are re-excited and reoriented. This excitation is axial, so solely the photoinduced polar order is erased. The fact that the birefringence is preserved during the irradiation with the single frequency beam clearly shows that the axial terms of the excitation probability are at the origin of the birefringence related to the seeding process.

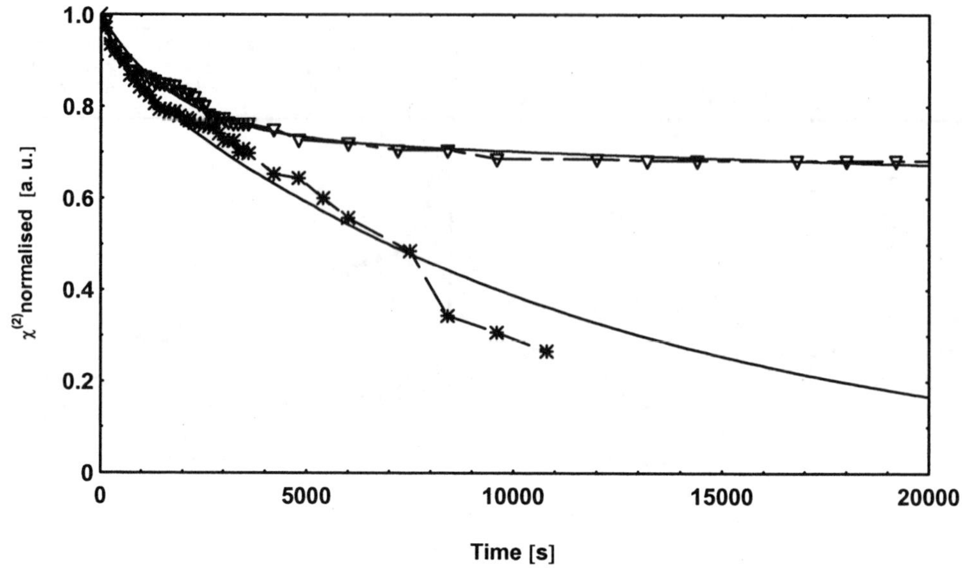

Fig.6. Decay of the normalized $\chi^{(2)}$ of the non-reticulated sample under constant illumination with fundamental beam (triangles) and green beam (stars). Solid lines are fittings to the eq. (1).

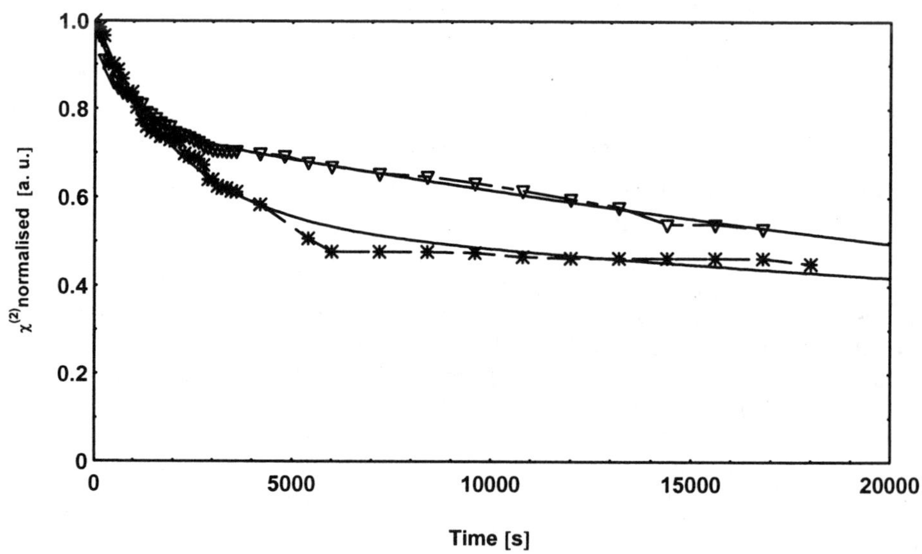

Fig.7. Decay of the normalized $\chi^{(2)}$ of the reticulated sample under constant irradiation with 1,06 μm beam (triangles) and with 532 nm beam (stars). Solid lines are fittings to the eq.(1).

As it can be seen from fig.6, after the seeding process is stopped, two time domains of the decay of the second order nonlinear susceptibility $\chi^{(2)}$ can be observed: an initial very fast decay occurring within a few seconds after interruption of the preparation

process, and a second, much slower, multiexponential one. As the hyperpolarizability β of DR1 is different in the *cis* and *trans* isomers [15], the first very rapid decay contains a contribution, which corresponds to lifetime of the *cis* form (produced through the photoinduced *trans-cis* isomerization in the seeding phase), which comes back to the *trans* form without any net orientation [13, 15]. The second component originates from an orientational scattering of the polarly oriented molecules of the *trans* form inside the matrix. The first fast decay of the thermodynamically unstable *cis* form is a thermally induced reaction. The second one (*trans* isomer orientational scattering) depends on the matrix scattering constant (rigidity of the polymer matrix), laser repetition rate, the way the dye molecule exists in the polymer host, etc.
It was found that the decay experimental curves could be fitted with the biexponentional function, given by the following equation [16]:

$$y(t) = A \cdot \exp(-t / \tau_A) + B \cdot \exp(-t / \tau_B) \quad (1)$$

where *y(t)* is the normalized value of the $\chi^{(2)}$ at point *t* after the seeding is stopped: $y(t) = \frac{|\chi^{(2)}(t)|}{|\chi^{(2)}_{\max}|}$. The first term depicts the faster decay process with the time constant τ_A, the second term - describes the slower one, with the τ_B time constant. A and B are contributions to the saturated value of $\chi^{(2)}$ (*y* at time zero) and $A + B = 1$. The parameters (τ_A , τ_B , *A*, *B*) obtained through the fitting procedure are summed up in Table 1.

	Infrared decay				***Green decay***			
Sample	**τ_A [s]**	***A***	**τ_B [s]**	***B***	**τ_A [s]**	***A***	**τ_B [s]**	***B***
Non-reticulated	1750	0,26	203218	0,74	580	0,1	11876	0,9
Reticulated	753	0,23	46233	0,77	2124	0,45	73617	0,55

Table 1. Parameters derived from the fitting of the decay of the photoinduced $\chi^{(2)}$ curves to the eq.(1).

The value of the τ_A could be attributed to the thermal *cis-trans* back reaction, which in different polymer systems may vary from seconds to minutes [16, 17]. The τ_B correspond to the *trans* isomer orientational scattering inside the matrix. The value of the *A* parameter indicates that the loss of the polar order originates mostly from the reorientation of the *trans* isomer inside the matrix, however the observed stabilization of the signal on a certain level indicates that a large fraction of the chromophore molecules stay fixed in the oriented position. The thermal cross-linking reaction may be induced by the seeding beams themselves – the sample heats as it absorbs light. This could be tested by increasing the laser repetition rate, but we could not perform this experiment with our experimental set-up (we already use the maximal repetition rate of the laser). The reticulation procedure strongly influences the rigidity of the polymer matrix, as it occurs that for the polymer system studied, the reticulation leads to the increase of the matrix rigidity. The reaction of crosslinking performed before poling hinders the mobility of the dye molecules, which leads to the decrease of the photoinduced nonlinearity. In the reticulated sample some of the bounds which attach the molecules to the epoxy group are weakened, molecules are excited and then re-attached – that is probably why we observe photoinduced SHG. Surely, not all dye molecules are re-attached and that is why we observe fast decay of the induced orientation. However, the process of reticulation appears much more accented in the non-reticulated sample and concerns bigger population of photochromic molecules. The non-reticulated polymer studied has a very low T_g, lower than the crosslinked one, and as the sample is seeded, the reticulation reaction progresses, but the randomization process goes on simultaneously, and the final result is the competition between these two. The photoinduced polar order is preserved via the reticulation, as the polarly oriented molecules are fixed in their positions, but it is also destroyed by molecular orientational scattering of non-fixed molecules. The fact that the saturation of the second order susceptibility $\chi^{(2)}$ is not attained after 5 hours of the seeding process indicates that the dynamical equilibrium is not established. The reticulation reaction through light absorption has a significant contribution to the temporal stability of the photoinduced order, as there stays a part of nonlinear optical moieties cemented in the matrix and difficult to reorient. Similar experiment was performed on the high T_g (180°C) polyurethane polymer grafted system by Si et al. [18]. They observed that photoinduced crosslinking resulted in better stability of the induced orientation, when compared to the case when the sample was seeded with green seeding beam 10 times weaker and at lower laser repetition rate, however lower than the stability of the orientation induced when the sample was oriented in conditions of low power and low repetition rate, but with simultaneous heating of the sample to 85°C.

We also tested the dark decay process in both samples and it turned out to be much longer than the photostimulated one, which indicates that the orientational scattering is slow in both samples. It should be also noticed that the photoinduced polar orientation is erased in the probing process by the strong laser fundamental ω beam. However the observed decay of the SH signal reveals very good temporal stability even under the influence of this optical erasure, in comparison to the DR1-MMA grafted system, in which the loss of one order of magnitude in the value of the SH signal is observed after two hours of constant illumination with fundamental beam [8].

The photostimulated relaxation erasure is the major obstruction of the use of such systems. The stability of the induced polar order would increase if the dye molecules were fixed after or during orientation process, with the reduction of their free volumes. The thermal cross-linking reaction is induced by the seeding beams themselves – the sample heats as it absorbs light. We also tested the stability of the polar orientation induced by Corona poling method. The sample was placed at 45° incidence angle, in the *s*-incidence beam polarization configuration, as in the Corona poling method dye molecules are oriented perpendicularly to the surface of the film. The SH signal generated by the sample remained stable over two hours of constant irradiation with strong fundamental beam. The same test was performed under the irradiation with the second harmonic beam and the SH signal generated by the corona poled sample was also constant.

7. Conclusions

Structural and dynamic studies of the induced noncentrosymmetry were performed in a reticulable polymer system with azobenzene chromophores. Growth and decay kinetics of the photoinduced second order susceptibility $\chi^{(2)}$ was tested through photoinduced SHG in the samples. From fitting of the experimental results to the theoretical equation, parameters representing the *cis-trans* thermal backreaction and orientation scattering of *trans* molecules with net orientation were obtained. In particular good photostability of the induced polar orientation was demonstrated, presumably due to the fraction of molecules, which remained fixed in their positions via photoinduced reticulation reaction.

Erasure mechanisms following an intense monochromatic irradiation appear to be the major limitation to the use of such systems, as the decay process determines their potential application to optical storage for instance. Yet, experiments performed recently by some research groups with crosslinkable polymers [18] turned out to be an attractive way of improving the photo-stability of the photoinduced polar order. Another method could be the self-seeding of the optically induced orientation [19].

8. References

[1] Charra, F., Kajzar, F., Nunzi, J.-M., Raimond, P., and Idiart, E. (1993), Light-induced second harmonic generation in azo-dye polymers, *Opt. Lett.* **18,** 941-943.

[2] Churikov, V. M., Hsu, C.-C. (2001), Dynamics of photoinduced second order nonlinearity in dimethylamino-nitrostilbene polymer thin films, *Opt. Comm.* **190,** 367-371.

[3] Branger, C., Lequan, M., Lequan, R. M., Large, M., Kajzar, F. (1997), Polyurethanes containing boron chromophores as sidechains for nonlinear optics, *Chem. Phys. Lett.* **272,** 265-270.

[4] Fiorini, F., Nunzi, J.-M., Charra, F., Kajzar, F., Lequan, M., Lequan, R. M., Chane-Ching, K. (1997), Light-induced orientation of a low absorbing phosphine oxide azo-dye/PMMA copolymer: towards a trade-off between transparency and photoinduced non-linearity, *Chem. Phys. Lett.* **271,** 335-340.

[5] Xu, G., Yang, Q. G., Si, J., Liu, X., Ye, P., Li, Z., Shen, Y. (1999), Application of all-optical poling in reversible optical storage in azopolymer films, *Opt. Comm.* **159,** 88-92.

[6] Xu, G., Liu, X., Si, J., Ye, P., Li, Z., Shen, Y., (1999), Optical poling in a crosslinkable polymer system, *Appl. Phys. B* **68,** 693-696.

[7] Xu, G., Si, J., Liu, X., Yang, Q., Ye, P., Li, Z., Shen, Y. (1998) Permanent optical poling in polyurethane via thermal crosslinking, *Opt. Comm.* **153,** 95-98.

[8] Fiorini, C., Charra, F., Nunzi, J.-M. and Raimond, P. (1997), Quasi-permanent all-optical encoding of noncentrosymmetry in azo-dye polymers, *J. Opt. Soc. Am. B* **14,** 1984-2003.

[9] Bosc, D., Foll, F., Boutevin, B., Rousseau, A. (1999), Synthesis of a novel difunctional NLO azo-dye chromophore and characterizations of crosslinkable copolymers with stable electrooptic properties, *Journal of Applied Polymer Science* **74,** 974-982.

[10] Buisson , J. P. , Mevellec, J. Y. , Zeraoui, S., Lefrant, S. (1991), Comparative study of the vibrational properties of polymers with phenyl ring, *Synth. Met.* **41,** 287-290.

[11] Mevellec, J. Y., Quillard, S., Boyer, M. I., Buisson, J. P. (2001). Structural effects on the vibrational behaviour of *N,N*-Diphenyl-1,4-phenylenediamine, *J. Raman Spectrosc.* **32,** 701-709.

[12] Apostoluk, A., Chapron, D., Sahraoui, B., Gadret, G., Fiorini, C., Raimond, P., Nunzi, J.-M. (2002), Novel real-time monitoring technique of the all-optical poling process, *Nonlinear Materials : Growth, Characterization, Devices and Applications*, Lowenthal, D. D. and Kalisky, Y. Y. (eds.), San Jose, CA, Proc. of SPIE, vol. 4628, 39-45.

[13] Loucif-Saibi, R., Nakatani, K., Delaire, J. A., Dumont, M., Sekkat, Z. (1993), Photoisomerization and second harmonic generation in disperse red one-doped and -functionalized poly(methyl methacrylate) films, *Chemistry of Materials* **5,** 229-236.

[14] Aoki, H., Ishikawa, K., Takezoe, H. , Fukuda, A. (1996), Photoinduced destruction of polar structure in dye-pendant polymers studied by second-harmonic generation, *Jpn. J. Appl. Phys.* **35,** 168-174.

[15] Rau, H. (1989). In: Rabek, J. F. (ed.), Photoisomerization of azobenzenes, *Photochemistry and photophysics*, CRC Press Inc., Boca Raton, Vol.2, Chap. 4
[16] Xu, G., Si, J., Liu, X., Yang, Q. G., Ye, P., Li, Z., Shen, Y. (1999), Comparison of the temperature dependence of optical poling between guest-host and side-chain polymer films, *J. Appl. Phys.* **85,** 681-685.
[17] Sekkat, Z., Wood, J., Knoll, W., Volksen, W., Miller, R. D., Knoesen, A. (1997), Light-induced orientation in azo-polyimide polymers 325°C below the glass transition temperature, *J. Opt. Soc. Am. B* **14,** 829-833.
[18] Si, J., Kitaoka, K., Mitsuyu, T., Ye, P., Hirao, K.(1999), Thermosetting enhancement of the light-induced polar orientation stability of molecules in polymers, *J. Appl. Phys* **12,** 8018-8022.
[19] Kitaoka, K., Si, J., Mitsuyu, T., Hirao, K. (2000), Photoinduced highly efficient second-harmonic generation in azo-dye-doped thick films, *Jpn. J. Appl. Phys.* **39,** 363-365.

CHARACTERIZATION OF NONLINEAR OPTICAL PARAMETERS OF METAL-DOPED POLYVINYLPYRROLIDONE NANOSTRUCTURES AND ORGANIC DYE MOLECULES

R. A. GANEEV, A. I. RYASNYANSKY, R. I. TUGUSHEV, T. USMANOV
NPO Akadempribor, Akademgorodok, Tashkent 700143, Uzbekistan

Abstract

The results of third-order nonlinear susceptibility studies of Fe- and Zn-doped polyvinylpyrrolidone (PVP) in processes of third harmonic generation (THG) of Nd:YAG laser radiation are presented. Nonlinear susceptibilities ($\chi^{(3)}(-3\omega;\omega,\omega,\omega)$) of PVP:Fe and PVP:Zn were calculated to be 5×10^{-13} esu and 3×10^{-13} esu respectively. Third harmonic conversion efficiencies in these metalocomplexes were measured to be 8×10^{-7} and 5×10^{-7} respectively. The influence of high-frequency Kerr nonlinearities in PVP:Fe was negligible on the $I_{3\omega}(I_{\omega})$ dependence. The optical limiting studies of cobalt-doped PVP have shown that this process was due to self-defocusing at the wavelength of 1064 nm and due to reverse saturable absorption and self-defocusing at the wavelength of 532 nm. The results of nonlinear optical characteristics (n_2, $\chi^{(3)}(-\omega;\omega,\omega,-\omega)$, β) measurements of investigated media are discussed.

We also present the theoretical calculations of nonlinear susceptibilities, responsible for the frequency conversion and self-interaction effects, and of nonlinear refractive indices of organic dyes. The values of third- and fifth-order nonlinear susceptibilities responsible for the THG of Nd:YAG laser radiation are calculated on the basis of free electron model. In particular, the nonlinear susceptibilities $\chi_m^{(3)}(-3\omega;\ \omega,\ \omega,\ \omega)$ of paraterphenyl and pentacene were calculated to be 2.4×10^{-30} esu and 1.7×10^{-30} esu respectively. The relation between the length of conjugated bonds of organic dyes and their third- and fifth-order nonlinear susceptibilities has been analysed. Calculations of third-order Kerr-induced nonlinear susceptibilities and nonlinear refractive indices of organic dyes are presented.

1. Introduction

The investigations of optical and nonlinear-optical properties of organic compounds possessing double conjugated bonds in the processes of harmonic generation, optical limiting, optical switching, etc., are of great interest during last time. Such type of compounds covers a number of arenes (naphthalene, antrathene etc.), fullerenes, polymers, etc. π-electrons delocalization in such kind of organic molecules with double conjugated bonds is responsible for their high nonlinearities. For example, delocalized π-

F. Charra et al. (eds.), Organic Nanophotonics, 367–384.

electrons in arenes cause the high values of transition dipole moments [1] and also responsible for high values of fullerenes' nonlinear susceptibilities, observed both at harmonic generation [2-4] and wave conjugation [5] processes. Metal injection into the polymer structure leads to the variations of their both optical and nonlinear-optical characteristics [6]. Note an interest on metal-doped polymers with conjugated stabilisers (in particular, polyphenylenevinylene and polyvinylpyrrolidone [7]). Polyvinylpyrrolidone (PVP) has an advantage of accepting of high concentrations of various guest molecules and atoms without the loss of its good optical properties (in particular the negligible scattering). PVP can be used as a stabiliser for nanostructures such as metal clusters that makes such polymer also interesting for the stable suspensions preparation for nonlinear-optical investigations [8].

The investigation of optical limiting (OL) in various materials opens new opportunities of their application as laser switching systems for protection of eyes and sensitive registrars from intense laser radiation. The mechanisms that cause OL have different origin. Excited state induced nonlinearity is one of the main sources of optical nonlinearities, which has potential applications of OL [9]. Fullerene C_{60} with a variety of central metal atoms is perhaps the most famous material known as possessing strong reverse saturable absorption (RSA). The considerable interest on organometalic molecules doped with such metals as ruthenium, gold, nickel, and metallic clusters was due to high second- and third-order nonlinearities they posses. Some of such structures posses the RSA that makes them attractive for possible applications as optical limiters.

Gaseous molecular systems are of a great interest due to their possible applications as the nonlinear media for the frequency conversion of laser radiation in VUV and XUV ranges [10-14]. The organic dye vapours are the attractive molecular systems for these purposes. Such molecules with double-conjugated bonds and delocalized π- electrons seems to be the effective nonlinear materials. It was shown previously that the nonlinear susceptibilities of some organic molecules with double-conjugated bonds such as organic dyes are comparable with the resonant nonlinear susceptibilities of atoms [15]. The nonlinear-optical characteristics of organic dye vapours were studied in [16] and various frequency conversion schemes were offered. Third harmonic generation due to fifth-order nonlinearity in naphthalene vapours was investigated in [17]. However, there are some problems that not yet considered so far, such as the phase-matching conditions at frequency conversion in different channels, the signs of dispersion at the nonlinear-optical processes, the influence of various nonlinear processes on frequency conversion efficiency, etc.

Below we present our studies on nonlinear-optical parameters of PVP solutions doped with iron, zinc and cobalt using third harmonic generation (THG) of picosecond Nd:YAG laser radiation and Z-scan measurements. Nonlinear susceptibilities ($\chi^{(3)}(-3\omega,\omega,\omega,\omega)$), nonlinear refractive indices (n_2) and conversion efficiencies (η) of THG in these media were determined and the influence of Kerr nonlinearities on THG conversion efficiency was discussed. The studies on optical limiting in PVP doped with different concentrations of cobalt using picosecond infrared radiation (λ = 1064 nm, t=35 ps) are presented and compared with that in visible spectral range (λ = 532 nm).

Further, the results of theoretical calculations of the nonlinear-optical nonlinearities of organic dye vapours and solutions are presented. The values of third- and fifth-order nonlinear susceptibilities responsible for THG of Nd:YAG laser radiation in organic dye vapours and solutions have been calculated in the frames of free electron (FE) model. We analyse the dependence between the length of double conjugated bonds of dye molecules and their third- ($\chi^{(3)}(-3\omega;\omega,\omega,\omega)$) and fifth- ($\chi^{(5)}(-3\omega;\omega,\omega,\omega,\omega,-\omega)$) order susceptibilities. The calculations of Kerr-induced third-order nonlinear susceptibilities ($\chi^{(3)}(-\omega;\omega,-\omega,\omega)$) and n_2 of organic solutions are also presented. These results are compare with the experimental data of Z-scan measurements of $\chi^{(3)}(-\omega;\omega,-\omega,\omega)$ and n_2 of naphthalene and paraterphenyl solutions.

2. Third harmonic generation of laser radiation in Fe- and Zn- doped polyvinylpyrrolidone

Output characteristics of Nd:YAG laser radiation were as follows: pulse duration 35 ps, pulse energy 1 mJ, $\lambda = 1064$ nm. Laser radiation was focused by 25-cm focal length lens into the quartz cells with metallopolymer solutions. Pump radiation was registered by a calibrated photodiode. Converted radiation ($\lambda = 354.7$ nm) was filtered by UV filter from pump radiation and was directed to the spectrograph and registered by a photomultiplier tube. The laser radiation was converted to second and third harmonics in *KDP* crystals and the calibration of registrar apparatus was made for determination of the absolute values of conversion efficiencies.

The aqueous PVP solutions doped with iron (4.25 wt. %) and zinc (0.85 wt. %) were used. 1 g of metallopolymers was dissolved in 100 ml of distilled water at a room temperature. Investigated solutions were kept in 1-mm (in the case of Z-scan experiments) and 2-mm (in the case of THG experiments) cells. Our interest in application of PVP as a matrix was due to its stabilising properties. Our nanoparticles had a longer lifetime in respect to colloidal aqueous solutions, which fall in the sediments after some period of aggregation.

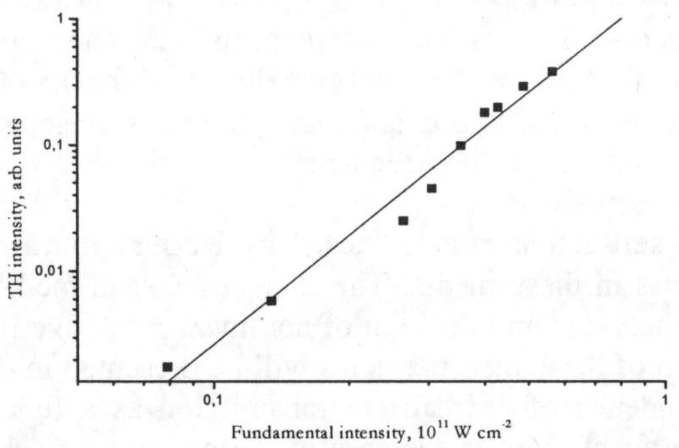

Figure 1. The dependence of TH intensity versus pump radiation intensity for PVP:Fe solution

Fig. 1 shows the dependence of third harmonic (TH) intensity on pump radiation intensity in the case of iron-doped PVP solution (PVP:Fe). The slope of $I_{3\omega}(I_\omega)$ dependence was 2.85 that was close to the theoretical predictions and remained approximately constant in the whole intensity range. The same features were in the case of zinc-doped PVP solution (PVP:Zn). The experiments took place at the intensities below the breakdown threshold of solutions ($I_{thr} = 10^{11}$ W cm^{-2}). The maximal values of conversion efficiencies for PVP:Fe and PVP:Zn were measured to be 8×10^{-7} and 5×10^{-7}

respectively. No THG was observed in pure PVP and other metal-doped PVP solutions, particularly in PVP doped with cobalt and manganese. Thus the role of π-bonds in PVP was negligible in respect to the influence of metallic nanoparticles on the nonlinear-optical properties of metal-doped PVP structures. The possible explanation of the fact that cobalt- and manganese-doped PVP solutions does not exhibit THG is that their surface plasmon resonances are situated far from two- or three-photon resonances with pump radiation frequency (in respect to iron and zinc nanoparticles) .

One of the main factors, which govern the conversion efficiency of laser radiation frequency in isotropic media, is the self-interaction associated with a change of refractive index of medium in an electromagnetic field. This effect alters the phase conditions of a nonlinear interaction between the waves and also results in a redistribution of the intensity of confined wave beams. The influence of the self-interaction has been analysed previously by numerical and analytic methods. As a rule, an analysis has been made without full account for variation of the field along a transverse coordinate in a nonlinear medium [18,19].

The analysis of harmonic generation was used to determine the nonlinear susceptibilities responsible for harmonic generation. Analysis was first provided by a method taking into account the self-action effects caused by the Kerr nonlinearity [20]. In the case of weak focusing of fundamental radiation (medium length L much less than confocal parameter of fundamental beam b), there is following simple equation for third harmonic intensity taking into account the variations of the fields along a transverse coordinate (r) in a nonlinear medium:

$$I_{3\omega} = \gamma^2 L^2 I_{10}^3 \exp(-6k_1 r^2 / b) \frac{\sin^2 \Delta(L,r)}{\Delta^2(L,r)} \quad (1)$$

Here $\gamma=24\pi^3\chi^{(3)}(-3\omega,\omega,\omega,\omega)/(n_1^{3/2}n_3^{1/2}c\lambda_1)$; $\Delta(L,r)=2b/L + \alpha - \beta$; $\alpha = L\Delta k$ is the normalised phase-mismatching; $\Delta k = k_1 - 3k_3$, $\beta = 72\pi^3 L\Delta\chi_k I_{10}\exp(-2k_1r^2/b)/(n_1^2 c\lambda_1)$; $\Delta\chi_k = \chi^{(3)}(-\omega,\omega,-\omega,\omega)/2 - n_1\chi^{(3)}(-3\omega,3\omega,\omega,-\omega)/n_3$ is the difference of Kerr nonlinearities, responsible for refractive indices changes at the wavelengths of fundamental and harmonic radiation caused by fundamental radiation; λ_i, k_i and n_i are the wavelengths, the wave numbers and the refractive indices at the frequency of i-radiation, I_{10} is the maximal intensity at the plane of beam waist.

We estimated the influence of self-action effects caused by Kerr nonlinearities on phase characteristics of THG process in these media. The standard Z-scan method [21] was applied in order to determine the value and the sign of nonlinear refractive index of solutions. More Detailed description of these measurements will be presented in the next Chapter. Fig. 2 presents the dependence of normalised transmission as a function of PVP:Fe position respectively to the focal point in the case of closed aperture scheme at the wavelength of 1064 nm. As one can see, the investigated solution demonstrates self-defocusing of laser radiation. In this case, the observed nonlinear process was due to the Kerr nonlinearities' influence. We neglect the influence of thermal lens on the process of self-defocusing. The studies of nonlinear absorption in open aperture scheme showed an absence of this process at investigated wavelength.

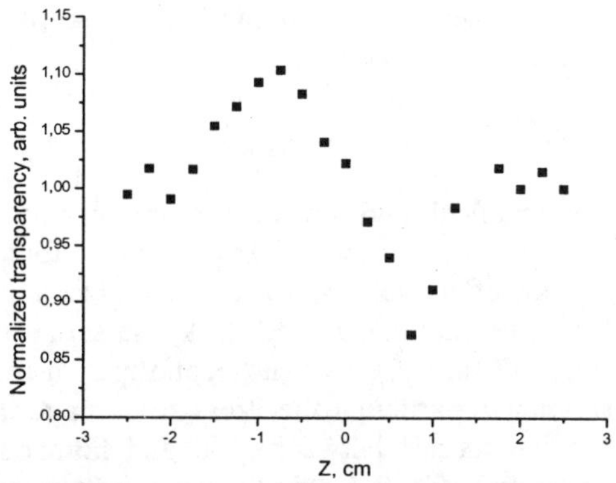

Figure 2. Normalized transmittance as a function of sample's (PVP:Fe) position with respect to the focusing point (Z=0) in closed aperture Z-scan scheme.

The value of nonlinear refractive index of PVP:Fe at the wavelength of λ = 1064 nm was measured to be $n_2 = -6{,}7\times10^{-13}$ esu. n_2 of pure PVP was of the order of 10^{-14} esu. Following to the theoretical calculations, this nonlinear variation of refractive index does not influences on $I_{3\omega}(I_\omega)$ dependence at the intensities used. The same conclusion was made in our previous THG investigation in colloidal platinum and copper [22]. Note that the deviation from the cubic dependence observed in fullerene-doped polyimide films [4] was due to considerable nonlinear variations of the refractive index. The same deviation was observed previously in gaseous media [23,24] when the high-frequency Kerr effect led to the phase mismatching. The reverse case takes place in the media with positive sigh of n_2 when the phase parameter $\sin^2(\Delta kL/2)/(\Delta kL/2)^2$ increases with the growth of pump intensity. This peculiarity can be used for THG optimisation in the media with normal dispersion [24].

The $\Delta k=k_3-3k_1=0$ conditions should be fulfilled to achieve the phase matching of THG in the case of plane waves interaction. The confocal parameter of radiation focused by a 25-cm focal length lens was 20 mm whereas the length of nonlinear medium was 2 mm, i. e. the conversion process took place in plane waves conditions. It is well known that above-mentioned phase-matching conditions (Δk=0) take place when the frequency of generated harmonic is located in the range of anomalous dispersion of medium [25]. The investigated solutions posses the normal dispersion in the range of λ=354.7 nm, so the phase-matching conditions were not fulfilled for observed process. The value of nonlinear susceptibility $\chi^{(3)}(-3\omega;\omega,\omega,\omega)$ in this case was estimated from relation [26] taking into account a negligible influence of Kerr nonlinearities:

$$\left|\chi^{(3)}(-3\omega,\omega,\omega,\omega)\right|^2 = \frac{\lambda_1^2 n_3 n_1^3 c^2 \eta}{256\sqrt{3}(\ln^3 2)\pi^5 L^2 I_\omega^2}\left(\frac{\sin(\Delta kL/2)}{\Delta kL/2}\right)^{-2} \qquad (2)$$

where I_ω is the pump radiation intensity, η is the conversion efficiency, c is the light velocity. Third-order nonlinear susceptibilities responsible for the harmonic generation in PVP:Fe and PVP:Zn determined from Eq.2 were calculated to be 5×10^{-13} esu and 3×10^{-13} esu respectively.

Our previous studies of nonlinear-optical characteristics of colloidal metals [22] and fullerenes [4,26] and their exhibition of THG have shown that the concentration of active particles in solutions (Pt and Cu nanoclusters in aqueous solutions, C_{60} and C_{70} in toluene) seems to be crucial factor of frequency conversion efficiency. The concentration of nanoparticles should be chosen in appropriate manner for the optimisation of conversion processes, so that the linear absorption at fundamental and harmonic frequencies and another limiting factors (nonlinear refraction, nonlinear absorption, etc.) did not prevented the growth of the conversion efficiency.

3. Optical limiting in cobalt-doped polyvinylpyrrolidone

Previously it was shown that the chemical and physical environment used to synthesise and process poly(p-phenylenevinylene) by the soluble precursor route in the matrix of PVP has a remarkable effect on the effective conjugation length of the resulting polymer, which directly affects the third order nonlinearity [27]. At the same time, the thermal response of such bulk polymers (in the case of nanosecond and microsecond radiation) can lead to a confusing behaviour. In particular the stress-strain patterns in the polymer matrix generates stress at the boundaries between cold and hot areas. The refraction index is changing in that case both due to medium density alternation and stress-induced birefringence (elastooptic effect). An application of organometalic structures in dissolved phase disposes of the last process and makes possible to take in to account only thermal process due to medium expansion (for long pulses) and high-frequency Kerr effect (in the case of short pulses).

Pure PVP was dissolved in methylenechlorid: acetone at the ratio of 1:4. This solution then was mixed with cobalt chalogenide solution in acetone at room temperature and inert gas atmosphere. The next stages of preparation of organometalic compound were the separation of one from the solution and the drying in vacuum conditions. The formula of these organometalic complexes is $[(C_6H_9NO)_2 \text{ Me } X_2]_n$, where X is the halogen atom, Me is the metal atom (Co in our case), n is the polymerisation rate of PVP.

The aqueous solutions of PVP doped with different concentrations of cobalt (2, 5.3, 6.2 and 13.5 weight %) were used. 1 g of organometalic compound was dissolved in 100 ml of distilled water at room temperature. The spectroscopic measurements of cobalt-doped PVP solutions were carried out on spectrophotometer SF-26. The absorption spectra of organometalic polymer solutions near resonant line (525 nm) are presented in Fig. 3.

Investigations of organometalic solutions were carried out at the wavelengths of 1064 nm and 532 nm applying the standard Z-scan technique [21]. Experimental set-up has been described in detail elsewhere [26]. The Nd:YAG laser generated picosecond pulse train was used in our experiments. Single pulse (τ = 35 ps) extracted from the train was focused by 25-cm focal length lens (FL, Fig. 4). The

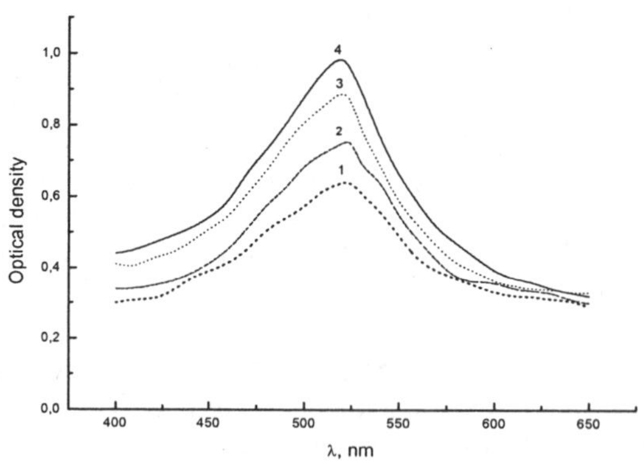

Figure 3. Absorption spectra of aqueous PVP solutions with different cobalt concentration. 1- 2%, 2 - 5.3%, 3 - 6.2%, 4 - 13.5%.

beam waists were 150 μm and 100 μm for fundamental and second harmonic radiation respectively. The laser pulse energy was measured by calibrated photodiode and was registered by digital voltmeter. 1-mm quartz cells (QC) with organometalic solutions were moved through the focal point by movable table (MT).

1-mm aperture (A), which transmitted 3% of laser radiation, was fixed at the distance of 150 cm from the focal plane (closed aperture scheme). The photodiode (PD_2) was kept behind this aperture for measurement of output radiation. We used the ratio of transmitted radiation registered by PD_2 and incident radiation registered by PD_1 as a normalised transmittance to avoid the influence of laser radiation non-stability on measured results. Closed aperture scheme allowed to determine the sign and the magnitude of n_2 of organometalic polymer solutions. The aperture was removed (open aperture scheme) for measurements of nonlinear absorption coefficient (β). The detector was kept at such distance from the sample that allowed measure all transmitted radiation to determine the dependence of sample's transmission on the intensity of laser radiation.

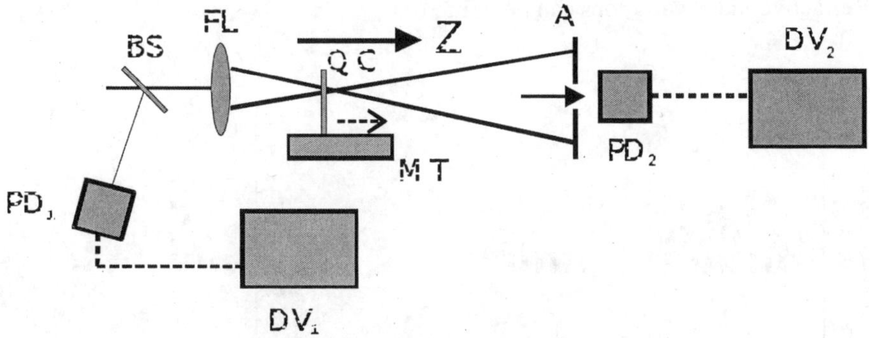

Figure 4. Experimental setup. FL: focusing lens; QC: quartz cell with organometalic solution; $PD_{1,2}$: photodiodes FD24K; $DV_{1,2}$: digital voltmeters V4-17; A: aperture; BS: beam splitter; MT: movable table.

An attention was given to prevent optical damage of investigated medium. The intensities of optical damage of polymer structures were estimated to be 4×10^{11} W cm^{-2} and 10^{11} W cm^{-2} at the wavelengths of the fundamental and second harmonic radiation respectively. The intensities used in the experiments not exceeded 8×10^{10} W cm^{-2} and 10^{10} W cm^{-2} respectively.

Fig. 5,a shows the dependencies of normalised transmittances at the wavelength of 1064 nm in closed aperture scheme of organometalic polymers doped with different cobalt concentrations. Each point corresponds to the averaging of 20 measurements. The same experiments were carried out at λ=532 nm (Fig. 5,b). These dependencies were measured at different radiation intensities at λ = 1064 nm ($I_o = 4 \times 10^{10}$ W cm^{-2}) and λ = 532 nm ($I_o = 3 \times 10^9$ W cm^{-2}). The peculiarity of Z-scan measurements is that it makes possible to determine the sign of nonlinear refractive index and character of nonlinear process (in this particular case the sign was negative that corresponded to the self-defocusing). Note the sign and concentration dependence persistence at two wavelengths studied. The appreciable increase of T(Z) dependence with cobalt concentration growth was observed.

We neglected the thermal lens influence on investigated process. It is well known that the thermal effect appearance in the medium connected with the acoustic wave spreading. The time required for density reduction is determines by the ratio of radius of beam cross-section (d) to the velocity of sound in the dielectric (V_s).

$$\tau = \frac{d}{V_s} \quad (3)$$

Taking into account our experimental conditions (d = 75 μm for 1064 nm and 50 μm for 532 nm, V_s = 1500 m/s), this time is equal to 50 ns and 33 ns for the fundamental and second harmonic radiation respectively that is three orders longer than pulse duration. Note the low repetition rate of the laser (0.4 Hz) in these experiments that also restricts the thermal lens appearance. Thus the observed effect of self-defocusing in our experiments was due to Kerr-induced nonlinearities.

Refractive index variations can be written as

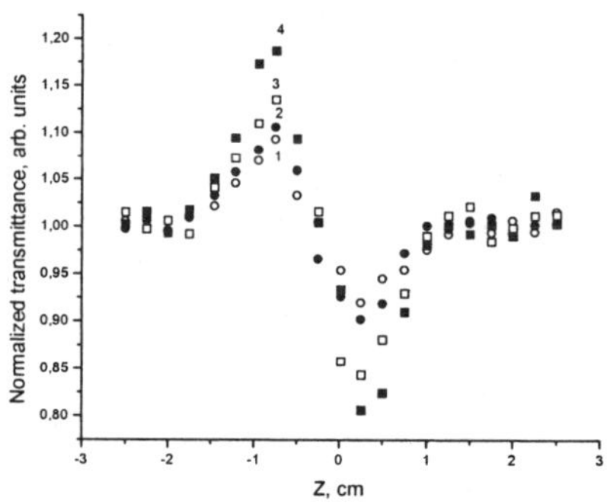

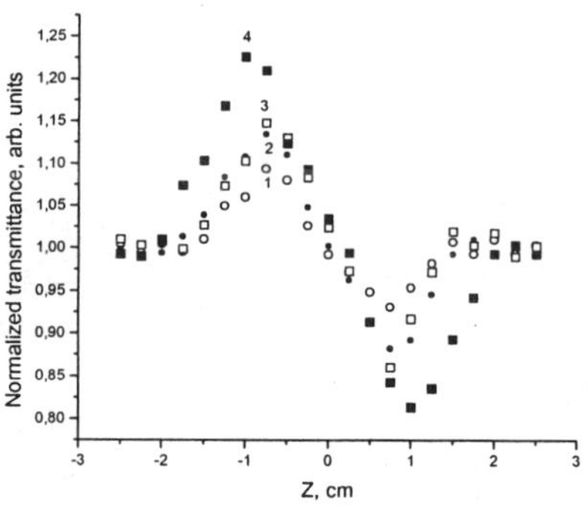

Figure 5. Normalized transmittances as a functions of the cell's position in closed aperture scheme at the wavelengths of (a) 1064 nm ($I_o = 4 \times 10^{10}$ W cm^{-2}) and (b) 532 nm ($I_o = 3 \times 10^9$ W cm^{-2}). 1- 2%, 2 – 5.3%, 3 – 6.2%, 4 – 13.5%.

$$\Delta n = \gamma I_\omega = n_2 \frac{|E_\omega|^2}{2} \quad (4)$$

Here I_ω and E_ω are the intensity and electric field of electromagnetic wave. Nonlinear refractive indices (n_2 and γ) representing the same process in different units are connecting each other by the relation n_2 [esu units] = ($c \times n$ / 40π) γ [SI units]. Here c is the light velocity and n is the linear refractive index of the medium.

For determination of nonlinear refractive indices of investigated medium we used the equation [21]

$$\gamma = \frac{\alpha\lambda}{0.404(1-S)^{0.25}\,2\pi I_o[1-\exp(-\alpha L)]}\Delta T_{p-v} \qquad (5)$$

where ΔT_{p-v} is the normalised difference between peak and valley in T(Z) dependence in closed aperture scheme, I_o is the radiation intensity at the focal point, S is the aperture's transmittance (part of radiation, propagated through the aperture), λ is the radiation wavelength, L is the sample's thickness, α is the linear absorption coefficient.

Nonlinear refractive index of 13.5% cobalt-doped PVP solution was measured to be -1.3×10^{-12} esu at λ=1064 nm, whereas this parameter at λ=532 nm was -9.5×10^{-12} esu.

Nonlinear absorption studies have shown that this process was observed only at the wavelength of 532 nm (Fig. 6). There was not noticeable nonlinear absorption at the wavelength of 1064 nm, in spite of radiation intensity we used (up to 8×10^{10} W cm^{-2}).

Normalized transmittance in the case of open aperture scheme is determined as [28,29]

$$T(z) = \frac{\ln(1+q_o)}{q_o}, \qquad (6)$$

where

$$q_o = \beta\frac{I_o[1-\exp(-\alpha L)]}{\left(1+\left(\frac{z}{z_o}\right)^2\right)\alpha}. \qquad (7)$$

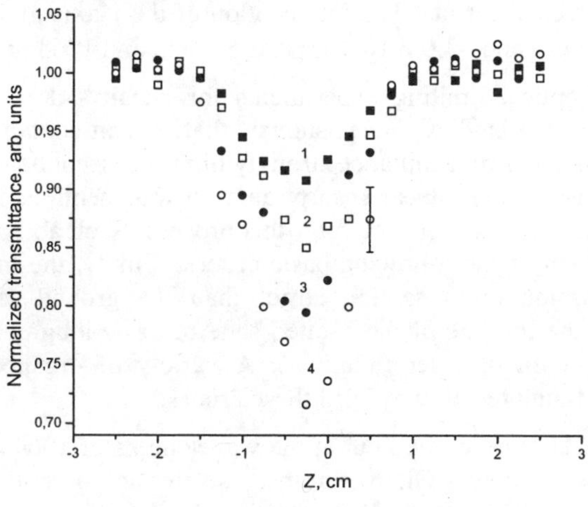

Figure 6. Normalized transmittances as the functions of cell's position in open aperture scheme at the wavelength of 532 nm. 1- 2%, 2 - 5.3%, 3 - 6.2%, 4 - 13.5%.

Here z_o is the diffraction length of the laser beam, $z_o = kw_o^2 / 2$, k is the wave number, $k = 2\pi / \lambda$, w_o is the beam radius at the focal point.

Using (4) and (5), one can calculate the nonlinear absorption coefficient of investigated medium. Transmittance decreasing at the wavelength of 532 nm was due to RSA that manifested in a number of organometalic and polymer structures in visible range [30]. Nonlinear absorption of PVP solution with 13.5 % cobalt concentration was measured to be 3.2×10^{-9} cm W^{-1}. The shot-to-shot fluctuations in the normalised transmission did not exceeded 3%. On this assumption we can estimate the highest limit of nonlinear absorption coefficient to be 2×10^{-11} cm W^{-1} at the wavelength of 1064 nm.

Third-order nonlinear susceptibilities in media with nonlinear absorption and refractive index can be considered to be a complex quantity

$$\chi^{(3)} = \mathrm{Re}\,\chi^{(3)} + i\,\mathrm{Im}\,\chi^{(3)}, \tag{8}$$

where the imaginary part is related to the nonlinear absorption coefficient through

$$\mathrm{Im}\,\chi^{(3)} = \frac{n^2 \varepsilon_o c \lambda \beta}{2\pi}, \tag{9}$$

and the real part is related to γ through

$$\mathrm{Re}\,\chi^{(3)} = 2n^2 \varepsilon_o c \gamma\ . \tag{10}$$

Here ε_o is the permittivity of free space. For the presentation of $\chi^{(3)}$ in esu units we used the relation $\chi^{(3)}$ [esu units] = $(9\times10^8 / 4\pi)\ \chi^{(3)}$ [SI units]. Nonlinear refractive indices, nonlinear absorption coefficients and both real and imaginary parts of susceptibilities are presented for investigated organometalic polymers in Table 1. Samples with different cobalt concentration were tested and their third-order susceptibility values were derived from Eqs. (8-10). Typical results for γ are shown in Fig. 7. The solid lines are linearly fitted to the experimental results. The interception of the y-axis gives γ of the pure PVP that are -1.8×10^{-12} esu and -3.6×10^{-13} esu for 532 nm and 1064 nm respectively.

The results of optical limiting experiments for cobalt - doped PVP solutions are presented below. Our studies have shown that cobalt-containing PVP–stabilized suspensions of high linear transmittance strongly limit picosecond laser pulses at 532 nm. One of the mechanisms of nonlinear absorption, as it was mentioned, is the RSA. A five-level model [31] has been used to interpret this process. Such absorption in the medium requires the fulfilment of the following basic criteria. Firstly, the material must have an excited-state absorption cross section larger than the ground-state absorption cross section. Secondly, the lifetime of the excited state must be long enough (in comparison with the pulse duration of laser radiation). A variety of organometalic and organic materials has been found recently to fulfil these criteria.

Investigation of OL was carried out at the wavelengths of 1064 nm and 532 nm. The solutions did not show any sigh of nonlinear scattering for both wavelengths at the intensity levels used. In spite of the absence of nonlinear absorption in infrared spectrum region, the cobalt-doped PVP

TABLE 1. Nonlinear optical characteristics of cobalt-doped PVP solutions.

Organometalic polymer	λ, nm	n_2, esu	Re $\chi^{(3)}$, esu	β, cm W^{-1}	Im $\chi^{(3)}$, esu
PVP (2% Co)	1064	-4.5×10^{-13}	-1.09×10^{-13}	-	-
PVP (5.3% Co)	1064	-7.3×10^{-13}	-1.62×10^{-13}	-	-
PVP (6.2% Co)	1064	-9.1×10^{-13}	-2.02×10^{-13}	-	-
PVP (13.5% Co)	1064	-1.3×10^{-12}	-2.88×10^{-13}	-	-
PVP (2% Co)	532	-2.4×10^{-12}	-5.4×10^{-13}	7.1×10^{-10}	2.7×10^{-13}
PVP (5.3% Co)	532	-5.4×10^{-12}	-1.2×10^{-12}	1.4×10^{-9}	5.4×10^{-13}
PVP (6.2% Co)	532	-6×10^{-12}	-1.6×10^{-12}	2.1×10^{-9}	7.8×10^{-13}
PVP (13.5% Co)	532	-9.5×10^{-12}	-2.1×10^{-12}	3.2×10^{-9}	1.2×10^{-12}

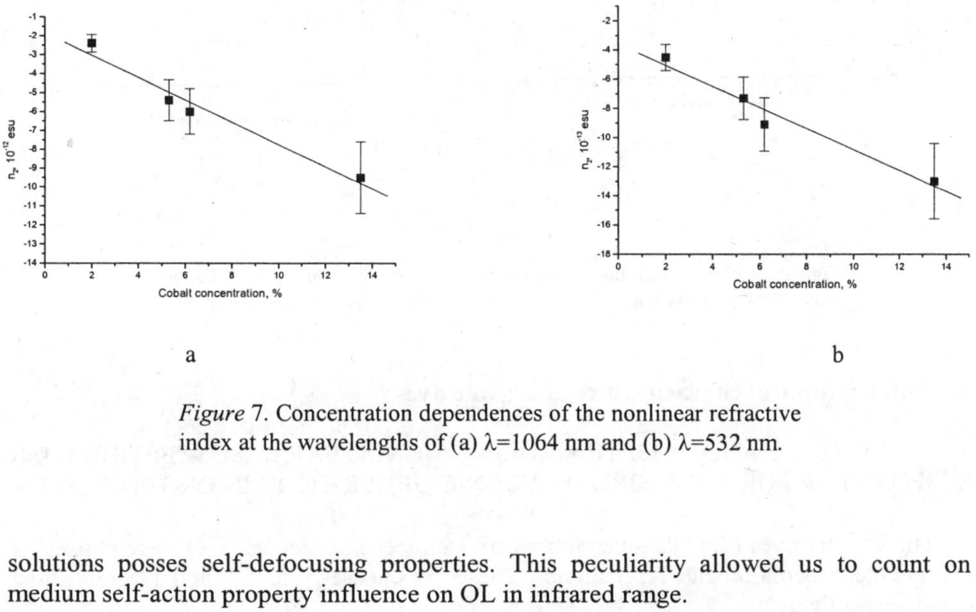

Figure 7. Concentration dependences of the nonlinear refractive index at the wavelengths of (a) λ=1064 nm and (b) λ=532 nm.

solutions posses self-defocusing properties. This peculiarity allowed us to count on medium self-action property influence on OL in infrared range.

Fig. 8,a shows the dependencies of normalised transmittance of cobalt-doped PVP solutions as the functions of laser radiation intensity in the case of closed aperture scheme (λ = 1064 nm). In this case the cell with solution was kept in the region of lowest transmittance, i. e. in the region of the valley (see Fig. 5).

Another picture took place at the wavelength of 532 nm (Fig. 8,b). In that case OL was attributed to the self-defocusing and nonlinear absorption. Nonlinear absorption in this case can be explained by the influence both two-photon absorption and RSA (at high intensities). The absence of information about spectral characteristics and lifetimes of excited singlet and triplet states of investigated organometalic solutions does not allow us to make the conclusion about relationship between these processes. Optical limiting threshold at both wavelengths was found to be lower than that of C_{60} [26].

Our results show that the OL threshold decreases with the incorporation of additional cobalt atoms. In these experiments we have shown that excited state absorption can be significant and needs to be considered in the case of even weakly absorbing organometalic clusters.

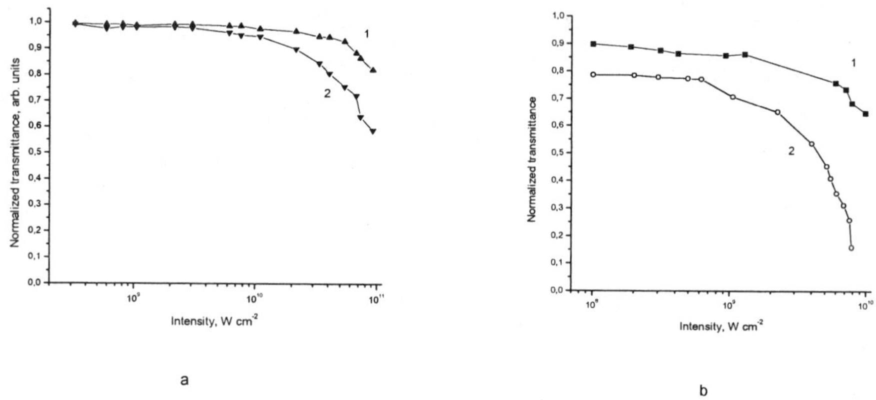

Figure 8. Normalized transmittances as the functions of laser radiation intensity at the wavelengths of (a) 1064 nm and (b) 532 nm. 1- 2%, 2 – 13.5%.

4. Nonlinear optical characteristics of organic dyes

4.1. THIRD- AND FIFTH-ORDER NONLINEAR SUSCEPTIBILITIES RESPONSIBLE FOR THE THIRD HARMONIC GENERATION IN ORGANIC DYES

The organic dyes (naphthalene, anthracene, tetracene and pentacene) were chosen as the nonlinear media for the THG studies. These dyes posses j=5, 7, 9 and 11 conjugated bonds respectively.

Nowadays, three quantum-mechanics methods are used for characterisation of the molecules with delocalized π-electrons. They are the method of delocalized valent bonds, the method of molecular orbitals and the free-electron method. We used FE model for the calculations of nonlinear susceptibilities of organic dyes. FE model describes organic molecule systems with double conjugated bonds on the base of an assumption that the σ-electrons of carbon atoms form the molecular skeleton of certain geometrical shape. Due to positive charge the skeleton creates the potential field under the action of which all π-electrons are moving along this conjugated structure. Although FE model is very simple for double conjugated bond description, it enables to calculate their basic optical and nonlinear optical characteristics.

The description of π-electrons based on FE model was used by many authors (see, for example, [32]). Kuhn in [33] interpreted and predicted the basic spectral properties of organic dyes. Some of them, for example the symmetrical cyanide dyes, can be presented in a framework of one dimensional model, in which π-electrons' behaviour can be compared with free electrons in potential well.

For the determination of organic dye molecules' dipole moments we calculated the length of conjugated bond (L) in accordance to the relationship [34]

$$L = d\,(j + \delta)\,, \tag{11}$$

where j is the quantity of double conjugated bonds, d is the distance between conjugated atoms, δ is the correction parameter, which is equal 0 for all above mentioned dyes and equal 1 for paraterphenyl. In particular, for naphthalene $j = 5$, d = 0.14 nm, $\delta = 0$. In that case L for naphthalene is equal to 0.7 nm.

The energy values E_n of discrete states can be described by the following equation

$$E_n = \frac{n^2\pi^2\hbar^2}{8m_0L^2} \tag{12}$$

Here m_o is the electron mass.

The ground state of naphthalene equal to n = 5 (E_n = 36628.17 cm^{-1}), and first 5 states are occupied by 5 electron couples. The transitions take place only from higher states with n > 5. The transition frequencies determine by the relationship:

$$\omega_{mn} = \frac{1}{\hbar}\left(E_m - E_n\right) = \frac{\hbar\pi^2}{8m_0L^2}\cdot(m^2 - n^2) \tag{13}$$

Transition dipole moments were calculated in accordance to [34]:

$$d_{mn} = \begin{cases} 0, & \text{when } (m\pm n) \text{ are even} \\ \dfrac{4eL}{\pi^2\left[\dfrac{1}{(m-n)^2} - \dfrac{1}{(m+n)^2}\right]^{-1}} & \text{when } (m\pm n) \text{ are odd} \end{cases} \tag{14}$$

where e is the electron charge.

Third-order nonlinear susceptibility of a single molecule (for THG of Nd:YAG laser radiation, λ=1064 nm, $\omega+\omega+\omega\Rightarrow 3\omega$ process; $\chi_m^{(3)}(-3\omega;\omega,\omega,\omega) = \chi^{(3)}(-3\omega;\omega,\omega,\omega)/N$, N is the medium concentration) was calculated in accordance to [25]:

$$\chi_m^{(3)}(-3\omega;\omega,\omega,\omega) = \frac{1}{4\hbar^3} \times \sum_{n,m,l}\left[\frac{d_{gn}d_{nm}d_{ml}d_{lg}}{(\omega_{ng}-\omega)(\omega_{mg}-2\omega)(\omega_{lg}-3\omega)}\right], \qquad (15)$$

using calculated energy states, transition frequencies and transition dipole moments.

The value of nonlinear susceptibility $\chi_m^{(3)}(-3\omega;\omega,\omega,\omega)$ of naphthalene was calculated to be 0.42×10^{-34} esu. In isotropic media the difference frequency generation ($\omega + \omega + \omega + \omega - \omega \Rightarrow 3\omega$) is also realises alongside with $\omega + \omega + \omega \Rightarrow 3\omega$ process. Fifth - order nonlinear susceptibility $\chi_m^{(5)}(-3\omega;\omega,\omega,\omega,\omega,-\omega)$ for the $\omega+\omega+\omega+\omega-\omega\Rightarrow 3\omega$ process was also calculated to be 1.3×10^{-46} esu using the same technique. Reported earlier experimental results on THG of picosecond Nd:YAG laser radiation in four- ($\omega + \omega + \omega \Rightarrow 3\omega$) and six-photon ($\omega + \omega + \omega + \omega - \omega \Rightarrow 3\omega$) processes in naphthalene vapours ($C_{10}H_8$) [17] have shown prevailing opportunity of the latter one.

Calculation results of third- and fifth-order nonlinear susceptibilities of THG process as the functions of the length of conjugated bonds are presented in Table 2. It should be noted the large values $\chi_m^{(3)}(-3\omega;\omega,\omega,\omega)$ of paraterphenyl and pentacene (2.35×10^{-30} esu and 1.66×10^{-30} esu respectively) that are comparable with $\chi_m^{(3)}(-3\omega;\omega,\omega,\omega)$ of alkali metal vapours. The calculated results have shown that the growth of the length of double conjugated bonds leads to the increase of nonlinear susceptibilities.

4.2. THIRD-ORDER KERR-INDUCED NONLINEAR SUSCEPTIBILITIES OF ORGANIC DYES

The investigations of wave interaction processes in various media due to refractive index modulation are of great interest, as it was mentioned previously. On the one hand, the refractive index modulation in the field of intense electromagnetic wave (Kerr effect) limits the frequency conversion of laser radiation in isotropic media due to infringement of optimal phase-matching conditions [24], but, on the other hand, it allows the generation of odd harmonics in the media with positive dispersion [35].

TABLE 2. Third- and fifth-order nonlinear susceptibilities responsible for the THG in organic dyes. λ = 1064 nm.

Organic dye	L, nm	$\chi_m^{(3)}(-3\omega;\omega,\omega,\omega)$ esu	$\chi_m^{(5)}(-3\omega;\omega,\omega,\omega,\omega,-\omega)$ esu
Naphthalene	0.7	0.42×10^{-34}	1.3×10^{-46}
Anthracene	0.98	2.8×10^{-33}	6×10^{-44}
Tetracene	1.26	7.4×10^{-31}	2.85×10^{-40}
Pentacene	1.54	1.66×10^{-30}	9.6×10^{-39}
Paraterphenyl	1.4	2.35×10^{-30}	7×10^{-38}

So, it is necessary to take into account the refraction index modulation, which influences on effective harmonic generation of laser radiation in such media. It should be noted another important peculiarity. The investigation of optical limiting in various materials opens new opportunities of their application for the protection from intense laser radiation. Optical limiting is based both on the nonlinear absorption and on the processes of refractive index variations. Below we present the theoretical calculations of third-order nonlinear susceptibilities and nonlinear refractive indices of a number of organic dyes. Third-order Kerr-induced nonlinear susceptibility of single molecule ($\chi_k^{(3)}$ (-ω; ω, -ω, ω) = $\chi^{(3)}$(-ω; ω, -ω, ω) / N) for the process ω + ω – ω → ω was calculated in accordance to [36]

$$\chi^{(3)}(-\omega;\omega,-\omega,\omega) = \hbar^{-3}\sum_{ijk} d_{gi} d_{ij} d_{jk} d_{kg} (C_{gi,gj,gk} - C_{ig,jg,kg}) \qquad (16)$$

where

$$C_{gi,gj,gk} = \left[(\omega_{kg} - \omega)^{-1} + (\omega_{kg} + \omega)^{-1}\right](\omega_{ig} - \omega)^{-1}(\omega_{jg} - 2\omega)^{-1} \pm$$
$$(1 - \delta_{gj})\left[(\omega_{kg} - \omega)^{-1} + (\omega_{kg} + \omega)^{-1}\right]\omega_{ig}^{-1}(\omega_{jg} \pm \omega)^{-1} +$$
$$\delta_{gj}\left[(\omega_{kg} - \omega)^{-1} + (\omega_{kg} + \omega)^{-1}\right](\omega_{ig} \pm \omega)^{-1}$$

Calculated $\chi_k^{(3)}$ and n_2 / N values of various organic dyes are presented in Table 3. The relation between $\chi^{(3)}$(-ω; ω, -ω, ω) and n_2 in the case of the absence of nonlinear absorption was determined as follows [25]

$$\chi^{(3)}(-\omega;\ \omega,\ -\omega,\ \omega) = n_2\, n_1 / 3\pi \qquad (17)$$

We measured $\chi^{(3)}$(-ω; ω, -ω, ω) and n_2 of some organic dye solutions using Z-scan method for the comparison with theoretical calculations. Laser radiation (λ = 1064 nm, E = 2 mJ, t = 35 ps) was focused by 25-cm focal length lens into the cell (*L*=5 mm) with paraterphenyl and naphthalene solutions in toluene. Solutions concentrations were 20 mg/ml and 50 mg/ml respectively.

In Fig. 9, T(Z) dependence of paraterphenyl solution is shown. Nonlinear optical characteristics ($\chi^{(3)}$(-ω; ω, -ω, ω) and n_2) of paraterphenyl were measured from the relationships, which connect these parameters with T(Z) dependence, solution concentration, linear absorption and laser radiation intensity [21]. Measured nonlinear optical characteristics of paraterphenyl were $\chi_k^{(3)}$ (-ω; ω, -ω, ω) = - (1 ± 0.5)×10^{-30} esu and n_2 = - (3 ± 1.5)×10^{-10} esu respectively at a concentration of N = 5×10^{19} cm^{-3}. The same measurements have been performed for naphthalene solution ($\chi_k^{(3)}$ (-ω; ω, -ω, ω) = 5×10^{-34} esu and n_2 = 2×10^{-14} esu. The last medium possesses the self-focusing features at the investigated wavelength in comparison to the paraterphenyl possessing the self-defocusing features.

TABLE 3. Kerr-induced nonlinearities and nonlinear refractive indices of organic dyes. $\lambda = 1064$ nm.

Organic dye	$\chi_k^{(3)}(-\omega; \omega, \omega, -\omega)$, esu	n_2/N, esu
Naphthalene	1.54×10^{-33}	9×10^{-33}
Anthracene	1.08×10^{-31}	6×10^{-31}
Tetracene	5×10^{-29}	3×10^{-28}
Pentacene	-2.46×10^{-29}	-1.5×10^{-28}
Paraterphenyl	-4.15×10^{-30}	-2.5×10^{-29}

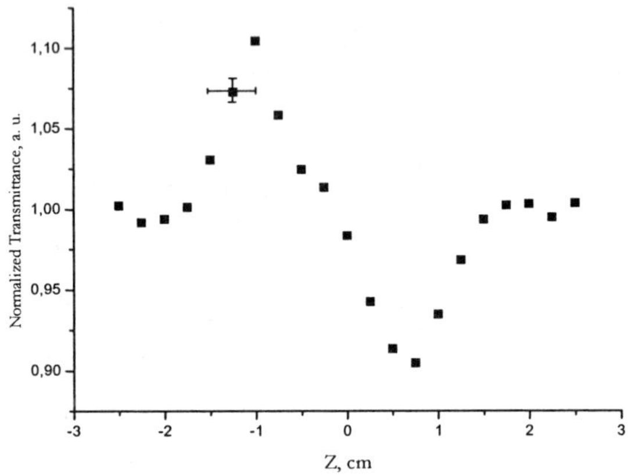

Figure 9. The normalized transmittance dependence on the cell position for paraterphenyl solution in toluene.

These experiments were performed at fixed energy gates of laser pulse. We have performed a set of experiments at other levels of pulse energy to be sure that n_2 is independent on the intensity of laser radiation and the self-focusing and self-defocusing were due to Kerr-induced nonlinearities.

Let us consider the signs of nonlinear susceptibilities of paraterphenyl (negative sign) and naphthalene (positive sign). The sign of $\chi_k^{(3)}$ is determined by the sign of frequency detuning from the resonance. On the basis of Eqs. (13-15) we obtained that for paraterphenyl molecule the quasi-resonant conditions occur (two-photon resonance for the wavelength used). This fact is in a good agreement with experimental results reported in [37,38] Reverse case was observed for naphthalene molecule. It should be noted that for naphthalene molecule at the wavelength of $\lambda = 1064$ nm, the four-photon resonance conditions fulfil. This fact was discussed in [17]. Using Eqs. (12-16) we obtained, that the signs of $\chi_k^{(3)}$ are positive for the naphthalene and negative for the paraterphenyl in a good correlation with experimental results.

The comparison of experimental results and theoretical calculations of Kerr-induced nonlinear susceptibilities and nonlinear refractive indices of investigated organic dyes have shown the agreement within a factor of four. Some discrepancies of $\chi_k^{(3)}(-\omega; \omega, -\omega, \omega)$ and n_2 measurements and calculations can be explained by the inexactitude of dipole moments calculations.

Our calculations have shown that some of investigated dyes (tetracene, paraterphenyl and pentacene) posses the considerable value of Kerr-induced nonlinearity, which can influence on the frequency conversion in such media. Declination from phase-matching conditions can be observed due to refractive index variations at pump and generated frequencies.

As it was mentioned above, the media with high value of Kerr-induced nonlinearities are of great interest due to their possible application as the optical limiters of laser radiation. The pentacene ($n_2 = - 1.5\times10^{-28}$ esu) can be of especial interest for this purpose. Such medium with high-expressed tendency of self-defocusing can be used as an effective optical limiter.

5. Conclusions

In conclusion, we investigated the nonlinear-optical characteristics of aqueous solutions of iron- and zinc-doped PVP in the processes of third harmonic generation of Nd:YAG laser radiation. Nonlinear susceptibilities ($\chi^{(3)}(-3\omega;\omega,\omega,\omega)$) of PVP:Fe and PVP:Zn were calculated to be 5×10^{-13} esu and 3×10^{-13} esu respectively. Third harmonic conversion efficiencies in these metalocomplexes were measured to be 8×10^{-7} and 5×10^{-7} respectively. The influence of high-frequency Kerr nonlinearities in PVP:Fe was negligible on the $I_{3\omega}(I_\omega)$ dependence.

We have also studied the nonlinear optical properties of aqueous solutions of PVP doped with different cobalt concentrations. Nonlinear absorption investigations have shown the absence of last process in infrared range, whereas in visible range the nonlinear absorption was considerable. It was shown that the main parameter influenced on the nonlinear process of picosecond radiation interaction with such media is the cobalt concentration in PVP. Optical limiting experiments showed that this process took place due to self-defocusing at the wavelength of 1064 nm and as a result of self-defocusing and particular influence of nonlinear absorption at the wavelength of 532 nm. The results of nonlinear optical characteristic measurements of investigated medium are presented.

Finally, we have presented the theoretical calculations of nonlinear susceptibilities, responsible for the frequency conversion and self-interaction effects, and also of nonlinear refractive indices of organic dyes. The values of third- and fifth-order nonlinear susceptibilities responsible for the THG of Nd:YAG laser radiation are calculated on the basis of FE model. In particular, the nonlinear susceptibilities $\chi^{(3)}(-3\omega; \omega, \omega, \omega)$ of paraterphenyl and pentacene were calculated to be 2.35×10^{-30} esu and 1.66×10^{-30} esu respectively. The relation between the length of conjugated bonds of organic dyes and their third- and fifth-order nonlinear susceptibilities has been analysed. Calculations of third-order Kerr-induced nonlinear susceptibilities and nonlinear refractive indices of organic dyes are presented. It was shown an importance of Kerr-induced nonlinearity influence on phase-matching conditions of the frequency conversion in tetracene and pentacene. The opportunity of some of organic dyes applying as the optical limiters was considered due to their high negative value of n_2 (-1.5×10^{-28} esu for pentacene).

Analysing our results one can see a good agreement of theoretical calculations provided by FE model with experimental measurements of nonlinear optical characteristics of organic dyes. These results allow calculate the fundamental nonlinear optical parameters of such materials using simple apparatus of FE model.

Acknowledgements

R.A.G., R.I.T. and T.U. acknowledge the partial support of the Scientific and Technology Centre in Ukraine under agreement Uzb-29. A.I.R. acknowledges the support of the State Committee of Science and Technology of Uzbekistan (grant 7/01).

References

1. Ganeev R.A., Kamalov S.R., Kodirov M.K., Malikov M.R., Ryasnyansky A.I., Tugushev R.I., Umidullaev S.U., and Usmanov T., (2000) *Tech. Phys. Lett.* **26**, 7.
2. Meth J.S., Vanherzeele H., and Wang Y., (1992) *Chem. Phys. Lett.* **197**, 26.
3. Neher D., Stegeman G.I., and Tinker F.A., (1992) *Opt. Lett.* **17**, 1491.
4. Ganeev R.A., Ryasnyansky A.I., Kamanina N.V., Kodirov M.K., and Usmanov T., (2001*) J. Opt. B* **3**, 88.
5. Knize R.J., and Partanen J.P., (1992) *Phys. Rev. Lett.* **686**, 2704.
6. Pan G.S., Kesavamoorthy R., and Asher S.A., (1998) *J. Am. Chem. Soc.* **120**, 6525.
7. Samoc A., Samoc M., and Luther-Davies M., (1997) *Proc. SPIE* **3147**, 166.
8. Sun Y.-P., Riggs J.E., and Rollins H.W., (1999) *J. Phys. Chem. B.* **103**, 77.
9. Tutt L.W., and Boggess T., (1993) *Prog. Quantum Electron.* **17**, 299.
10. Innes K.K., Stoicheff B.P., and Wallace S.C., (1976) *Appl. Phys. Lett.* **29** 715.
11. Ashfold M.N.R., Heryet C.D., and Prince J.D., (1986) *Chem. Phys. Lett.* **131** 291.
12. Frazer D.J., Hutchinson M.H.R., Marangos J.P., Shao Y.L, Tisch J.W.G., and Castillejo M., (1995) *J. Phys. B* **28,** L739.
13. Linga C., L`Huillier A., and Wahlstrem C.G., (1996) *J. Phys. B* **29,** 3292.
14. Hay N., De Nalda R., Halfmann T., Mendham K.J., Mason M.B., Castillejo M., and Marangos J., (2000) *Phys. Rev. A* **62,** 041803 (R).
15. Lukinykh V.F., Myslivets S.A., Popov A.K., and Slabko V.V., (1985) *Appl. Phys. B* **38**,143.
16. Aleksandrovsky A.S., Karpov S.V., Myslivets S.A., Popov A.K. and Slabko V.V., (1993*) J. Phys.* B **26,** 2965.
17. Ganeev R.A., Kamalov S.R., Kodirov M.K., Malikov M.R., Ryasnyansky A.I., Tugushev R.I., Umidullaev S.U., and Usmanov T., (2000) *Optics Commun.* **184,** 305.
18. Mahon R., and Yun M.Y., (1980) *Opt. Lett.* **5,** 279.
19. Mittleman D.M., Douglass D.C., and Henis Z., (1996) *J. Opt. Soc. Am.* **13,** 170.
20. Kulagin I.A., and Usmanov T., (1998) *Quantum Electron.* **28,** 1092.
21. Sheik-Bahae M., Said A.A., Wei T., Hagan D.J., and Van Stryland E.W., (1990) *IEEE J. Quantum Electron.* **26,** 760.
22. Ganeev R.A., Ryasnyansky A.I., Kodirov M.K., Kamalov S.R., and Usmanov T., (2001) *J. Phys. D* **34,** 1602.
23. Zych L.J., and Young J.F., (1978) *IEEE J. Quantum Electron.* **14,** 147.
24. Ganeev R.A., Kulagin I.A., Begishev I.A., Redkorechev V.I., and Usmanov T., (1996) *Nonlin. Opt.* **16,** 109.
25. Reintjes J.F., (1984) *Nonlinear optical parametric processes in liquids and gases,* Academic Press, London.
26. Ganeev R.A., Ryasnyansky A.I., Kodirov M.K., and Usmanov T., (2000) *Opt. Commun.*, **185,** 473.
27. Samoc A., Samoc M., and Luther-Davies B., (1997), *Proc. SPIE* **3147**, 166.
28. Rakovich J.P., Artemyev M.V., Rolo A.G., Vasilevskiy M.I., and Gomes M.J.M., (2001) *Phys. Stat. Sol. (b)* **224,** 319.
29. Capple P.B., Staromlynska J., Hermann J.A., and Mckay T.J., (1997) *J. Nonlin. Opt. Phys. Matter.* **6,** 251.
30. Perry J.W., Mansour K., Marder S.R., Perry K.J., Alverez D., and Choong L., (1994) *Opt. Lett* **19,** 625.
31. Tutt L.W., and McCahon S.W., (1990) *Opt. Lett.* **15,** 700.

NOVEL HETEROCYCLE-BASED TWO-PHOTON ABSORBING DYES

A. ABBOTTO,[a] L. BEVERINA,[a] R. BOZIO,[b] S. BRADAMANTE,[c] A. FACCHETTI,[a] C. FERRANTE,[b] G. A. PAGANI,[a] D. PEDRON,[b] R. SIGNORINI[b]

[a]*Department of Materials Science, University of Milano-Bicocca, Via Cozzi 53, I-20125, Milano, Italy, e-mail: alessandro.abbotto@unimib.it;*

[b] *Department of Physical Chemistry, University of Padova, Via Loredan 2, I-35131 Padova, Italy*

[c] *CNR – Institute of Molecular Science and Technologies, Via Golgi 19, I-20133 Milano, Italy*

Abstract. The synthesis and nonlinear optical characterization of two novel series of heteroaromatic-based quadrupolar chromophores are described. The new systems present a A-π-D-π-A and a D-π-A-π-D general framework, respectively, where A is a π-deficient heteroaromatic ring (pyridine, quinoline, benzothiazole) and D a π-excessive 5-membered heterocycle (pyrrole, furan, 3,4-ethylenedioxythiophene). Systems exhibit large two-photon absorption (TPA) activities in the femtoseconds regime (TPA cross section as high as 150 GM with 150-fs laser pulses). Their TPA-based optical limiting activity is also shown.

1. Introduction

Organic molecules which can simultaneously absorb two or more photons of the same energy to be promoted to its excited state have recently been the subject of much research work in the scientific community due to the growing interest in advanced photonic applications such as two-photon laser scanning fluorescence microscopy,[1] optical limiting,[2] three-dimensional optical storage,[3] microfabrication,[4] and upconverted-lasing.[5] Many of these applications take advantage of the fact that the two-photon absorption (TPA) activity scales quadratically with the intensity of the incident laser radiation. Some of these fields are further based on the fluorescence properties of the two-photon excited molecule. One further important aspect is the fact that biological tissues and organic coatings are linearly transparent to the fundamental wavelength of the incident radiation, providing increasing penetration. To meet market criteria for these applications, molecules with large TPA cross-sections σ are required. However, it is only in very recent times that very efficient TPA dyes[6] and a few first studies on structure/property relationships[7] have been reported. It is now apparent that a successful design strategy involves quadrupolar systems having a general structure of the type D-π-A-π-D or A-π-D-π-A.

The vast majority of the known efficient systems is based on substituted conjugated benzene rings, relying on the electronic properties of primary organic simple

F. Charra et al. (eds.), Organic Nanophotonics, 385–393.

functionalities. The lack of a systematic use of heteroaromatic compounds is surprising if one considers that π-deficient and π-excessive heterocycles[8] may act as efficient acceptor and donor moieties, respectively.[9] Furthermore, the intrinsic tunable electron nature of the heteroaromatic rings (even when unsubstituted) makes these systems particularly appropriate in order to finely control electronic and optical properties. We have recently reported the design of new heterocyle-based push-pull dyes[9,10] and prepared new NLO-phores with large second-order NLO[11] and frequency-upconverted lasing activities.[5]

In the past years we have undergone a systematic study aimed at elucidating and quantitatively rank the electron-acceptor properties of π-deficient heteroaromatics rings, including azines, azoles, triazoles, and purines.[12] In this way we have experimentally ascertained that heterocyclic azines, such as pyridine and quinoline, and 1,3-azoles, such as benzothiazole, are strong acceptor groups. On the other hand, the widely known π-excessive nature of five-membered monoheterocycles (furan, thiophene, pyrrole),[8] as established by their σ_R values and oxidation potential,[13] makes these systems particularly suitable as tunable donor moieties. Pyrrole is the stronger π-electrondonor group among these rings. Our method of choice was NMR multinuclear analysis of α-heterocyclic carbanions and applications of ^{13}C (Eq. 1)[14] and ^{15}N (Eq. 2)[15] shift/π-charge relationships for experimentally mapping π-electron distribution and ranking the resonance electron-withdrawing power.

$$\Delta\delta^{13}\text{C} = -160\ \Delta q^{\pi}_{\text{C}} \quad (1)$$
$$\Delta\delta^{15}\text{N} = -366\ \Delta q^{\pi}_{\text{N}} \quad (2)$$

Relationships (2) and (3) correlate the shift variation for each carbon or nitrogen site with that of the π-electron density between two systems (e.g. a sample and a reference compound), provided that such atoms have the same hybridization in the two sites (each for system) that are to be compared. In this manner, we can quantitatively determine the effect of different acceptor and donating heterocyclic groups on the π-system of a conjugated chromophore and correlate the nature of the heteroaromatic rings with the π-electron structure and, eventually, with nonlinear optical properties.

We herein report the design and synthesis of two series of novel heterocycle-based TPA dyes, namely belonging to the general structure A-π-D-π-A and D-π-A-π-D, respectively. We then describe the characterization of their π-electron structure as obtained by multinuclear NMR analysis and linear optical (absorption and emission) characterization. Finally, two-photon absorption properties and TPA-based optical power limiting activities are shown.

2. Results and Discussion

2.1. DESIGN AND SYNTHESIS OF CHROMOPHORES.

Starting from the acceptor and donor properties of simple 6- and 5-membered heterocycles, as outlined above, we designed two series of systems: a) symmetric and asymmetric quadrupolar chromophores of the type A-π-D-π-A or A-π-D-π-A’, respectively (Scheme 1); 2) symmetric quadrupolar chromophores of the type D-π-A-π-D (Scheme 2). In the latter series, the weak donor properties of thiophene were increased by using the 3,4-ethylenedioxythiophene (EDOT) ring.

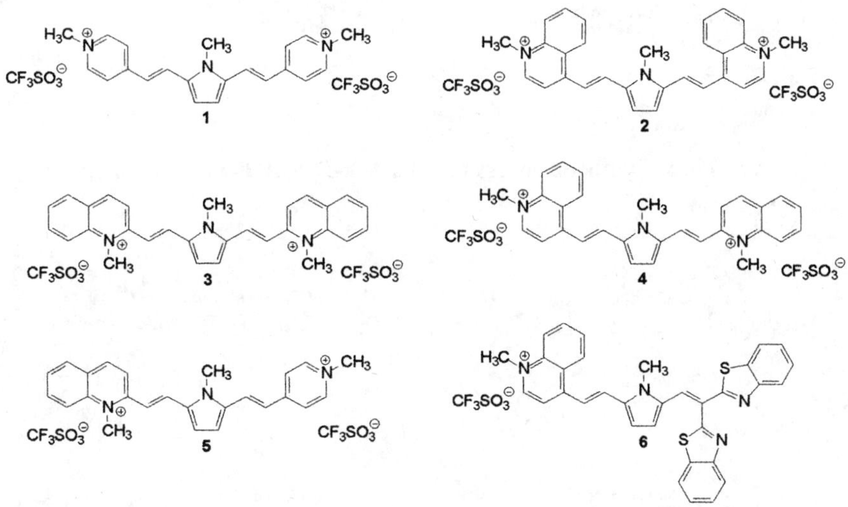

Scheme 1. Quadrupolar heterocycle-based dyes of the type A-π-D-π-A or A-π-D-π-A'.

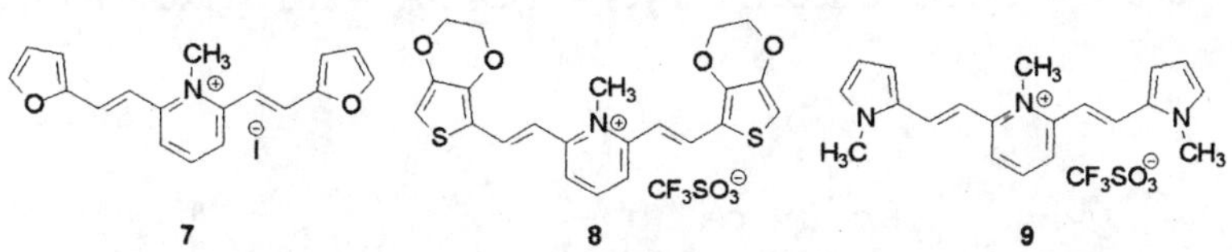

Scheme 2. Quadrupolar heterocycle-based dyes of the type D-π-A-π-D.

The synthetic access to symmetric A-π-D-π-A dyes **1-3** was quite general and involved, in the first step, the Knoevenagel-type condensation of an azinium salt with *N*-methyl-2-pyrrolecarbaldehyde, subsequent Vilsmeier formylation to give the corresponding A-π-D aldehyde, and finally condensation with a second equivalent of the azinium salt (Scheme 3). Unfortunately, this protocol could not be applied to access asymmetric compounds **4** and **5** due to the formation of a complex mixture of symmetric and asymmetric products, the isolation of which is not viable because of their saline nature. We therefore designed the indirect route described in Scheme 4. Conversely, the asymmetric compound **6** was obtained by condensation in the last step with bis-2-benzothiazolylmethane. In this case, the poor solubility of **6** in the reaction solvent allowed the isolation of the desired compound in good yields from the mixture of products.

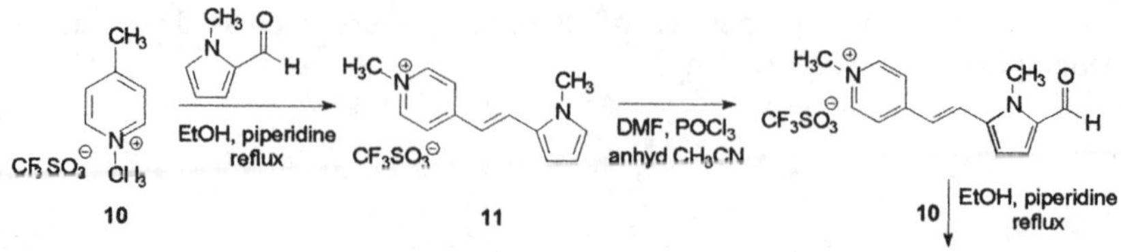

Scheme 3. Synthetic access to dyes A-π-D-π-A **1-3** (exemplified for **1**).

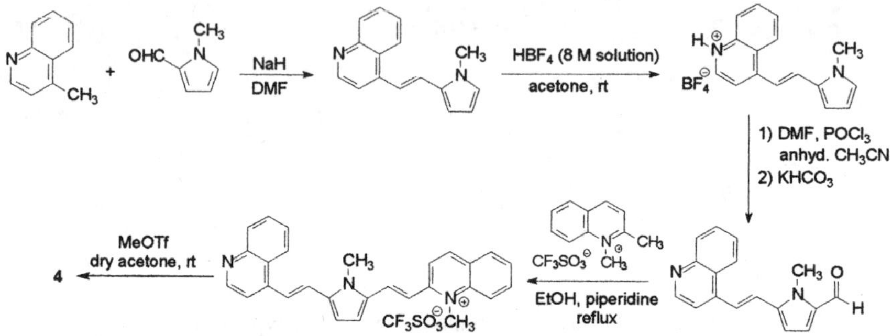

Scheme 4. Synthetic access to dyes A-π-D-π-A' **4** and **5** (exemplified for **4**).

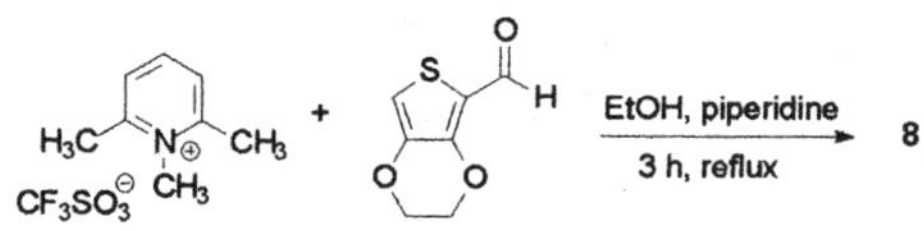

Scheme 5. Synthetic access to dyes D-π-A-π-D **7-9** (exemplified for **8**).

2.2. MULTINUCLEAR NMR ANALYSIS OF THE π-ELECTRON STRUCTURE.

We used ^{13}C and ^{15}N NMR spectroscopy and applied chemical-shift/π-electron-density relationships (1) and (2) in order to experimentally investigate, in a quantitative manner, the π-electron distribution of the ground state of TPA dyes in solution (DMSO) and quantitatively rank, in terms of intramolecular π-charge transfer, donor and acceptor strength of heterocyclic moieties. In this way, a correlation between π-electron ground state properties and NLO response can be drawn and used for the design of new optimized TPA dyes.

We show, as an example, the multinuclear NMR investigation applied to symmetric A-π-D-π-A dyes **1-3** and to the asymmetric A-π-D-π-A' dye **6**. Electron distribution of the π-framework of TPA dyes is shown in terms of site-by-site variation of π-electron density with respect to the reference system **11**, the π-electron structure of which was previously investigated in detail.[10b] Dipolar chromophore **11** was also shown to possess a very large two-photon pumped cavity lasing activity.[5] The ^{13}C and ^{15}N NMR shifts of molecules **1** – **3** and **6**, together with those of the reference system **11**, are collected in Table I. Table II shows relative π-electron distributions with respect to the reference system.

TABLE I. $^{13}C^a$ and $^{15}N^b$ NMR shifts (ppm) of compounds **1-3**, **6**, and **11** in DMSO.

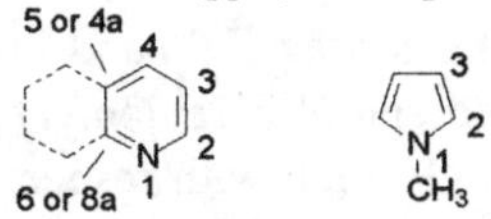

compd	azine ring positions 1	2	3	4	5 or 4a [c]	6 or 8a [d]	bridge positions α [e]	β [e]	pyrrole ring positions 1	2	3
11	188	144.30	122.10	153.33	122.10	144.30	117.40	129.50	154	130.40	109.90
1	190	144.52	122.67	152.46	122.67	144.52	121.18	128.04	154	135.56	113.47
2	179	147.08	115.61	151.92	125.92	138.84	117.86	129.96	156	137.21	114.97
3	173	155.36	120.77	142.74	127.48	139.25	117.21	133.51	157	137.17	116.31
6 [f]	-	147.00	115.12	151.98	125.88	138.80	117.69	130.05	-	136.02	114.53

[a] Relative to Me_4Si (0.0 ppm). [b] Relative to liquid NH_3 (0.0 ppm), 380.23 from neat nitromethane. [c] Position 5 of a pyridine ring or 4a of a quinoline ring. [d] Position 6 of a pyridine ring or 8a of a quinoline ring. [e] Positions relative to the azinium ring. [f] ^{15}N NMR shifts could not be recorded due to the poor solubility of **6** in DMSO.

TABLE II. Variation of the local π-electron densities Δq^{π} (millielectrons) for each *i*th position on going from the reference compound **11** to the cromophores **1-3** and **6**.[a,b]

compd	azine ring positions 1	2	3	4	5 or 4a [c]	6 or 8a [d]	bridge α [e]	β [e]	pyrrole ring positions 1	2	3
1	-5	-1	-4	5	-4	-1	-24	9	0	-32	-22
2	25	-17	41	9	-24	34	-3	-3	-5	-43	-32
3	41	-69	8	66	-34	32	1	-25	-7	-42	-40
6	-	-17	44	8	-24	34	-2	-3	-	-35	-29

[a] Positive values correspond to an increment of π-electron density. [b] Obtained from Eqs. (1) and (2) for carbon and nitrogen atoms, respectively. [c] Position 5 of a pyridine ring or 4a of a quinoline ring. [d] Position 6 of a pyridine ring or 8a of a quinoline ring. [e] Positions relative to the azinium ring.

We can use the variation of the π-electron density on the pyrrole ring, present as a median donor group in all of the four dyes, to compare the effect of different acceptor terminal azinium moieties on the π-system. Replacement of the 4-pyridinium with the 4-quinolinium group on going from **1** to **2** (same linkage position to the ethenylic spacer) resulted in a decrease of a total of 47 π-millielectrons on the pyrrole ring. The quinoline ring is therefore a stronger π-acceptor group, consistently with what we have previously found in our general investigation on π-deficient heteroaromatics.[12] Comparison between dyes **2** and **3** shows that the 2-quinolinium is slightly more electron-withdrawing than the 4-quinolinium substituent, again in agreement with our previous findings on 2- and 4-pyridine.[12] Finally, in system **6** we look at the effect of replacing a *N*-alkyl azinium with a not quaternized azine group. Although the electron-withdrawing capacity is decreased, as expected, the presence of two not quaternized benzothiazolyl rings is sufficient to provide **6** with the same total electron-withdrawing capacity of the terminal groups in azinium derivatives **1-3**. This result is important in view of the fact that a non-saline nature of the terminal groups and the presence of the azine as a free base can be used to build structures (including nanostructures) by means of known techniques. In addition, the coordinating properties of the bis-benzothiazolylmethyl fragment towards transition metal ions[16] can be exploited to prepare hybrid organic-inorganic TPA systems.

2.3. LINEAR OPTICAL CHARACTERIZATION.

All of the new compounds present a strong CT band in the visible region of the spectrum and are transparent in the visible red and NIR region. The CT band of D-π-A-π-D dyes **7** – **9** is hypsochromically shifted by ca. 100 nm with respect to the A-π-D-π-A counterpart. In addition, they all show a fluorescence emission, which is needed in TPA-based important applications such as fluorescence imaging[1] and upconverted lasing.[5] Table III collects absorption and emission data relative to the charge transfer (CT) band in DMSO for molecules **1** - **3** and **6** - **9**. As an example, the linear absorption and fluorescence spectrum of **6** in DMSO is shown in Figure 1.

TABLE III. Linear absorption and emission properties of compounds **1-4** and **6-9**.

Compd	λ_{max}(abs) [nm] [a]	λ_{max}(em)[nm] [a]	Φ [a,b]	ε [$M^{-1}cm^{-1}$] [c]
1	524	620	0.14	72100
2	603	701	-	75000
3	569	657	-	-
4	556	688	-	-
6	559	669	0.004[c]	51400
7	414	518	-	41400
8	436	536	-	35100
9	471	556	-	48100

[a] In DMSO. [b] Fluorescence quantum yield determined relative to 9,10-diphenylanthracene (Φ = 0.90 in cyclohexane). [c] Extinction coefficient.

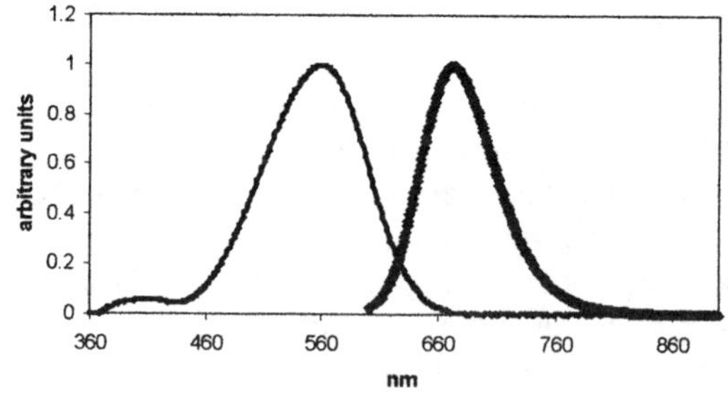

Figure 1. Linear absorption and emission spectra of compound **6** in DMSO.

2.4. TWO-PHOTON ABSORPTION PROPERTIES.

TPA cross sections of the new dyes were measured by open aperture Z-scan experiments performed with a femtosecond (fs) laser source. Open aperture Z-scan allows one to measure the nonlinear absorption of the samples by recording the transmittance as a function of the input intensity. To this end, the sample position is moved along the propagation direction (z-axis) of a focused beam, the energy of which is kept fixed, while the intensity increases by approaching the focal plane. Far from the focus position only the linear transmittance contributes to the signal, while in the vicinity of the focus the intensity increases very rapidly and the non linear absorption process generates a dip in the transmittance. The dip is symmetric with respect to the focal plane position.

The fs-laser measurement is able to discriminate between coherent two photon processes and multiphoton sequential absorption implying excited states populated by non-radiative processes. The latter does not contribute to the measured signal when fast (~100 fs) pulses are used, since most non-radiative processes occur on a picosecond (ps) or nanosecond (ns) timescale. Therefore, fs data provide a more accurate evaluation of the true TPA activity. Figure 2 shows the Z-scan data of dye **1** in DMSO. The normalized transmittance (i.e. $I(z)/I(\infty)$, with $I(\infty)$ the linearly transmitted intensity far from the focal plane) is reported as a function of the sample position (z). The normalized transmittance of the solvent is flat, indicating that nonlinear absorption from the solvent does not occur at 0.35 μJ, the upper limit of used pulse energies. Conversely, system **1** shows a deep dip typical of nonlinear absorption.

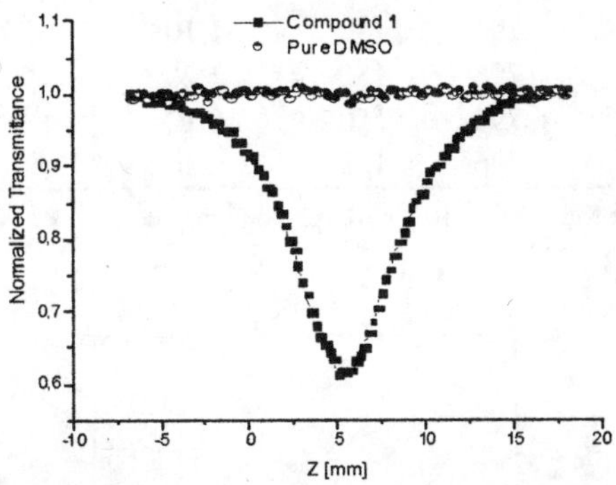

Figure 2. Z-scan experimental data of compound **1** in DMSO ($3x10^{-2}$ M) (squares) and of pure DMSO (circles) in a 1-mm cell, using an input intensity laser beam of 0.35 μJ.

Table IV reports the fs TPA coefficients of compounds **1** and **6** – **9** compared to those of the most efficient reported systems of comparable molecular weight measured via fs nonlinear transmission or Z-scan methods. It can be easily seen that some of the novel chromophores overcome by nearly one order of magnitude the TPA performances of AF50 and AF250, which are commonly used as TPA benchmarks.[17] The TPA coefficient of the *N,N*-diphenylamine-dithienothiophene derivative TP-DTT-TP is larger[6c] but was obtained close to the maximum TPA wavelength.[18] In fact, it should be considered that the our TPA measurements were carried out at the non-optimized 790-nm wavelength position. Indeed, recent results demonstrate that the 790-nm laser source is far away from the peak position of the TPA spectrum of **1** and that a significant enhancement of the σ value is expected at optimized parameters. Finally, we compare the performances of our dyes to that of a branched system, PRL-701,[6b] which exhibits one of the largest σ value so far known in the fs regime. Although our systems well compare to PRL-701, it should be noted that the high value of the latter originates from its trimeric structure, where a cooperative effect enhances σ by more than six times with respect to the 'monomer' system (σ = 87 GM). Multi-branched structures based on heteroaromatic donor and acceptor units have been developed in our laboratories and will be reported in due course. As an example of TPA-based applications, we show in Figure 3 the optical limiting properties of chromophore **1** obtained in the fs time scale.

TABLE IV. TPA cross section coefficient of compounds **1** and **6-9** in DMSO and comparison with literature data.

compd	Pulse laser		β	σ
	λ [nm]	FWHM [fs]	[cm^4/GW] ×10^{21}	[GM]^a
1	790	150	0.08	119
6	790	150	0.02	150
7	790	150	0.01	10
8	790	150	0.03	40
9	790	150	0.07	103
AF50[17]	796	150	1.10	22
AF250[17]	796	150	1.20	30
TP-DTT-TP[6c]	796	160	0.12	270
PRL-701[6b]	796	173	2.4	600

^{a} 1 GM (Goppert-Mayer) = 1x10^{-50} cm^4 s photon^{-1} molecule^{-1}.

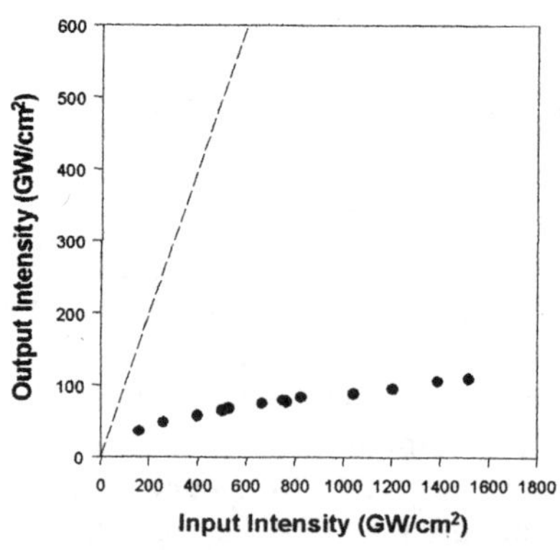

Figure 3. Optical power limiting properties of **1** in DMSO for a fs laser source. The straight line corresponds to the linear transmission of the solution.

In conclusion, the novel heterocycle-based TPA dyes presented in this work are high-ranked among the most important TPA systems so far known. The expected[6,17] much larger ns multiphoton absorption, due to an additional contribution of the linear absorption of the two-photon allowed excited state, makes these chromophores very attractive for a number of applications operating in the ns regime.

3. Acknowledgment

We thank Prof. G. Musumarra, University of Catania, Italy, for providing us with a sample of compound 7. This work was supported by grants from CNR-Progetto Finalizzato MSTA II and MIUR (Grant 2001034442).

4. References

1. Denk, W., Strickler, J. H., and Webb, W. W. (1990) *Science* **248**, 73-76.
2. Ehrlich, J. E., Wu, X. L., Lee, L. Y. S., Hu, Z. Y., Rockel, H., Marder, S. R., and Perry, J. W. (1997) *Opt. Lett.* **22**, 1843-1845.
3. Day, D., Gu, M., and Smallridge, A. (1999) *Opt. Lett.* **24**, 948-950.
4. Zhou, W., Kuebler, S. M., Braun, K. L., Yu, T., Cammack, J. K., Ober, C. K., Perry, J. W., and Marder, S. R. (2002) *Science* **296**, 1106-1109.
5. Abbotto, A., Beverina, L., Bozio, R., Bradamante, S., Ferrante, C., Pagani, G. A., and Signorini, R. (2000) *Adv. Mater.* **12**, 1963-1967.
6. (a) Albota, M., Beljonne, D., Bredas, J. L., Ehrlich, J. E., Fu, J. Y., Heikal, A. A., Hess, S. E., Kogej, T., Levin, M. D., Marder, S. R., Mccordmaughon, D., Perry, J. W., Rockel, H., Rumi, M., Subramaniam, C., Webb, W. W., Wu, I. L., and Xu, C. (1998) *Science* **281**, 1653-1656. (b) Chung, S. J., Kim, K. S., Lin, T. H., He, G. S., Swiatkiewicz, J., and Prasad, P. N. (1999) *J. Phys. Chem. B* **103**, 10741-10745. (c) Kim, O. K., Lee, K. S., Woo, H. Y., Kim, K. S., He, G. S., Swiatkiewicz, J., and Prasad, P. N. (2000) *Chem. Mater.* **12**, 284-286. (d) Drobizhev, M., Karotki, A., Rebane, A., and Spangler, C. W. (2001) *Opt. Lett.* **26**, 1081-1083. (e) Mongin, O., Porres, L., Moreaux, L., Mertz, J., and Blanchard-Desce, M. (2002) *Org. Lett.* **4**, 719-722.
7. Zojer, E., Beljonne, D., Kogej, T., Vogel, H., Marder, S. R., Perry, J. W., and Bredas, J. L. *J.* (2002) *Chem. Phys.* **116**, 3646-3658.
8. Katritzky, A. R. (1983) *Handbook of Heterocyclic Chemistry*, Pergamon Press, Oxford.
9. Bradamante, S., Facchetti, A., and Pagani, G. A. (1997) *J. Phys. Org. Chem.* **10**, 514-524.
10. Abbotto, A., Bradamante, S., and Pagani, G. A. (2001) *J. Org. Chem.* **66**, 8883-8892.
11. (a) Abbotto, A., Bradamante, S., Facchetti, A., and Pagani, G. A. (1997) *J. Org. Chem.* **62**, 5755-5765. (b) Abbotto, A., Bradamante, S., Facchetti, A., Pagani, G. A., Ledoux, I., and Zyss, *J.* (1998) *Mat. Res. Soc. Symp. Proceed.* **488**, 819-822.
12. Abbotto, A., Bradamante, S., Facchetti, A., and Pagani, G. A. (1999) *J. Org. Chem.* **64**, 6756-6763, and references cited therein.
13. Zotti, G., Schiavon, G., Berlin, A., and Pagani, G. (1991) *Synth. Met.* **40**, 299.
14. Bradamante, S., Pagani, G. A. (1984) *J. Org. Chem.* **49**, 2863-2870.
15. Abbotto, A., Bradamante, S., and Pagani, G. A. (1993) *J. Org. Chem.* **58**, 449-455.
16. Abbotto, A., Bradamante, S., Facchetti, A., and Pagani, G. A. (2002) *J. Org. Chem.* **67**, 5753-5772.
17. Swiatkiewicz, J., Prasad, P. N., and Reinhardt, B. A. (1998) *Opt. Commun.* **157**, 135-138.
18. Lee, K. -S., Lee, J. -H., Kim, K. -S., Woo, H. -Y., Kim, O. -K., Choi, H., Cha, M., He, G. S., Swiatkiewcz, J., Prasad, P. N., Chung, M. -A., and Jung, S. -D. (2001) *Nonlinear Opt.* **27**, 87-100.

3D CHARACTERIZATION OF MOLECULAR PHOTO-ORIENTATION: APPLICATION TO ALL-OPTICAL POLING

Ngan NGUYEN THI KIM and Michel DUMONT
Laboratoire de Photonique Quantique et Moléculaire (CNRS, UMR8537)
ENS de Cachan, 61 Av. Président Wilson, 94235 Cachan cedex, France

Abstract

The symmetry of molecular angular distributions obtained by All-Optical Poling (AOP) has been widely studied, but not the equilibrium between the different tensorial orders. Particularly the photoinduced anisotropy (quadrupolar contribution) is in competition with the non-centrosymmetric contributions (dipolar and octupolar) responsible for the generation of $\chi^{(2)}$. For that reason, we developed theoretical and experimental tools for studying the dynamics of the 3D orientational order, by measuring simultaneously the SHG and the trichroism at several wavelengths. We also introduced a third excitation beam, to obtain a 3D pumping.

1. Introduction

Light is a powerful tool for the micro structuration of initially isotropic and homogeneous materials, such as functionalized polymers. Focussing light, or illuminating with an interference pattern, permits a spatial organization of optical properties (bulk modulation of absorption and of the index of refractive, or thickness modulation of thin films), while the polarization of pumping beams induces an orientational order of active molecules and produces unusual symmetries of the susceptibility tensor. In this paper, we focus our attention on photo-orientation of optically active molecules in amorphous polymer films, submitted to a spatially uniform illumination.

Many materials become anisotropic (birefringent and dichroic), when they are illuminated by a linearly polarized light beam. In some cases, a strong laser excitation, in the transparency spectral region of the material, induces a positive anisotropy (the index increases in the direction of light polarization), because of the optical Kerr effect torque, which align anisotropic molecules, parallel to the electric field of light [1]. Often, in the case of photochromic molecules, a weak illumination in the absorption band produces a strong negative anisotropy. It is the Photo-Induced Anisotropy (PIA), or Weigert effect, which is known for a long time [2-4] and which have been explained by a two step mechanism. The first mechanism is a selective optical pumping, or angular hole burning (AHB): the polarized pump beam produces an anisotropic depletion of the

F. Charra et al. (eds.), Organic Nanophotonics, 395–404.

ground state, according to a $\cos^2\theta$ law (θ is the angle between the transition dipole of molecules and the polarization of light). The second mechanism is the angular redistribution (AR): molecules achieve a small random rotation during the excitation and relaxation cycle. After several cycles, molecules are accumulated perpendicularly to the polarization of light (negative anisotropy). A good PIA efficiency requires a reversible photoinduced structural modification of molecules, which generates the rotation and which permits many cycles. Azodyes are the most efficient molecules, because of the reversible trans-cis photoisomerization of the N=N double bound.

AHB and AR lead to an angular distribution which has the symmetry of the pumping light, but with the opposite sign. This is true whatever is the geometry of excitation. Particularly interesting is the All-Optical Poling method (AOP), for creating a non-centrosymmetric distribution and consequently $\chi^{(2)}$ non linear optical properties [5, 6]. The material (resonant at 2ω) is coherently pumped by the fundamental frequency (ω) and the second harmonic of a strong pulsed laser. The interference of these two beams excites molecules, with a probability which is not centrosymmetric. For instance, rod-like molecules, pumped by linearly polarized beams, are excited with a probability proportional to $\mathscr{E}_{2\omega}^{*}\mathscr{E}_{\omega}^{2}\cos^3\theta$. Beside this non-centrosymmetric term, two centrosymmetric anisotropic pumping terms are resonant: the one-photon absorption at 2ω ($|\mathscr{E}_{2\omega}^{2}|\cos^2\theta$) and the two-photon absorption at ω ($|\mathscr{E}_{\omega}^{4}|\cos^4\theta$). With other polarization configurations a more sophisticated tensorial order can be produced, for example, quadrupolar order can be induced by contra-rotating circular polarizations [7, 8].

The efficiency of AOP is limited by PIA, which occurs simultaneously, particularly because of the two centrosymmetric terms. The greatest part of molecules is accumulated in a symmetric arrangement, perpendicularly to the polarization of beams. The equilibrium between the two processes depends on the relative power of both beams, but also on the symmetry of molecules (for instance, there is no PIA effect for pure octupolar molecules, without quadrupole) and on the different dynamical parameters (lifetimes, angular diffusion), which determine the saturation of the pumping cycle. For that reason, we are developing theoretical and experimental tools in order to study the dynamics of the 3D orientational order for different types of molecules and for different pumping beams configurations. Another interesting idea is to pump molecules with a third beam polarized perpendicularly to the polarization of the AOP beams [9].

2. Modelization

In the case of cigar-like molecules pumped with an axial symmetry, PIA (also photoassisted electrical poling) has been described by a theoretical model and its transient behavior has been studied [3, 4, 10, 11]. The model has been generalized to AOP and to any kind of symmetry, in the basis of irreducible tensors [8, 12], but, the complexity of the problem forbids a detailed calculation of transients. Here we recall this model and we simply give the asymptotic angular distribution ($t=\infty$), for 3D pumping of cigar-like molecules. The angular distribution of the trans state, $n_T(\Omega) = n_T(\varphi, \theta, \psi)$, obeys to the general equation:

$$\frac{dn_T(\Omega)}{dt} = -Pr_{TC}(\Omega)\, n_T(\Omega) + \int R^{P}_{CT}(\Omega'\to\Omega)\, Pr_{CT}(\Omega')\, n_C(\Omega')d\Omega' + \frac{1}{\tau_c}\int R^{R}_{CT}(\Omega'\to\Omega)\, n_C(\Omega')d\Omega' + \left(\frac{dn_T(\Omega)}{dt}\right)_{Diff} \tag{1}$$

The cis distribution, $n_C(\Omega)$, obeys to a similar equation. Pr_{TC} (Pr_{CT}) is the probability of trans→cis (cis→trans) photoisomerization. $R_{AB}(\Omega'\to\Omega)$ is the probability of rotation from Ω' to Ω during the A→B transition (superscript P or R refer to photoinduced or to spontaneous isomerization). The last term in (1) is the thermal orientational diffusion. Assuming that molecules are resonant at 2ω, the AOP probability of optical pumping contains three components: one-photon absorption at 2ω, two-photon absorption at ω and the interference term. Each term is expressed as the scalar product of the imaginary part of a polarizability tensor ($\boldsymbol{\alpha}''_{\Omega}$, $\boldsymbol{\beta}''_{\Omega}$, $\boldsymbol{\gamma}''_{\Omega}$) by a field tensor of the same rank, built as a tensorial product of the complex amplitude of fields, $\mathscr{E}_{\omega}$ and $\mathscr{E}_{2\omega}$:

$$\begin{aligned}
Pr(\Omega) \propto\ & 2\boldsymbol{\alpha}''_{\Omega}(-2\omega;2\omega)\bullet\Re e\left(\vec{\mathscr{E}}^{\,*}_{2\omega}\otimes\vec{\mathscr{E}}_{2\omega}\right)+\left(2\boldsymbol{\beta}''_{\Omega}(-2\omega;\omega,\omega)+\boldsymbol{\beta}''^{*}_{\Omega}(-\omega;-\omega,2\omega)\right)\\
&\bullet\Re e\left(\vec{\mathscr{E}}^{\,*}_{2\omega}\otimes\vec{\mathscr{E}}_{\omega}\otimes\vec{\mathscr{E}}_{\omega}\right)+\boldsymbol{\gamma}''_{\Omega}(-\omega;-\omega,\omega,\omega)\bullet\Re e\left(\vec{\mathscr{E}}^{\,*}_{\omega}\otimes\vec{\mathscr{E}}^{\,*}_{\omega}\otimes\vec{\mathscr{E}}_{\omega}\otimes\vec{\mathscr{E}}_{\omega}\right)\\
=\ & \sum_{Kqq'}(-1)^{q}\, D^{K}_{q'q}(\Omega)\left[\boldsymbol{\alpha}''^{K}_{q'}\,{}^{(2)}E^{K}_{-q}+\boldsymbol{\beta}''^{K}_{q'}\,{}^{(3)}E^{K}_{-q}+\boldsymbol{\gamma}''^{K}_{q'}\,{}^{(4)}E^{K}_{-q}\right]
\end{aligned} \tag{2}$$

$\boldsymbol{\alpha}''_{\Omega}$, $\boldsymbol{\beta}''_{\Omega}$ and $\boldsymbol{\gamma}''_{\Omega}$ are expressed in the laboratory frame: they depend on the orientation Ω of the molecule under consideration. In the last row, tensors have been written in the spherical irreducible representation and susceptibilities have been expressed in the molecular axes, with the help of the Wigner rotation operators $D^{K}_{q'q}(\Omega)$, according to Brasselet and Zyss [8]. If the Kleinman symmetry applies, K is 0 and 2 for α, 1 and 3 for β and 0, 2 and 4 for γ. Because of the optical resonance other components exist, such as $\beta^{K=2}$, but, if the writing field tensors, ${}^{(n)}E$, are symmetrical the non-symmetrical components of β are not excited. In this formalism, at the first order, $\chi^{(2)J}_{m}\propto[\boldsymbol{\beta}^{J}\bullet\boldsymbol{\beta}^{J}]\,{}^{(3)}E^{J}_{m}$. In order to take into account multi-cycles interaction (*i.e.* saturation of the optical pumping) the angular distributions could be develop on Wigner function [12], in the same way as they have been developed on Legendre polynomials, in the case of full axial symmetry [3, 4]. With this general formalism, the number of components and of parameters is so large that it should be very difficult to calculate the transient effects. Fortunately, with some further approximations (negligible angular diffusion in trans and isotropic pumping from cis to trans), equation (1) has a simple steady state solution:

$$n_T(\Omega)=\frac{N_C}{8\pi^2\tau_c\,Pr(\Omega)}\quad;\quad n_C(\Omega)=\frac{N_C}{8\pi^2}\;;\quad N_C=N\left(1+\frac{1}{\tau_c}\int\frac{d\Omega}{8\pi^2\,Pr(\Omega)}\right)^{-1} \tag{3}$$

Within this approximation, we have developed a computer program, which calculates all the Cartesian components of the susceptibility tensors, for any kind of molecules and any kind of optical excitation. Here we only show the results for 3D optical pumping of polar cigar like molecules, with the geometry of the experiment presented below, *i.e.* a X linearly polarized Yag laser and a YZ circularly polarized Ar laser (θ is the angle with the axis X):

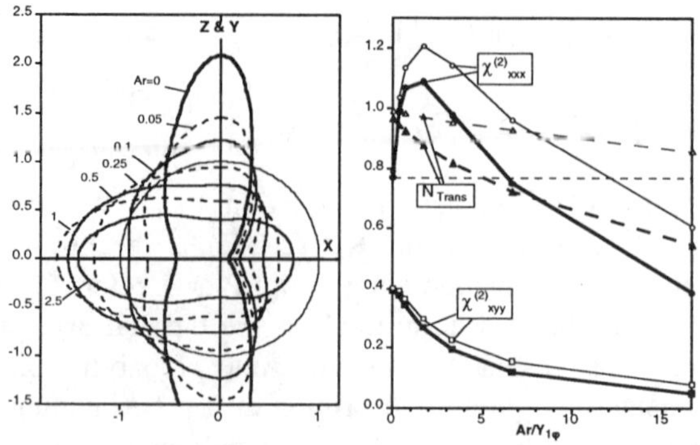

Figure1. Simulation of All-Optical Poling of cigar-like molecules, induced by an *X* linearly polarized Yag laser ($\omega + 2\omega$) and a *ZY* circularly polarized Argon laser, for $t = \infty$. Arbitrary parameters have been used in equation 4 ($Y_{1\varphi}$=0.15, $Y_{interf.}$=0.15, $Y_{2\varphi}$=0.10, e=0.7 i.e. $\sigma_{//} = 8\sigma_{\perp}$). *Left:* Angular distribution in trans state, as a function of the Ar laser intensity. *Right:* $\chi^{(2)}_{xxx}$, $\chi^{(2)}_{xyy}$ (arbitrary units) and the trans population N_T, as a function of the relative intensity of the Ar laser. Bold curves correspond to the angular distribution of the left graph. Tiny curves are obtained by dividing all pumping terms by 5.

$$\Pr(\theta) \propto Y_{1\varphi}\left(\frac{1-e}{3} + e\cos^2\theta\right) + Y_{interf}\cos^3\theta + Y_{2\varphi}\cos^4\theta + Ar\left(\frac{1-e}{3} + e\frac{\sin^2\theta}{2}\right) \quad (4)$$

In fig. 1, we chose arbitrary parameters, only for demonstration. The non-centro-symmetric term (Y_{interf}) is probably overestimated. *Ar* represents the argon laser intensity. The ellipticity of molecules, $e = (\sigma_{//} - \sigma_{\perp})/(\sigma_{//} + 2\sigma_{\perp})$, has a critical effect on the angular distribution. Fig. 1 also shows the variations of N_T, $\chi^{(2)}_{xxx} \propto \int n_T \cos^3\theta \, d\Omega$ and $\chi^{(2)}_{xyy} \propto \int \frac{1}{2} n_T \cos\theta \sin^2\theta \, d\Omega$, versus the relative intensity of the Ar beam. When Ar increases, $\chi^{(2)}_{xxx}$ first grows (pumping of molecules in YZ plane) but reaches a maximum and then decreases because of the reduction of the total population N_T. Reducing the intensity of all beams increases the asymptotic $\chi^{(2)}$. In [12] the angular distribution for octupolar AOP is shown.

3. Experimental Setup

The sample is a polymer film, spin-coated on a microscope glass slide. In order to allow multidirectional pumping and probing, the sample is sandwiched between two half-cube silica prisms (fig. 2). Nujol mineral oil is used as an index matching liquid. The sample can be excited simultaneously by an argon laser beam and by collinear 1064 and 532 nm beams from a pulsed Nd-Yag laser. Taking into account the index of refraction of the polymer, the incidence angles of both beams are calculated so that they cross at right angle in the film, at 45° of the normal to the sample. With the MMA-DR1 copolymer, this cannot be absolutely perfect, since the dispersion is quite important between 1064 and 532 nm. The external incidence angles are of the order 15°. For the experiments presented hereafter, both Yag beams (propagating along Z, in the sample) are linearly polarized in the plane of incidence (X) while the Ar laser beam, propagating along X, is circularly polarized in the YZ plane. The setup allows another interesting configuration,

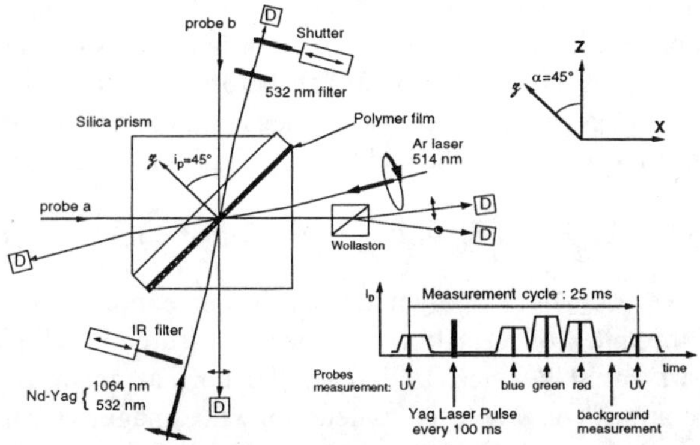

Figure 2. 3D Experimental setup for All-Optical Poling and anisotropy measurements.

for octupolar AOP: the Yag beams can be circularly polarized (co- or contra-rotating in XY plane) and the Ar beam linearly polarized, parallel to Z. In both configurations, X, Y and Z will be the principal axes of the susceptibility tensor. The index dispersion in the prism produces a fast rotation of the relative phase of ω and 2ω waves, so that there are several $\chi^{(2)}$ fringes in the beam diameters (in fact a $\chi^{(2)}$ hologram is recorded).

The probe beams *a* and *b* are chosen to propagate along X and Z, respectively, in the silica prisms, so that they are incident at 45° on the sample. With a linear analyzer, on probe *b*, and a Wollaston prism on probe *a*, absorption is measured for three polarization directions. The s polarization of probe *a* is rigorously Y. p polarizations of probe *b* and *a* are approximately X and Z, respectively (probes are not exactly orthogonal). The probe beams are produced by a Xe arc lamp and a monochromator with several entrance slits (up to 6). A rotating slit-wheel selects successively each chosen wavelength with a repetition rate of 25 ms [14]. The signal from each detector is integrated during 1 ms, for each wavelength. Except the Wollaston prism, all optical elements are made of fused silica, which allows near UV probing. After a few tens of seconds, when the variation rate of all signals slows down, the probe beams are switched on during a small fraction of time only, in order to avoid pumping by these beams.

The Yag laser has a repetition rate of 100 ms and is synchronized on the rotating wheel (fig. 2). The growing of $\chi^{(2)}$ is monitored by interrupting the 532 nm beam by an IR filter, for 1 s every 30 s, and by opening the shutter of the photomultiplier, which measures the second harmonic generated by the sample [6]. During the relaxation periods, the fundamental Yag beam is interrupted, except during short reading intervals.

4. The Optical Density at Large Incidence

For each probe beam (i = sa, pa or pb), the measured optical density is:

$$OD_i(t) = \log I_{0,i} - \log I_{T,i}(t) = 2h \log(e)\, \Im m\big(k_{z,i}(t)\big) \tag{5}$$

Where $I_{0,i}$ and $I_{T,i}$ are the incident and transmitted intensities. h is the thickness of the sample and $k_{z,i}$ is the component of the complex wave vector, perpendicular to the sample. With the above experimental geometry, two problems appear. The first one is that it is practically impossible to measure $I_{0,i}$, since it is very difficult to replace the sample by an uncoated glass plate, like it is easy to do in 2D ordinary experiments. If the complex index of refraction, $\tilde{n}_0$, of the isotropic material (before pumping) is known, it is possible to measure the thickness, by a previous measurement of the normal absorption spectrum and to calculate the k_z, for the incidence i_p inside the silica prism:

$$k_z = \frac{2\pi}{\lambda}\sqrt{\tilde{n}_0^2 - \beta^2} \qquad \text{with } \beta = n_{silica} \sin i_p \tag{6}$$

In the case of the MMA_DR1 copolymer, used in our experiments, the real and the imaginary part of the index of refraction have been measured by ellipsometry. In other cases, the real part could be measured from the normal absorption spectrum and Kramers-Kronig relations could be used to determine the imaginary part. Then, from (5) and (6), the value of the optical density of the isotropic film, under the incidence i_p, can be calculated and used to normalize the experimental measurements:

$$OD_i(t) = OD_{calc}(0) + \log I_{T,i}(0) - \log I_{T,i}(t) = 2h \log(e)\, \Im m\left(k_{z,i}(t)\right) \tag{7}$$

The second problem is to establish the relation between the variations of optical density and the anisotropy of the film. With $\beta_b = -\beta_a = \beta$, one can show that (j = a or b):

$$\begin{aligned} k_{z,sa} &= \frac{2\pi}{\lambda}\sqrt{\tilde{n}_Y^2 - \beta_a^2} \\ k_{z,pj} &= \frac{2\pi}{\lambda}\,\frac{\beta_j \cos\alpha \sin\alpha\left(\tilde{n}_X^2 - \tilde{n}_Z^2\right) + \tilde{n}_X\tilde{n}_Z\sqrt{\sin^2\alpha\, \tilde{n}_X^2 + \cos^2\alpha\, \tilde{n}_Z^2 - \beta_j^2}}{\sin^2\alpha\, \tilde{n}_X^2 + \cos^2\alpha\, \tilde{n}_Z^2} \end{aligned} \tag{8}$$

$\tilde{n}_X$, $\tilde{n}_Y$, and $\tilde{n}_Z$ are the three principal indices of the polymer. $\alpha = \{z,Z\}$ is the angle between the normal to the film and the propagation axis of the Yag beam. For a rigorous expression of the transmission it is also necessary to take into account the reflection losses, which are important for s polarization. In the present form of the experimental setup, $\alpha = i_p = 45°$. For a phenomenological interpretation of the preliminary results presented in the next paragraph, it is sufficient to use a first order development, by assuming a small anisotropy:

$$\tilde{n}_i \approx \tilde{n}_0\left(1 + \tilde{\delta}_i\right), \text{ with } \tilde{\delta}_i = \delta\chi_{ii}^{(1)}/\left(2\tilde{n}_0^2\right) \qquad (i = X, Y, Z) \tag{9}$$

$$\begin{aligned} k_{z,sa} &= \frac{2\pi}{\lambda}\sqrt{\tilde{n}_0^2 - \beta^2}\left(1 + \frac{\tilde{n}_0^2}{\tilde{n}_0^2 - \beta^2}\tilde{\delta}_Y\right) \\ k_{z,pj} &= \frac{2\pi}{\lambda}\left(\beta_j\left(\tilde{\delta}_X - \tilde{\delta}_Z\right) + \sqrt{\tilde{n}_0^2 - \beta^2}\left(1 + \frac{\tilde{n}_0^2}{\tilde{n}_0^2 - \beta^2}\,\frac{\tilde{\delta}_X + \tilde{\delta}_Z}{2}\right)\right) \end{aligned} \tag{10}$$

$$OD_m = OD_{p,a} + OD_{p,b} + OD_{s,a} \propto \Im m\left(\sqrt{\tilde{n}_0^2 - \beta^2} + \frac{\tilde{n}_0^2}{\sqrt{\tilde{n}_0^2 - \beta^2}}\left(\tilde{\delta}_X + \tilde{\delta}_Y + \tilde{\delta}_Z\right)\right) \tag{11}$$

$$\delta_{XZ}OD = OD_{p,b} - OD_{p,a} \propto \beta\, \Im m\left(\tilde{\delta}_X - \tilde{\delta}_Z\right) \tag{12}$$

As $\tilde{n}_0$ is complex, the variations of absorption do not give directly the variations of the imaginary part of $\chi^{(1)}$, but, within the limit of small anisotropy, they are linear functions of the components of $\delta\chi^{(1)}$. Therefore, they express correctly the modifications of the angular distribution of molecules. Let us notice that, if the imaginary part of $\tilde{n}_0$ is small and if $\beta^2 = n_0^2/2$ (propagation at 45°), $\delta OD_{p,b} \propto \Im m(\delta_X)$ and $\delta OD_{p,a} \propto \Im m(\delta_Z)$

5. Experimental Results

5.1 . TWO-PHOTON INDUCED ANISOTROPY

Before beginning AOP experiments, it is interesting to show the anisotropy induced separately by each of the three pump beams. Experimental curves presented in figure 3 have been recorded in a classical 2D experiment (normal pumping and probing, without prisms [14]), with a DR1 doped PMMA film (10% w). In all cases, the pump is X polarized, so that one can assume that $OD_Z = OD_Y$. It appears that the photoinduced anisotropy has a similar behavior with two-photon and one-photon absorption. The numerical comparison of the three experiments is difficult, because in the present state of the setup, the power densities are not measured with enough precision. Nevertheless, a careful observation shows that, within the experimental uncertainty, the ratio of the slopes of OD_X and OD_Y, at the beginning of pumping, is compatible with the values calculated from the hole burning theory, for one-photon absorption ($\delta n_T \propto \cos^2\theta \rightarrow$ ratio = 3) and for two-photon absorption ($\delta n_T \propto \cos^4\theta \rightarrow$ ratio = 5).

In all cases, in the beginning of pumping, the anisotropy is positive at 374 nm (where $\sigma_T \approx 0.28\,\sigma_C$ [13]), since cis molecules are oriented parallel to the pump polarization, by the AHB process (a bump in cis, complementary to the hole in trans). After

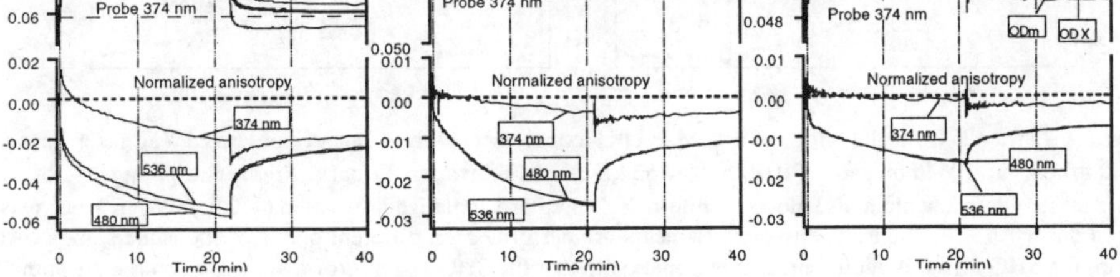

Figure 3. Photoinduced anisotropy in a DR1-doped PMMA film. In the left part, the pump is an argon laser (3mW/cm²,), in the right part, a Nd-Yag laser (100ps pulses, 10 Hz, ≈ 1.5 mJ, ∅ ≈ 0.4mm) and in the center, the second harmonic of the same laser (≈ 3 μJ, ∅ ≈ 0.6mm). In the three cases, the pump polarization is X and the optical density is recorded simultaneously with X and Y polarizations, at 374 nm ($\sigma_T \approx 0.28\,\sigma_C$) and at 480 nm ($\sigma_T \approx 2.2\,\sigma_C$). The variations of $OD_m = (OD_X + 2\,OD_Y)/3$ are proportional to the population of cis. The normalized anisotropy is $(OD_X - OO_Y)/(3OD_{m(t=0)})$.

some time, the anisotropy changes of sign, because the large increase of the alignment of trans, by AR, dominates the decreasing anisotropy of cis. When the pump is switched off, the anisotropies, at 374 nm and that at 480 nm ($\sigma_T \approx 2.2\,\sigma_C$), show a complementary fast variation, corresponding to the cis ›trans relaxation of positively aligned molecules. We never observed that in the past, because we always pumped up to saturation, where anisotropy of cis is destroyed, according to equation (3).

5.2. 3D ALL-OPTICAL POLING

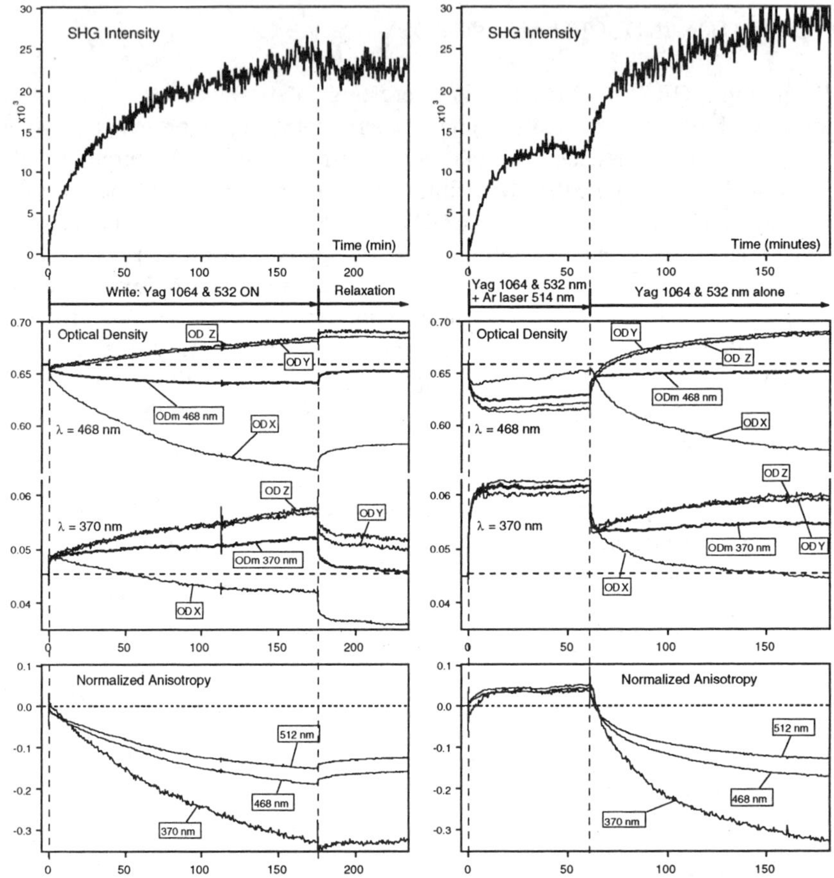

Figure 4. ***Left:*** All-Optical Poling, of a MMA_DR1 copolymer, with X linearly polarized Yag laser beams (≈ 200 mJ/cm^2 at 1064 nm and ≈ 30 µJ/cm^2 at 532 nm) and relaxation in dark. ***Right:*** the poling by the X polarized beams is maintained 3 hours, while a YZ circularly polarized Ar laser (≈ 0.5 mW/cm^2) beam is applied during the first hour. The two experiments are performed on different points of the same sample. At the top, the SHG signal is represented: it is remarkable that the SHG signal grows higher, when the Ar pumping has been applied. Below, the optical density, measured with the three probe beams, is shown for two wavelengths. OD_Y is obtained with the s polarization of probe *a* and rigorously corresponds to the Y polarization. OD_X and OD_Z, from the p polarization of *b* and *a* probes, respectively, correspond only approximately to X and Y polarizations. The variations of $OD_m=(OD_X+OD_Y+OD_Z)/3$ are proportional to the population of the cis state. All optical densities have been normalized according to (7). The lowest graphs show the normalized anisotropy, $[OD_X-(OD_Y+OD_Z)/2]/OD_{m(t=0)}$.

Fig. 4 shows one of the first simultaneous records of SHG and 3D anisotropy obtained with the AOP setup shown in fig. 2. The sample is a copolymer of methyl-methacrylate (MMA 70%) and MMA-DR1 (30 %). On the left part of Fig. 4 (without Ar laser), one can observe that the dichroism grows like in usual photoinduced anisotropy experiments: as pump beams are X polarized, OD_X decreases, while OD_Y and OD_Z increase slowly after a short decrease. As it is well known, this behavior corresponds to the hole burning process, during the first seconds and to the angular redistribution process, subsequently. Beside a possible experimental uncertainty, the small difference between OD_Y and OD_Z could be due to the fact that OD_Z differ slightly from a Z polarized probe (Cf. §4). The SHG signal grows slowly, like the anisotropy, while the populations of cis and trans states rapidly reaches a steady state (OD_m). This is the proof (probably the first direct one) that the AOP of DR1 results from the rotation of molecules due to the redistribution process, and not only from hole burning.

The right part of fig. 4 represents an AOP experiment with a 3D pumping. The Yag laser beams have the same polarization and the same pulse energy, as on the left part of fig. 4. The Ar laser beam is YZ circularly polarized. The anisotropy from the Ar laser is slightly dominant over that from the Yag laser: this should be favorable for a large non-centrosymmetry. Nevertheless, after one hour, the SHG signal is significantly smaller than that obtained with the Yag laser alone (but, the small decrease, observed on Fig. 4, after 50 min, seems meaningless). This is certainly due to a too large decrease of the trans population: this decrease is estimated to 10%. When the Ar laser is switched off, the poling grows higher than in the first experiment without Ar pumping. This reproducible phenomenon could be the consequence of the 3D pumping (reduction of the number of molecules in Z and Y directions), but it could also be due to an increase of the molecular mobility induced by the Ar pumping, like it has been observed in photoassited electrical poling (increase of the free volume) [11,14].

In a 2D geometry, we tried to reproduce the experiment of reference [9], by switching on the Ar pumping after the stabilization of the orientation induced by Yag beams (Ar and Yag polarization were perpendicular). We almost always observed a decrease of SHG. At very low Ar power, when the Yag induced anisotropy was just canceled, we apparently saw a very small increase of SHG, at the limit of noise. Probably, the net increase observed in [9] could be obtained by reducing drastically the intensity of the three beams.

6. Conclusion

The simultaneous measurement of SHG and of 3D variations of absorption, at several wavelengths, is a powerful tool, for studying the dynamics and the saturation behavior of All-Optical Poling. The preliminary results are promising, but some experimental improvements are necessary for quantitative analyses, particularly the precision of beam crossing and of power density measurements. Different symmetries of molecules and of optical excitation will be studied experimentally and theoretically. We have clearly demonstrated the role of the angular redistribution in AOP of DR1. It will be interesting to do the same experiment with molecules which are not photoisomerizable and for which a small poling occurs, presumably from a photodegradation hole burning.

We tried to increase the efficiency of AOP by pumping with a third laser beam, which is supposed to recycle the molecules being oriented perpendicularly to the direction of non-centrosymmetry. We always observed a decrease of $\chi^{(2)}$ when the third pump was applied: the theory suggests that the effect of the reduction of population overcomes that of the enhancement of the non-centrosymmetric order. Experiments must be pursued with weaker intensity of the three beams, but the time needed to complete the poling will increase. The role of the reverse cis→trans photoisomerization deserve some attention. The improvement of $\chi^{(2)}$ after the removal of the third beam (Fig.4), is not clearly understood.

The 3D characterization of photoinduced anisotropy will be of great interest in non-amorphous materials, like liquid crystal polymers. Indeed, the symmetry imposed by cooperative effects and boundary conditions is often in competition with the symmetry of light. The transient reorientation often follows a 3D trajectory.

References

1 Canva M., Le Saux G., Georges P., Brun A., Chaput F. and Boilot J.P. (1992) *Optics Lett.*, **17**, 218-220

2. Neporent B.S., Stolbova O.V. (1961) *Opt. Spectrosc.*, **10**, 146 and (1963) *Opt. Spectrosc.*, **14**, 331. Makushenko A.M., Neporent B.S., Stolbova O.V. (1971) *Opt. Spectrosc.* **31**, 295 and **31**, 397.

3. Dumont M. and Sekkat Z. (1992) Dynamical study of photoinduced anisotropy and orientational relaxation of azo-dyes in polymeric films. Poling at room temperature, *SPIE proceedings*, **1774** p 188-199.

4. Sekkat Z. and Dumont,M. (1993) Photoinduced orientation of azo dyes in polymeric films. Characterization of molecules angular mobility, *Synth. metal*, **54**, 373-381

5. Charra F., Kajzar F., Nunzi J.M., Raimond P. and Idiart E. (1993) Light-induced second-harmonic generation in azo-dye polymers, *Optics Lett.* **18**, 941.

6. Fiorini C., Charra F., Nunzi J.M. and Raimond P. (1995) Photoinduced non centrosymmetry in azo-dye polymers, *Nonlinear Optics* **9**, 339.
Etilé A.C.,Fiorini C., Charra F., Nunzi J.M. (1997) phase-coherent control of molecular polar order in polymers using dual-frequency interfernces between circularly polarize beams,Phys. Rev. **A 56**,3888

7. Nunzi J.M., Charra F., Fiorini C. and Zyss, J.(1994) Transient optically induced non centrosymmetry in a solution of octupolar molecules, *Chem.Phys. Lett.* **219**, 34-354.

8 Brasselet S.and Zyss J. (1997) Control of the polarization dependence of optically poled nonlinear polymer films, *Opt. Lett.*, **22**, 1464-1466 and (1998) Multipolar molecules and multipolar fields: probing and controlling the tensorial nature of nonlinear molecular media, *J. Opt. Soc. Am.*,**B 15**, 257-288

9 Xiuqin Yu, Xiaoxia Zhong, Qu Li, Shouyu Luo, Yingli Chen Yu Sui and Jie Yin (2001) Method of improving optical poling efficiency in polymer films, *Optics Letters* **26**, 220-222.

10. Dumont M., Hosotte S., Froc G. and Sekkat Z. (1993) Orientational manipulation of chromophores through photoisomerization, *SPIE proceedings*, **2042**, 2-13 . and Dumont M., Froc G. and Hosotte S. (1995) Alignment and orientation of chromophores by optical pumping, *Nonlinear Optics* , **9**, 327-338.

11. Dumont M., and El Osman A. (1999) On spontaneous and photoinduced orientational mobility of dye molecules in polymer film, *Chemical Physics*, **245**, 437-462;

12 Dumont M. (2000) Modelization of the angular redistribution in optical ordering processes in dye containing polymers, *Nonlinear Optics*, **25**, 195-200 and (2001) New developments in optical ordering of NLO Dyes in Polymers, *SPIE proceedings,* **4461**,149-163

13 Loucif-Saibi R., Nakatani K., Delaire J., Dumont M. and Sekkat Z. (1993) Photoisomerization and second harmonic generation in disperse red one-doped and -functionalyzed Poly(mathyl methacrylate) films, *Chemistry of materials*, **5**, 229

14. El Osman A. and Dumont M. (1998) Dynamical and spectroscopic study of photoinduced orientation of dye molecules in polymer films, *SPIE proceedings*, **3417**, 36-46

CONTROL OF LIGHT EMISSION PROPERTIES OF ELECTROLUMINESCENT DIODES BY SURFACE PATTERNING

LICINIO ROCHA,[1] CÉLINE FIORINI-DEBUISSCHERT,[1] CHRISTINE DENIS,[1] PASCAL MAISSE,[1] PAUL RAIMOND,[1] BERNARD GEFFROY[1] AND JEAN-MICHEL NUNZI[2]
[1]*CEA-Saclay, Bat. 451*
DECS/SE2M/Laboratoire des Composants Organiques
91191 Gif-sur-Yvette cedex, France
[2]*Laboratoire POMA, UMR-CNRS 6136, Université d'Angers, 2 Boulevard Lavoisier, F49045 Angers cedex, France*

1. Introduction

The scattering properties of structures with periodicities in the optical wavelength scale have been widely employed to control the light propagation and generation of thin luminescent films. An example of the control on radiative properties of thin films through periodic modulations is the well known DFB (Distributed FeedBack) laser. Such devices consist of layers exhibiting a periodical modulation of their optical constants (refractive index and/or gain) which enable controlled modification of light emission and propagation resulting in lasing without mirrors. Light propagation and generation is then modified following interaction with the resulting waveguide grating.
Grating definition is a critical technological step that is commonly provided via lithographic methods. However such methods are not always suitable in the case of organic materials since they make use of solvents and developers that can dissolve the organic layers. Moreover they require an expensive equipment, multiple fabrication steps and so don't take advantage of the ease of processing of organic materials. For full exploitation of organic materials in optoelectronic devices new approaches have been considered with the aim of reducing the fabrication steps and costs (see for example [1],[2],[3]).
Here we propose a simple and original one step method of surface patterning based on laser controlled molecular motion in azo-dye functionalised polymer films. Surface relief gratings, with controlled pitch and modulation amplitude, are induced at room temperature[4],[5] through irradiation of the film surface with an interference pattern. The interference pattern is produced by the superposition of two coherent light beams at a frequency in the absorption band of the azo-dye. Surface deformation results from mass transport following successive photo-isomerisation cycles of the azo-dye molecule.
As an application of the method, the case of the organic light emitting diodes (OLEDs) is considered. The goal was to improve the external efficiency of OLEDs since, in

F. Charra et al. (eds.), Organic Nanophotonics, 405–412.

classical systems, 80% of the light generated inside the active layer is trapped in the diode structure and then lost for the user. We demonstrate that a part of the light guided can be out coupled by modulating the diode surface.

2. Photo-induced patterning of diazobenzene functionalised polymers

Azo-dye molecules have been the subject of many studies since the first observation in the 80's[6] of birefringence and dichroism optically induced in polymer thin films. It has been shown that under resonant excitation with linearly polarised light, the chromophores tend to orient their axis in a direction orthogonal to the excitation beam polarisation, reaching then a stable position. The key process behind axial reorientation is the *trans-cis* isomerisation of the azo-dye. Azo-dyes exist indeed in two different conformations (figure 1) and following optically induced isomerisation from the *trans* to the *cis* form and thermal relaxation back to the *trans* form, a rotation of the molecule axis occurs. After several isomerisation and relaxation cycles, the molecules reach a position perpendicular to the excitation polarisation, minimizing the interaction with the excitation.

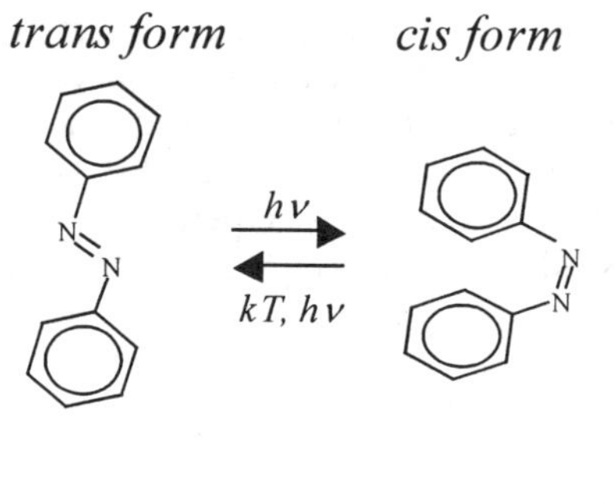

Figure 1: isomerisation of an azo-dye molecule

More recently, large surface relief gratings[4],[5] have been realised following excitation of an azo-dye polymer with an interference pattern between two Argon laser beams. The efficiency of the process has been shown to depend on the total energy sent onto the film. An important feature of the phenomenon is the polarisation dependence of the deformation efficiency that can be 10 times higher in the case of a polarisation orthogonal to the interference fringes than in the case of a polarisation along the fringes. The process is reversible so that the gratings can be erased by heating the polymer film above its glass transition temperature. Although there is still no fully satisfying modelling of the process, it is now well admitted that the principal cause leading to the film surface deformation is related to mass transport following *trans-cis* isomerisation instead of purely thermal effects such as ablation or swelling. It has been demonstrated that migration occurs from the high intensity regions of the interference pattern to the

low intensity regions, the resulting surface relief grating corresponding then to the negative of the intensity figure[7].
A simple microscopic model[8] accounting for the main features is described hereafter. The model is based on the fact that the rotational diffusion is accompanied by translation[9] of the molecule (figure 2). Following photo-isomerisation and thermal relaxation, the chromophores migrate along the polarisation direction, leaving the high intensity to the low intensity regions and tracking in their motion the polymer chains into which they are covalently bound resulting in surface deformation.

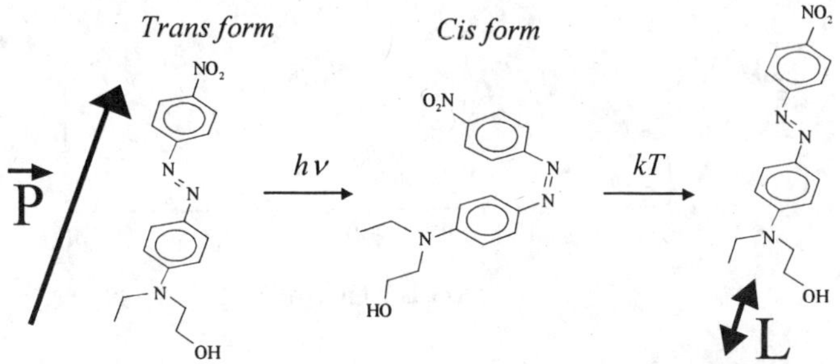

Figure 2: "molecular creeping" model : photo-isomerisation followed translation of an azo-dye molecule

The grating profile depends then on the intensity figure sent onto the film surface, the direction of migration can be controlled with the polarisation, and the modulation amplitudes are controlled by the exposure time. A typical curve of the modulation dynamic is represented in figure 3 which evidences an increase of the modulation amplitudes with the irradiation time which is followed by saturation.
The period Λ of the resulting grating corresponds to the fringes spacing of the interference pattern and is given by :

$$\Lambda = \frac{\lambda_p}{2n\sin\theta} \qquad (1)$$

where λ_p is the excitation wavelength, θ the incidence angle of the two interfering beams and n is the index of the medium where the θ angle is considered (n=1 for θ in air as represented in figure 2).

The grating period can be controlled by the incidence angle θ, for a given excitation wavelength. The smallest periods are obtained for grazing incidences (θ=90°) corresponding to a period $\Lambda=\lambda_p/(2n)$. As an example, using a polymer functionalised with an azo-dye absorbing in the ultraviolet region, gratings with periods of 200nm have been realised using the 364nm line of an Argon laser.
However reaching small periods leads to a decrease of the migration efficiency : for example, the modulation amplitudes we have recorded for periods of Λ=200nm are

limited to 20nm whereas they can be as high as the initial film thickness for periods close to 1μm.

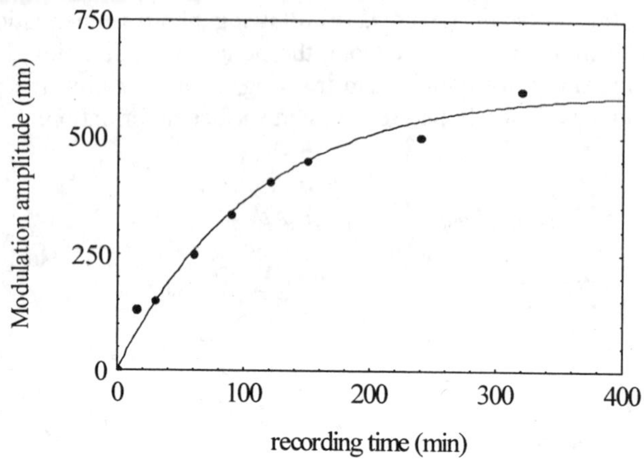

Figure 3 : modulation amplitude growth in function of the irradiation time. Points correspond to measurements performed with an AFM. The solid line is a theoretical fit to the points.

The influence of parameters related to the material such as the polymer functionalisation degree, or related to the experimental configuration such as the interference pattern contrast and light polarisation projection have been studied[10],[11],[12]. The mechanisms responsible for the decrease of the modulation amplitudes at lower periods (<400nm) are under investigation.

3. Application to the control of OLED emission

Surface relief gratings are recorded on spin-coated copolymer films of the azo-dye molecule Disperse Red 1 [4-(N-(2-hydroxyethyl)-N-ethyl-)amino-4'-nitroazobenzene] (DR1) covalently bound to poly(methylmethacrylate) (PMMA) (see insert in figure 4). The gratings are produced by illuminating the film surface with the interference pattern obtained by the superposition of two beams from an Argon laser at 514.5nm with a total intensity of 300mW.cm^{-2}. The experimental set-up is represented in figure 4 with a typical image of the modulated surface recorded with an atomic force microscope (AFM).

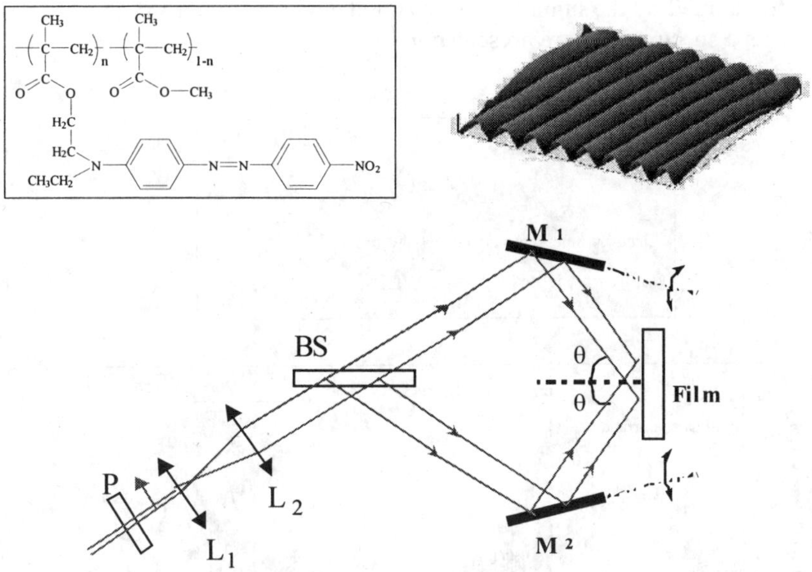

Figure 4 : Chemical formula of the (DR1-MMA) copolymer. Typical AFM scan of a film surface following irradiation with a two beams interference pattern. Experimental set-up. P : polariser, L_1-L_2 : afocal system, BS : beam splitter, M_1-M_2 : co-rotating mirrors.

The realisation of an optically pumped laser based on a distributed feedback configuration has recently been demonstrated using an irradiated film of an azo-dye copolymer as a modulated substrate[13]. As a first step towards the realisation of an electrically pumped laser based on the same kind of geometry, we have implemented the modulation technique described in this paper to optimise the emission properties of OLEDs.

An OLED is typically composed of an electroluminescent layer sandwiched between two electrodes, one of which is transparent (ITO). The multi-layer structure of the diodes provides confining effects that decrease the external efficiency of the device by trapping a part of the light generated inside the active layer. Theoretical calculations confirmed by measurements with an integration sphere show that almost 80% of the light emitted is guided in the diode layers. For the applications involving OLED, such as flexible displays, the improvement of the external efficiency reveals to be a key parameter to consider.

As a first step, we investigated the emission of a widely employed electroluminescent material (Alq3) doped with rubrene evaporated onto the azo-dye modulated copolymer previously recovered by an aluminium electrode (see fig. 5). Only a part of the substrate surface is modulated which enables realisation of two diodes on the same substrate for direct comparison between a modulated and a non-modulated diode.

The evaporation of a transparent electrode onto the organic layer being a critical technological step, the experiments were first performed using optical excitation of the Alq3 doped rubrene layer with an ultraviolet lamp.

The spectra emitted by the luminescent layer and collected by means of an optical fibre coupled with a spectrometer are presented in figure 5.

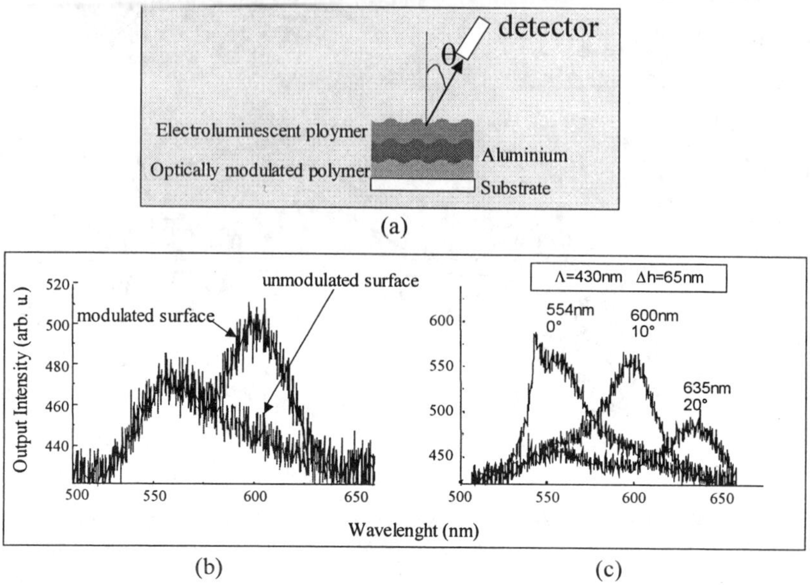

Figure 5: transversal view of a modulated sample(a) and modification of the light emission through structuration of the emitting layer at a given observation angle(b) and for different angles of observation(c). The grating period was Λ = 430 nm, with modulation amplitude Δh = 65 nm. The electroluminescent layer was the reference material Rubrene doped Alq3

In figure 5b are shown the spectra emitted by both the modulated and the non-modulated area of the Alq3 doped rubrene layer, collected in a direction θ (figure 5a) and excited in the same conditions. In the case of the signal collected from the modulated part of the surface, an additional band peaking at 600nm appears in the spectrum. The additional band corresponds to a part of the light initially guided inside the layers that is extracted by propagation in the grating[14]. The grating diffraction effect is more clearly evidenced in figure 5c where we can see a tuning of the additional band for different angles of observation, indicating that a different part of the trapped energy can be outcoupled. Indeed, gratings are dispersive in such a way that modes with different propagation constants will be diffracted into different directions according to :

$$k_0 \sin\theta = \pm\beta_g \pm pK \qquad (2)$$

where k_0 is wave vector and θ the angle of propagation of the outcoupled mode, β_g is the propagation constant of a trapped mode in the layers, K is the grating vector, and p is an integer corresponding to the diffraction order.

The integration of the spectra emitted in all directions of a plane defined by the normal of the device surface and the grating axis, shows that the quantity of light extracted from the modulated part of the device is more than twice the energy extracted from the non-modulated part.
The next step was the realisation of an electrically pumped device. The deposition of a transparent ITO electrode on the top of the device being critical for the organic layers, the experiments were realised with a more conventional diode structure emitting through the substrate. The inversion of the layers stack resulting in a modification of the waves confinement, a change in the diffraction properties of the gratings have been observed. Experiments towards the improvement of the energy localisation inside the layers are under way and the first results are promising.

4. Conclusion

We have made use of a simple one step method to pattern, in the optical wavelength scale, the surface of polymer thin films in view of photonic applications. The method is based on a photo-induced mass transport through repeated photo-isomerisation of azo-dye molecules. The parameters of the resulting grating (the period and the modulation amplitudes) can be easily controlled by adjusting the interference pattern parameters and the exposure time. Modulation as high as the initial film thickness have been recorded.
As an application of the method, we have considered the case of organic light emitting diodes. We have shown that it is possible by this method to increase significantly the OLED emission. As a first step, photoluminescence was investigated : integration of the light emitted in all directions results in a total emitted intensity twice greater for a modulated device than for a classical non-modulated one. The increase of the diode external efficiency is, in the present case, due to diffraction onto the grating of the light initially trapped inside the diode layers. The results reported in this paper deal with one dimensional gratings allowing control on the emission in the direction defined by the grating axis. The second step will be the insertion of bi-dimensional gratings[15] in order to control the diodes emission properties not only in one direction, but in all directions of the plane of the device which should lead to a even more significant increase of the light extracted.

References

1. Schäffer, E., Thurn-Albrecht, T., Russell, T. P. and Steiner, U. (2000) Electrically induced structure formation and pattern transfer, *Nature* **403**, 874-877
2. Rogers, J. A., Meier, M. and Dodabalapur, A. (1998) Using printing and molding techniques to produce distributed feedback and Bragg reflector resonators for plastic lasers, *Appl. Phys. Lett.* **73**, 1766-1768
3. Chou, S. Y., Zhuang, L. And Guo, L. (1999) Lithographically induced self-construction of polymer microstructures for resistless patterning, *Appl. Phys. Lett.* **75**, 1004-1006

4. Kim, D. Y., Tripathy, S. K., Li, L. and Kumar, J. (1995) Laser-induced holographic surface relief gratings on nonlinear optical polymer films, *Appl. Phys. Lett.* **66**, 1166-1168
5. Rochon, P., Batalla, E. and Natansohn, A. (1995) Optically induced surface gratings on azoaromatic polymer films, *Appl. Phys. Lett.* **66**, 136-138
6. Todorov T., Nikolova L., Tomova N. (1984) Polarization holography. I. A new high efficiency organic material with reversible photoinduced birefringence, *Appl. Opt.*, **23**, 4309-4312
7. Bian, S., Li, L., Kumar, J., Kim, D. Y., Williams, J. and Tripathy, S. K. (1998) Single laser beam-induced surface deformation on azobenzene polymer films, *Appl. Phys. Lett.* **73**, 1817-1819
8. Lefin, P., Fiorini, C. and Nunzi, J.-M. (1998) Anisotropy of the photo-induced translation diffusion of azobenzene dyes in polymer matrices, *Pure Appl. Opt.* **7**, 71-82
9. Hall D. B., Dhinojwala A., Torkelson J. M. (1997) Translation-rotation paradox for diffusion in glass-forming polymers: the role of the temperature dependence of the relaxation time distribution, *Phys. Rev. Let.*, **79**, p.103-106
10. Fukuda T., Matsuda H., Shiraga T., Kimura T., Kato M., Viswanathan N. K., Kumar J., Tripathy S. K. (2000) Photofabrication of Surface Relief Grating on Films of Azobenzene Polymer with Different Dye Functionalization, *Macromolecules*, **33**, 4220-4225
11. Fiorini C., Prudhomme N., Etilé A.-C., Lefin P., Raimond P., Nunzi J.-M. (1999) All-optical Manipulation of Azo-dye Molecules, *Macromol. Symp.*, **137**, 105-113
12. Hubert, C., Malcor, E., Maurin, I., Nunzi, J.-M., Raimond, P. and Fiorini, C. (2002) Microstucturing of polymers using a light-controlled molecular migration process, *Appl. Surface Science* **186**, 29-33
13. Rocha, L., Dumarcher, V., Denis, C., Raimond, P., Fiorini, C. and Nunzi, J.-M. (2001) Laser emission in periodically modulated polymer films, *J. Appl. Phys.* **89**, 3067-3069
14. Lupton, J. M., Matterson, B. J., Samuel, I. D. W., Jory, M. J. and Barnes, W. L. (2000) Bragg scattering from periodically microstructured light emitting diodes, *Appl. Phys. Lett.* **77**, 3340-3342
15. Hubert, C., Fiorini-Debuisschert, C., Maurin, I., Nunzi, J.-M. and Raimond, P. (2002) Spontaneous patterning of hexagonal structures in an azo-polymer light-controlled mass transport, *Adv. Mater.* **14**, 729-732

PHOTON STATISTICS OF A SINGLE PHOTON SOURCE

F. TREUSSART, R. ALLÉAUME, V. LE FLOC'H, L.T. XIAO and J.-F. ROCH
Laboratoire de Photonique Quantique et Moléculaire, ENS Cachan, 61 avenue du Préseident Wilson, 94235 Cachan Cedex, France*

J.-M. COURTY
Laboratoire Kastler Brossel†, UPMC case 74, 4 place Jussieu, 75252 Paris Cedex 05, France

Abstract. We report on the realization of a single photon source relying on the pulsed excitation of a single organic molecule at room temperature, and the study of its intensity fluctuations over 4 orders of magnitude of the observation time scale. On time scale of a few excitation periods, sub–poissonian statistics is clearly observed and the probability of multiphoton events is 10 times smaller than for equivalent Poissonian pulses. An excess of noise appears on longer timescale, due to the blinking produced by the molecular triplet state.

Key words: single molecule spectroscopy, nonclassical electromagnetic field, quantum cryptography

1. Introduction

The production of a regular stream of photons on demand has attracted much of interest over the past few years. Such a triggered Single Photon Source (SPS) source could replace attenuated laser pulses in order to fully secure quantum key distribution (QKD) protocols [1]. An ideal single photon source (SPS) should produce light pulses containing exactly one photon per pulse delivered at the place of interest with 100% efficiency. For any given measurement time T, this source would emit exactly $N = T/\tau_{\rm rep}$ photons, so that the standard deviation $\Delta N \equiv \sqrt{\langle N^2\rangle_T - \langle N\rangle_T^2} = 0$ ($\langle\ \rangle_T$ has to be understood as a mean value over a set of measurements lasting

* Laboratoire du CNRS, UMR 8537, associé à l'École Normale Supérieure de Cachan.
† Laboratoire de l'École Normale Supérieure et de l'Université Pierre et Marie Curie, associé au CNRS.

F. Charra et al. (eds.), Organic Nanophotonics, 413–422.

T). Such a source would then be virtually free of intensity fluctuations, therefore corresponding to perfect intensity squeezing [2].

While there have been impressive realizations of SPS relying on the strong coupling of a single atom and a high finesse cavity [3, 4], a solid state SPS is more desirable for practical applications. A first category of solid state SPS realized relies on optically [5–8] or electrically [9] pumped semiconductors nanostructures. A second category uses the fluorescence of a single organic molecule embedded in a crystal host. This system behaves indeed like a two-level emitter as shown in numerous remarkable experiments [10]. The population of the molecule is efficiently transferred to its excited state by an adiabatic passage. The molecular transition frequency is swept with the Stark effect induced by an electric field, around the laser frequency, so that the molecule is coherently driven into its excited state. It then emits a single photon as shown by Brunel et al.[11]. However, in this experiment, the single photon production efficiency was limited to about 70%, due to technical limitations and background photons. Moreover, due to the cryogenic temperature required in all the previously described experiments, the collection efficiency of photons was hardly higher than a few 10^{-3}.

A higher collection efficiency (about 5%) can be easily achieved in setup operating at room temperature. The excitation scheme then relies on the pulsed saturated emission of a 4–level fluorophore [12, 13]. When the pulse duration τ_{p} is much shorter than the dipole radiative lifetime τ_{rad}, such a single emitter can then emit only one single photon per excitation pulse. Two types of emitters have been considered, organic molecules and Nitrogen–Vacancy (NV) color centers in diamond [14, 15]. Lounis and Moerner [16] realized the first room temperature SPS relying on the fluorescence of a single organic molecule, terrylene, imbedded in a crystal host of p–terphenyl. This host–guest system was identified to be very photostable even at room temperature [17, 18]. However, due to crystal absorption and scattering by defects, this system yields a rather low signal to background ratio of 5–10, which is responsible for multiphoton detection events and then for deviation from the perfect SPS. We were able to achieve a higher signal to background ratio (more than 20) by replacing the crystal host by a thin polymer film [19]. Similar sample structures were used to realized a SPS relying on the fluorescence of NV centers in diamond nanocrystals also immobilized in a thin polymer layer [20].

In the following, we first describe how we realized our molecular based SPS and compare it to equivalent attenuated laser pulses. We then analyse the intensity noise properties of this source in the time domain and quantify its departure from a perfect SPS.

2. Experimental setup and data processing

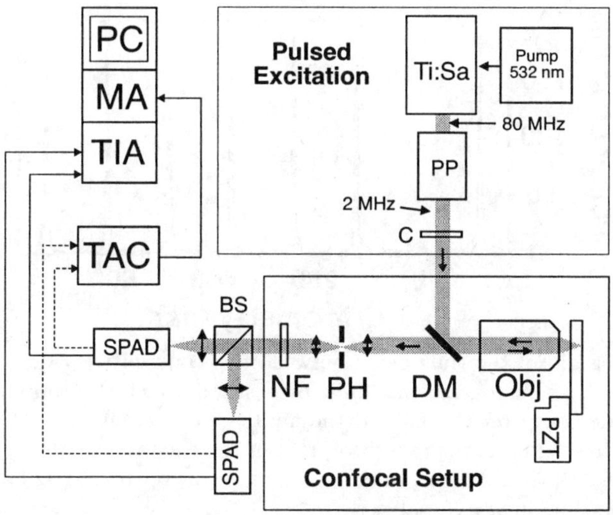

Figure 1. Experimental setup. PZT: x, y, z piezo translation stage; Obj: $\times 60$, $NA = 1.4$ objective; DM: dichroic mirror; NF: notch filter. Pulse excitation is achieved by a titanium dopped sapphire femtosecond laser (Ti:Sa); PP: pulse picker; C: single path propagation in a doubling crystal ($LiIO_3$); BS: non polarizing beamsplitter; SPADs: single photon counting avalanche photodiodes; TAC: time-to-amplitude converter, TIA: time interval analyser; MA: multichannel analyser, PC: computer.

The experimental setup is depicted on Fig.1. The samples are made of cyanine dye $DiIC_{18}(3)$ molecules dispersed at a concentration of about 1 molecule per 10 μ m^2 into a 30 nm thick polymethylmetacrylate (PMMA) layer, spincoated over a microscope coverplate. The fluorescence from the single molecule is excited and collected by the standard technique of scanning confocal optical microscopy [21]. The molecules are non–resonantly excited at 532 nm, with femtosecond pulses generated by a Ti:Sapphire laser and frequency doubled by single pass propagation inside a $LiIO_3$ crystal. The repetition rate, initially at 82 MHz, is divided by a pulsepicker. The energy per pulse E_p is finally adjustable by an electro–optic modulator. The pulse duration is about $\tau_p \approx 100$ fs. The excitation light is reflected by the dichroic mirror of an inverted microscope, and then focused by an oil–immersion objective to form a spot of $\approx 1\mu\text{m}^2$ surface area. The fluorescence light from the sample, spectrally centered on 570 nm (50 nm FWHM), is collected by the same objective and then focused into a 30 μm diameter pinhole. After recollimation, a holographic notch filter removes the residual pump light. A standard Hanbury Brown and Twiss setup is then used to split the beam and detect single photons on two identical avalanche photodiodes. Glass filters are placed onto each arm to suppress parasitic crosstalk [22] between the two photodiodes.

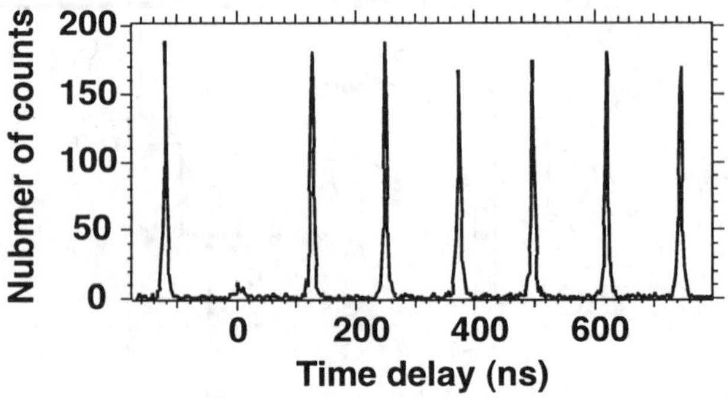

Figure 2. Histogram of the time delay between two consecutively detected photons on each photodiode. This histogram was built by a standard Start–Stop method using the time to amplitude converter (TAC) and the multichannel analyser (MA). In the case of the molecule addressed here, the repetition rate of the excitation pulses was chosen to be 8 MHz, whereas in all the following data presented it was set to 2 MHz for some technical limitations of the time interval analyser.

In order to identify rapidly single molecule emission, we first measure the interphoton time histogram by the standard Start–Stop technique with a time–to–amplitude converter [11]. When a single emitter is addressed, there is virtually no event registered at $\tau = 0$, since a single photon cannot be simultaneously detected on both sides of a beamsplitter [23]. The histogram presented on Fig.2 shows a peak pattern at the pulse repetition period $\tau_{\rm rep}$. As explained in Ref.[11], the central peaks' area allow one to infer the probabilities $P_{\rm S}(2)$ for the source (S), of giving two photocounts per excitation pulse, where these counts are due to deviations from the ideal SPS. Nevertheless, this technique can hardly be used to extract the intensity fluctuations on timescale longer than a single pulse. We have therefore chosen to record every photodetection time event with a two–channel time interval analyser computer board (TIA).

In a typical experiment, we first raster scan the sample at low excitation energy per pulse (0.5 pJ). When a single molecule is located, we apply the excitation energy ramp $E_{\rm p}(t)$ shown on inset of Fig.3, and simulteaneously record the fluorescence counts on a 1 ms integration time[1] Fig.3 displays the fluorescence counting rate R vs. $E_{\rm p}$. The large intensity fluctuations are due to triplet state excursion of the molecule. If this state is not taken into account, the molecular energy levels can be modelised by a 2-level system

[1] This progressive excitation method appears to yield a higher photostability than by shinning upfront the full pump power onto the molecule. This interesting behavior is under study.

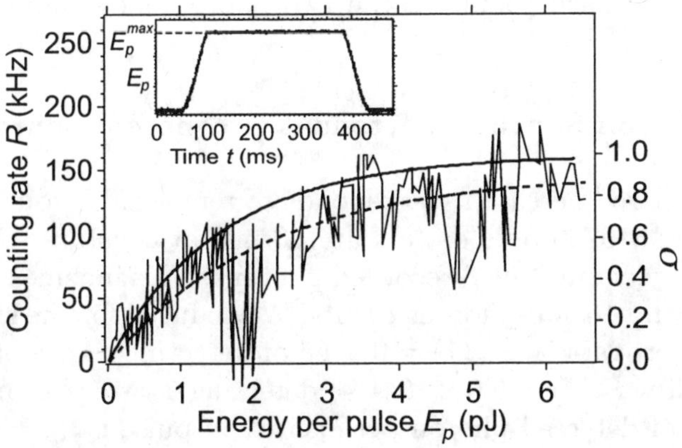

Figure 3. Photon counting rate R vs. energy per pulse, for a single molecule. The inset shows the excitation ramp $E_{\rm p}(t)$, with $E_{\rm p}^{\rm max} = 13$ pJ in this case. The record of the saturation curve was limited by the photobleaching of the dye. The dashed curve is a fit of the raw data according to eq.(1) and the solid line is a fit after correction of triplet state excursion. The right scale shows the excited state population σ.

assuming a very fast non-radiative relaxation between the two higher and lower energy states. The excited state population σ at the time $\tau_{\rm p}$ after the pulse arrival is then

$$\sigma = \frac{E_{\rm p}/E_{\rm sat}}{(1+E_{\rm p}/E_{\rm sat})}\left(1 - e^{-\frac{\tau_{\rm p}}{\tau_{\rm rad}}\left(1+\frac{E_{\rm p}}{E_{\rm sat}}\right)}\right). \tag{1}$$

The data $R(E_{\rm p})$ are fitted by the function $R = R_0 \times \sigma$ in a two steps procedure. After a first fit of the raw data, the points below this fit by more than one standard deviation, which are attributed to triplet state excursion, are removed. The fit of the remaining set of data yields $R_0 = 160 \times 10^3$ counts/s and $E_{\rm sat} = 5.6 \times 10^{-5}$ pJ.

In order to optimize the number of emitted photons and avoid rapid photobleaching, we set $E_{\rm p}^{\rm max}$ to 5.6 pJ. Such a value would correspond to $\sigma = 97\%$, for the molecule studied in Fig.3.

During the constant maximum pumping energy period of the excitation ramp, 10^4 detection events are typically recorded before photobleaching. Thanks to the high stability of the period of the pulsed laser, this set of times can be synchronized on an excitation timebase. This procedure consists in minimizing the time delays between the detection and excitation events over the whole set of measurements, with the laser pulse excitation period and a constant emitter-to-detector propagation delay as the two free parameters. We finally end–up with a table which includes, for each excitation pulse i, the number of photocounts $n_i = 0, 1, 2$ and the time delay δt_i between the excitation pulse and the recording of the detected photon. Photons which are delayed by more than $10 \times \tau_{\rm rad}$ are considered to come from the dark counts of the two photodiodes, and are therefore rejected.

3. Single Photon Source vs attenuated coherent source

The data considered hereafter corresponds to a molecular source (S) which survived during 319769 periods (about 160 ms) at constant saturation excitation power, yielding 14928 recorded photons. This includes 14896 single photon events and 16 two-photons events. We deduced for the single photon source (S) the probability $P_S(1) = 0.0466$ of detecting a one–photon pulse, and a probability $P_S(2) = 5.0 \times 10^{-5}$ of detecting a two–photon pulse. The mean number of detected photon per excitation pulse is $\overline{n}_S = 0.0467$. The real source is considered as the superposition of an attenuated ideal SPS with an overall quantum efficiency η, and a coherent source simulating the background which adds a mean number of detected photon per pulse γ. From the measured values of $P_S(1)$ and $P_S(2)$, we infer $\eta \approx 0.0445$ and $\gamma \approx 2.2 \times 10^{-3}$. This leads to a signal-to-background ratio of about 20, similar to the one previously achieved under CW excitation [19].

We also compared experimentally our SPS to an experimental reference source (R) made of attenuated pump laser pulses, with approximately the same mean number of detected photons per pulse. We checked that this source is a very good approximation of a coherent source and measured that the ratio of two-photon events for the reference $P_R(2)$ and the SPS source $P_S(2)$ is $P_S(2)/P_R(2) \approx 0.10$. The number of two photons pulses is therefore 10 times smaller for our SPS than for the reference poissonian source (R). Since the presence of two photons is an open door to eavesdroppers in QKD, such an efficient single photon soure allows one to considerably improve the performance of QKD schemes, as recently shown by A. Beveratos et al. [24].

4. Intensity fluctuations of the molecular SPS

In order to quantify the fluctuation of the intensity of the molecular SPS, and measure its non classical features, we now consider the number $N(T)$ of photons detected during the observation time T, and then introduce the Mandel parameter [25]

$$Q(T) \equiv \langle(\Delta N)^2\rangle_T / \langle N \rangle_T - 1$$

. In this definition, $\langle\rangle_T$ is the mean value over the largest set of measurements lasting T in which the finite time sequence of photons acquired can be splitted. We can also define a Mandel parameter $Q_s(T)$ for the number of photons *emitted* by the source in the same period of time T. In the case of an ideal SPS, one simply has $Q = \eta \times Q_s$ [26]. Since $Q_s = -1$

for ideal single photon emission, $Q(T) = -\eta$ for any value of T. Such a negative value of Q is the signature of the intensity noise reduction, i.e. of a sub–poissonian counting statistics. Let us point out that due to the existence of a photodetection deadtime, the triggered reference source (R), which simulates the equivalent coherent source, also yields a sub–poissonian counting statistics. One can indeed show that a coherent source (C) having the same mean number $\overline{n}_C = \overline{n}_S$ of photons per pulse than our SPS, would give $Q_C(T) = -\overline{n}_C/2 = -0.0231$, for any T.

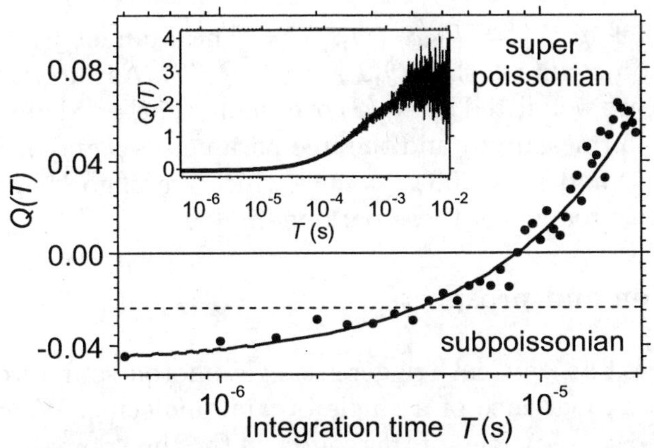

Figure 4. Direct measurement of Mandel parameter $Q(T)$ over 4 orders of magnitude of the recording time T. Dashed line shows $Q(T)$ for the equivalent coherent source (C). Inset shows $Q(T)$ for longer time of observation. The solid curve is a fit to the data by a model accounting for intermittency in the SPS emission. On long timescales, large fluctuations of $Q(T)$ are due to the finite number of available photocounts.

Fig.4 shows that we did observe subpoissonian intensity fluctuations of the SPS ($Q(T) < Q_C(T)$) on timescales from $T = 1 \times \tau_{\rm rep}$ to $T \approx 8 \times \tau_{\rm rep}$, with the minimum value $Q(\tau_{\rm rep}) = -0.0445$ achieved on a single pulse timescale as explained above. Albeit still limited by the quantum efficiency η, this direct measurement of the Mandel parameter $|Q(\tau_{\rm rep})|$ on a single pulse time scale is already larger by more than one order of magnitude than previous measurements realized in experiments with either a single atom [27] or a single trapped ion [28].

When we now consider the number of detected photons on timescales larger than 10^{-5} s, the intensity fluctuations exhibit a superpoissonnian behaviour ($Q(T) > 0$) as shown on the inset of Fig.4. To account for these obervations, we developped a simple model relying on the intermittency of the SPS emission due to the presence of the molecular dark triplet state. In this model, the molecule is either available for fluorescence (ON state), or it is in its triplet state which does not fluoresce (OFF state). Let us note p, the probability per unit of time to make a ON $\rightarrow$ OFF transition, and

$q = 1/\tau_T$ the one to make the reverse OFF $\rightarrow$ ON transition, where τ_T is the lifetime of the triplet state. Note that $p\tau_{rep} = \mathcal{P}_{ISC}$ is the intersystem crossing probability per excitation pulse. From measured values at the single molecule level with $DiIC_{18}(3)$ cyanine dye [29], $p\tau_{rep} \approx 10^{-4} \ll 1$ and $q\tau_{rep} \approx 2.5 \times 10^{-3} \ll 1$. In this limiting case, the Mandel parameter of the source is

$$Q_s(k\tau_{rep}) = \frac{2 \times \mathcal{P}_{ISC}}{\beta^2}\left\{1 - \frac{1}{k\beta}\left[1 - (1-\beta)^k\right]\right\} - 1 \tag{2}$$

where $\beta \equiv (p+q)\tau_{rep} = \mathcal{P}_{ISC} + \tau_{rep}/\tau_T$. The Mandel parameter of the detected photon counts is then $Q(T) = \eta \times Q_s(T)$. As shown on Fig.4, our data for $Q(T)$ are well fitted by eq.(2) over more than 4 orders of magnitude, with $\eta = 0.0445$ (measured) and the free parameters p and q. The fit yields $p\tau_{rep} \approx 2 \times 10^{-4}$ and $\tau_T \approx 250\,\mu s$, values which are in good agreement with a previous measurement of these parameters [29].

5. Conclusion and prospects

We have realized an efficient triggered single photon source relying on the control of the fluorescence of a single organic molecule. We characterized the intensity noise properties of this source in the time domain and observed the signature of non classical photocount statistics.

The record of every photocount allows one to deduce all moments of the photon statistics in order to measure any kind of characteristic function like the second order correlation function $g^{(2)}(\tau)$ and Q. Such a direct statistical analysis can give a deep insight into the photophysical properties of molecules, like conformational changes [30, 31] or resonant energy transfer [32].

Acknowledgements

We are grateful to Carl Grossman, Philippe Grangier, Robert Pansu and Joseph Zyss for help and stimulating discussions. The experimental setup was built thanks to the great technical assistance of A. Clouqueur, J.-P. Madrange and C. Ollier. This work is supported by France Telecom R&D and ACI "jeunes chercheurs" of the Ministère de la Recherche et de l'Enseignement Supérieur.

References

1. N. Gisin, G. Ribordy, W. Tittel, and H. Zbinden. Quantum cryptography. *Rev. Mod. Phys.*, **74**,145, 2002.

2. Serge Reynaud. Introduction à la réduction du bruit quantique. *Ann. Phys. Fr.*, **15**,63, 1990.
3. S. Brattke, B.T.H. Varcoe, and H. Walther. Generation of photon number states on demand via cavity quantum electrodynamics. *Phys. Rev. Lett.*, **86**,3534, 2001.
4. A.Kuhn, M. Hennrich, and G. Rempe. Deterministic single–photon source for distributed quantum networking. *Phys. Rev. Lett.*, **89**,067901, 2002.
5. J. Kim, O. Benson, H. Kan, and Y. Yamamoto. A single-photon turnstile device. *Nature*, **397**,500, 1999.
6. P. Michler, A. Kiraz, C. Becher, W.V. Schoenfeld, P.M. Petroff, Lidong Shang, E. Hu, and A. Imamoğlu. A quantum dot single-photon turnstile device. *Science*, **290**,2282, 2000.
7. C. Santori, M. Pelton, G. Solomon, Y. Dale, and Y. Yamamoto. Triggered single photons from a quantum dot. *Phys. Rev. Lett.*, **86**,1502, 2001.
8. E. Moreau, I. Robert, J.-M. Gérard, I. Abram, L. Manin, and V. Thierry-Mieg. Single-mode solid-state single photon source based on isolated quantum dots in pillar microcavities. *Appl. Phys. Lett.*, **79**,2865, 2001.
9. Z. Yuan, B. E. Kardynal, R.M. Stevenson, A.J. Shields, C.J. Lobo, K. Cooper, N.S. Beattie, D.A. Ritchie, and M. Pepper. Electrically driven single-photon source. *Science*, **295**,102, 2002.
10. Ph. Tamarat, A. Maali, B. Lounis, and M. Orrit. Ten years of single molecule spectroscopy. *J. Phys. Chem. A*, **104**,1, 2000.
11. C. Brunel, B. Lounis, P. Tamarat, and M. Orrit. Triggered source of single photons based on controlled single molecule fluorescence. *Phys. Rev. Lett.*, **83**,2722, 1999.
12. F. De Martini, G. Di Giuseppe, and M. Marrocco. Single–mode generation of quantum photon states by excited single molecules in a microcavity trap. *Phys. Rev. Lett.*, **76**,900, 1996.
13. R. Brouri, A. Beveratos, J.-P. Poizat, and P. Grangier. Single-photon generation by pulsed excitation of a single dipole. *Phys. Rev. A*, **62**,063817, 2000.
14. C. Kurtsiefer, S. Mayer, P. Zarda, and H. Weinfurter. *Phys. Rev. Lett.*, **85**,290, 2000.
15. R. Brouri, A. Beveratos, J.-Ph. Poizat, and Ph. Grangier. Photon antibunching in the fluorescence of individual colored centers in diamond. *Opt. Lett.*, **25**,1294, 2000.
16. B. Lounis and W. E. Moerner. Single photons on demand from a single molecule at room temperature. *Nature*, **407**,491, 2000.
17. L. Fleury, B. Sick, G. Zumofen, B. Hecht, and U. Wild. High photo-stability of single molecules in an organic crystal at room temperature observed by scanning confocal optical microscopy. *Molecular Physics*, **95**,1333, 1998.
18. L. Fleury, J.-M. Segura, G. Zumofen, B. Hecht, and U. Wild. Nonclassical photon statistics in single–molecule fluorescence at room temperature. *Phys. Rev. Lett.*, **84**,1148, 2000.
19. F. Treussart, A. Clouqueur, C. Grossman, and J.-F. Roch. Photon antibunching in the fluorescence of a single dye molecule embedded in a thin polymer film. *Opt. Lett.*, **26**,1504, 2001.
20. A. Beveratos, S. Kühn, R. Brouri, T. Gacoin, J.-P. Poizat, and P. Grangier. Room temperature stable single photon source. *Eur. Phys. J. D*, **18**,191, 2002.
21. S. Nie and R.N. Zare. Optical detection of single molecules. *Annu. Rev. Biophys. Biomol. Struct.*, **26**,567, 1997.
22. C. Kurtsiefer, P. Zarda, S. Mayer, and H. Weinfurter. Optical detection of single molecules. *J. Mod. Opt.*, **48**,2039–2047, 2001.
23. P. Grangier, G. Roger, and A. Aspect. *Europhys. Lett.*, **1**,173, 1986.

24. A. Beveratos, R. Brouri, T. Gacoin, A. Villing, J.-P. Poizat, and P. Grangier. Single photon quantum cryptography. to appear in *Phys. Rev. Lett.*, 2002.
25. L. Mandel. Sub–poissonian photon statistics in resonance fluorescence. *Opt. Lett.*, **4**,205, 1979.
26. J.A. Abate, H.J. Kimble, and L. Mandel. Photon statistics of a dye laser. *Phys. Rev. A*, **14**,788, 1976.
27. R. Short and L. Mandel. Observation of sub-poissonian photon statistics. *Phys. Rev. Lett.*, **51**,384, 1983.
28. F. Diedrich and H. Walther. Nonclassical radiation of single stored ion. *Phys. Rev. Lett.*, **58**,203, 1987.
29. J.A. Veerman, M.F. Garcia-Parajo, L. Kuipers, and N.F. Van Hulst. Time-varying triplet state lifetimes of single molecules. *Phys. Rev. Lett.*, **83**,2155, 1999.
30. H. Yang and X.S. Xie. Statistical approaches for probing single–molecule dynamics photon–by–photon. to appear in *Chem. Phys.*, 2002.
31. V. Barsegov and S. Mukamel. Probing single molecule kinetics by photon arrival trajectories. *J. Chem. Phys.*, **116**,9802, 2002.
32. A. Berglund, A. Doherty, and H. Mabuchi. Photon statistics and dynamics of fluorescence resonance energy transfer. *Phys. Rev. Lett.*, **89**,068101, 2002.

SINGLE MOLECULE OPTICALLY CONTROLLED CURRENT SWITCH: BEYOND THE ELECTROSTATIC APPROACH

S. NEŠPŮREK[a,b], J. SWORAKOWSKI[c], P. TOMAN[a]
[a]*Institute of Macromolecular Chemistry, Academy of Sciences of the Czech Republic, Heyrovský Sq. 2, 162 06 Prague, Czech Republic*
[b]*Faculty of Chemistry, Technical University of Brno, Purkyňova 118, 612 00 Brno, Czech Republic*
[c]*Institute of Physical and Theoretical Chemistry, Wroclaw University of Technology, Wyb. Wyspianskiego 27, 50-370 Wroclaw, Poland*

Abstract

A scheme of optically controlled current switch is put forward based on electrical properties of a conjugated polymer chain ('molecular wire') containing chemically attached photochromic side groups. Upon changing the dipole moment of the photochromic units by light of a suitable wavelength, the ionization energy of the polymer segments containing the dipolar species and energies of electrostatic interactions of charge carriers on the molecular wire and the dipoles are locally modified. These changes result in the formation of a potential wall structure and in the creation of local states for charge carriers on the molecular wire, and hence in a decrease of the on chain carrier mobility. According to *ab initio* quantum chemical calculations, a suitable system for the molecular switch seems to be poly[methyl(phenyl)silylene] molecular wire with photochromic 6-nitro-1',3',3'-trimethylspiro[2H-1-benzopyran-2,2'-indoline] moieties chemically attached to the phenyl ring via –CO–CHOH– spacers.

1. Introduction

The architecture of molecular-scale electronic devices can be designed starting from molecular segments whose properties have been known from experiment and/or suitable theoretical models. Such an approach can be employed in designing a molecular switch – one of basic components for the construction of memories and logic elements. Some examples of molecules or nanostructures that might be used as switching units have recently been demonstrated (e.g., [1-8]).

A pioneering construction of a molecular switch, put forward by Carter [1], was based on the electron tunneling principle. The transmission coefficient of an electron travelling along a 'molecular wire' (e.g., a conjugated polymer chain) and approaching a finite series of periodic potential barriers would be unity if the electron energy matched

F. Charra et al. (eds.), Organic Nanophotonics, 423–436.

a pseudostationary energy level of the barriers. The tunnel switch can be turned off by either changing a barrier height or the depth of a potential well, which can be controlled by the dipole moment of side groups attached directly to the polymer backbone. However, such a switch has a disadvantage: the energy of the transmitted electron must exactly match the energy levels of the barriers formed on the molecular wire. In reality, any distribution of geometries of the chain segments should lead to a dispersion of the positions of energy levels throughout the polymer chain thus resulting in a sharp decrease of the transmission coefficient of the electron. Thus the idea of the Carter's switch is likely to fail in confrontation with the situation in real polymer chains. To avoid this disadvantage, another type of the switch was proposed [2]: a photochromic unit incorporated within the molecular wire. In the stable form, the unit fitted the polymer conjugation being thus a part of the structure in which a charge carrier was delocalized. Exposure to light of a suitable wavelength should change the chemical structure of the photochromic unit, resulting in the interruption of the conjugation and, consequently in the localization of the charge on the molecular wire. However, the chemical synthesis of these types of structures is complicated, hence this type of the switch has never been realized in practice.

Recently, Chen et al. [6] proposed a new type of the electrical switch based on a two-step reduction process. As the voltage is increased, the molecular wire initially undergoes a one-electron reduction on the site of donor-acceptor substituted unit, thereby supplying a charge carrier for electron flow through the system. A further increase in voltage causes a second electron reduction resulting in the current blockade. The negative resistance behavior of this type of switch was demonstrated on a molecule containing a nitroamine redox center 2'-amino-4,4'-di(ethynylphenyl)5'-nitro-1-benzenethiolate. However, recent studies of Donhauser et al. [8] showed that the switching in such a system is due to conformational changes in the molecules or bundles rather than to electrostatic effects of charge transfer. Thus, the basic philosophy, physical background, and material engineering of molecular switches are still open problems.

In this paper we present an alternative approach to the construction of the molecular switch. The idea, presented in our earlier papers [9,10], is based on results obtained from model calculations performed on a model molecular lattice, involving only electrostatic charge-dipole interactions [11-14]. The switch would consist of a molecular wire with suitably attached side groups containing reversible photochromic moieties.

2. The architecture of the switch

As was mentioned in the Introduction, the proposed molecular switch consists of a molecular wire with photochromic side units chemically attached via a spacer [9,10]. The schematic picture is given in Fig. 1.

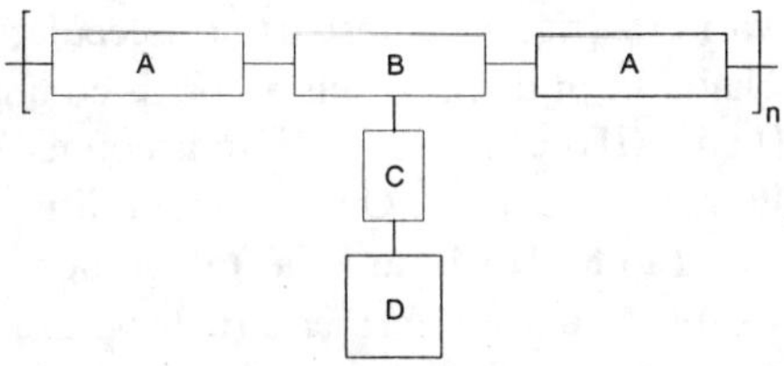

Figure 1. A scheme of a molecular current switch. See the text for a further discussion.

Scheme 1. Chemical structures of some materials suitable for the construction of a molecular switch. The σ–conjugated polymer (II) in which R^1 = phenyl, R^2 = methyl, will be referred to as poly[methyl-(phenyl)silylene] (PMPSi).

(I)
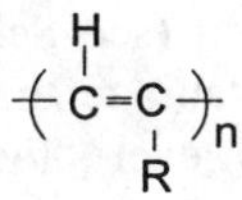

(II)
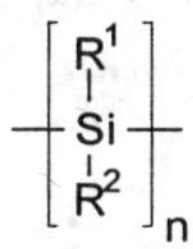

(III)
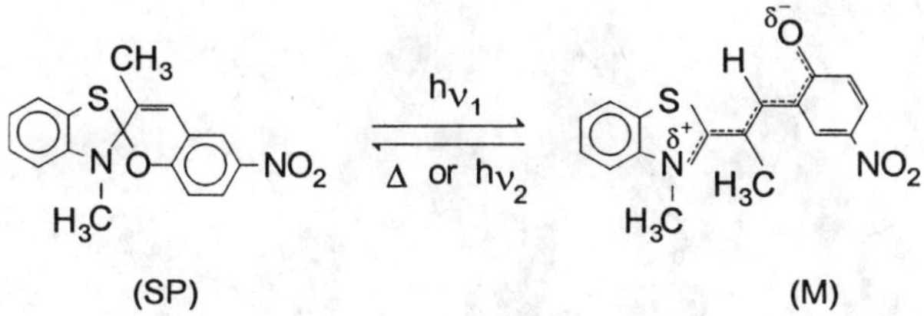

(IV)
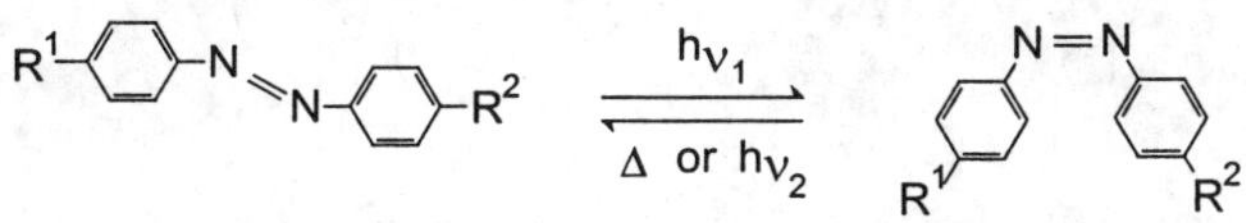

(V) $-CO-(CH_2)_n-O-$

(VI) $-CH=N-NH-$

(VII) $-CO-CHOH-$

(VIII)
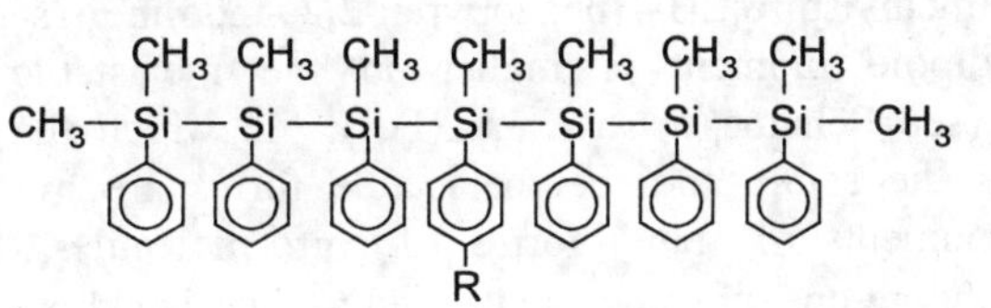
(7MPSi-DNPH)

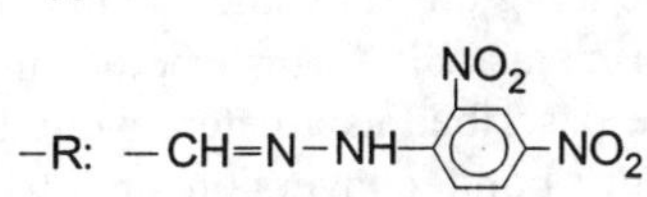

The A-B-A sequence of blocks shown in the figure represents a molecular wire, which is realized by a conjugated polymer chain. In principle, both π– or σ–conjugated polymer chains (see Scheme 1, structures (I) and (II)) can be taken into account. Thus, the charge injected into the chain can simply move through it. On-chain mobilities in π-conjugated carbon polymers are usually higher (up to 7×10^{-5} $m^2V^{-1}s^{-1}$) than those in σ-conjugated silicon ones (up to 2×10^{-5} $m^2V^{-1}s^{-1}$) [15] but the former structures are more sensitive to conformational disorder and structural defects in the backbone. In σ-conjugated polymers, this problem can be overcome by a prolongation of alkyl side groups. The chemical modification results in higher barriers to conformations other than the *all-trans* one and, consequently, in a lower conformational disorder and an increase of overall electronic coupling. In this paper, functioning of the switch will be discussed taking as an example σ-conjugated poly[methyl(phenyl)silylene] (see Scheme 1, (II), R^1 = phenyl, R^2 = methyl). The methyl and phenyl groups stabilize the chain. This polymer is a good hole transporting material [16]. The HOMO level ("the edge of the valence band" in the solid state terminology) consists of mixed σ and π orbitals, as follows from Fig. 2.

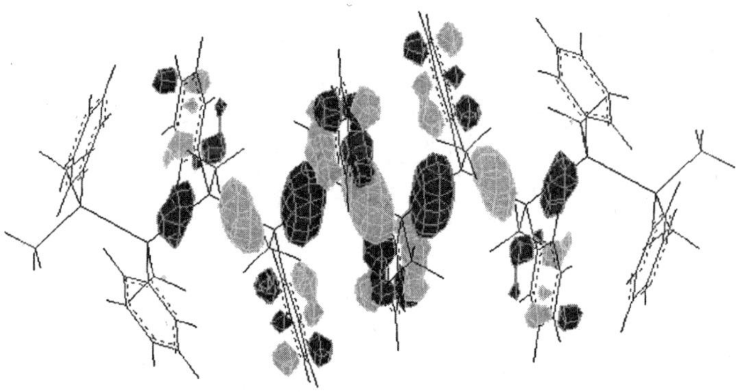

Figure 2. Shape of the HOMO orbital of deca[methyl(phenyl)silylene]. Positive and negative values of the wavefunction are indicated with gray and black colors, respectively.

The block D in Fig. 1 is a photochromic group, the magnitude and/or orientation of its dipole moment differing as much as possible between the stable and the metastable form. There are many photochromic materials that can be used for this purpose, see e.g. [17,18] and references therein. Two photochromic molecules will be mentioned in this context: 6-nitro-1',3',3'-trimethylspiro[2H-1-benzopyran-2,2'-indoline] (see Scheme 1, structure (III)), whose dipole moment is higher in the metastable form, and a substituted azobenzene (see Scheme 1, structure (IV)), in which the orientations of the dipole moments of the stable and the metastable forms are mutually nearly orthogonal. The dipole moments of stable forms of spiro-molecules (SP) amount typically to (2÷7) D. Depending on substituents attached to their backbones, the dipole moments of the respective metastable forms (merocyanines, M) may exceed 10 D. The overall mechanism may be described as a heterocyclic bond cleavage and ring opening, but there are also cis-trans bond conversion and triplet-triplet contributions to the mechanisms. It should be noted that steric factors play an important role in the photochromic response: since the spirocarbon atom in the stable form of the molecule constitutes an intersection of two nearly perpendicular planes of the molecule, formation of the metastable form, after the initial C-O bond cleavage, requires a 90° twist. The photochromism of azobenzene is based on trans-cis isomerization. The schemes of the photochromic reactions are shown in Scheme 1, reactions (III) and (IV).

The spacer C can be realized by several chemical structures, e.g. as those given in Scheme 1, (V)-(VII), its length, shape and other properties can easily be tailored to needs resulting from, e.g., model quantum chemical calculations. We must consider two important requirements for the spacer C: its chemical structure and its length. There are two limiting factors for the chemical structure: the spacer can be "quasi-neutral" from the point of view of the charge transfer, like the spacer (V) presented in Scheme 1, or it can simply transfer the charge during the formation of the dipole moment in the side group (e.g., spacer (VI) in Scheme 1). In the former case, the electrostatic contribution of the dipolar side-group based on the charge-dipole interactions is the most important factor which influences the behavior of the charges on the molecular wire. In the latter case, the situation is more complex: a charge redistribution during the photochemical transformation is possible which can influence the ionization energy of the polymer segment with the polar side group.

The two cases are compared in Fig. 3. The situation was modeled on a derivative of the hepta[methyl(phenyl)silylene (7MPSi, see Scheme 1, (VIII)). The derivative, hereafter referred to as 7MPSi-DNPH, contained a chemically attached 2,4-dinitrophenyl-hydrazono-methyl unit (–R, abbreviated DNPH). The reference case, where the charge redistribution was impossible, was 7MPSi with m-dinitrobenzene (DNB) placed in the same position as the dinitrobenzene group in 7MPSi-DNPH. The

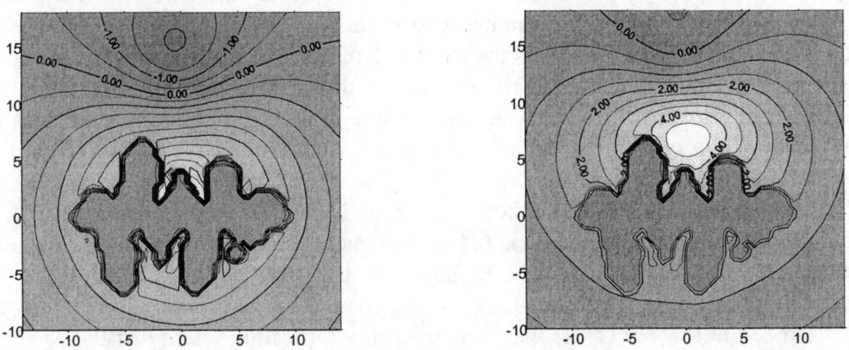

Figure 3. Differential potential map of 7MPSi-DNPH and 7MPSi (left), and of 7MPSi:DNB and 7MPSi (right). The distances on the axes are expressed in 0.1 nm units, the potential differences are given in Volts.

results of the calculations are given in Fig. 3. The map on the left shows the differences of electrostatic potentials between 7MPSi-DNPH and 7MPSi molecules. It is evident that the largest potential difference in this case is located on the Si oligomer chain. The map on the right shows similar differences between 7MPSi:DNB and 7MPSi. Here, the largest differences are positioned outside the oligomer chain; note that the dinitrophenyl group is located in the half-circle in the uppermost part of the figure. The influence of the dinitrobenzene (DNB) dipole on the electronic structure of the Si chain is therefore expected to be weaker. This is also demonstrated in Fig. 4 where the potential differences calculated in the centers of Si-Si bonds are presented. While the difference of the electrostatic potentials reaches up to about 0.3 V (curve 1) for the system 7MPSi-DNPH / 7MPSi, it amounts to ca. 0.14 V for the system 7MPSi:DNB / 7MPSi (curve 2). The magnitude of the electrostatic contribution can be determined by model calculations [11,12,14,19,20]. The calculations performed on a model molecular lattice demonstrate that polar dopants locally modify the polarization energy thus creating new traps for charge carriers in the vicinity of the dipole. The dipolar species with the dipole moment 10 D can form traps 0.5 eV deep on a neighboring site (a nearly linear dependence between the dipole moment and trap depth was found).

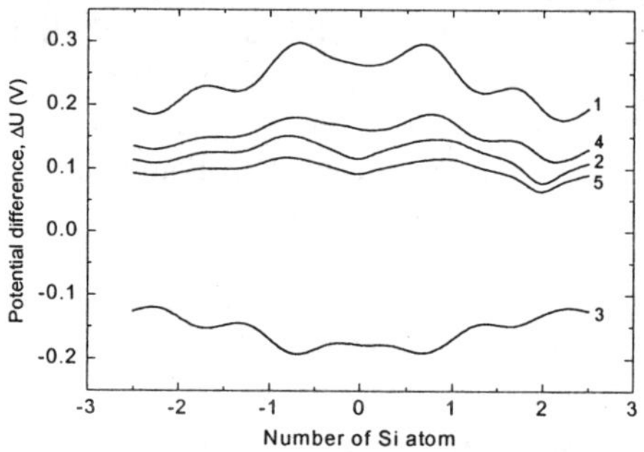

Figure 4. Potential differences in the centers of Si-Si bonds for the system 7MPSi-DNPH and 7MPSi (curve 1), and 7MPSi:DNB and 7MPSi (curve 2). The curve 3 shows the potential difference for the opposite orientation of the dipole (cf. curves 2 and 3), curves 4 and 5 show the potential difference for shorter (-0.1 nm) and longer (+0.1 nm) distance of the dipole from the polymer chain.

Thus, changes of the electrostatic potential on the molecular wire are larger if a "conjugated" (and suitably oriented) spacer is used for the attachment of the polar side group. However, it should be mentioned that in this case the localization of the HOMO orbital is changed: it is shifted from the main polymer chain towards the side group. This is associated with the charge redistribution and with the change of the ionization energy of the polymer segment containing the dipolar side group. The effect is demonstrated on the example of the system consisting of the penta[methyl(phenyl)silylene] (5MPSi) and spiropyran (Scheme 1, (III)). It comes from quantum chemical calculations performed by the *ab initio* Hartree-Fock method (cf. Table 1) that the dipole moment of a free-standing spiropyran (SP) molecule is 6.4 D, increasing to 8.3 D upon the photochemical transformation to merocyanine (M). If, however, the same molecule is attached to the silicon wire with the –CO–CHOH– spacer, the change of the dipole moment was found to be from 8.1 D to 18.9 D. At the same time, the ionization energy of the Si segment with side group changes from 7.9 eV (SP) to 6.8 eV (M); it should be pointed out that the ionization energy of the unsubstituted oligosilylene is 7.7 eV. These changes are accompanied by an important charge redistribution on the side photochromic unit during the photochemical transformation (cf. Table 1). The results presented in Table 1 convincingly demonstrate that the chemical structure of the photochromic side group *and* of the spacer strongly influence the ionization energy. Consequently, chemical traps can be created on the chain. The charge-dipole interactions (electrostatic term) additionally contribute to the effect.

Table 1. Basic characteristics of the spiropyran photochromic unit and changes in charge densities during the photochromic conversion SP → M. Note that the side group is attached to the Si wire through the Si atom No. 63.

Molecule

(SP)	(M)
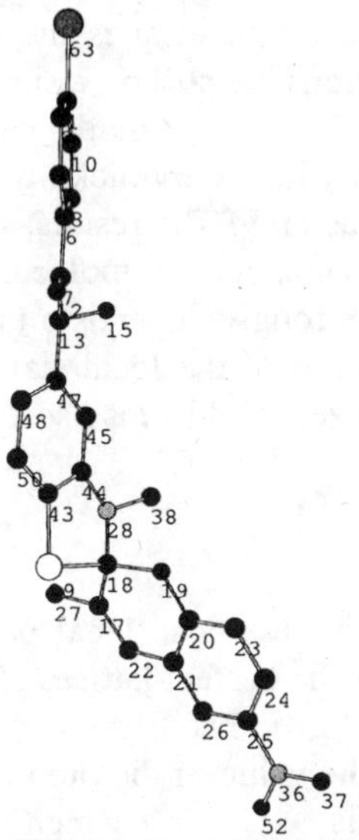	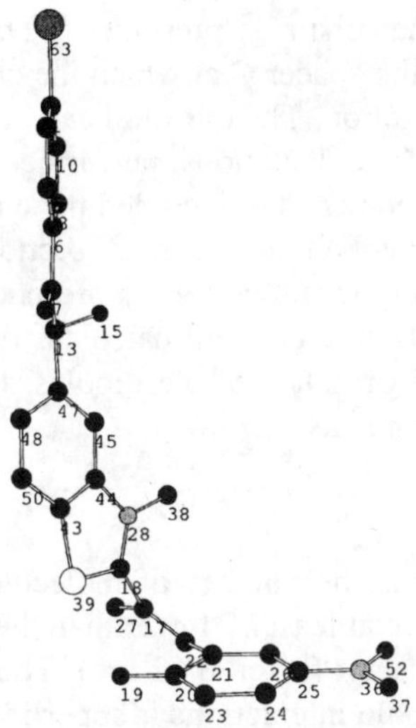

Dipole moments *p* of side group units (in D)

	SP	M
Free-standing molecule	6.4	8.3
Chemically attached to 5MPSi	8.1	18.9

Ionization energy *I* of the polymer segment with side group (in eV)

SP	M
7.9	6.8

5MPSi: Mulliken atomic charges with hydrogens summed into heavy atoms (in unit charges)

No.	Symbol	SP	M
	Charges on the atoms with the largest increase of the atomic charge upon transformation		
39	S	0.26	0.59
38	C	0.27	0.36
20	C	0.48	0.56
45	C	-0.04	0.03
22	C	0.06	0.12
28	N	-0.71	-0.66
48	C	0.04	0.09
	Charges on the atoms with the largest decrease of the atomic charge upon transformation		
18	C	0.33	0.17
21	C	-0.07	-0.22
23	C	-0.02	-0.12
19	O	-0.72	-0.79
43	C	-0.23	-0.29
52	O	-0.47	-0.52
25	C	0.15	0.10

Another problem, which should be solved, is that of the length of the spacer. From the point of view of chemical synthesis it is difficult to attach chemically the photochromic group directly to the polymer chain. The energy of electrostatic (charge-dipole) interactions depends on the distance between the interacting particles (see also Fig. 4, where the sensitivity of the potential difference to the distance of the dipole from the polymer chain is presented). Therefore, the question arises what is the maximum length of the spacer C at which the charge-dipole interactions are still on a suitable level of the operation. The question can be partly answered on the basis of model calculations [12,14]. The calculations, performed on a model molecular lattice mimicking properties of anthracene crystal, included the electrostatic interactions only. The results, which will also be invoked in the next section, showed that the number of molecules whose energies are perturbed by the presence of a polar dopant strongly depends on the dipole moment. Hence one can calculate the average cross-section of the local states created due to the presence of the dipoles. For this purpose the zero-field cross section of the trap is defined as

$$S = (n_{kT})^{2/3} a^2 \tag{1}$$

where n_{kT} is the number of molecules adjacent to the dipole whose local polarization energies are at least kT lower than the value calculated for the perfect lattice, and a is the lattice constant (taken 0.62 nm). The result is shown in Fig. 5. The cross-section of the charge-dipole interactions is superlinearly dependent on the value of the dipole moment. For p = 10 D, the cross section is nearly 10 nm^2 (radius ~ 1.7 nm) which represents about 12 C–C bonds. Thus, with a sufficiently high dipole moment of the side group after its photochromic transformation there is enough place for necessary chemical connecting groups. Clearly, the above result has been obtained with the electrostatic component only taken into account, i.e. with a 'totally neutral' spacer linking the photochromic moiety to the chain. One should expect, however, that the presence of any conjugated spacer would result in an increase of the 'cross-section' of the side group.

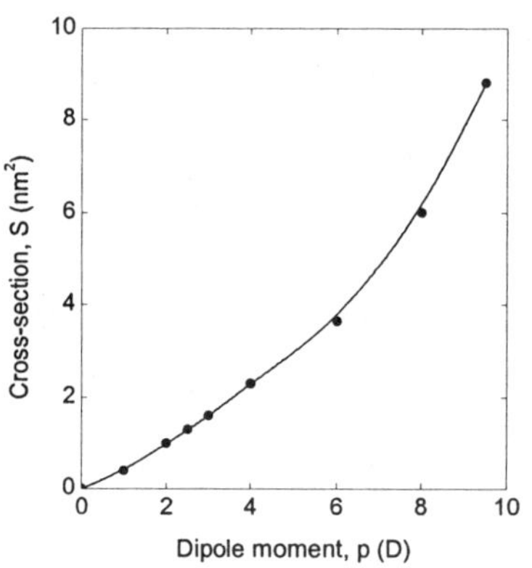

Figure 5. The zero-field cross-section of the trap created in a non-polar model molecular lattice by a polar dopant as a function of the dipole moment of the polar species.

3. The function of the switch

As early as in the fifties, Lyons [21,22] demonstrated that in perfect and nearly-perfect molecular crystals, energies of bands for excess charge carriers, measured with respect to the vacuum level, are determined by ionization energies and electron affinities of constituent molecules, and by energy of electrostatic interactions of a carrier (momentarily residing on a given molecule) with electrically neutral surrounding molecules. The latter parameter is referred to as the polarization energy B ($B < 0$).

In the following we will briefly demonstrate the basic ideas of the model (because of hole transport in PMPSi, which is considered in this work as model molecular wire, we will concentrate on the positive charges). A hole can be localized on crystal sites (molecules) [23,24] having local values of the ionization energy (I_c^{loc}) smaller than the respective value in a perfect crystal (I_c). The trap depth E_t^h is then given as

$$E_t^h = I_c - I_c^{loc} \quad . \tag{2}$$

The values of I_c and I_c^{loc} are related to the molecular value I_g by the simple formula $I_c = I_g + B$. Thus traps will be created on (or in the vicinity of) defects causing suitable local changes of I_c^{loc} and/or B.

The problem considered in this paper can be described within a suitably modified Lyons's approach. It should be pointed out that the magnitudes of the molecular ionization energy and electron affinity strongly depend on the chemical structure of the guest molecules or on the chemical structure of the polar side group of the molecular wire and the chemical spacer between them. Thus, if there are some segments on the molecular wire with a lower ionization energy (e.g. due to the specific side groups) than the respective value of neighboring segments, then the former segments will operate as localization centers. If the ionization energy I_c^{loc} of the molecule is greater than the respective value in a perfect crystal I_c, the molecule in a 3D lattice will act as a scattering center.

Considering the behavior of molecular-scale systems, one should, however, introduce an important distinction. Individual molecular wires should behave as one-dimensional (1D) systems, where there is no scattering. In this case, both the increase and decrease of the ionization energy result in the formation of local states on the molecular wire. A typical example of the formation of these types of traps represents a PMPSi wire with SP side group during the photochemical transformation (see the ionization energies in Table 1; note that the ionization energy of PMPSi wire without side groups is 7.7 eV).

Another difference that should be considered is the effect of the environment. In 3D (macroscopic) systems, it enters *via* the polarization energy. In molecular-scale devices, this effect should be re-defined; apparently, to within a reasonable approximation, it would be sufficient to take into account the charge-permanent dipole interactions.

Apart from traps created because of changes of I_c^{loc} due to local modifications of the chemical structure of the polymer, one should consider formation of traps (or a modification of their parameters) due to electrostatic interactions of polar side groups with charge carriers travelling on the chain. Here, the problem is equivalent to some extent to the case of 'dipolar traps' considered in our earlier papers. The calculations [11,12,14] demonstrate, in agreement with results of experiments performed by several groups [13,19,20,25-27], that parameters of such traps (their depths, cross-sections etc.) depend on the dipole moment of the dopant, as well as on the concentration and mutual orientation of the dipoles. The electrostatic contribution can further increase or decrease (depending on the orientation of the dipole) the local ionization energy and new local electronic states can be formed or the localization energy of existing local states can be modified. It follows from the model calculations of the polarization energy performed on 3D model crystal [11,12,14] that local states are formed on the adjacent molecules in the vicinity of the dipole. In 1D polymer chain the dipolar species in the side groups influence the local states on the molecular wire.

The main results concerning the electrostatic contribution can be summarized as follows: (i) the presence of polar species results in local perturbations of the polarization energy on adjacent molecules, the number of affected molecules depending on the dipole moment of the guest; (ii) the depth of traps formed on molecules adjacent to the polar dopant or on the chain in the vicinity of a polar side group is a nearly linear function of its dipole moment; (iii) the density-of-states (DOS) function is broadened in the presence of the dipole [13,19,20,25-27], the broadening parameter increasing with increasing dipole moment.

Due to local changes of the ionization energy of polymer segments which contain polar groups and due to electrostatic charge-dipole interactions, new local states are formed and the on-chain mobility of charge carriers decreases. If the side group is a photochromic one capable of changing its dipole moment under the irradiation, one can expect a modulation of the on-chain mobility of charge carriers by light. This seems to be the case for PMPSi wire with SP side groups: the results of calculations given in Table 1 show that the open form of spiropyran (merocyanine) generates local centers capable of trapping holes. On the other hand, results obtained for PMPSi wire with azobenzene side groups indicate that the switching cannot be realized, probably due to a small change of the dipole moment under irradiation: the calculations performed for a 5MPSi segment yield the ionization energies equal to 7.8 eV and 7.8 eV, and the dipole moments amounting to 7.7 D and 6.2 D before and after the photochemical transformation, respectively.

There are several open questions, which need further investigations. Namely, the charge redistribution, orbital occupation and localization, and geometry of the side group during the photochromic conversion, which can influence the ionization energy, should be elucidated.

4. Charge carrier mobilities influenced by polar guest molecules

A molecular switch, as its name implies, would be a molecular-scale device operating on a single molecule or an assembly of a few molecules. It can be shown, however, that macroscopic assemblies of molecules, whose architecture was discussed in the preceding sections, will behave in a similar way. It should be noted that the macroscopic systems will behave as (more or less anisotropic) 3D solids. It would be instructive to assess the effect of traps associated with the presence of polar impurities in such systems. Two limiting cases can be considered: nearly-perfect crystals with a low concentration of polar species, and highly disordered (amorphous or partly crystalline) molecular systems.

In nearly-perfect solids, the trap-controlled mobilities follow the equation [28]

$$\mu = \mu_o \left[1 + x_{dip} \exp\left(-\frac{E_t}{kT} \right) \right]^{-1} \tag{3}$$

where μ_o is the drift mobility of the carriers in a perfect solid, x_{dip} is the mole fraction of the polar species, and E_t is the depth of trap created due to the presence of the dopant ($E_t < 0$). As comes from simple estimates, even for moderate dipole moments of the impurity, one should expect a significant decrease of the mobility.

In disordered solids, more relevant to the matter of this paper, the photochemical modulation of carrier mobility is more difficult to assess since, apart from the modification of the trap depths (cf., e.g., [29]), one should also expect a modification of the shape of the density-of-states functions. The importance of the latter effect can be estimated employing the model put forward by Bässler [30,31]. According to the model, DOS in disordered solids may be approximated by a Gaussian function

$$N(E) = \frac{N_{tot}}{(2\pi\sigma^2)^{1/2}} \exp\left(-\frac{(E - E_m)^2}{2\sigma^2} \right) \tag{4}$$

N_{tot} standing for the total density of states, E_m for the energy of the distribution maximum, and the variance σ being a measure of the disorder; typically, $\sigma \approx 0.1$ eV. The zero-field drift mobility parameter follows a non-Arrhenius temperature dependence

$$\mu_o(\sigma, T) = \mu_{o,o} \exp\left[-\left(\frac{2\sigma}{3kT} \right)^2 \right], \tag{5}$$

$\mu_{o,o}$ being usually identified with the mobility in a parent perfect material (on-chain mobility in the case of polymers). The effect of dipolar species has been rationalized invoking a dipolar contribution in the DOS function: the width of DOS is a superposition of van der Waals and dipolar components (σ_{vdW} and σ_{dip}, respectively)

$$\sigma^2 = \sigma_{vdW}^2 + \sigma_{dip}^2 \tag{6}$$

with the latter component related to the permanent dipole moment of the polar species and its mole fraction [25,26]

$$\sigma_{dip} = \frac{\kappa x_{dip}^{b} p}{\varepsilon a_o^2} \tag{7}$$

Here, κ and b are constants, ε is the relative electric permittivity of the medium, and a_o is the inter-site distance. A similar dependence has been found in polysilylenes as has been experimentally verified [13,19,20,27]. Taking the values reported in Ref. [27], one may approximate the concentration dependence of the dispersion parameter in diluted systems ($x_{dip} < 0.02$) by the following equation

$$\sigma^2 = 8.65\times10^{-3} + 0.23 p^2 x_{dip}^2 \tag{8}$$

σ in eV, p in D). Assuming the equation to hold for PMPSi doped with 2 mole % of spiropyran and taking the dipole moments from Table 1, one may estimate that the presence of the stable (closed) form results in broadening of σ by about 20 %, whereas the increase of σ due to the presence of the metastable (open) form amounts to 44 %. The mobility ratio $\mu^{(high\ p)}/\mu^{(low\ p)}$ at 300 K for the same difference dipole moments would amount to 0.18. Although not very low, the ratio should easily be decreased by a chemical modification of the side group: if the dipole moments of the stable and metastable form amount to 2 and 10 D, the mobility ratio would be equal to ca. 3×10^{-3}.

5. Conclusion

The estimates point to the possibility of a controlled modulation of charge carrier mobilities in ordered and disordered molecular materials, by a light-induced reversible photochemical reaction. Moreover, one may envisage the construction of a molecular-scale switch acting on the same principle. Discussing the architecture and function of the molecular switch, it is important to mention the role of the spacer. With a neutral spacer fixing the geometry of the photoactive side group, the electrostatic contribution of the dipole to the function of the switch is dominant. With spacer electronically connecting the side group with the main chain, i.e., allowing for the charge redistribution, the trapping of a charge carrier on the side group becomes more probable if only its ionization energy is lower than that of the main chain. The electrostatic contribution is then superimposed on the main term. In the former case, one deals with a dipolar trap created *on the chain* due to electrostatic interactions of the carrier with the dipole of the side group, whereas the latter case is equivalent to creation of a chemical trap mainly *on the side group*.

6. Acknowledgements

The research was supported by the Grant Agency of the Academy of Sciences of the Czech Republic (grants No A1050901 and AV0Z4050913) and by the Polish State Committee for Scientific Research (grant No 4 T09A 132 22).

7. References

1. Carter, F.L. (1982) Conformational switching at the molecular level, in: F.L. Carter (Ed.) *Molecular Electronic Devices*, M. Dekker, New York, pp. 51-71.
2. Carter, F.L. (1984) The molecular device computer: point of departure for large scale cellular automata, *Physica* **10D**, 175-194.
3. Nešpůrek, S. and Sworakowski, J. (1994) Electroactive and photochromic molecular materials for wires, switches and memories, *IEEE Eng. Med. Biol.* **13**, 45-57.
4. Tour, J.M., Kozaki, M., and Seminario, J.M. (1998) Molecular scale electronics: a synthetic/computational approach to digital computing, *J. Am. Chem. Soc.* **120**, 8486-8493.
5. Tans, S.J., Verschueren, A.R.M., and Dekker, C. (1998) Room-temperature transistor based on a single carbon nanotube, *Nature* **393**, 49-52.
6. Chen, J., Reed, M.A., Rawlett, A.M., and Tour, J.M. (1999) Large on-off ratios and negative differential resistance in a molecular electronic device, *Science* **286**, 1550-1551.
7. Gittins, D.I., Bethell, D., Schiffrin, D.J., and Nichols, R.J. (2000) A nanometre-scale electronic switch consisting of a metal cluster and redox-addressable groups, *Nature* **408,** 67-69.
8. Donhauser, Z.J., Mantooth, B.A., Kelly, K.F., Bumm, L.A., Monnell, J.D., Stapleton, J.J., Price D.W., Jr., Rawlett, A.M., Allara, D.L., Tour, J.M., and Weiss, P.S. (2001) Conductance switching in single molecules through conformational changes, *Science* **292**, 2303-2307.
9. Nešpůrek, S. and Sworakowski, J. (2001) Molecular current modulator consisting of conjugated polymer chain with photochemically attached photoactive side groups, *Thin Solid Films* **393**, 168-176.
10. Nešpůrek, S., Toman, P., Sworakowski, J., and Lipinski, J. (2002) Photochromic molecules as control units in molecular switches, in Graja, A., Bulka, B.R., and Kajzar, F. (Eds.), *Molecular Low Dimensional and Nanostructured Materials for Advanced Applications,* Kluwer Acad. Publishers, Dordrecht, pp. 37-47.
11. Sworakowski, J. and Nešpůrek, S. (1998) Contribution of dipolar species to the formation of local states for charge carriers in molecular materials, *Polish J. Chem.* **72**, 163-171.
12. Sworakowski, J. (2000) Effect of dipolar arrays on the localization of charge carriers in molecular materials, *IEEE Trans. Diel. Electr. Insul.* **7**, 531-536.
13. Nešpůrek, S., Pfleger, J., Brynda, E., Kmínek, I., Kadashchuk, A., Vakhnin, A., and Sworakowski, J. (2001) Poly(silylene)s: Effect of polar acceptor side groups on the charge carrier photogeneration and transport, *Mol. Cryst. Liq. Cryst.* **355**, 191-216.
14. Sworakowski, J. and Nešpůrek, S. (2002) Dipolar species in molecular materials: Towards modulation of charge carrier mobilities by electrostatic interactions, in Graja, A., Bulka, B.R., and Kajzar, F. (Eds.), *Molecular Low Dimensional and Nanostructured Materials for Advanced Applications,* Kluwer Acad. Publishers, Dordrecht, pp. 25-35.
15. Grozena, F.C., Siebbeles, L.D.A., Warman, J.M., Seki, S., Tagawa, S., and Scherf, U. (2002) Hole conduction along molecular wires: σ-bonded silicon versus π bond-conjugated carbon, *Adv. Mater.* **14,** 228-231.
16. Nešpůrek, S. and Eckhardt, A. (2001) Polysilylenes: charge carrier photogeneration and transport, *Polym. Adv. Technol.* **12**, 427-440.

17. Dürr, H. and Bouas-Laurent, H. (Eds.) (1990) *Photochromism. Molecules and Systems*, Elsevier, Amsterdam.

18. Crano, J.C. and Guglielmetti, R.J. (Eds.) (1999) *Organic Thermochromic and Photochromic Compounds*, Kluwer, New York.

19. Valerián, H., Brynda, E., Nešpůrek, S., and Schnabel, W. (1995) Effect of polar additives on charge carrier transport in polysilylenes, *J. Appl. Phys.*, **78**, 6071-6078.

20. Nešpůrek, S., Valerián, H., Eckhardt, A., Herden, V., and Schnabel, W. (2001) Charge carrier transport in poly[methyl(phenyl)silylene]: the effect of additives, *Polym. Adv. Technol.* **12**, 306-318.

21. Lyons, L.E. (1957) Photo- and semi-conductance in organic crystals. Pt. V. Ionized states in molecular crystals, *J. Chem. Soc.* 5001-5007.

22. Gutmann, F. and Lyons, L.E. (1967) *Organic Semiconductors*, J. Wiley, New York, Ch. 6.

23. Sworakowski, J. (1970) On the origin of trapping centres in organic molecular crystals, *Mol. Cryst. Liq. Cryst.* **11**, 1-11.

24. Silinsh, E.A. (1980) *Organic Molecular Crystals. Their Electronic States*, Springer Verlag, Berlin, Ch. 2&3.

25. Dieckmann, A., Bässler, H., and Borsenberger, P.M. (1994) An assessment of the role of dipoles on the density-of states function of disordered molecular solids, *J. Chem. Phys.* **99**, 8136-8141.

26. Young, R.H. and Fitzgerald, J.J. (1995) Effect of polar additives on charge transport in a molecularly doped polymer: evaluation of disorder models, *J. Chem. Phys.* 102, 9380-9391.

27. Nešpůrek, S., Sworakowski, J., and Kadashchuk, A. (2001) The influence of dipolar species on charge carrier transport in a linear polysilicon, *IEEE Trans. Diel. Electr. Insul.* **8**, 432-441.

28. Hoesterey, D.C. and Letson, G.M. (1963) The trapping of photocarriers in anthracene by anthraquinone, anthrone and naphthacene, *J. Phys. Chem. Solids* **24**, 1609-1615.

29. Wolf, U., Bässler, H., Borsenberger, P.M., and Gruenbaum, W.T. (1997) Hole trapping in molecularly doped polymers, *Chem. Phys.* **222**, 259-267.

30. Bässler, H. (1981) Localized states and electronic transport in single component organic solids with diagonal disorder, *Phys. Stat. Sol. (b)* **107**, 9-54.

31. Bässler, H. (1993) Charge transport in disordered organic photoconductors. A Monte Carlo simulation study, *Phys. Stat. Sol. (b)* **175**, 15-56.

Fluorescent Photochromic Diarylethene Oligomers

H. CHO, E. KIM*

Advanced Materials Division Lab 7, Korea Research Institute of Chemical Technology, P. O. Box 107, Yusong, Daejeon, 305-600, South Korea

1. Introduction

Fluorescent photochromic materials collect strong interest for their capability of controlling fluorescence by photochromic conversion of materials. In particular fluorescent diarylethenes, which shows reversible change in fluorescence intensity with photochromic reaction, are useful for light probe and non-destructive optical readout system.[1] Since diarylethene molecules do not show strong fluorescence (fluorescence quantum yields lower than 3 %), they have been modified by substituting a fluorescent chromophore such as anthracene, 2,4,5-triphenylimidazole, diphopyrines, or tungstene complex.[2-4]

Highly fluorescent diarylethene could be prepared by attaching the diarylethene unit to a fluorescent bis (phenylethynyl)-anthracene, however, it shows poor photochromic color contrast due to large absorption at 400-500 nm in both open(**O**) and closed (**C**) form.[5] Thus we embarked on syntehsis of a new type of diarylethene polymer having fluorophore bridge between the diarylethene unit. As a fluorophore bridge we introduce p-phenylene vinylene (PPV) derivatives since they have absorption shorter than 400 nm and show strong fluorescence.[6] Furthermore PPV has a planar structure that may induce planarity (anti parallel) in diarylethene unit. To minimize interaction between the fluorophores, we designed an alternating copolymer of photochromophore and fluorophore, as described below:

F. Charra et al. (eds.), Organic Nanophotonics, 437–446.

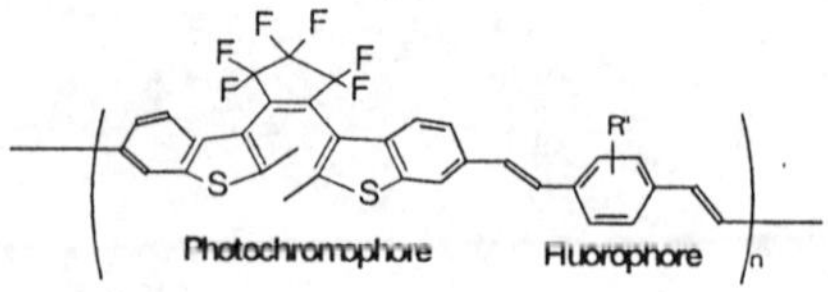

Here we report fluorescence properties of diarylethenes substituted with a vinyl and a styryl and diarylethene oligomers linked by PPV.

2. Experimental

2.1 MATERIALS

Potassium *tert*-butoxide (1.0 M solution in THF) was purchased from Aldrich and used as received. Triphenyl phosphine was purchased from Aldrich and recrystallized from hexane. Solvents were dried and purified by fractional distillation over sodium/benzophenone and handled in a moisture-free atmosphere. Column chromatography was performed using silica gel (Merck, 250-430 mesh). 1,2-Bis(2-methylbenzo[b]thiophene-3-yl)hexafluorocyclopentene (BTF6), 1-(6-formyl-2-methylbenzo[b]thiophene-3-yl),2-(2-methylbenzo[b]thiophene-3-yl)hexafluoro cyclopentene (CHOBTF6), 1-(6-vinyl-2-methylbenzo[b]thiophene-3-yl),2-(2-methylbenzo[b]thiophene-3-yl)hexafluorocyclopentene (VNBTF6), 1-(6-(2-phenylethenyl)-2-methylbenzo[b]thiophene-3-yl), 2-(2-methylbenzo[b]thiophene-3-yl)hexafluorocyclopentene (PEBTF6) 1,4- bis(triphenylphosphonionmethyl)benzene dibromide were synthesized according to the literature.[7-11] 2,5-Bis(2'-carbomethoxyethyleneoxy)-1,4-xylenebis(triphenylphosphonium bromide) was synethsized from 2,5-bis(2'-carbomethoxyethyleneoxy)-1,4-bis(bromomethyl)benzene and triphenylphosphine using the procedure in ref. 11.

2.2 SYTHESIS OF DIARYLETHENE OLIGOMER (BTF-PPV)

CHOBTF6 (0.7 g, 1.3 mmol) and 1,4-bis(triphenylphosphonionmethyl)-benzene dibromide (1.05 g, 1.3 mmol) were dissolved in $CHCl_3$ (10 mL) at room termperature. To the above solution was added dropwise a solution of *t*-BuOK(0.75 g, 6.6 mmol) dissolved in 10 mL of EtOH. After stirring for 24 hours, solvent was removed, to give yellow solids. Then, the product was redissolved in dichloromethane and precipitated in methanol. The product yield (BTF-PPV) was 62 %. M_n, 2702 ; M_w, 3806 (M_w / M_n =1.40). ^{1}H NMR (300 MHz, $CDCl_3$) δ 2.22 (s, 3H), 2.50 (s, 3H), 6.16 (m, 4H), 7.06-7.19 (4H), 7.58-7.84 (m, 4H), 8.06-8.14 (m, 2H).

Similary CHOBTF6 (1.05 g, 2 mmol) was reacted with 1,4-bis(triphenylphosphonionmethyl)-2,5-di(2'-carbomethoxyethyleneoxy)benzene dibromide salt (1.98 g, 2 mmol) in 20 mL of $CuCl_3$ and t-BuOK (1.12 g, 10 mmol) in ethanol (15 mL), to give BTF-EOPPV(Yield : 48%) : M_n, 3967 ; M_w, 7673 (M_w / M_n =1.93).

2.3 CHARACTERIZATION

^{1}H NMR spectra were recorded using a Bruker AM-300 spectrometer. The UV-visible spectra were recorded on a Shimadzu UV-3100 spectrophotometer with baseline corrections and normalizations carried out using Origin 6.0 software (Microcal, Software INC.). The FT-IR spectra were measured on a Nicolet DX-5B spectrometer in KBr pellets. The molecular weights of the oligomers were determined by gel permeation chromatography (GPC) analysis relative to polystyrene calibration (Waters high-pressure GPC assembly Model M590 pump, μ-styragel columns of 10^5, 10^4, 10^3, 500, and 100 Å, refractive index detectors) in a THF solution. Emission spectra were made on a dilute (~10^{-5} M) solution collected on a FluoroMax-2 (ISA Instruments S.A Inc) fluorometer utilizing a lock-in amplifier system uner Ar atmosphere at 25 °C. Flourescences quantum yields were determined by using quinine sulfate solution as a reference.[12]

3. Results and Discussion

3.1 FLURESCENCE PROPERTIES OF DIARYLETHENES SUBSTISUTED WITH VINYL AND 2-PHENYLETHENE

Photochromism of diarylethenes arises from the photo cyclization/ring opening of hexatriene(C1-C6) /cyclohexdiene, respectively, in diaryethene. In order to investigate the effect of substituent on the photochromism and photoluminescence property we studied absorption and emission property using three types of diarylethene, BTF6, VNBTF6, and PEBTF6, of which chemical structures are drawn below:

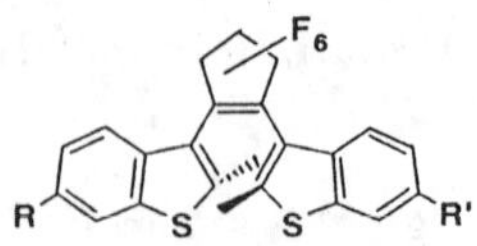

R, R': H (BTF6), R: $CH=CH_2$ R': H (VNBTF6), R, R' : $CH=CH-C_6H_5$ (PEBTF6)

Figure 1 shows the absorption spectral change of VNBTF6, and PEBTF6 in $CHCl_3$ by photo irradiation, compared to unsubstituted BTF6. Upon irradiation with UV light, the colorless solution of the diarylethene solution turned deep red, in which characteristic absorption maxima were observed at visible range. The red color indicates formation of the ring-closed isomer (**C**). Upon visible (λ > 540 nm) light irradiation, the red color disappeared, indicating the formation of initial open form (**O**). The λ_{max} of **O** form of PEBTF6 was observed at 313 nm, which is shifted toward longer wavelength as compared with those of VNBTF6 and BTF6. Similarly, the λ_{max} at 545 nm for **C** isomer at visible region was also shifted toward longer wavelength compared to those of VNBTF6 and BTF6.

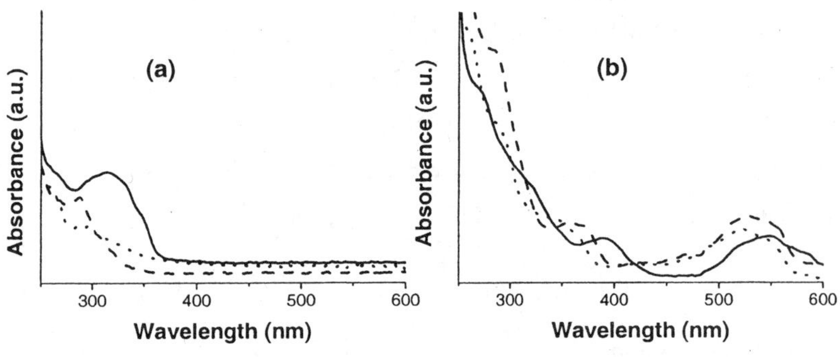

Figure 1. UV-visible spectra of BTF6 (dot), VNBTF6 (dash), and PEBTF6 (solid) in open form (a), and those in closed form (b). Concentration of diarylethene in chloroform : (1×10^{-5} M).

This result indicates that the ground and excited state of PEBTF6 in both **C** and **O** were stabilized by 2-phenyl ethenyl substitutent, possibly due to longer π-conjugation. Figure 2 shows photoluminescence spectra for VNBTF6 and PEBTF6 dissolved in $CHCl_3$ (1x 10^{-5} M). The emission intensity of diarylethene derivatives of **O** form in Figure 2 was stronger than that of **C** form, excited at the λ_{max} of open form. This result correlates well with previous studies for other diarylethene derivatives.[1-4]

Interestingly emission intensity of PEBTF6 was stronger than those of BTF6 and VNBTF6. The λ_{max} of the emssion band (λ_{em}) was red shifted as the the conjugation length increassed. Thus λ_{em} for PEBTF6 in **O** form was red shifted by over 60 nm compared to BTF6. Similary λ_{em} for PEBTF6 in **C** form was red shifted by over 40 nm compared to BTF6. Fluorescence quantum yield (Φ_F) was increased in PEBTF6 in both **O** and **C** forms. However, Φ_F of VNBTF6, was low. This result indicates that substituting vinly only can not increase Φ_F. On the otherhands, 2-phenylethene group increase Φ_F by a factor of 2.5, compared to BTF6. Spectral data and fluorescence quantum yield (Φ_F) are summarized in Table 1.

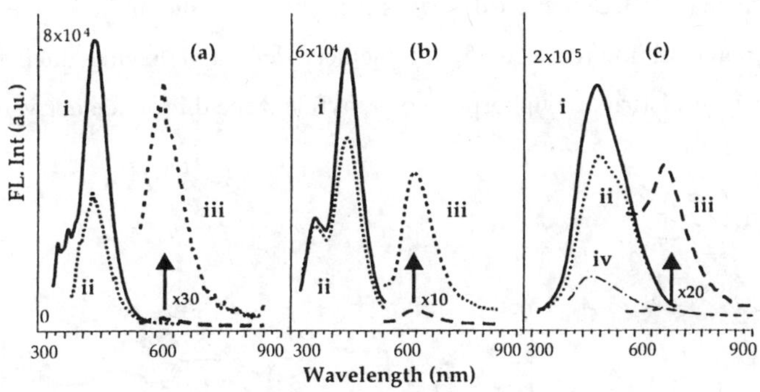

Figure 2. Photoluminescence spectra of BTF6 (a), VNBTF6 (b), and PEBTF6 (c) in chloroform solution, open form (**i**, excited at λ_{max} of open form) and in closed form (**ii** - **iv**) excited at λ_{max} of closed form in UV region (**ii**), that in visible region (**iii**), and at 389 nm (**iv**).

3.2 DIARYLETHENE OLIGOMERS

As described above substituting vinly only cannot increase Φ_F but it was increased when diarylethene was substituted by 2-phenylethene. This results led us to synthesize diarylethene oligomers linked by a phenylenevinylene, to increase Φ_F.

TABLE I. Sepctral data and fluorescence quantum yield (Φ_F) of diarylethene derivatives

Compds	state	λ_{ex} nm	λ_{em} nm	stokes' shift eV	Φ_F (%)
BTF6	open	297	432	1.3	2.2
	closed	297	432	1.3	0.6
	closed	520	621	0.39	0.1
VNBTF6	open	290	441	1.47	1.5
	closed	290	444	1.47	1.1
	closed	522	646	0.46	0. 09
PEBTF6	open	313	491	1.44	5.1
	closed	313	491	1.44	3.6
	closed	389	478	0.59	0.8
	closed	545	664	0.41	0.1

Two types of diarylethene oligomers were synthesized by changing substituents of p-phenylenevinylene (PPV). Scheme 1 describes the synthetic route of diarylethene oligomers by Wittig polycondensation reaction of a diarylethene derivative having aldehyde groups (**1**, CHOBTF6) and 1,4-bis(triphenylphosphoniomethyl)benzene dibromide derivatives (**2**).

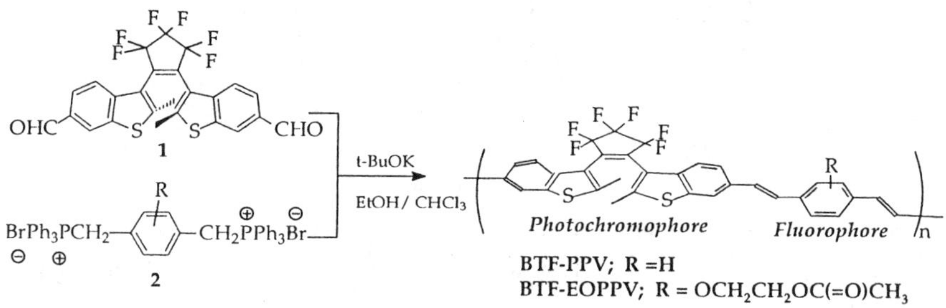

Scheme 1. Synthetic route of diaylethene oligomers linked with a fluorescent builder

Figure 3 (a) shows UV-visible spectral change of BTF-PPV dissolved in $CHCl_3$ by photo irradiation. Upon irradiation with 325 nm light, the colorless solution of the open form of BTF-PPV solution turned deep red, in which characteristic absorption maxima were observed at 381 and 555 nm. The red color indicates formation of the ring-closed (**C**) form. Upon visible (λ> 540 nm) light irradiation, the red color disappeared indicating the formation of initial open (**O**) form. The λ_{max} of the **O** form of BTF-PPV was observed at

340 nm, which is red shifted by ~30 nm compared to that of the monomeric type (PEBTF6, compare with Figure 1). Similarly, the λ_{max} at 565 nm for closed isomer at visible region was also red shifted by 20 nm compared to that of PEBTF6. This result indicates that the enegy levels of the singlet-excited state for both **C** and **O** form of BTF-PPV are stablized by increase in the π-conjugation length.

Interestingly, photochromic conversion of BTF-PPV was fast and quantum yield of photocyclization (Φ_{pc}) of BTF-PPV in $CHCl_3$ solution was determined as 0.38.[13] Φ_{pc} of BTF-PPV was comparable to that of tetramer (0.40 in hexane) bridged by ethynylene group.[14] Such high Φ_{pc} of BTF-PPV could be due to the formation of high **ap** isomer. Figure 3(b) shows UV-Vis spectra of BTF-EOPPV in chloroform solution (1x 10^{-5} M) before and after irradiation with UV. The absorption spectrum of the **O** form for BTF-EOPPV showed two bands with lower intensity at 330 nm and maximized at 396 nm (λ_{max}). Surpringly, BTF-EOPPV does not show photochromic property when the solution was excited with a UV light. As shown in Figure 3(b) the spectrum after UV irradiation was almost overlaped with the spectrum of the sample before UV irradiation.

A chloroform solution of BTF-PPV (1x 10^{-5} M) showed strong fluorescence bands when excited with a UV light. Figure 4 shows photoluminescence spectra of BTF-PPV solution excited at absorption maxima. Emission band at 450 nm was observed from the **O** form of BTF-PPV, when excited with a UV light. The fluorescence intensity of BTF-PPV was much higher than those of monomeric diarylethenes (VNBTF6 and PEBTF6).

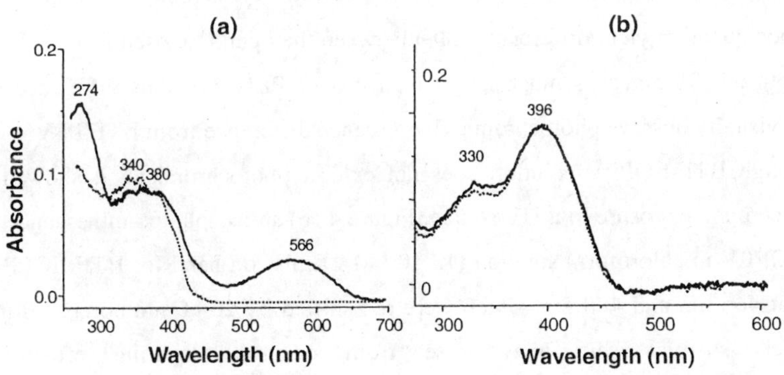

Figure 3. (a) UV-visible spectra of BTF-PPV in chloroform solution: before (dotted line) and after UV irradiation (solid line). (b) UV-visible spectra BTF-EOPPV in chloroform solution : before (dotted line) and after irradiation with a 325 nm laser (solid line).

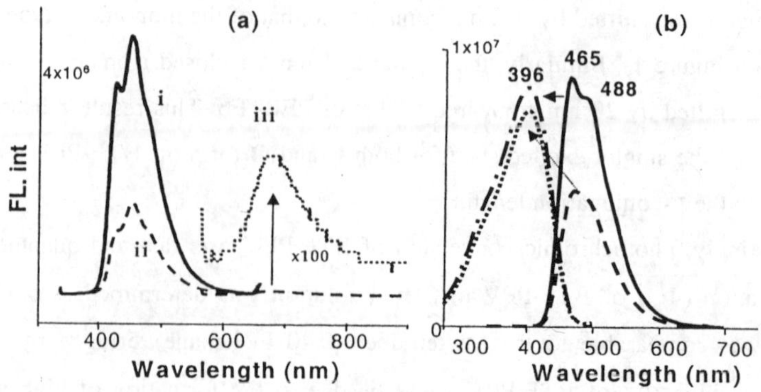

Figure 4. (a) Photoluminescence spectra of BTF-PPV, open form (**i**, excited at ëmax of open form) and in closed form (**ii** and **iii**) excited at ëmax of the closed form in UV region (**ii**), and that in visible region (**iii**). (b) Photoluminescence and excitation spectra of BTF-EOPPV. Excited at 330 nm (dashed), 390 nm (solid line).

The Φ_F for **O** form of BTF-PPV excited at 381nm was determined as 0.53, which is quite high as compared to that of other photochromic diarylethenes. Thus both photochromic response and fluorescence quantum yields of BTF-PPV were increased by incorporating PPV unit, compared to those of monomeric diarylethenes. However, Φ_F of **O** form of BTF-PPV was much decreased when excited at 339 nm, indicating that photochromic cyclization and photoluminescence process are competitive in this state. Φ_F of **C** form of BTF-PPV, excited at 381 nm was lower than that of the **O** form. Such result correlates well with the previous observation that the closed isomers do not emit significantly.[4,15] Importantly the fluorescence quantum yield difference ($\Delta\Phi_F$) between the open *vs* closed form of BTF-PPV, excited both at 381 nm, was much larger than that of PEBTF6. This difference is large enough to visually observe photochromic fluorescence difference from BTF-PPV.

Although BTF-EOPPV solution does not exhibit photochromism, it showed strong fluorescence upon exposure to a UV source. Figure 4 (b) shows photoluminescence spectra of BTF-EOPPV in chloroform solution (1x 10^{-5} M). Emission band for BTF-EOPPV were observed at 465 nm and 488 nm, which were red shifted by 20-40 nm compared to BTF-PPV. Such spectral shift may arise from electron donating effect of 2-carbomethoxyethyleneoxy group. Excitation spectra for BTF-EOPPV in Figure 4 (b) mirrored the absorption band at 390 nm (Figure 3(b)). The results of the steady-state fluorescence study for diarlyethene oligomers are summarized in Table 2.

TABLE II. Fluorescence property of diarylehene oligomers dissolve in chloroform.

Oligomer	State of diaryethene	λ_{ex}	λ_{em}	Stoke shift, eV	Φ_{fl}
BTF-PPV	Open	339	450	0.902	0.16
	open	381	450	0.50	0.53
	Closed	339	453	0.920	0.07
	Closed	381	454	0.523	0.13
	Closed	556	674	0.390	0.002
BTF-EOPPV	Open	330	465, 488	1.090	0.55
	"	396		0.590	0. 59

The fluorescence intensity of BTF-EOPPV was increased compared to that of BTF-PPV. Therefore singlet exited state molecules of BTF-EOPPV decays back to ground state with light emission, without photocyclization. Significant intramolecular charge transfer between the hexafluorocyclopentene (acceptor) and the other part of molecules (benzothiopene and PPV) could occur to yield higher content of parallel isomer compared to BTF-PPV. ^{1}H NMR spectrum for BTF-EOPPV indicates that high content of **p** isomer. Further systematic studies are in progress to understand the photophysical behavior of BTF-EOPPV.

4. Conclusion

A new type of fluorescent diarylethene polymer was synthesized by connecting diarylethene unit with PPV, through Wittig polycondensation reaction. Such connections of PPV with diarylethene unit are an effective strategy for obtaining dual functionality of photochromism and fluorescence with high Φ_F. The dual functionality was controllable by the wavelength of excitation light source and highly affected by the substituent of PPV ring.

Acknowledgement.

This work was supported by MOST of Korea under National Research Laboratory program.

References and Notes

1. Irie, M. and Sayo, K. (1992) Solvent effects on the photochromic reactions of diarylethene derivatives *Journal of Physical Chemistry* **96**, 7671.

2. (a) Tsivgoulius, G. M. and Lehn, J. -M. (1996) Photoswitched and functionalized oligothiophenes: synthesis and photochemical and electrochemical properties *J. Chem. Eur.* **2**, 1399. (b) Takeshita, M. and Irie. M. (1998) Reversible fluorescence intensity change of a diarylethene *Chem. Lett.* 1123.

3. Ern, J., Benz, A. T., Martin, H. -D., Mukamel, S., Tretiak, S., Tsyganenko, K., Kuldova, K., Trommsdorff, H. P. and Kryschi, C. (2001) Reaction dynamics of a photochromic fluorescing dithienylethene *J. Phys. Chem. A*, **105**, 1741.

4. Yagi, K., Soong, C. F. and Irie, M. (2001) Synthesis of fluorescent diarylethenes having a 2,4,5-triphenylimidazole chromophore *J. Org. Chem.* 66, 5419.

5. Kawai, T., Sasaki, T. and Irie, M. (2001) A photoresponsive laser dye containing photochromic dithienylethene units *Chem. Comm.* 711.

6. (a) Burroughes, J. H., Bradley, D. D. C., Brown, A. R., Marks, R. N., MacKay, K., Friend, R. H., Burn, P. L. and Holmes, A. B. (1990) Light-emitting diodes based on conjugated polymers *Nature* **347**, 539. (b) Hsieh, B. R., Yu, Y., Forsythe, E. W., Schaaf, G. M. and Feld, W. A. (1998) A New Family of Highly Emissive Soluble Poly(*p*-phenylene vinylene) Derivatives. A Step toward Fully Conjugated Blue-Emitting Poly(*p*-phenylene vinylenes) *J. Am. Chem. Soc.* **120*(1)*** 231.

7. Irie, M.; Miyatake, O.; Uchita, K. and Eriguchi, T. (1994) Photochromic Diarylethenes with Intralocking Arms *J. Am. Chem. Soc*, 9894

8. Cho H., Cho S. Y., Kim Y., Jeon D. and Kim E. (2002) Fluorescence property of photochromic diarylethene derivatives , *J. Korean Society for Imaging Science*, in press.

9. Cho S. Y., Yoo M., Shin H. -W., Ahn K. -H. , Kim Y. -R., and Kim E. (2002) Preparation of Diarylethene Copolymers and Their Photo-induced Refractive Index Change *Organic Materials*, in press.

10. Kim M. –S., Kawai T. and Irie M (2000) Synthesis and photochromism of amorphous diarylethene having stynyl substituents, *Mol. Cryst. And Liq. Cryst.* **345**, 251.

11. Wang S., Xiao S., Li Y., Shi Z., Du C., Fang H. and Zhu D. (2002) Synthesis and characterization of new C60-PPV dyads containing carbazold moiety, *Polymer* **43**, 2049.

12. (a) Demas J. N. and Crosby G. A Measurement of photoluminescence quantum yields. Review (1971) *J. Phys. Chem* **75**, 991 (b) Ng S. C., Ma Y. F., Chan H. S. O and Dou Z. L (1999) Syntheses and characterization of electrically conductive and fluorescent poly[3-(-bromoalkyl)thiophenes] *Synth. Metals* **100**, 269

13. Mejiritski, A., Polykarpov, A. Y., Sarker, A. M. and D. C. Neckkers (1997) Determined from the changes of absorption due to ring-cyclization reactions according to, *J. Photochem. Photobiol* **108**, 289.

14. Kaieda, T., Kobatake, Miyasaka, S., Murakami, H. M., Iwai, N., Nagata, Y., Itaya, A. and Irie, M. (2002) Efficient Photocyclization of Dithienylethene Dimer, Trimer, and Tetramer: Quantum Yield and Reaction Dynamics *J. Am. Chem. Soc.* **124**, 2015.

15. Fernandez-Acebes, A. and Lehn, J. (1998) Optical switching and fluorescence modulation in photochromic metal complexes M. *Adv. Matr.* **10**, 1519.

THE SEARCH FOR HIGHLY POLAR BETAINIC TYPE MOLECULES FOR ELECTRO-OPTICAL APPLICATIONS

O. NEILANDS[a], I. MUZIKANTE[b]
[a]*Department of Organic Chemistry, Riga Technical University, 14/24 Azenes Street, LV-1048, Riga, Latvia*
[b]*Institute of Physical Energetics, 21 Aizkraukles Street, LV-1006, Riga, Latvia*

Abstract. We propose a novel class of NLO active materials – dipolar intramolecular salts (betaines), comprising in molecule directly connected electron donor five member β-dicarbonyl compound anion moiety and electron acceptor heterocyclic N-onium cation moiety. The betaines are different in principle from commonly used push-pull dipolar molecules. Calculations have shown a strong localization of HOMO and LUMO. PIET from HOMO to LUMO calls giant change of dipole moment. The calculated hyperpolarizabilities are remarkable. Betaines are being commonly to synthesize. It is a good possibility for fine tuning of their NLO properties. Preparing of betaine solution in polymer matrices (host-guest systems) and formation of films is the best method for investigation of betaines NLO properties. Obtained films is being subjected to corona poling and then measured the surface potential and changes by irradiation. It is urgently needed to exclude contact with air or exclude lighting in the processing of betaines films due to easy photo oxidation.

1. Introduction

Molecular nonlinear optics (NLO) has attracted increasing interest over the past ten years. NLO materials are highly promising for applications in various fields including telecommunications, optical data storage and processing, optical power limitation and others. The strategy of search of novel NLO molecules is based usually on the design of dipolar push-pull chromophores (molecules combining an electron-donating (so-called donor – D) group and an electron withdrawing (so-called acceptor – A) group, connected by a conjugated system (spacer) [1,2]. "Direct polar conjugation" between D and A groups realizes in many cases. The term "direct polar conjugation" means the constitution of molecule can be described by two mesomeric formulas: D-CH=CH-A ↔ $^{+}$D=CH-CH=A^{-}. The main groups A and D, and spacers of a dipolar push-pull system are depicted in Fig.1. It is important to remark recently synthesized [1] push-pull molecules comprising donor ferrocene (b) or tetrathiafulvalene (d) do not belong to "direct polar conjugation" family.

A new concept of octupolar nonlinearities of two- and threedimensional molecules has recently been proposed [3]. Research of octupolar structures has mainly focused on organic chromophores with threefold symmetry, for example, the dye "crystal violet". Metal complexes offer many possibilities for the design of threedimensional octupolar molecules. Enhanced molecular hyperpolarizability has been observed for octahedral

F. Charra et al. (eds.), *Organic Nanophotonics*, 447–462.

and tetrahedral metal complexes. A model of a tetrahedral metal complex is depicted in Fig. 2. The acceptor A is a metal atom, but four donor groups D are bound by four spacers in a tetrahedral configuration.

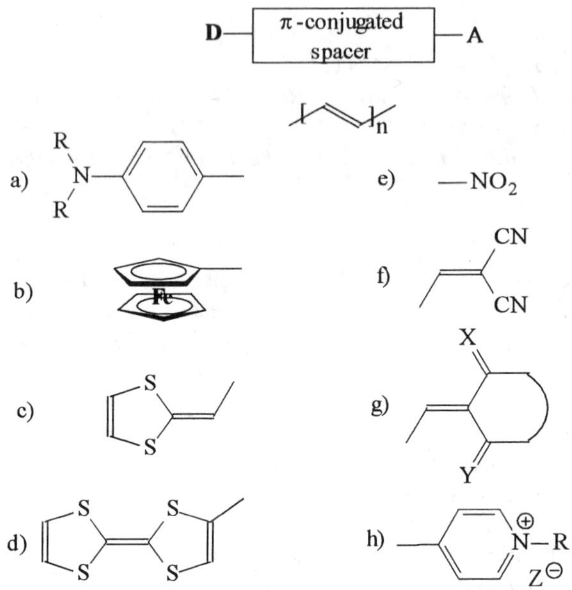

Figure 1. The main types of push-pull dipolar molecules

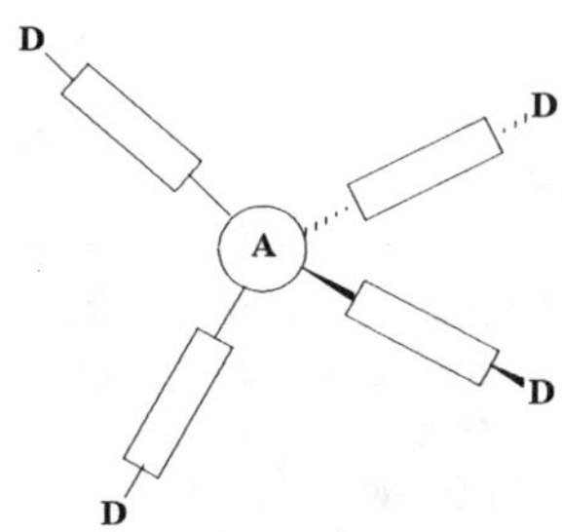

Figure 2. Model of a tetrahedral octupolar molecule

2. Dipolar intramolecular salts (betaines) – different from push-pull dipolar molecules class of NLO materials

Our approach of search for NLO materials is different of commonly used [4-6]. We aimed to synthesize dipolar intramolecular salts – betaines, comprising in molecule directly connected electron donor anion part and acceptor N-onium cation part. We can see in Fig. 3 the general constitution of our objects of interest.

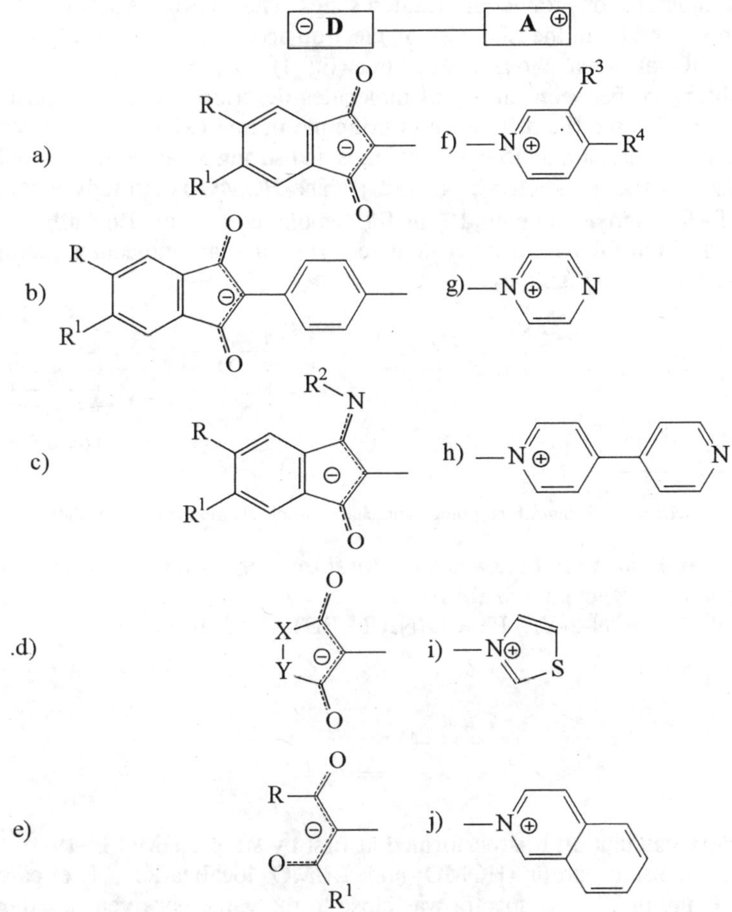

Figure 3. The general constitution of dipolar intramolecular salts (betaines)

Fine tuning of acceptor and donor properties of betaines is easily to realize. For example, pyrazinium cation (f) or isoquinolinium cation(j) are much more stronger acceptors as pyridinium cation (a). Respectively, open chain β-dicarbonyl compound anion (e) is a weaker donor as indan-1,3-dione anion (a).

Simple quantum chemical calculations [6] showed that molecules of betaines are strong polar, the electric dipole moments of the ground state are equal 4 – 16 D depending on the space between D and A centers. The dipole moment changes by excitation considerably. Photo induced intramolecular electron transfer (PIET) from HOMO to LUMO calls remarkable change of the electric dipole moment in the excited state, sometimes the dipole moment changes to opposite. The difference between values of dipole moments in ground and excited states is unusually large. The topology of frontal molecular orbitals, HOMO and LUMO is the cause of the large difference

between dipole moments of ground and excited states. The HOMO is strongly localized on electron donor part D, but the LUMO – on electron acceptor part A. Overlapping of HOMO and LUMO takes place to a small degree (fig. 4).
What is the difference between push-pull molecules described on fig. 2 and betaines depicted on fig. 3? We see the difference in principle in non-existence of "direct polar conjugation" between donor and acceptor moieties and strong localization of LUMO on acceptor moiety (in fig. 3 moieties f, g, h, i, j). The HOMO is strongly localized on anion part of β-dicarbonyl compound (in fig.3 moieties a – e). Recently, a strong localization of HOMO has been calculated for dipolar molecules comprising tetrathiafulvalene moiety[1] (Fig. 2, d).

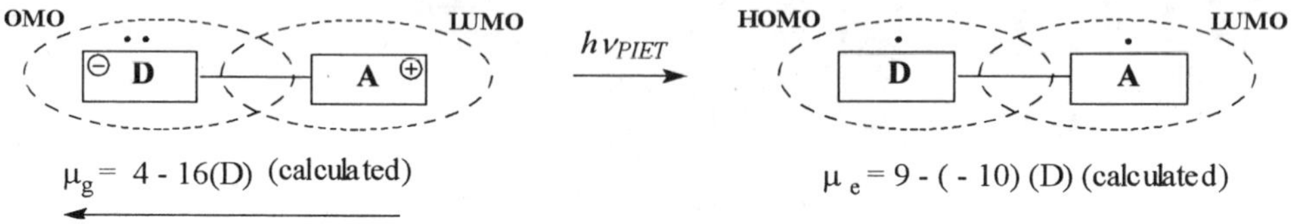

Figure 4. Scheme of frontal MO topology and change of dipole moments by excitation

3. Characterization of main betaines by orbital topology, dipole moments, UV VIS spectra and hyperpolarizability

3.1. N- (INDAN-1,3-DION-2-YL) PYRIDINIUM BETAINE (IPB) (1)

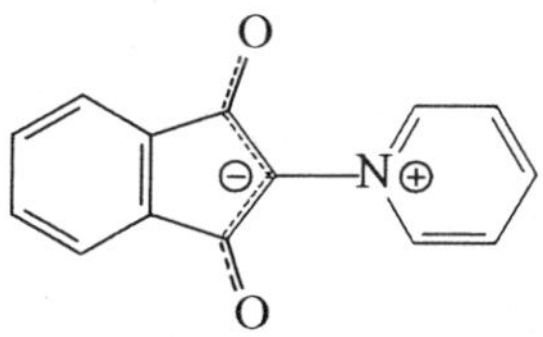

1

Quantum chemical calculation was performed at first by simple HMO in 1970 [7]. It has been shown an unusual strong HOMO and LUMO localization. The calculated absorption maximum in UV Vis spectra was close to the value observed in experiment (412 nm in chloroform). This absorption was called as "intramolecular electron transfer band". Further calculations[8] confirmed the possibility of an effective photo induced electron transfer (PIET). The present calculation (AM1, ZINDO/S) will demonstrate (Fig. 5) the topology of HOMO and LUMO of IPB, values of electric dipole moments in ground μ_g and excited μ_e states, and the difference $\Delta\mu = \mu_g - \mu_e$, absorption maximum λ, polarizability α_o and hyperpolarizability β_o.
The constitution of IPB molecule is very simple. Despite of this simplicity IPB molecule theoretically possess unexpected NLO properties due to strong localization of HOMO and LUMO, comparable with hyperpolarizability of 4-nitro-4'-dimethylaminostilbene.
The electric dipole moment IPB molecule measured in dioxane solution is as high as 4.48 (D)[10].

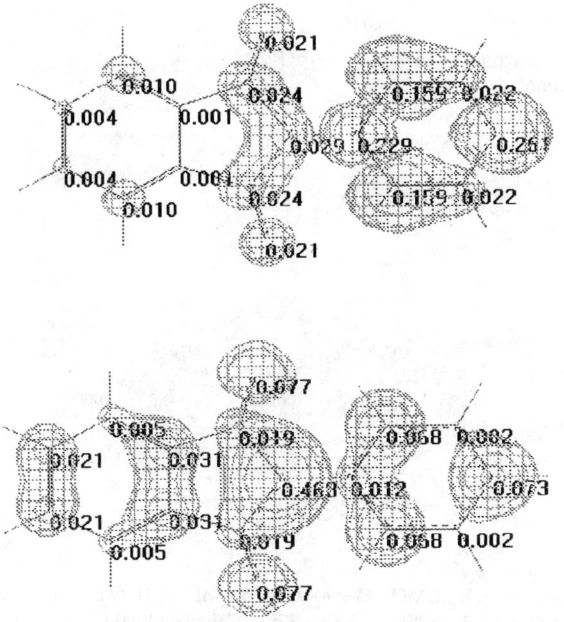

Figure 5. HOMO (below) and LUMO (above) topology of IPB (**1**). Calculated electron density on pyridinium moiety for HOMO – 23%, for LUMO – 85%. Calculated values of μ , λ , α and β: $\mu_g = 4.2$ (D) , $\mu_e = -3.3$ (D) , $\Delta\mu = 7.5$ (D); $\lambda = 435$ nm (experim. 392 nm in ethanol, 431 nm in hexane ($\varepsilon = 33500$)); $\alpha_o = 48$ (10^{-24} esu), $\beta_o = 19$ (10^{-30} esu)[9].

3.2. N-[4'-(INDAN-1,3-DION-2-YL) PHENYL] PYRIDINIUM BETAINE (IPPB) (2)

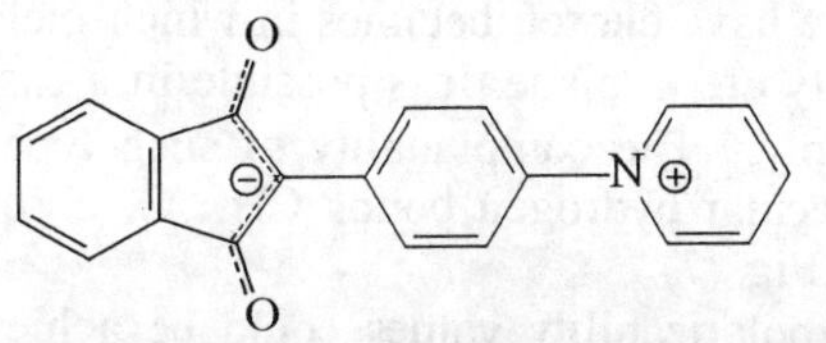

2

The molecule of IPPB (**2**) is longer as molecule of IPB and distance between anion moiety and pyridinium cation is higher as 6 Å. Therefore this betaine has very large dipole moment in the ground state. We can see in Fig. 6 the strong localization of HOMO on indandione moiety and LUMO on pyridinium moiety[9].

The IPPB molecule characterizes with unusual high calculated hyperpolarizability, and experimental has been observed remarkable batochromic shift (143 nm) of absorption maximum by change of polar solvent methanol to chloroform.

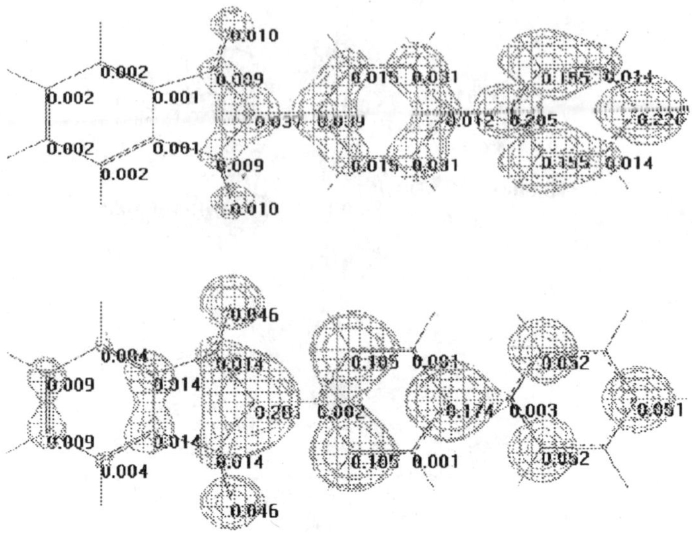

Figure 6. HOMO (below) and LUMO (above) topology of IPPB (**2**). Calculated electron density on pyridinium moiety for HOMO – 16%, for LUMO – 77%. Calculated values of μ , λ , α and β: $\mu_g = 13$ (D) , $\mu_e = 1$ (D) , $\Delta\mu = 12$ (D); $\lambda = 615$ nm (experim. 430 nm in methanol, 573 nm in chloroform); $\alpha_o = 106$ $(10^{-24}$ esu), $\beta_o = 120$ $(10^{-30}$ esu)[9].

3.3. HOW TO FIND IPB TYPE MOLECULES POSSESSING HIGH HYPERPOLARIZABILITY?

We have performed quantum chemical cakculations (AM1, ZINDO/S) for a line of betainic type molecules comprising donor and acceptor sustituents in different positions of betainic molecule [6, 9]. We have chosen betaines in which molecules pyridinium ring and indandione anion moiety are in plane. It is possible in a case of five member ring β-diketone anion system only. The complanarity of such type betainic molecules is stabilized by two intramolecular hydrogen bonds CH…O, observed by X-Ray [11] and NMR [12] measurements (see Fig. 7).

The enhancement of hyperpolarizability values could be achieved by increasing of donor properties of anion system and increasing of acceptor properties of pyridinium moiety. We can see on Fig. 7 that dimethylaminogroup (donor) increases β_o value when is located in anion moiety only. Cyanogroup (acceptor) contrary to dimethylaminogroup increases β_o value when is located in pyridinium moiety.

Betaine **4** is of especially interest. This is a derivative of heterocyclic five membered β-dicarbonyl compound – pyrazolidine-3,5-dione. It means not only indandione derivatives, but also betaines of heterocyclic five membered dicarbonyl compounds could be useful for NLO materials.

Betaine **4** is of especially interest. This is a derivative of heterocyclic five membered β-dicarbonyl compound – pyrazolidine-3,5-dione. It means not only indandione derivatives, but also betaines of heterocyclic five membered dicarbonyl compounds could be useful for NLO materials.

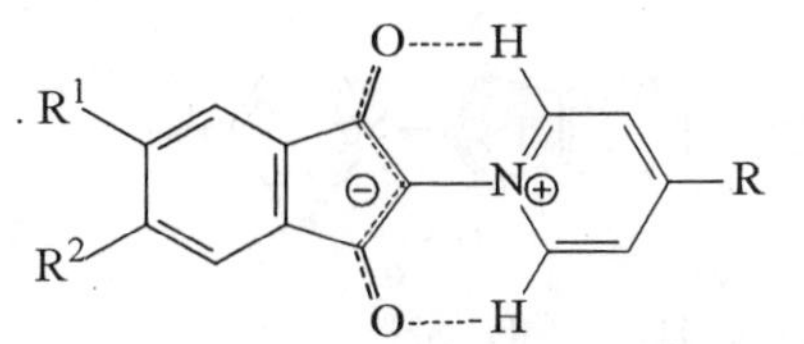

3

R	R^1	R^2	β_o (10^{-30} esu)
CN	H	H	30
$(CH_3)_2N$	H	H	17
H	$(CH_3)_2N$	H	31
H	CH_3O	CH_3O	29

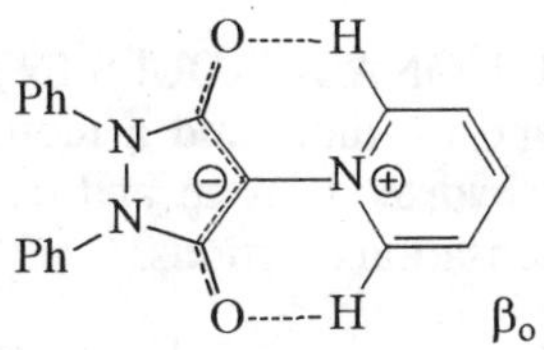

4

$\beta_o = 20$

Figure 7. Calculated hyperpolarizabilities of some substituted IPB molecules **3** and N-(1,2-diphenyl-pyrazolidine-3,5-dion-4-yl) pyridinium betaine (**4**). The intramolecular hydrogen bonds are shown.

4. Syntheses of dipolar intramolecular salts (betaines)

4.1. SYNTHESIS OF INDAN-1,3-DIONE BETAINES

Synthesis of IPB system have been performed by use of 2 different methods. The first method is based on phthalic anhydride or substituted phthalic anhydrides **5** condensation with N-(carboxymethyl) pyridinium salts **6** in the presence of acetanhydride and triethylamine [13,14]

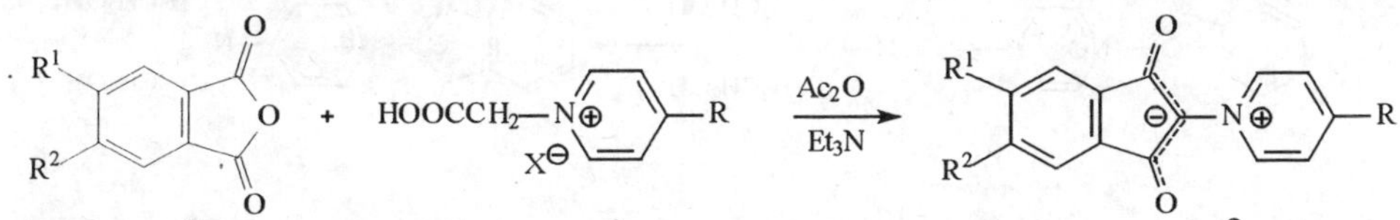

5 **6** **3**

This method is very simple and useful for many substituted phthalic anhydrides, cinchomeronic and quinolinic acid anhydrides [15 - 18]. Reaction failed in the case of aminosubstituted phthalic anhydrides.

Pyromellitic dianhydride (7) is a highly active compound, easily reacts with salts **6** and transforms in IPB-dicarboxylic acid anhydride (**8**)[19]. Mentioned anhydride is an important starting material for synthesis of oligomers.

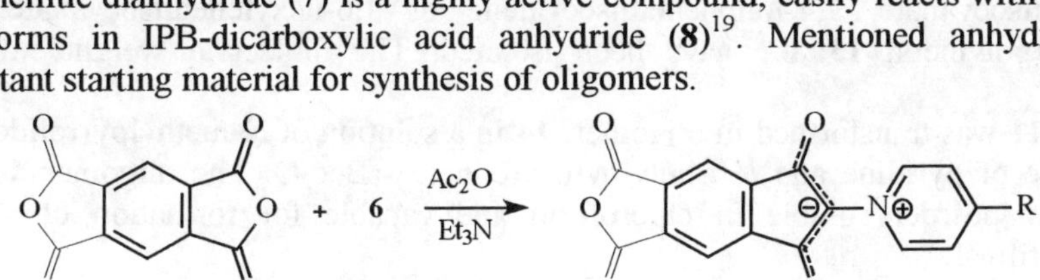

7 **8**

The second method is based on transformations of 2-dicyanomethyleneindan-1,3-dione oxide (**9**) in reactions with N-heterocycles (pyridine, isoquinoline, thiazole, imidazole, pyrazole)[20 - 24]. This is an important method for preparation of chloro-, amino-, mercapto-, hydroxysubstituted IPB **10**.

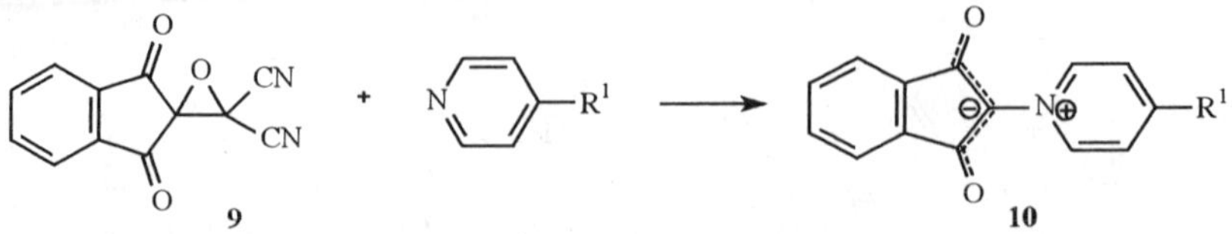

R^1 = Cl, NH_2, $N(CH_3)_2$, OH, SH

4.2. TRANSFORMATIONS OF FUNCTIONAL GROUPS IN BETAINES **8** AND **10**

Betainanhydride **8** interacted with butane-1,4-diole and a monoester **11** was isolated. This monoester comprises in molecule hydroxyl group and carboxyl group, therefore could be used as monomer for polycondensation reactions.

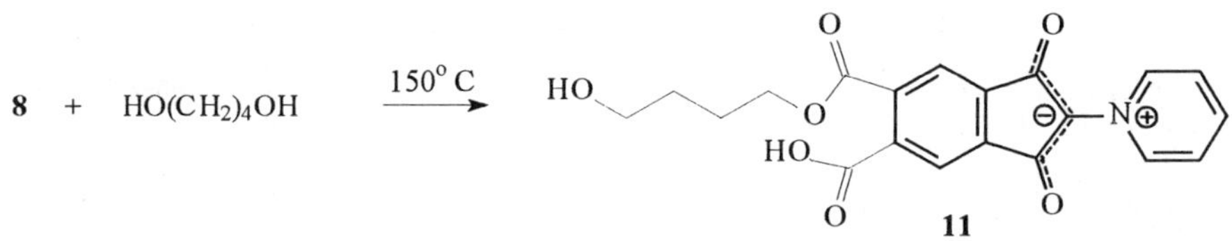

Chlorosubstituted IPB (**10**, R^1 = Cl) contains in molecule a mobile chlorine atom capable for interaction with nucleophilic reagents. A substitution take place by reaction with amines, for example,with diethanolamine. Di(hydroxyethyl)amino-IPB **12** resulted in a high yield [25, 26]. This compound is useful for synthesis of polyuretanes.

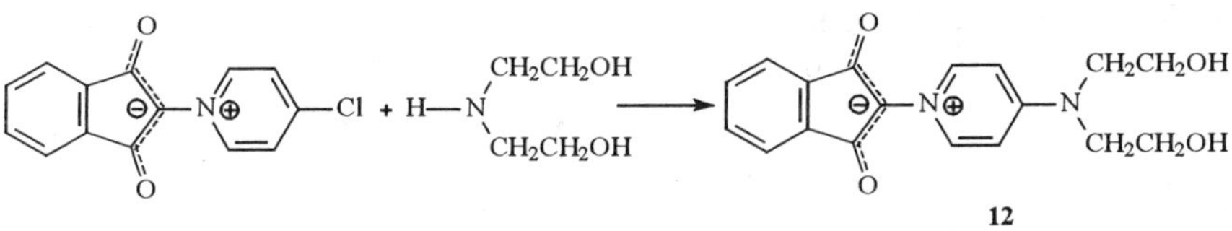

4.3. OLIGOMERIZATION AND POLYMERIZATION OF BIFUNCTIONAL BETAINES **3**

We was in success to produce oligo(poly)uretanes[25,26] and oligo(poly)esters, comprising IPB moieties. Dihydroxyethylamino-IPB (**12**) was reacted with 1,4-phenylenediisocyanate, 2,4-toluylenediisocyanate or 1,6-hexylenediisocyanate, and oligomers (polymers) **13a,b,c** have been isolated. The molecular weight was not measured.

Monoester **11** was transformed in oligomer **14** in a solution of N-methylpyrrolidone in the presence of pyridine and *tert*-butylpyrocarbonate - Boc_2O. The oligomer **14** is a deep yellow powder soluble in chloroform and capable for formation of yellow transparent films.

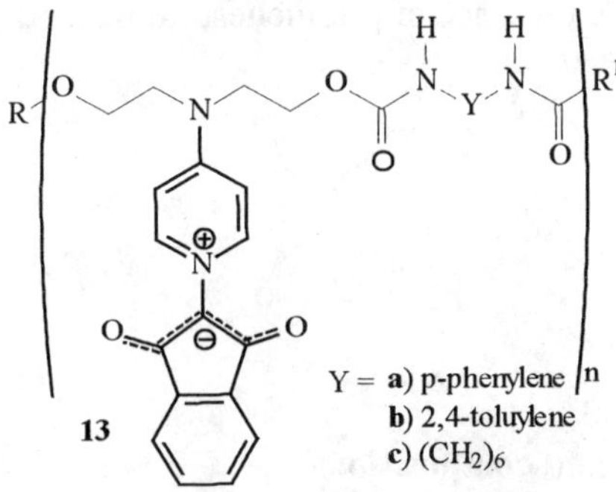

13

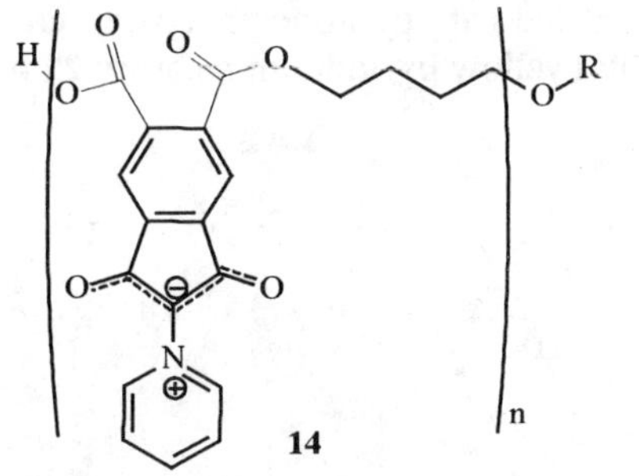

14

4.4. SYNTHESIS OF N-[4'-(INDAN-1,3-DION-2-YL) PHENYL] PYRIDINIUM BETAINE (IPPB) (2) AND SUBSTITUTED ONES

The starting compound for synthesis is being 2-(4'-aminophenyl)indan-1,3-dione (**15**, R = H) or substituted compounds. The aminogroup can be transformed in pyridinium ring by two ways. At first, arylamines react easily with N-2,4-dinitrophenylpyridinium salts and transform in N-arylpyridinium salts[27]. Secondly, substituted pyrilium salts interact with arylamines, and reaction results in substituted N-arylpyridinium salts [27]. The arylamine in our case is compound **15**.

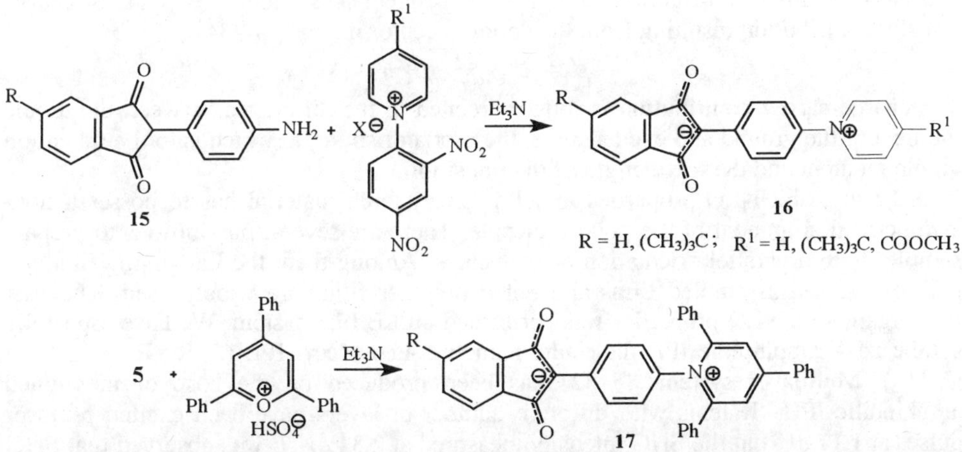

15 16

R = H, $(CH_3)_3C$; R^1 = H, $(CH_3)_3C$, $COOCH_3$

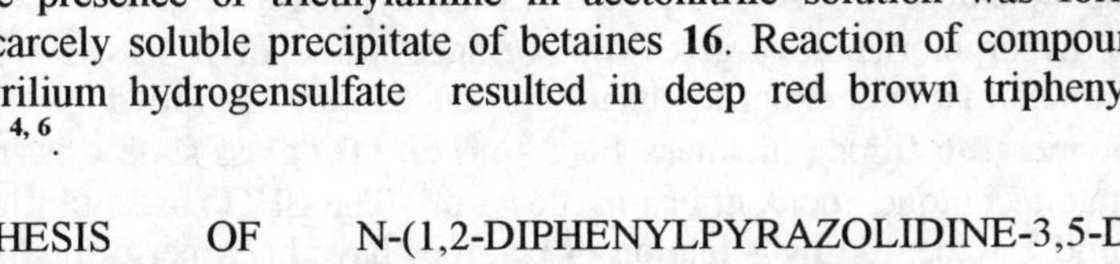

17

By interaction of compounds **15** with N-(2',4'-dinitrophenyl)-4-substituted pyridinium salts in the presence of triethylamine in acetonitrile solution was formed a deep coloured scarcely soluble precipitate of betaines **16**. Reaction of compounds **15** with triphenylpyrilium hydrogensulfate resulted in deep red brown triphenylpyridinium betaines **17** [4, 6].

4.5. SYNTHESIS OF N-(1,2-DIPHENYLPYRAZOLIDINE-3,5-DION-4-YL) PYRIDINIUM BETAINE (4) AND SUBSTITUTED ONES

1,2-Diphenylpyrazolidine-3.5-dione (**18**) easily interacts with phenyliodosyldiacetate (**19**) and transforms in phenyliodonium betaine **20**. This phenyliodonium betaine reacted with pyridine or substituted pyridines in the presence of p-toluenesulfonic acid , and yellow pyridinium betaines **21** have been isolated [28].

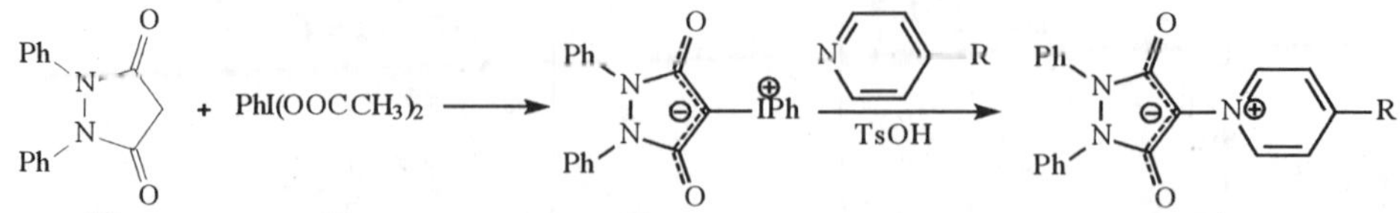

5. Observed NLO effects of betaines solutions and compositions

Dipolar betaines are easily preparable compounds as we can see from material above. Experimental investigations of NLO properties are not so many. The high photoconductivy of dipolar betaines in polycrystallinic evaporated films was investigated much more[29].

As a result of the asymmetry of charge distribution, the IPB molecule possesses in the ground state a considerable permanent dipole moment equal to 4.5D. Such molecule from point of view of SHG are regarded as 1-D since the hyperpolarizability β vector is oriented parallel to the ground state dipole moment μ_g of the molecule and directly proportional to the diference $\Delta\mu=\mu_g-\mu_e$. The hyperpolarizability β of D-π-A compouds consists of additive contributions from the independent substituent effects and of charge transfer contribution resulting from the donor-acceptor interaction [34]

$$\beta = \beta_{add} + \beta_{CT}$$

The charge carrier contribution is roughly related to the difference between the dipole moment of the ground and excited state, the oscilator strength, which is to the transition dipole moment and the wavelength of the transition.

In order to exibit NLO properties in solid state, a bulk material has to possessa non-symmetrical alignment of the polar molecule. There are several possibilities to prepare samples with a parallel orientation of molecules. Among them the Langmuir- Blodgett (LB) films, self-assembled films and poled polymer films are widely used. The first investigation on NLO properties was performed on LB film system. We have especially synthesized amphiphile IPB molecule, namely 5-heptadecyl-IPB (**3**, R =R^2 = H, R_1 = $C_{17}H_{35}$). Multilayer systems (8–32) has been produced on the base of mentioned amphiphilic IPB. Systems with different number of layers have been excited by laser pulses at 1.17 *eV*, but the SHG intensity measured at 2.34 *eV*. It was observed that SHG intensity of different samples was a function of the square of the layers number[18].

It was investigated chloroform solutions of IPB **1** by Hyper-Rayleigh scattering method. Evaluation of hyperpolarizability gives an averaged value of β 138.10 $^{-30}$ esu[30]. It should be comment that later experiments to reproduce this value failed.

The poled polymer **13b** films (thickness 1 – 3 μm) on ITO glass slide was investigated by Maker fringe technique, incident beam 1064 μm. The SHG susceptibility d_{33} = 25 pm/V was found[26]. Recently investigation of betaines have been performed in polymer (polymethylmetacrylate, PMMA) matrices (host-guest systems). A solution of PMMA and IPB or substituted derivatives has been evaporated by spin-coating method and obtained IPB/PMMA films have been subjected to a high-electric field corona poling at softening temperature of polymer. The arrangement of the polar molecules in the film is expected to be clarified through the surface potential measurements. The surface potentials are governed by dipole moment of the ordered polar molecules in the film and are related to chemical constitution, packing densities, molecular orientation of the molecules. The surface potential can be related to the dipole moment normal to the plane of the film $\mu_x = \varepsilon\varepsilon_0 A U_S$, where A is the average area occupied by the molecule [35]. Surface potential U_S studies provide useful information regarding both structural and electronic properties of oriented films [35, 36].

5.1. EXPERIMENTAL

In our case the host-guest polymer films with betaine molecules were chosen. The poled polymer films are promising due to preparing sufficient quality and sizes films for optical applications. Optically transparent polymer PMMA with molecular weight 440000 was used. IPB and it derivative molecules were dissolved in PMMA at concentration of 1wt%. The chloroform as solvent was used to deposit casted films onto gold and ITO electrodes. The thickness of the film was between 0.14 and 1.1 μm.
The high-electric field corona poling was used to generate dipole orientation of betaine molecules in polymer matrix. The corona poling of PMMA and IPB/PMMA films was performed according to Ref.[37]. A high voltage of 4 - 8kV was applied to a tungsten wire (diameter 25 μm) at the distance of 1 cm from the sample surface. During the poling procedure, which continued 10 min, the corona current was constant of 1 μA. The poling temperature of IPB/PMMA film was constant in the range 50-95°C. The poling temperature of PMMA film was about 110°C. The samples were corona poled through a mask with diameter 1cm and cooled to room temperature at applied electric field maintaining a constant current.

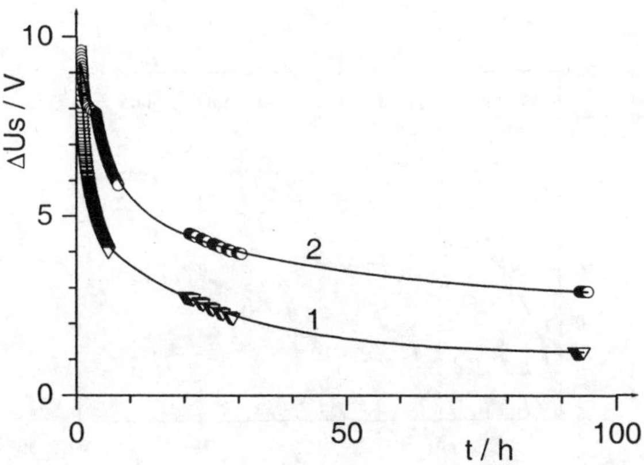

Figure 8. Decay of surface potential of PMMA (∇) and IPB/PMMA (O) films after corona poling fitted with stretched exponential functions (1) and (2) respectively.

The changes of the value of the molecular dipole moment of betaine molecules change the surface potential ΔU_S of the host-guest films. The measurements of changes of the surface potential were made by the Kelvin probe technique [38]. The surface potential was measured as the difference between the surface potentials of the grounded bottom electrode and the film.

The surface potential of the host-guest film changed going from unpoled to poled region of the film. In the case of IPB/PMMA films the value of the surface potential increased from some mV in un-poled region to several volts in the poled region. The unpoled polymer is isotropic due to random distribution of polar molecules in polymer matrix. The total dipole moment of the film is close to zero and the value of surface potential is negligible. The increase of the surface potential would have to been due to orientation of polar molecules in the polymer. For applications, it is important that the highly poled order is maintained for a considerable time. So the efficiency and persistence of alignment and studies of relaxation rates are of considerable interest. As shown in Fig. 8, a decay of the surface potential after poling both of PMMA and IPB/PMMA films has been observed. The decay of the surface potential is well represented by the sum of two exponential terms $U_S = A\exp[-t/\tau_1) + B\exp(-t/\tau_2)$, where τ_1 and τ_2 represent the time constants of two relaxation process, a faster and slower one. The initial faster decay of the surface potential in the case of IPB/PMMA host-guest was τ_1~6 hours. In the case of pure PMMA film the decay was faster of τ_1~1.4 hours [31]. In common with the surface potential the intensity of SHG signal is due to polar molecules aligning in the polymer matrix. According to studies in the decay of non-linear optical properties [39], the initial faster decay of the surface potential may be related to the rotation of polar molecules in the polymer matrix. At the same time decay of a surface charge, which arose during corona poling, decreases both the intensity of SHG signal and the surface potential.

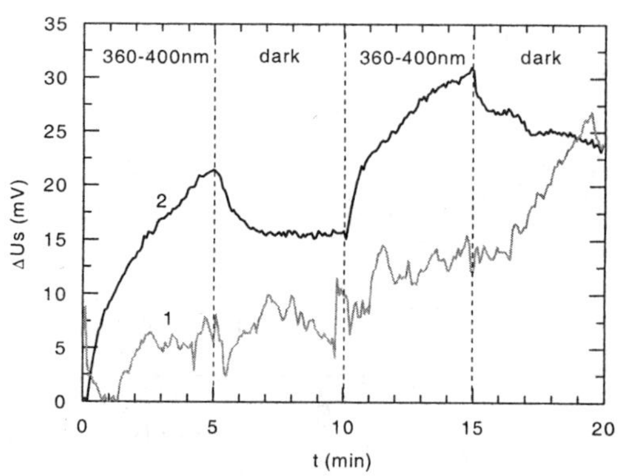

Figure 9. Dependence of the surface potential on irradiation within 360-400nm spectral region of un-poled (1) and poled (2) IPPB (**17**)/PMMA host guest film. The sample was poled at 6.2kV, 1μA, 75°C for 10 min.

The slower process with τ_2=43 hours for IPB/PMMA and τ_2=23 hours for PMMA samples may be due to thermal relaxation of the polymer matrix. The time constant of the decay of SHG coefficient d_{33} was measured in poled host-guest PMMA films. In the case of Disperse red 1 as guest molecules the time constants are of the same order τ_1~0.4 hours and τ_2=10 hours [40].
The changes of the surface potential of both IPB and IPPB host-guest films were examined under irradiation with light of selected wavelength. In the case of un-poled samples the switching process was negligible. While in the case of poled samples the value of ΔU_S was in the range of 95mV for IPB/PMMA and of 15mV for IPPB/PMMA films (see Fig. 9.). The response time was $\tau_{res} \approx 7$min. The values of ΔU_S and τ_{res} remained the same after several repeated cycles of alternating irradiation.

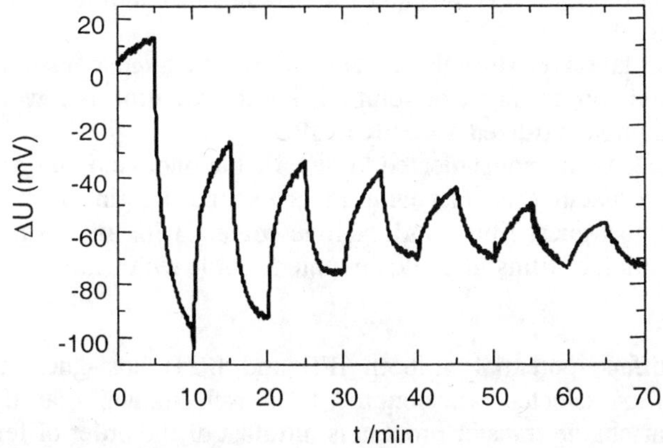

Figure 10. Dependence of the changes of the surface potential of N-(1,2-diphenylpyrazolidine-3,5-dion-4-yl)-pyridinium betaine (**4**) in PMMA matrix on alternate irradiation with light at 380 nm. The changes of surface potential decrease on irradiation and increase in darkness.

The host-guest films of betaine molecules are unstable to irradiation with light due to the photochemical oxidation. Hence as shown in Fig. 10, the photoinduced changes of the surface potential at PIET were partly reversible and decreased with irradiation time. The changes of surface potential decreased on irradiation with light centered at 380 nm. In darkness the reversible process was observed. The changes of the surface potential due to irradiation decreased with increasing to number of irradiation cycles. The decrease of the changes of the surface potential might be described by exponential function with time constant of 6min.
We have often observed that yellow solution of IPB and derivatives discolored by keeping in sunlight.
Nearly all betaines are unstable on irradiation in the PIET band in solutions and host-guest polymer films in the presence of air oxygen. We propose the photochemical oxidation by air oxygen takes place. Experimental investigations conformed interaction of IPB with oxygen on irradiation resulted in decomposition products: phthalic acid (**23**), 2,2-dihydroxyindan-1,3-dione (**24**) [33] and other products.

of IPB with oxygen on irradiation resulted in decomposition products: phthalic acid (**23**), 2,2-dihydroxyindan-1,3-dione (**24**) [33] and other products.

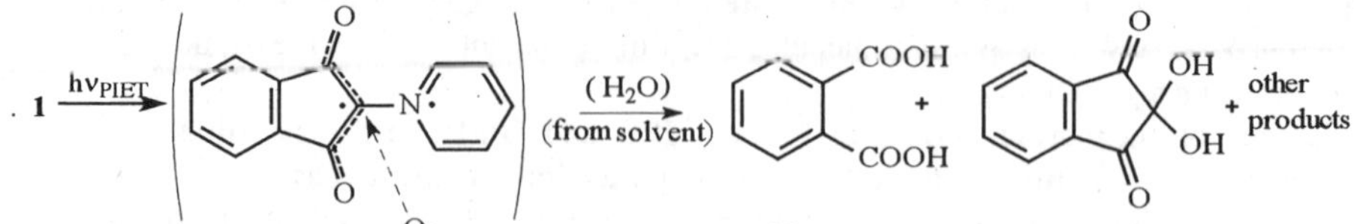

22 **23** **24**

IPB in crystals are not subjected to photochemical oxidation. We explain this easy photochemical oxidation in solutions by the excited state **22** peculiarities, namely high single electron density on 2-carbon atom of indan-1,3-dione moiety (see Fig. 5) and higher life-time of excited state. IPB molecules in crystal form a dimer[11] what is stable to photochemical oxidation.

The rate of photo oxidation is strongly dependent on chemical constitution of IPB derivatives, solvent and concentration of solution. Photo oxidation is lowered in water solutions. Betaine **4** is photo oxidized as easily as IPB.

The "longer" betaines **2**, **16** are not subjected to remarkable photo oxidation.

It is urgently needed to exclude contact with air or exclude lighting in the process of preparing of betaines host-guest films, and exclude contact with air oxygen in optical and electrical investigation of films. It could be quite feasible technically.

5.2. DISCUSSION

Why changed the surface potential of both IPB and IPPB host-guest films under irradiation with light of selected wavelength? It is well-known, that the optically induced intramolecular charge transfer process is ultrafast of the order of femtoseconds [41]. In our case the response time is of several minutes. The Kelvin probe technique is too slow to follow ultrafast processes.

We propose, in our case the following responses to irradiation caused by fast intramolecular changes of dipole moment in betaine molecules were observed. The surface potential is a macroscopic response of the film. Switching of single molecules yields only an extremely small change of the surface potential. We might monitor only the relation between molecules in the ground and excited states until the equilibrium state is achieved. The estimated change of the normal component of the dipole moment was of the order of 0.1D (for IPPB (17)/PMMA) and 0.2D (for IPB/PMMA), which was far less than theoretically calculated. It may be due to several reasons. One of them is that some of molecules are oriented by corona poling. Some molecules could form dimers due to large dipole moments. On the other hand, the orientation is calculated for isolated molecules, which should differ for molecules included in polymer matrix.

6. References

1. Gonzalez, M., Segura, J.L., Seoane, C., Martin, N., Garin, J., Orduna J., Alcala R., Villacampa, B., Hernandez V., and Navarrete, J.T.L. (2001) Tetrathiafulvalene derivatives as NLO-phores: synthesis,electrochemistry, Raman spectroscopy, theoretical calculations, and NLO properties of novel TTF-derived donor-π- acceptor dyads, *J. Org. Chem.* **66**, 8872-8882, and references therein.
2. Alain, V., Blanchard-Desce, M., Ledoux-Rak, I., and Zyss, J. (2000) Amphiphilic polyenic push-pull chromophores for nonlinear optical applications, *Chem. Commun.*, 353-354.
3. Renouard, T. Le Boze, H., Brasselet, S., Ledoux, I., and Zyss, J. (1999) Tetrahedral bipyridyl copper(I) complexes: a new class of non-dipolar chromophore for nonlinear optics, *Chem. Commun.*, 871-872, and references therein.
4. Neilands, O., and Utinans, M. (1997) Quantum chemical calculations, synthesis and properties of novel organic molecules with the photoinduced intramolecular electron transfer and giant change of the electric dipole moment in the exited state," *SPIE Proceedings* **2968,** 13-18.
5. Neilands O. (1998) From deep coloured 2-aryl-1,3-indandiones - organic semiconductors to photoconductors, non-linear optical materials and synthetic metals, *Latvian Journ.Physics and Techn. Sc.* 28-38.
6. Utinans, M., and Neilands O. (1999) Quantum chemical calculations, synthesis and properties of novel organic materials with the photoinduced intramolecular electron transfer and giant change of the electric dipole moment in the excited state," *Advanced Materials for Optics and Electronics* **9**, 19-25.
7. Neiland O. (1971) The characterization of π-electron structure of active methylene group betaine like derivatives. 3. 2-N-pyridiniumindandione-1,3-betaine, *Izv.AN Latvian SSR.. Ser.Khim.*, 85-91 (in Russian).
8. Kampar, V.E., Bruvers, Z.P., and Neiland, O.Y. (1981) The calculation of electron spectra in CNDO/S approximation and character of long wave absorption band of 2-N-pyridinium-1,3-indandione betaine, *Chem. Heterocycl. Compounds,* 658-661.
9. A. Jurgis (2002), unpublished results.
10. Kampars, V.E, Kacen, Y.Y., Mazheika, I.B., and Neiland, O.Y. (1979) The dipole moments in ground and excited states in some betaine like derivatives of 1,3-indandione, *Izv. AN.Latv. SSR, Ser.Khim.*, 541 - 546 (in Russian).
11. Kaminskii, V.F., Shibaeva, R.P, and Neiland, O.Y. (1976) The crystal and molecular structure of 2-N-pyridiniumindandione -1,3 betaine $C_{14}H_9NO_2$, *Zhurn. Strukturnoi Khim.*, **17**, 898-903 (in Russian).
12. Neiland, O.Y., Kampare, R.B., Prikule, D.E., and Liepinsh, E.E. (1979) On intramolecular hydrogen bond CH...O in 2-(N-pyridinium)-1,3-indandione betaines, *Khim. Geterocikl. Soedinenii,* 379-383 (in Russian).
13. Neiland, O.Y., and Pavars A.P. (1968) Method of preparing intramolecular salts of indandione-1,3 2-derivatives, USSR Pat. 228669, Byull. Izobret. 32.
14. Neiland, O.Y., and Pavars, A.P. (1966) Synthesis of 2-N-pyridinium and N-isoquinolinium betaines of indandione-1,3, *Zhurn. Organ. Khim.*, **6**, 634-635 (in Russian).
15. Pukitis G.G., and Neiland, O.Y. (1983) Method of preparing N-heteroaromatic onium betaines of 5-nitro-1,3-indandione," USSR Pat. 1023763.
16. Neiland, O.Y., and Krauze, A.A. (1980) Synthesis and properties of 2-(N-pyridinium)-4- and 5-aza-1,3-indandiones," *Khim. Heterocycl. Soedin.*, 1637 - 1641(in Russian).
17. Paulinsh, L.L., Pukitis, G.G., and Neiland, O.Y. (1985) Synthesis of strong electron donors in the series of 2-(N-pyridinium)-1,3-indandione betaines," *Izv. AN.Latv. SSR, Ser.Khim.*, 612 - 617 (in Russian).
18. Rutkis, M.A., Gerca., L.E , Silinsh, E.A., Neilands O.Y., Roze, M.P., Berzinsh, E.L., Klimkans, A.B. , and Larssons S. (1993) Langmuir-Blodgett films of indandione-1,3 pyridinium betaine. I Preparation, electronic structure, optical properties, *Advanced Materials for Optics and Electronics,* **2,** 319-330.
19. Neilands, O.Y., Shebenina E.B., and Pukitis, G.G. (1999) Investigation of reaction ability of substituted phthalic acids and anhydrides in a condensation reaction with N-carboxymethylpyridinium, substituted N-carboxymethylpyridinium and N-carboxymethylisoquinolinium salts and synthesis of novel N-(1H-indene- 1,3(2H)-dion-2-yl)pyridiniobetaines, *Khim. Heterocycl. Soedin.,*1647-1654 (in Russian).
20. Junek, H., Hermetter D., and Fischer-Colbrie,H. (1974) Ein Pyridinium Ylid des 1,3-Indandions, *Angew. Chem.* **86**, 380.
21. Neiland, O.Y., and Raiskuma, I.K. (1976) Method for preparation heterocyclic onium betaines of 1,3-indandione, USSR Pat. 537067, *Byull. Izobret. (44).*

22. Raiskuma, I.K., Pukitis, G.G., and Neiland, O.Y. (1985) Synthesis of N-heteroaromatic onium betaines of 1,3-indandione, *Izv. AN.Latv. SSR, Ser.Khim.*,612 - 617 (in Russian).
23. Neiland, O.Y., Prikule D.E., Adamsone, B.Y., Kampar, V.E., and Pukitis, G.G. (1980) Synthesis and investigation of properties amino- and carboxylderivatives of 2-(N-pyridinium)-1,3-indandione betaine," *Izv. AN.Latv. SSR, Ser.Khim.* 663 - 668 (in Russian).
24. Neiland O.Y., Raiskuma I.K., Kampare, R.B., and Kampar, V.E. (1982) 2-N- Imidazolyl- and 2-N-pyrazolyl-1,3-indandiones, *Khim. Heterocycl. Soedin.* 1082 – 1085 (in Russian).
25. Dubrovich O., Utinans, M., Zauls, V., and Neilands, O. (1999) Synthesis, Modelling and Characterization of Novel Non-Linear Optical Polymers Containing Indandionylpyridinium Betaine (IPB) Units," *Materials Science and Engineering C* **8/9**, 391-394.
26. Utinans, M., Dubrovich, O., Neilands,O., Zauls, V., and Muzikante I. (2000) NLO polymers containing as active chromophore indandionylpyridinium betaines units: synthesis, modelling and characterization , *in "Multiphoton and Light Driven Multielectron Processes in Organics: New phenomena, Materials, Application"*, ed. by F.Kajzar et al, NATO ASI Series, Kluwer Acad. Publ., 475-482.
27. Sliwa, W. (1985) The chemistry of N-substituted pyridinium salts, *Heterocycles*, **23,** 1513-1554.
28. Adamsone, B.Y., Neilands, O.Y., and Kampare, R.B. (1982) Iodonium derivatives of β-diketones. 21. Synthesis and properties of 1,2-diphenyl-3,5-pyrazolidinedione iodonium derivatives," *Zhurn. Organ. Khim.* **18**, 890-894 (in Russian)
29. Muzikante, I.,.Silinsh, E.A, Taure, L., and Neilands, O. (1998) Physical properties of thin evaporated films of N-(indan-1,3-dion-2-yl) pyridinium betaine and it derivatives, *Latvian Journ.Physics and Techn. Sc.*, **4**, 10-27.
30. Liberts, G., Zauls, V., and Neilands, O. (1997) Nonlinear optical properties of some indandione-1,3 betaine derivatives," *SPIE Procedings,* **2968**, 19-23.
31. Muzikante, I., Fonavs, E., Tokmakov, A., Stiller, B., Brehmer, L., Neilands, O., and Balodis, K. (2002) Studies of relaxation processes in dipolar dye-doped polymeric films," *Material Science &Engeneering C* (in press).
32. Muzikante, I., Neilands, O., and Markava, E. (2002) Optically induced electrical properties of new advanced organic materials, *SPIE Proceedings* (in press).
33. Kappe,Th., Khorkid-Zadeh,R., and Steininger,R. (1975) Mesoionische Sechsringheterocyclen. V Pyridinium -Ylide von Malonylheterocyclen, *Zeitschr. Naturforschung,* **30B**, 773-777.
34. Hissink, C.E. (1996) Silicon-bridged donor-acceptor compounds: synthesis and nonlinear optical properties, University Groningen, The Netherlands, 176 p.
35. Oliveira, O.N.Jr., Taylor, D.M., Lewis, T.J., Salvagno, S., and Stirling, C.J.M. (1989) Estimation of group dipole moments from surface potential measurements on Langmuir monolayers, *J.Chem.Soc., Faraday Trans. I*, **85**, 1009-1018.
36. Taylor, D.M. (2000), Developments in the theoretical modelling and experimental measurement of the surface potential of condensed monolayers, *Adv. Colloid and Interface Science* **87**, 183-203.
37. Mortazavi, M.A., Knoesen, A., Kowel, S.T, Higgins, B.G., and Dienes, A. (1989) Second-harmonic generation and absorption studies of polymer-dye films oriented by corona-onset poling at elevated temperatures, *J.Opt.Soc.Am.B* **6**, 733-741.
38. Vilitis, O., Fonavs, E., and Muzikante, I. (2001) A system for measuring surface potential by the Kelvin-Zisman vibrating capacitor probe, *Latv.J.Phys. Tech.Sciences* **5**, 38-56.
39. Suzuki, A., and Matsuoka Y. (1995) Relaxation processes in poled nonlinear optical polymer films, *J.Appl.Phys.* **77**, 965 – 969.
40. Casalboni, M., Sarcinelli, F., Pizzoferrato, R., D'Amato, R.., Furlani, A., and Russo M.V. (2000) Second-harmonic generation and absorption spectra of platinium organometallic complexes incorporated in PMMA films, *Chem.Phys.Lett.* **319**, 107-112.
41. Silinsh, E.A. (1997),Organic molecular materials for non-linear optics, *SPIE Proceedings*, **2968**, 2-12.

INTERMOLECULAR CHARGE TRANSFER MULTILAYERS FOR NLO APPLICATIONS

F. KAJZAR[A], Y. OKADA-SHUDO[B], C. MERITT[c] AND Z. KAFAFI[C]
[a]*CEA – DRT/LIST, DECS/SE2M/LCOF, Centre d'Etudes de Saclay, 91191 GIF SUR YVETTE CEDEX, France*
[b]*University of Electro-communications, Tokyo, Japan*
[c]*US Naval Research Laboratory, Washington, D. C. 20375, USA*

1. Introduction

It is well known that the centrosymmetric molecules and centrosymmetric materials don't exhibit, at least within the dipolar scheme, second order nonlinear optical properties useful fore a large class of practical applications, such as frequency conversion through second harmonic generation or parametric processes, light amplification, electro-optic modulation for signal transmission, generation of THz pulses, etc. Therefore a lot of effort was devoted to the design of the highly efficient molecules, with large molecular first hyperpolarizability β tensor and assembling them into a noncentrosymmetric bulk material with optimized $\chi^{(2)}$ susceptibility. The first approach is known as molecular engineering and the second as material engineering, respectively [1].

It was shown [2] that the 1D intramolecular charge transfer (CT) are characterized by large β values, with the β_{zzz} component enhanced in the charge transfer direction z. However these molecules exhibit also a large ground state dipole moment. Due to the usually strong dipole – dipole interaction the CT molecules align antiparallel in order to minimize the ground state energy. It leads to macroscopically centrosymmetric structures, despite the fact that the constituent molecules are noncentrosymmetric. As lot of practical applications are targeted in thin film configuration (integrated optics) much of effort was paid for building such noncentrosymmetric structures. Fundamentally four kind of approaches were used in getting noncentrosymmetric thin films

(i) single monocrystalline films
(ii) Langmuir-Blodgett layers
(iii) Self-assembly
(iv) Poled polymers

The first approach is costly and difficult. It is particularly difficult to obtain in this way monodomain thin films with low propagation losses, as required by device applications [3-4]. Lot of efforts was done with the fabrication of Langmuir-Blodgett thin films [5]. The technique allows to fabricate thin films with a good optical quality and well defined thickness. However the natural tendency in this technique is to form centrosymmetric Y

F. Charra et al. (eds.), Organic Nanophotonics, 463–474.

type films whereas noncentrosymmetric X and Z type are required for second-order NLO applications. Some molecules spontaneously arrange on a substrate giving a non centrosymmetric thin films. However the NLO susceptibilities of such films usually are small. Lot of successful work was devoted to the fabrication, characterization and device application of noncentrosymmetric thin films by using isotropic amorphous polymers with chemically bond highly nonlinear CT chromophores [6].

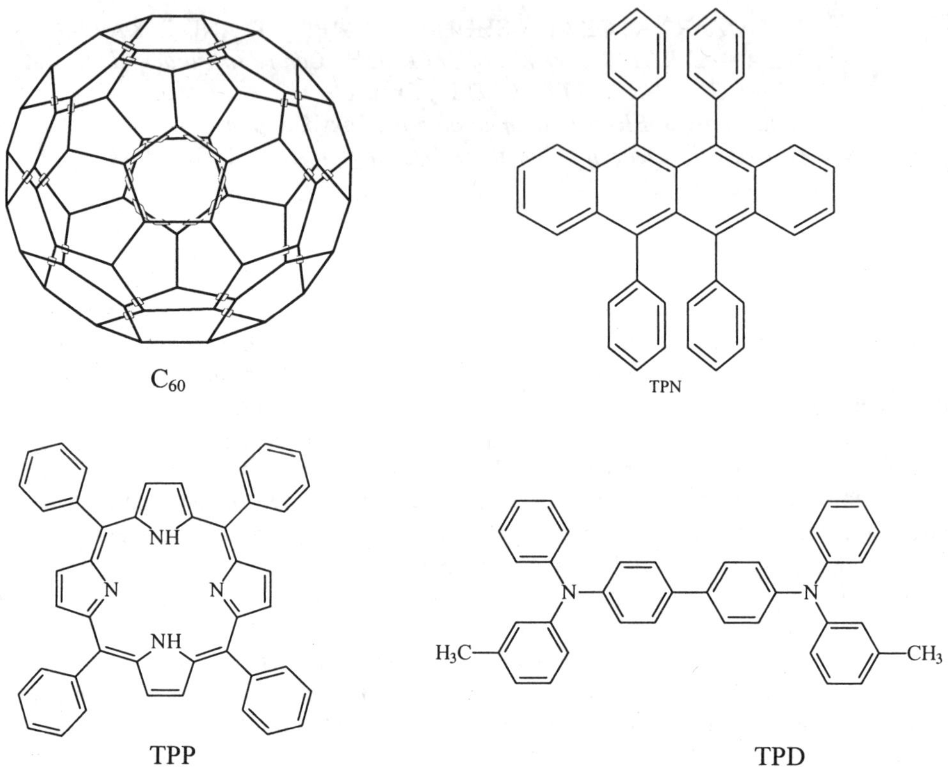

Figure 1. Chemical structure of molecules used as electron acceptors (C60) or electron donors (TPN, TPP, TPD)

In this paper we describe a new approach to the fabrication of noncentrosymmetric thin films for applications in devices based on second – order nonlinear optical properties. It consists on the fabrication of intermolecular charge transfer layers with the alternation of layers made from electron donating and electron accepting molecules, respectively. It is well known [7-10] that the sublimated thin films of C_{60} exhibit second order NLO properties despite the fact that presumably they are isotropic. The recent [10] more systematic and wave dispersed SHG measurements show that this is a bulk effect, and not an interface one. A resonant enhancement of SHG at 1.88 eV fundamental photon energy was observed [10]. This observation coincides well with the resonance observed in THG spectrum and interpreted as a two photon resonance [11].

The intrinsic SHG activity as well as the fact that C_{60} molecule can act as an electron donor or acceptor led us to explore the possibility of making the charge transfer complexes with the creation of a permanent dipole moment at the fundamental state and an enhanced quadratic nonlinear optical response with a good temporal and thermal stability. The fabrication of stable and efficient noncentrosymmetric thin films for quadratic NLO is surely an interesting challenge because of already mentioned large class of possible applications. The problems and the cost encountered with the fabrication of noncentrosymmetric structures through e.g. single crystal growth, epitaxy [3] or by using the Langmuir-Blodgett technique [5] led to an increasing interest in artificial structures such as the already mentioned poled polymers. The controlled charge transfer multilayer structures may offer an alternative solution, well argued by the controlled fabrication and expected enhanced stability. In this paper we report the preliminary results concerning the fabrication and second-order nonlinear optical properties of such intermolecular charge transfer complexes using C_{60} as electron accepting and 5, 10, 15,20 - tetraphenyl - 21H ,23H - porphine (TPP), 5, 6, 11, 12 - tetraphenylnaphathacene (rubrene ,TPN), and N,N'-bis-(3-methylphenyl)-N,N'-diphenyl - N,N' - bis (3 - methylphenyl) - 1, 1' - biphenyl - 4 - 4' - diamine (TPD); as electron donating molecules. The chemical structures of the used molecules are shown in Fig. 1. The NLO properties were studied by transverse optical second harmonic generation technique. A significant increase of SHG was observed in ABCABC.... type structures, where an effective charge transfer is expected with creation of a noncentrosymmetry in the direction perpendicular to the layers. The largest $\chi^{(2)}(-2\omega;\omega,\omega)$ susceptibility is obtained with multilayers containing rubrene (TPN). The SHG generation experiments performed on thin films with different thickness show that the nonlinear optical response comes from the bulk material and not from the interfaces. A quadratic dependence of SHG intensity on input power is also observed, showing that the charge transfer takes place at the ground state.

INTERMOLECULAR CHARGE TRANSFER STRUCTURES

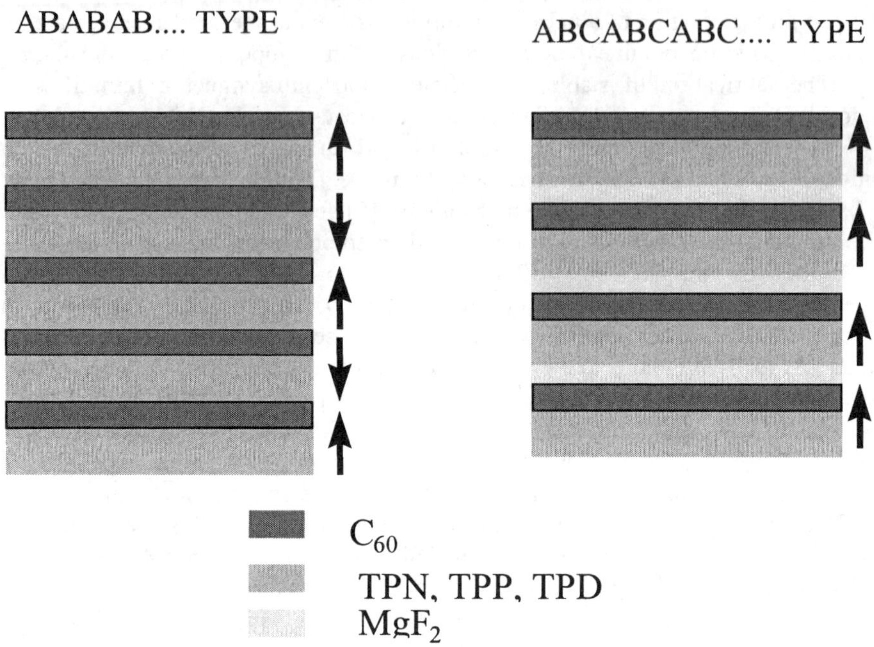

Figure 2. Schematic representation of two types of intermolecular charge transfer structures.

2. Multilayered structure fabrication and linear optical properties

High optical quality films of C_{60} with the electron donating molecules were prepared by sublimation under high vacuum on fused silica substrates. Two kinds of thin films were made: ABABAB ... and ABCABC...., (cf. Fig. 2), where the layer A is made from C_{60} molecules, layer B from electron donating (TPN, TPD or TPP) and C is a neutral layer (MgF_2 in the present case). Additionally the final structure was covered with an extra, protective layer of MgF_2. A total number of 25 AB or ABC type layers, with individual thickness of about 2 nm for each layer were made (cf. Fig. 2). Presumably ABAB.... type structures, despite possible intermolecular charge transfer, lead to macroscopically centrosymmetric structure while the ABCABC .... type structure to macroscopically non-centrosymmetric ones, respectively. For the sake of comparison composite thin films were also made by a simultaneous sublimation of C_{60} and electron donating molecules in with 1:1 molar ratio. The final thin film thickness was determined by examining the interference fringe patterns in the near infrared regions of the spectra where the films are transparent. They were verified independently with the use of an α-tallystep. The last measurements gave usually slightly different values for thin film thickness, especially for very thin films. This may be due to the inaccuracy in the used refractive index value.

Consequently in the data screening we have used for thin film thickness the values obtained with α-tallystep.

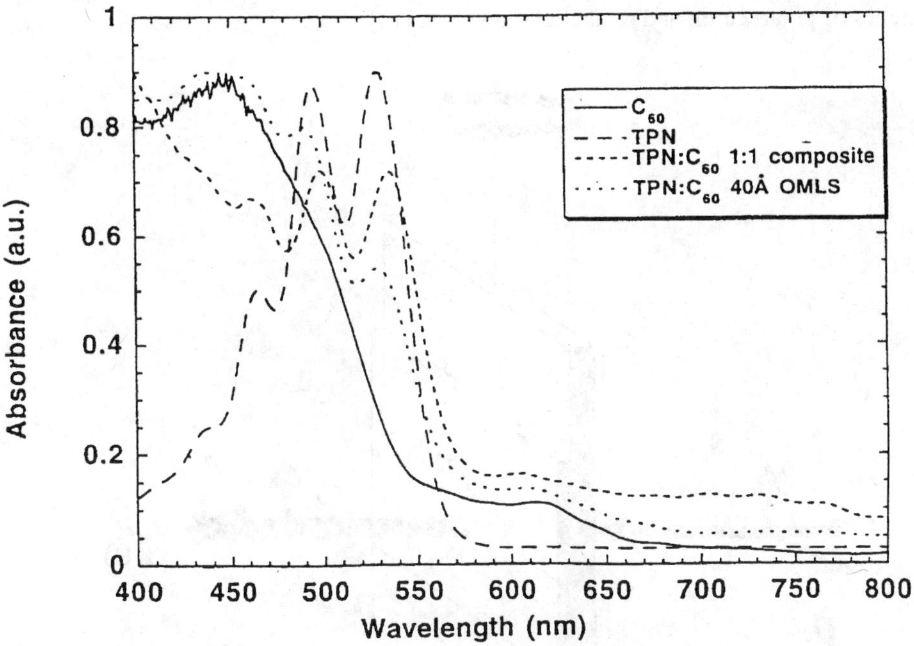

Figure 3. Optical absorption spectrum of TPN, TPN multilayers with C_{60} and TPN:C_{60} (1:1 molar ratio) composites

Figure 3 shows, as an example, the linear optical absorption spectra of the prepared thin films, measured with a Perkin-Elmer Lambda 9 Spectrophotometer for C_{60}, TPN, TPN :C_{60} 1 :1 composite and for a multilayer TPN :C_{60}. The observed weak absorption band, appearing in near IR (600 - 800 nm range) is assigned to an intermolecular charge transfer band. Its oscillator strength is the measure of the amount of charge transfer and consequently of the obtained second order NLO susceptibility.

3. Second harmonic generation measurements

The second-order NLO properties of deposited multilayers were studied by the transverse optical second harmonic generation technique. The principal advantage of this technique is the measure of the electronic response of material. As a matter of fact the response time in this technique is of the order of 10^{-15} s, as harmonic wave has to follow fundamental wave. This technique requires a change of the propagation length in material which is usually achieved by rotating thin film around an axis perpendicular to the beam propagation direction or by using wedge shaped samples and translating them. The SHG measurements can be also performed *in situ* allowing a real time study of the kinetics of polarization and relaxation processes [12].

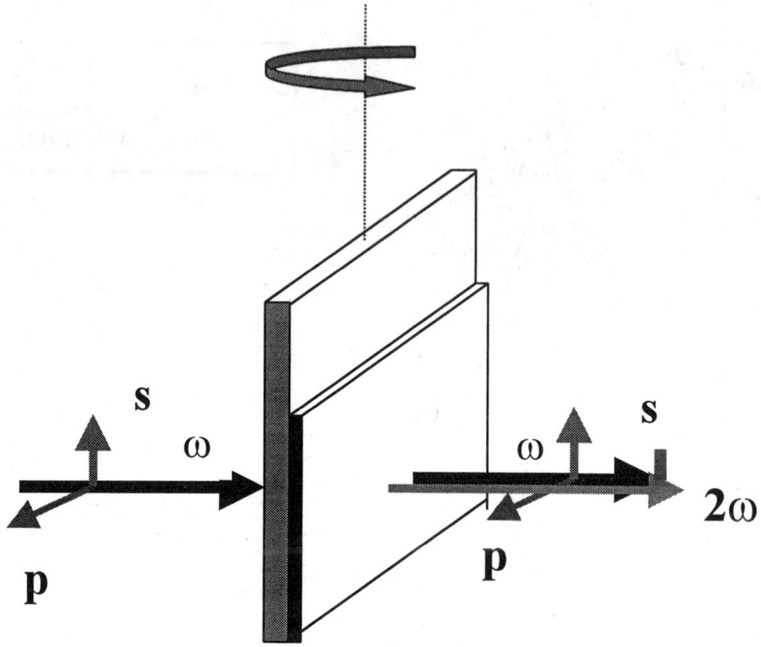

Figure 4. Schematic representation of different fundamental –harmonic beams polarization. At the output of nonlinear medium the fundamental beam is stopped.

The SHG measurements were performed using a Q-switched Nd : YAG laser operating at 1 0642 nm fundamental wavelength with 10 pps rate with pulse duration of 13 ns. The sample placed on a goniometer was rotated along an axis perpendicular to the beam propagation direction. Both input fundamental and output harmonic beam polarizations were controlled with Glan-Thompson polarizers as shown in Fig. 4. It allows to check the SHG efficiency as function of fundamental – harmonic beams polarization configuration. The SHG intensities were collected as function of incidence angle and are shown in Fig. 5 for different configurations of fundamental – harmonic beam polarizations for the multilayer ABCABC... type structure with TPN as electron donating molecule. It is seen that there is no SHG signal at normal incidence for all configurations. It appears when increasing the incidence angle, reaches a maximum at the incidence angle of about 55° and thereafter decreases. The largest SHG efficiency is 0observed for the p-p fundamental – harmonic beam configuration and a negligible signal for p-s and s-s configurations. This behavior is very similar to that observed in poled – polymers with ∞mm symmetry and with 2 nonzero $\chi^{(2)}$ $(-2\omega;\omega,\omega)$: diagonal $\chi^{(2)}zzz$ (= $\chi^{(2)}_{pp}$) and off-diagonal $\chi^{(2)}xxz$ (= $\chi^{(2)}_{sp}$), where Z axis is perpendicular to the thin film surface (or parallel to the poling direction) and X – axis is parallel to the substrate. It is what we expected as charge transfer should take place at the interface between deposited layers, i.e. perpendicular to the thin film substrate, leading to a non-centrosymmetry in this direction and isotropy in the thin film plane. Therefore in order to quantify the NLO susceptibilities we used the incidence angle θ dependencies for harmonic intensities for s-p and p-p fundamental – harmonic beam configurations derived for poled polymers (cf. e.g. [13]):

$$I_{2\omega}^{ffh}(\theta) = \frac{\pi^2}{2\varepsilon_0}\left|\frac{\chi_{ffh}^{(2)}(-2\omega;\omega,\omega)}{\Delta\varepsilon}\right|^2 \left|P_{fh}(\theta)\right|^2 \left|T_{fh}(\theta)(e^{i\Delta\varphi}-1)\right|^2 (I_\omega)^2 \tag{1}$$

where I_ω is the fundamental beam intensity, $\Delta\varepsilon$ is the dielectric constant dispersion at optical frequency

$$\Delta\varepsilon = n_\omega^2 - n_{2\omega}^2 \tag{2}$$

$P_{fh}(\theta)$ are projection factors:

$$P_{pp}(\theta) = \sin^2\theta_{2\omega}^p(\sin^2\theta_\omega^p + a\cos^2\theta_\omega^p) + 2a\sin\theta_\omega^p\cos\theta_\omega^p\cos\theta_{2\omega}^p \tag{3}$$

and

$$P_{sp}(\theta) = a\sin\theta_{2\omega}^p \tag{4}$$

where

$$a = \chi_{sp}^{(2)} / \chi_{pp}^{(2)} \tag{5}$$

The coefficients $T_{fh}(\theta)$ in Eq. (1) arise from transmission factors and boundary conditions (for details see e.g. [13]) and $\Delta\varphi$ is the phase mismatch between fundamental and harmonic beams given by

$$\Delta\varphi = \frac{4\omega l}{c}(n_{2\omega}\cos\theta_{2\omega} - n_\omega\cos\theta_\omega) \tag{6}$$

where $n_{\omega(2\omega)}$ are refractive indices of nonlinear medium at fundamental and harmonic frequencies, respectively.
The subscripts in Eqs. (2) – (4) denote beam frequencies and the superscripts its polarization, respectively.

The measured SHG intensities were fitted using Eqs. (1) – (5) and calibrated with similar measurements done on an y-cut α-quartz single crystal ($\chi^{(2)}_{pp}(-2\omega;\omega,\omega)$ =

1.0 pm/V [14]) done at the same conditions. Corrections for thin film absorption (cf. Fig. 3) at harmonic frequency were done in a standard way by introducing into Eqs. (1-6) imaginary refractive index at harmonic frequency:

$$n_{2\omega} = n_{2\omega}^{r} + i\kappa_{2\omega} \tag{7}$$

where the extinction coefficient $n_{2\omega} = n_{2\omega}^{r} + i\kappa_{2\omega}$ was determined from the linear optical absorption spectra of studied films. For the real part of refractive index the values determined by Ren et al [15] for C_{60} thin films were used. First the values of $\chi^{(2)}(-2\omega;\omega,\omega)$ were determined from s-p fundamental harmonic beam polarization configuration measurements ad then the obtained values were injected to corresponding formulas for p-p configuration in order to get the $\chi^{(2)}_{pp}(-2\omega;\omega,\omega)$ values.

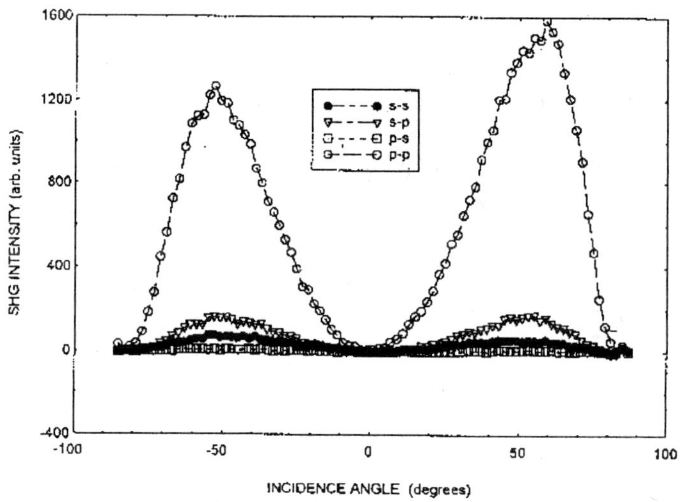

Figure 5. Incidence angle dependence of harmonic intensities for different fundamental – harmonic beam polarization and for the C_{60}/TPN/MgF_2 multilayers.

4. Results

As already mentioned the sublimated thin films of C60 exhibit SHG activity. Therefore it is important to take account of that when studying the NLO properties of composite systems containing this molecule. Also one has to be aware of any second order NLO activity of other components. Therefore careful measurements were done on thin film made from one molecule only. The measured SHG susceptibilities are listed in Table I. First of all, up to the dielectric breakdown, no SHG signal was observed from thin films made from pure TPN, TPP and TPD. One observes, indeed, an enhancement of $\chi^{(2)}_{pp}(-2\omega;\omega,\omega)$ tensor component, which in the case of C_{60}/TPN/MgF_2.... multilayer structure reaches a value close to 12 pm/V. If we correct it for the number of NLO C_{60} species it will give a value of about 35 pm/V to be compared with 3.2 pm/V the value we have got for a pure C_{60} thin film. It shows an almost one order of magnitude enhancement of the $\chi^{(2)}$ tensor due

Table I : Second order nonlinear optical susceptibilities of C_{60}, its composites and multilayers with TPN, TPP and TPD molecules acting as electron donors. The values are calibrated with α-quartz single crystal (d_{pp} = 0.5 pm/V [14])

Molecule	wavelength (μm)	χ^2_{sp} (pm/V)	χ^2_{pp} (pm/V)	Remark
C_{60}	1.064 1.064 0.69	1.7* 0.86	3.2 1.78 3.0	Ref. [8] Ref. [9] Ref. [10]
TPN	1.064	0^+	0^+	
TPP	"	"	"	
TPD	"	"	"	
MgF_2	"	"	"	
C_{60}:TPD multilayers	"	"	"	ABAB... type
C_{60}:TPD composite	"	"	"	~1:1
C_{60}:TPN multilayers	"	7.6	11.8	ABAB... type
C_{60}:TPN composite	"	2.6	5.0	~1:1
C_{60}:TPP multilayers	"	2.0±0.4	4.0±0.6	ABAB... type
C_{60}:TPP composite	"	2.0±0.4	4.2±0.6	~1:1
C_{60}:TPDmultilayers	"	2.6±0.4	2.2±0.4	ABCABC... type
C_{60}:TPN multilayers	"	4.4±0.6	9.2±1.2	ABCABC... type
C_{60}:TPP multilayers	"	2.6±0.4	5.4±0.6	ABCABC... type
α-quartz d_{11}	1.064		1.0	Choy and Byer [14]

* - estimated from the fit of $\chi^{(2)}_{pp}$
\+ - up to the limit of the dielectric breakdown

to the intermolecular charge transfer between layers of C_{60} and TPN. In order to check that this is not interface but a bulk effect we have done SHG experiments as function of the thin film thickness in four different thin films with the TPN donor. This molecule was chosen because it exhibits the largest $\chi^{(2)}$ susceptibility. Figure 6 shows the thickness dependence of the square root of SHG intensity observed at p-p fundamental harmonic beam configuration for these films. The observed linear behavior shows that the measured response is indeed due to the bulk contribution.

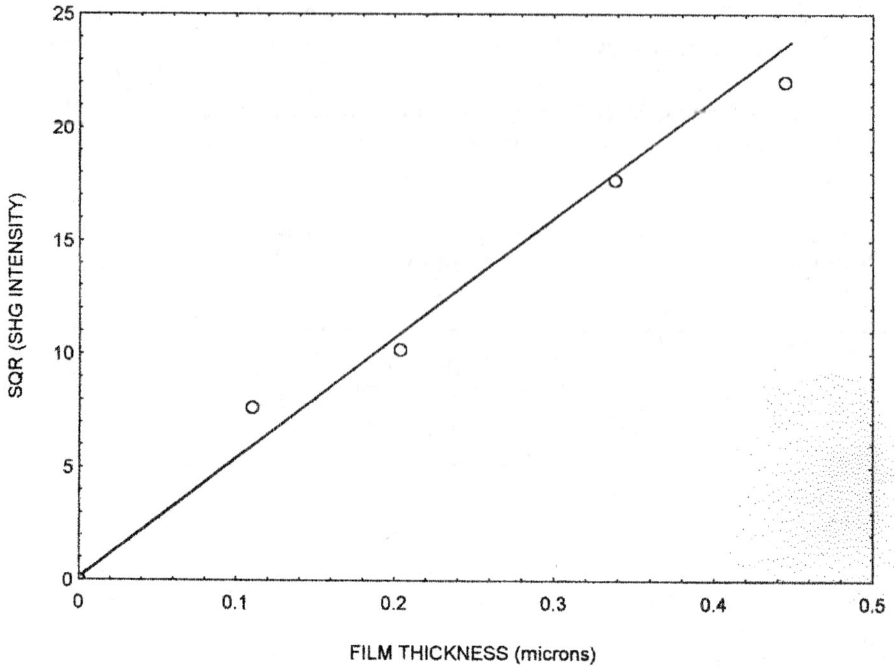

Figure 6. Square root variation of SHG intensity with thin film thickness. Solid line shows result of a least square fit.

Another important check of the intermolecular charge transfer origin of the observed increase of $\chi^{(2)}(-2\omega\,;\omega,\,\omega)$ susceptibility is the incidence power dependence of SHG intensity for a given thin film. By using different neutral density filters we have varied the incident fundamental beam intensity, again for a thin film made from the most effective C60/TPN/MgF_2 structure. Figure 7 shows the square root of SHG intensity in function of the fundamental beam intensity. Again an almost linear dependence is observed, showing that the SHG activity is due to the second order NLO effect and not to higher order effects.

5. Discussion

We observe a significant increase of $\chi^{(2)}(-2\omega\,;\omega,\,\omega)$ susceptibility in alternate ABCABC... type structures. This increase in the case of C_{60}/TPN/MgF_2 multilayers is by one order of magnitude, when compared to the number of active molecules.

Another important check of the intermolecular charge transfer origin of the observed increase of $\chi^{(2)}(-2\omega\,;\omega,\,\omega)$ susceptibility is the incidence power dependence of SHG intensity for a given thin film. By using different neutral density filters we have varied the incident fundamental beam intensity, again for a thin film made from the most effective C60/TPN/MgF_2 structure. Figure 6 shows the square root of SHG intensity in function of the fundamental beam intensity. Again an almost linear dependence is observed, showing that the SHG activity is due to the second order NLO effect and not to higher order effects.

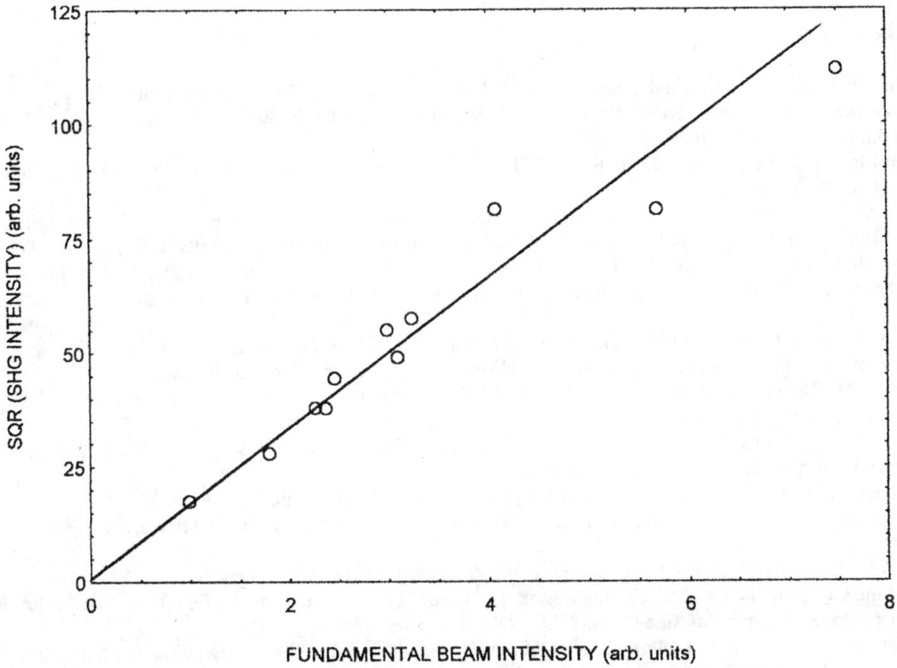

Figure 7. Square root dependence of SHG intensity on the fundamental beam intensity. Solid line shows result of a least square fit.

Conclusions

The present study shows that one can get efficient intermolecular charge transfer multilayer structures for second order NLO applications using alternate ABCABC.... Type layers with electron donating, electron accepting and neutral molecules. The value of the second-order NLO susceptibility of such structures depends on the chosen molecules. Between three electron donating molecules used in this study the most promising appears to be TPN (rubrene) and the less TPD. In the first case we observe one order of magnitude enhancement of $\chi^{(2)}(-2\omega;\omega,\omega)$ susceptibility as compared to the pristine C_{60} thin film, when corrected for the number of active molecules. This is a very interesting and promising result. It leads to the fabrication of controlled, temperature and temporal stable noncentrosymmetric thin films for second order NLO applications. As a matter of fact we didn't observe any change in NLO response of these structures over an extended period of storage in the dark.

By doing the SHG measurements as function of the thin film thickness we have shown that it is a bulk and not an interface effect. The SHG measurements performed in function of the incident light intensity show that the charge transfer takes place at fundamental state.

6. References

1. Chemla D. S., Oudar J. L. and Zyss J. (1981), L'Echo des Recherches (International Issue), 47.

2. Davydov B. L., Derkacheva L. D., Dunina V. V., Zhalotinskii M. E. Zolin V. F., Koreneva L. G. and Samoklina M. A. (1970), JETP Lett., **12**, 16

3. Le Moigne J., Oswald L., Kajzar F. and Thierry A. (1995), Oriented Thin Films for Nonlinear Optics, Nonl. Opt., **9**, 187-193.

4. Le Moigne J. 1996,, « *Epitaxy and Single Crystal Growth* », in Organic Thin Films for Waveguiding Nonlinear Optics, F. Kajzar and J. Swalen Eds., Gordon and Breach Publishers, Amsterdam pp 289 – 327.

5. Bosshard Ch. and Küpfer M. (1996), « *Oriented Molecular Systems* », in Organic Thin Films for Waveguiding Nonlinear Optics, F. Kajzar and J. Swalen Eds., Gordon and Breach Publishers, Amsterdam, pp 163 – 191.

6. Miyata S. and Sasabe H. (1997). (Eds*), Poled Polymers and their Applications to SHG and EO Devices*, Advances in Nonlinear Optics, vol. 4, Gordon and Breach Science Publishers, Amsterdam

7. Hoshi H., Nakamura N., Maruyama Y., Nakagawa T., Suzuki S., Shiromara H. and Achiba Y. (1991), *Japan. J. of Appl. Phys.,* **30,** L1397.

8. Wang X. K., Zhang T. G., Lin W. P., Shen Zhong Liu, Wong G. K. Kappes M. M., Chang R. P. H. and Ketterson J. B. (1992), *Appl. Phys. Lett.*, **60**, 810.

9. Kajzar F., Taliani C., Danieli R., Rossini S. and Zamboni R. (1993), *Synth. Metals,* **54** 21

10. Koopmans B., Anema A., Jonkman H. T., Sawatzky G. A. and Van der Woude F (1993), *Phys. Rev. B,* 48 2759.

11. Kajzar F., Taliani C., Danieli R., Rossini S. and Zamboni R., *Chem. Phys. Lett.* (1994), **217**, 418.

12. Dantas de Morais T., Noël C. and Kajzar, F. (1996), Electric Field Poling Effects on the Molecular Reorientational Dynamics of Side-Chain NLO Polymers, *Nonl. Optics,* 15, 315.

13. Swalen J. and Kajzar F. (1996), *Introduction*, in Organic Thin Films for Waveguiding Nonlinear Optics , F. Kajzar and J. Swalen Eds., Gordon and Breach Publishers, Amsterdam, pp. 1 – 44.

14. Choy M. and Byer R. L. (1976), *Phys. Rev. B*, **14,** 1693.

15. S. L. Ren, Y. Wang, A. M. Raon, E. McRae, J. M. Holden, T. Hager, KaiAn Wang, Wan-Tse Lee, H. F. Ni, J. Selegue and P. C. Eklund (1991), *Appl. Phys. Lett.,* **59,** 2678.

NOVEL MONOMERIC AMORPHOUS MATERIAL FOR SECOND-ORDER NLO: THE LOW-MOLECULAR-WEIGHT ORGANIC GLASSES

E. ISHOW[1], K. NAKATANI[1], J. DELAIRE[1], L. QIU[2]
[1]*Ecole Normale Supérieure de Cachan*
Laboratoire de Photophysique et Photochimie Supramoléculaire et Macromoléculaire, UMR CNRS 8531, 61 Av. Président Wilson, 94235 Cachan Cedex, France
[2]*On leave from the Institute of Photographic Chemistry, Academia Sinica, Beijing 100101, People's Republic of China*

1. Introduction

Low-molecular-weight (LMW) organic glasses have recently appeared as attractive materials for many optical applications as they combine the structural advantageous features of both molecular crystals and polymeric matrices.

They are based on bulky functionalized molecules with well-defined size which give reproducible homogeneous materials. Moreover, the absence of matrix allows a more straightforward understanding of the optical properties than in polymers. Their glassy state is induced by the sterically-demanding groups and can be achieved from melt powder or quick evaporation of concentrated solutions. As LMW glasses have only a weak tendency to crystallization, materials of pure compound can be elaborated. Hence a greater efficiency is expected compared to that of polymers severely limited by chromophore aggregation for high molar fractions of chromophores.

Contrarily to the case of single crystals where growing requires time and know-how, the manufacture of LMW glassy thin films showing a high optical quality involves a large choice of simple experimental processes: electrochemical deposition, concentrated solution spin-coating, vapor deposition under high vacuum. This latter process opens the way towards multilayer systems which are difficult to elaborate by the spin-coating technique except if a careful cross-solvent deposition is carried out for ruling out any dissolution of the previously spin-cast layers.

Most of the LMW-glasses applications have been mainly devoted to photorefractivity [1-3] for their high mixture ability and to organic light-induced electroluminescent devices [4-5] as hole transport layer materials. Curiously, to our knowledge, no specific studies have been carried out in the second-order nonlinear optics (NLO) domain where there is still a need for stably-poled and efficient materials. Only one work reported so far the use of LMW glasses for second harmonic-generation (SHG) but their low glass transition temperature (T_g) precluded subsequent developments at room temperature.[6]

In this paper, we want to present the synthesis and the optical properties of a novel molecule $DBABNO_2$ forming stable LMW glasses with T_g near 100 °C and exhibiting

F. Charra et al. (eds.), Organic Nanophotonics, 475–482.

both second-order NLO properties and photochromism whose coupling finds numerous applications into the photomodulation of NLO properties.

2. Synthesis and glassy state characterizations

2.1. PRECURSOR SYNTHESIS

The target molecule $DBABNO_2$ had to fulfil a threefold structural requirement, *i.e.* the presence of bulky substituents for RT glass-forming properties, a photoisomerizable moiety for photochromic behavior and a dipolar "push-pull" structure for second-order NLO-activity. An azobenzene-type precursor molecule DBAB, reported in the literature to form a photoisomerizable LMW glass characterized by $T_g = 68$ °C [7], satisfied the first two requirements. Besides the triarylamino group acting as a donor group, the nitro moiety was introduced as an electron-withdrawing group for building up the push-pull system $DBABNO_2$ (Figure 1). However, this small structural change led us to envisage another synthetic approach appearing more flexible and applicable to a larger range of acceptors and/or bulky group for varying the charge-transfer strength and the glass transition temperature.

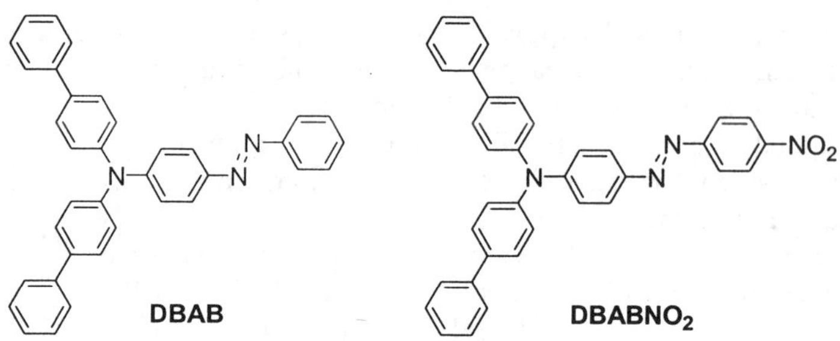

Figure 1. DBAB and $DBABNO_2$ forming photochromic LMW organic glasses

The three-step synthesis involved a phase-transfer coupling of a freshly prepared diazonium salt with triphenylamine. Bromination of the resulting azobenzene derivative and subsequent Suzuki coupling with phenylboronic acid provided $DBABNO_2$ in 66 % yield (for the last two steps).

2.2. LMW GLASSES OF $DBABNO_2$

Thin glassy films (0.3-0.5 µm-thick) were obtained by spin-coating a filtered 0.1 mol.L^{-1} $DBABNO_2$ chloroform solution onto a glass or ITO substrate; prior to use, films were dried under vacuum during 24 h. Glassy materials could also be made from melt polycrystalline powder which has been quickly cooled down to room temperature. The glassy character of the cooled melt was evidenced by differential thermal analysis (DTA) and powder wide angle X-Ray scattering and compared to that of the powder before melting. While the powder presented a sharp endothermic melting peak at T_m =

205 °C on the DTA curve, we only detected a slope change at T_g = 92 °C for the previously melt $DBABNO_2$ assignable to a glass transition. Neither subsequent crystallization nor melting features were observed on the DTA curves, which agrees with the stability of the amorphous state. Actually, some films remained glassy at room temperature over one year.

3. Photochromic properties

3.1. ABSORPTION SPECTRA

The absorption spectra of $DBABNO_2$ thin glassy films and solutions presented similar spectroscopic features: one broad band in the visible range and a second more intense one in the UV range (Figure 2). On the basis of comparative absorption studies of similar compounds as Disperse Red 1 and DBAB, the visible band was assigned to the donor-acceptor charge transfer transition, undergoing a 30 nm bathochromic shift when the solvent was changed from hexane to DMSO.

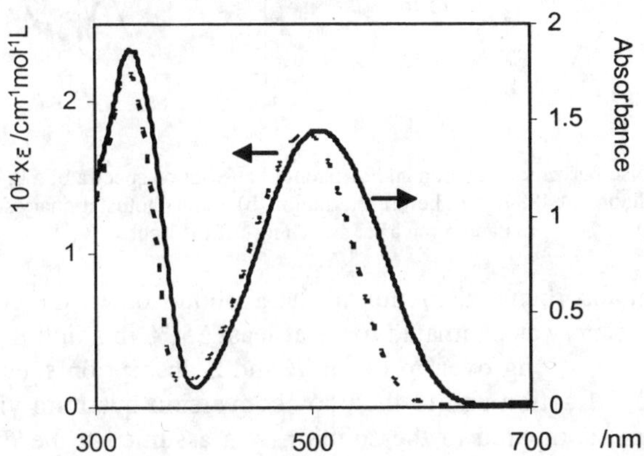

Figure 2. Absorption spectra. 0.3 µm-thick $DBABNO_2$ glassy film. $DBABNO_2$ toluene solution.

We can assume highly polar molecular surroundings within the thin films as the absorption maximum was found at λ_{max} = 509 nm close to 502 nm for a DMSO solution. On the other side, the UV band was attributed to mainly π–π* biphenyl-centered transitions (332 nm for thin films) which was confirmed by smaller spectroscopic changes upon varying the solvent polarity (312 nm and 330 nm for hexane and DMSO solutions respectively) and preliminary photoinduced anisotropy experiments.

3.2. PHOTOISOMERIZATION STUDIES

Photoisomerization studies were carried out in toluene solution and 0.4 µm-thick glassy films exhibiting both similar absorbance (around 1) for their visible absorption maximum. We used a CW argon ion laser operating at 488 nm (100 mW.cm^{-2}) to produce the $E \rightarrow Z$ photoreaction. Once the photostationary state has been reached, the pump laser was switched off so that the thermal back reaction $Z \rightarrow E$ could be investigated. The spectral changes of the samples were probed by a low-intensity xenon arc lamp; the light transmitted through the sample was recorded on a CCD camera coupled with a spectrograph. Through irradiation, the absorption maximum characteristic of the *E*-form centered at 493 nm for toluene solutions and 509 nm for thin films decreased in intensity with a 10 nm bathochromic shift (Figure 3). In the same time, a small increase around 400 nm and two isosbestic points - 345 and 415 nm for toluene solution, 370 and 430 nm for the film - were also observed.

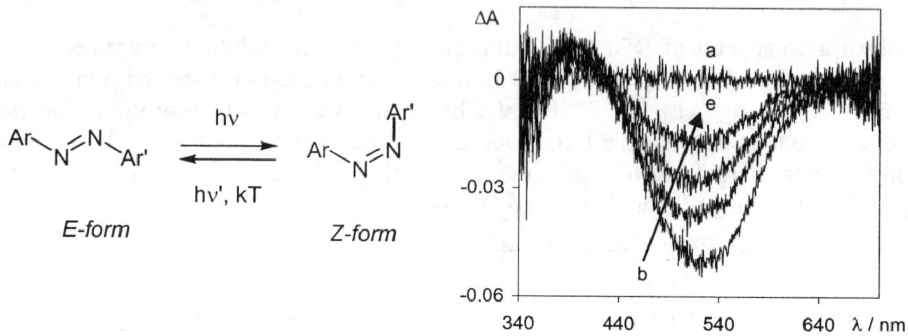

Figure 3. E-Z photoisomerization reaction and differential absorption spectra of a 0.4 μm-thick film of $DBABNO_2$ upon irradiation at 488 nm. a) before irradiation. b) at the photostationary state. c), d), e) 30 s, 2 mn and 5 mn after switching-off the light.

From maximum absorbance depletion of the *E*-photoisomer, the conversion yield at the photostationary state was estimated to be at least 35 % in solution and 7 % in spin-coated films. Actually, strong overlap of the *E* and *Z* absorption spectra precludes any direct access to the *Z* fraction and to the photoconversion quantum yield. The lower *Z* fraction for the films compared to the solution was assumed to be due to a restricted local free volume around the chromophores impeding their isomerization. Thermal relaxation of the *Z*-form was followed through the progressive recovery of the *E*-form absorbance and could be fitted by a first-order kinetics (Figure 4).

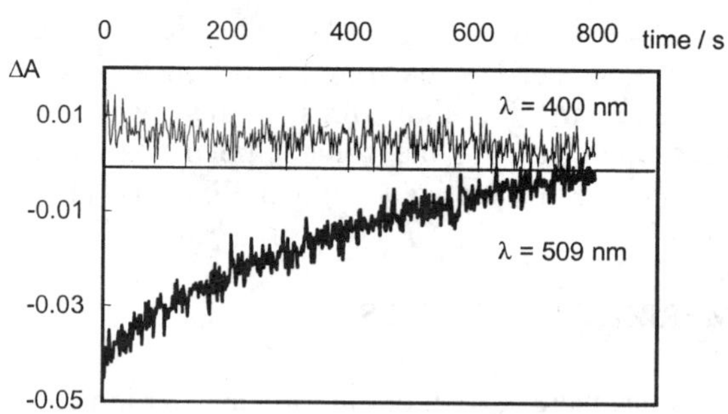

Figure 4. Evolution of the differential absorbance of an irradiated $DBABNO_2$ thin film at $\lambda = 400$ nm and 509 nm after light switching-off ($\lambda_{irr} = 488$ nm).

Use of a monoexponential law gave accurate fits. The rate constants k_r were found to be almost the same for toluene solution and glassy films, *i.e.* $1.95 \pm 0.01\times10^{-3}$ s^{-1} and $2.1 \pm 0.1\times10^{-3}$ s^{-1} respectively. These similar values let us think that the $E\rightarrow Z$ photoisomerization within the film only causes a small reorganization of the surroundings which still remain constraint. Thus, the *Z*-form is destabilized and quickly relaxes to the *E*-form. The thermal relaxation kinetics of photoirradiated $DBABNO_2$ films sharply contrasts with that of vacuum deposited thin films of DBAB which was decomposed into two components: one rate constant equal to 2×10^{-3} s^{-1} and the second one much slower equal to 5×10^{-5} s^{-1} weighing 95% of the whole decay.[7] This difference could originate either from the more polar character of $DBABNO_2$ since apolar azobenzene derivatives like DBAB are known to relax slower [8], or from the film preparation technique (spin-coating *vs* vacuum deposition). Work is currently under studies for investigating this second hypothesis.

4. Second-Order NLO Properties

4.1. EFISH STUDIES

4.1.1. *Experimental Setup*

In order to reduce second harmonic reabsorption and resonance effects, Electric Field Induced Second Harmonic (EFISH) and Second Harmonic Generation (SHG) experiments were carried out with a 1907 nm fundamental beam, Raman-shifted from the 1064 nm fundamental of a Q-switched Nd:YAG nanosecond laser (BM Industries; 10 Hz repetition rate, 10 ns pulse width) and focused in a 1 meter long hydrogen cell (50 bar).

Solutions of 2-methyl-4-nitroaniline in dioxane at various concentrations (0 to 3×10^{-2} $mol.L^{-1}$) allowed to calibrate the cell for EFISH measurements. Centrosymmetry of the sample or reference solutions was broken under a high-voltage pulse (around 6 kV applied on a 2 mm-thick cell during 5 μs) synchronized with the laser pulse. Instruments and calibration techniques have been more fully described elsewhere.[9]

The two-level model was adopted for resonance correction.[10] The off-resonance first hyperpolarisability β_0:

$$\beta_0 = \frac{6\pi^2 \mu_{01}^2 \Delta\mu}{h^2 \omega_1^2} \tag{1}$$

was deduced from β by mean of $\beta(-2\omega;\omega,\omega) = \beta_0 F(\omega,\omega_1)$ with the charge transfer transition moment μ_{01}, the charge transfer transition pulsation ω_1, the dipole moment difference $\Delta\mu$ between the ground and the excited states and the dispersion factor $F(\omega,\omega_1)$:

$$F(\omega, \omega_1) = \left[\frac{\omega_1^4}{(\omega_1^2 - 4\omega^2)(\omega_1^2 - \omega^2)} \right] \quad (2)$$

4.1.2. *Results*

The experimental value of $\mu\beta$ for $DBABNO_2$ was found to be 660×10^{-48} esu in dioxane. Energy minimization through semi-empirical AM1 computations (MOPAC) gave a vectorial β_0 almost parallel to the dipole moment vector μ_0 along the main donor-acceptor molecular axis. From the calculated ground state dipole moment μ_0(calc) = 7.9 D, the off-resonance first hyperpolarisability β_0 was deduced to be β_0(exp) = 51×10^{-30} esu.

For comparison, the first hyperpolarisability of the well-known DR1 which exhibits similar push-pull moieties was β_0(exp) = 47×10^{-30} esu.[11] The very slight difference could come from a greater electronic dispersion of the excited state over the conjugated π-electron system involving the biphenyl substituents.

4.2. SHG MEASUREMENTS ON THIN FILMS

4.2.1. Experimental Setup

Thermo-assisted poling (TAP) was carried out to break the bulk centrosymmetry of the thin LMW glassy $DBABNO_2$ films. This poling procedure consisted in applying a static electric field through corona discharge to a film at 100 °C (T_g = 92 °C). Once a steady-state dipole alignment was achieved, the film was slowly cooled down to room temperature while keeping the electric field on.

A 3 mm thick single crystal quartz (d_{11} = 0.3 pm/V) was taken as reference. The Maker fringe method was used for determining the second-order nonlinear coefficient of the materials (p polarization). Film thickness (0.35 μm) and refractive indices at ω and 2ω (n = 1.508 at λ =1907 nm and n_2 = 1.686 at λ = 950 nm) were determined by the interference method.[12]

4.2.2. Results

Neither the microcrystalline powder nor the spin-coated films before poling were active in SHG.

After the TAP process, poling and SHG relaxation of the sample were probed at 1907 nm. A fast decay of the second harmonic signal was observed during the first three hours followed by a slower one. d_{33} values were found to be 4 pm/V and 2 pm/V after one day and three days respectively as shown on Figure 5.

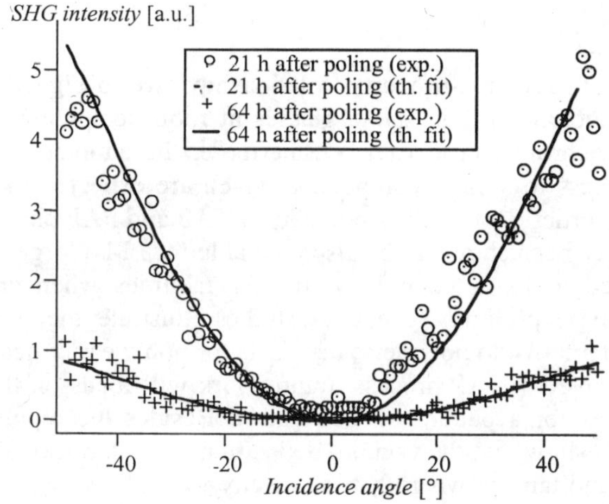

Figure 5. Maker fringe experiment on a glassy $DBABNO_2$ film (thickness: 0.35 μm) and fitted curve.

These significant values are in the same range of order as d_{33} for DR1-doped PMMA which was reported to be 8.4 pm/V at 1580 nm with comparable chromophore density (2.3×10^{26} m^{-3} for DR1 and 1.1×10^{27} m^{-3} for $DBABNO_2$ assuming a volumic mass equal to 1 kg.dm^{-3}).[13]

Taking advantage of the photoisomerization of the azobenzene moiety, photoassisted poling (PAP) procedure under corona discharge is an alternate poling procedure and can be carried out on LMW $DBABNO_2$ films at milder temperatures than TAP.

4.2.3. Photoassisted Poling

The PAP process works at room temperature and requires the simultaneous application of both a static electric field and an irradiation source to induce photoisomerization.[14] The corona poling with in-situ SHG monitoring was performed by applying a 4 kV voltage using a dc power supply to a stainless steel needle positioned normally to the film surface (ITO substrate) at a distance of 1 cm. The irradiation source was the same as that used in the photoisomerization studies except that the light was circularly polarized.

When the voltage was applied, film irradiation led to a larger increase of the SHG signal compared to that of a non-illuminated film. This difference has been reasonably attributed to a greater rotational mobility of the dipoles and consequently a better dipole alignment along the direction of the applied electric field when the irradiation is on. The rotational diffusion of the more compact *Z*-form generated through the photoisomerization reaction allows this phenomenon.

However, a large charge injection appeared on illumination which prevented an accurate determination of the PAP procedure efficiency. In fact, PAP on this material has to be optimized since the SHG intensity is about eight times lower than that obtained after TAP.

5. Conclusion

On the basis of structure-property relationship, we designed the first low-molecular weight material forming stable glasses at room temperature and exhibiting both NLO and photochromic properties. Furthermore, the adopted synthetic pathway provides a direct access to similar compounds which are expected to present varying NLO and glassy properties. Both poling procedures TAP and PAP classically applied to polymer matrices have been shown to be also available for LMW organic glasses which agrees with the need to have technology transfer facilities when change for novel materials occurs. These preliminary optical studies illustrate the possibility for the LMW to be a real alternative to polymeric materials for nonlinear optical applications.

In order to improve the PAP process, future work will focus on the elaboration of "sandwich" electrodes for a better voltage control. Besides the reported spin-coating technique, we plan to show that the vacuum deposition procedure can also be applied to the DBABX compound family (where X is an electron-withdrawing group) to form thin glassy films with high optical quality.

References

1. Lundquist, P.M., Wortmann, R., Geletneky, C., Twieg, R.J., Jurich, M., Lee, V.Y., Moylan, C.R. and Burland, D.M. (1996) Organic Glasses: a new class of photorefractive materials, *Science* **274**, 1182-1185.
2. Würthner, F., Yao, S., Schilling, J., Wortmann, R., Redi-Abshiro, M., Mecher, E., Gallego-Gomez, F. and Meerholz, K. (2001) ATOP Dyes. Optimization of a multifunctional merocyanine chromophore for high refractive index modulation in photorefractive materials, *J. Am. Chem. Soc.* **123**, 2810-2824.
3. Schloter, S., Schreiber, A., Grasruck, M., Leopold, A., Kol'chenko, M., Pan, J., Strohriegel, P., Zilker, S.J. and Haarer, D. (1999) Holographic and photoelectric characterization of a novel photorefractive organic glass, *Appl. Phys. B* **68**, 899-906.
4. Shirota, Y. (2000) Organic materials for electronic and optoelectronic devices, *J. Mater. Chem.* **10**, 1-25.
5. Tao, Y.T., Balasubramaniam, E., Danel, A., Jarosz, B. and Tomasik P. (2001) Organic light-emitting diodes based on variously substituted pyrazoloquinolines as emitting material, *Chem. Mat.* **13**, 1207-1212.
6. Eich, M., Looser, H., Yoon, D.Y., Twieg, R., Bjorklund, G. and Baumert, J.C. (1989) Second-harmonic generation in poled organic monomeric glasses, *J. Opt. Soc. Am. B.* **6**, 1590-1597.
7. Shirota, Y., Moriwaki, K., Yoshikawa, S., Ujike, T. and Nakano, H. (1998) 4-[Di(biphenyl-4-yl)amino]azobenzene and 4,4'-[bis(4'-*tert*-butylbiphenyl-4-yl)amino]azobenzene as a novel family of photochromic amorphous molecular materials, *J. Mater. Chem.* **8**, 2579-2581.
8. Dürr, H. and Bouas-Laurent, H. (1990) *Photochromism, Molecules and Systems*, Elsevier, Amsterdam.
9. Maltey, I., Delaire, J.D., Nakatani, K., Wang, P., Shi, X. and Wu, S. (1996) Hyperpolarisability of (donor)2-acceptor-type molecules determined by EFISHG, *Adv. Mater. Opt. Elect.* **6**, 233-238.
10. Oudar J. L. (1977) Hyperpolarisabilities of the nitroaniline and their relations to the excited state dipole moment, *J. Chem. Phys.* **67**, 446-457.
11. Katz, H.E., Singer, K.D., Sohn, J.E., Dirk, C.W., King, L.A. and Gordon, H.M. (1987) Greatly enhanced second-order nonlinear optical susceptibilities in donor-acceptor organic molecules, *J. Am. Chem. Soc.* **109**, 6561-6562.
12. Watanabe, T., Nalwa, H.S. and Miyata S. (1997) *Nonlinear Optics of Organic Molecules and Polymers*, CRC Press, Boca Raton, pp 57-92.
13. *ibid* p 419.
14. Sekkat, Z. and Dumont, M. (1992) Poling of azo dye doped polymeric films at room temperature, *Appl. Phys. B.* **54**, 486-489.

FUNCTIONAL POLYMERIC PHOTONIC CRYSTALS FOR INFRA-RED BANDS BY GUIDED COLLOIDAL ASSEMBLY TECHNIQUES

PRESTON B. LANDON, BOG G. KIM, RAY H. BAUGHMAN, ROBERT GLOSSER and ANVAR A. ZAKHIDOV
Nano Tech Institute, The University of Texas at Dallas
2601 North Floyd Rd. Richardson, TX 75083, USA

VLADIMIR KAMAEV and VALY VARDENY
Department of Physics, The University of Utah
115 South 1400 East, Salt Lake City, Utah 84112, USA

Abstract

A technique utilizing polyelectrolytes to assist in the self assembly of colloidal silica spheres into ordered crystallites is introduced. Techniques for coating silica spheres with poly(anilinesulfonic) acid, polypyrrole, and poly[2-methoxy-5-(2'-ethylhexyloxy)-1,4-phenylenevinylene] (MEH-PPV) are described with a self assemble method to guide them into face centered cubic (FCC) opals and into thin films with star shaped opalescent patterns. Inverse gold photonic crystals constructed from silica opal templates were constructed by the heat conversion of gold chloride to metallic gold and inverse gold/ polypropylene composite photonic crystals were fabricated. Inverse conjugated polymer/ polypropylene composite photonic crystals were constructed with MEH-PPV, poly(anilinesulfonic) acid and polypyrrole. The photoluminescence from MEH-PPV/ polypropylene photonic crystals was blue shifted compared to a uniform MEH-PPV film. Chalcogenide was infiltrated into 1.5μm opals and inverse chalcogenide photonic crystals were fabricated.

1. Introduction

Photonic crystals based on opal fcc lattices have proved as a very useful prototype for demonstrations of new phenomena, caused by the opening of a partial photonic band gap in the spectra of light. So we have demonstrated tunable intra-gap lasing, and new types of Braggariton excitations recently in opals. However the usual self-assembly methods can create opals with uncontrollable defects of structure and only with FCC lattices. In this paper we describe the geometry of guides to assemble colloidal spheres into FCC [1,2], tetragonal, body centered cubic and triclinic crystalline lattices. Inexpensive methods for the guided assembly of colloidal spheres into a tetragonal crystalline lattice are demonstrated. Many different methods for the assembly of

F. Charra et al. (eds.), Organic Nanophotonics, 483–492.

colloidal spheres have been investigated [3,4] in attempt to create synthetic opals with crystalline order for use as templates for photonic crystals.

Another heavily explored route to fabricate synthetic opals is the self assembly of colloidal spheres. This is used for the construction of FCC opals by controlling the surface charge, Debye length, and weight-to-weight (w/w) concentration of the colloidal spheres. The surface charge on colloidal spheres arises from the adsorption of ions by the surface of the spheres or by chemically modifying the colloidal surfaces [5, 6].

The engineering of colloidal particles [7,8], conjugated, polymers and synthetic opals are currently subjects of intense investigation for the fabrication of optoelectronic devices such as photonic crystals [9], thin-film transistors, organic light emitting diodes (OLED) [10], solar cells [11,12], pH sensors [13] and tunable lasers [14]. These new devices have great commercial application and will utilize the cost effective benefits of plastic manufacturing [11,12]. Recently, photonic crystals have been constructed from conjugated polymers by infiltrating synthetic opals with a monomer and initiating the polymerization process [15].

2. Guided Assembly of Colloidal Spheres

Guided assembly can be used to construct colloidal crystals with a face centered cubic (FCC) [1,2], tetragonal, triclinic or body centered cubic (BCC) crystalline lattice. A schematic illustration of the colloidal growth process utilizing guided assembly is in Figure 1.

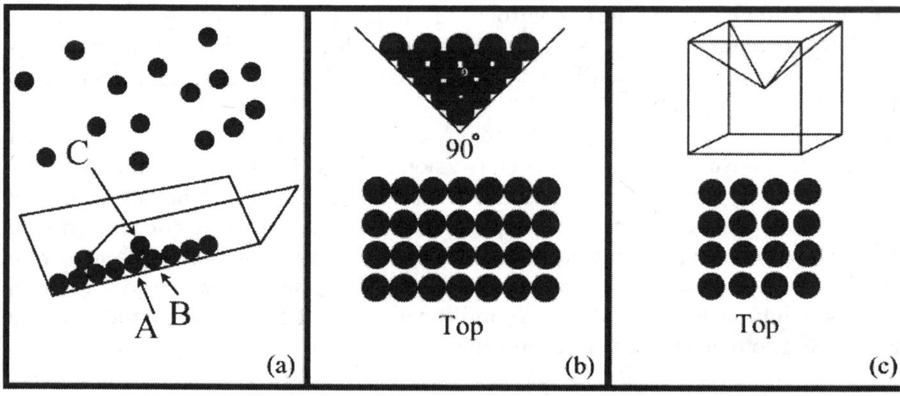

Figure 1. (a) Monodisperse colloidal spheres rain down into the valley shaped guide. The first row of spheres form a single file line at the bottom of the valley. The lowest gravitational potential energy for the next sphere is at point C between spheres A and B. The instability of this point causes the spheres to fall against the walls of the template where they form closed packed planes. (b) A 90° valley shaped guide templates colloidal spheres into a tetragonal crystal. (c) Four triangles joined at a common 70.53° point will guide colloidal spheres into the BCC lattice.

The FCC structure can be assembled in an inverted pyramid formed from four equilateral triangles, as well as in long valley shaped templates with an interior angle of either 70.53° [1,2] or 109.47°. The difference between the 70.53° and the 109.47°

templates is the direction along which the crystal structure grows. The tops of the 70.53° and the 109.47° templates display the (100) and the (110) direction of the FCC lattice, respectively. A tetragonal lattice can be formed in a valley with a 90° interior angle (Figure1b) and the triclinic lattice in any valley with an interior angle grater than 70.53° and less than 109.47° provided it is not equal to 90°. A BCC lattice can be assembled in a four sided inverse pyramid as shown (Figure 1c).

2.1 TECHNIQUES FOR FABRICATING COLLOIDAL GUIDES

The walls of a colloidal guide must be flat, smooth and come together at the bottom with a fine point relative to the size of the spheres assembled in the guide. We created tetragonal guides from two different materials. The first was plastic and the second was glass. A thick piece of Plexiglass was machined into a 3 x 2 x 1 cm block and a 90° valley .25cm deep was machined across the 1cm face with an end mill. The open ends of the guide were sealed with microscope slides and placed into an ultrasonic bath cleaner while in a small beaker. The ultrasonic cleaner was attached to a Variac and a small amount of 80μm (Figure 2a) and 30μm (Figure 2b) diameter polystyrene colloidal spheres were placed into the guide as a 5% w/w dispersion. The assembly process was viewed through a 70X zoom microscope. The voltage of the Variac was initially set on high, this helped the spheres move over any imperfections on the walls or in the bottom of the guide. The crystals observed while the spheres were submersed underwater were several times wider and had fewer defects than after they were dried. The reason is when the water dried across the face of the colloidal crystals the spheres migrated toward the ends of the guide, like tumble weeds in a wind tunnel, from cohesive forces. The cohesive force was overcome by centrifuging the spheres as they dried.

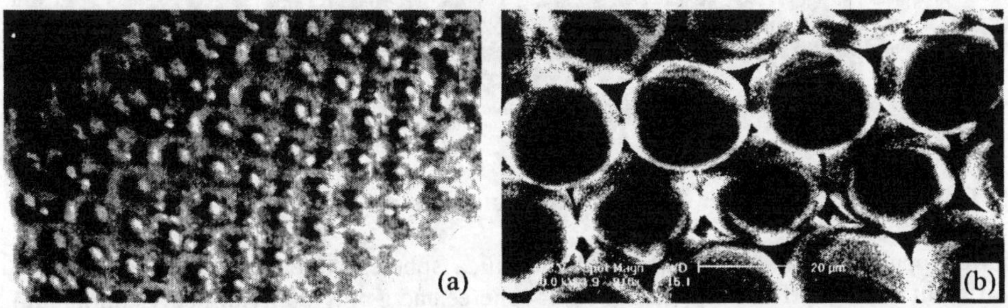

Figure 2. (a) A Polaroid image through a 70X zoom microscope of 80μm polystyrene latex colloidal spheres in a Plexiglass valley machined with an end mill, while spheres are still submersed underwater. (b) An SEM images of 30μm polystyrene spheres assembled into a tetragonal lattice in a Plexiglass guide.

Tetragonal guides constructed from glass have smoother walls than our machined ones allowing the assembly of smaller spheres. Two microscope slides were placed into a small AA-laboratory grade granite V-block and the slides were held together with a small drop of high viscosity gel super glue. A small drop of low viscosity superglue was run along the backside of the microscope slides to make a watertight seal. The inside of the glass V was carefully cleaned with acetone to remove any glue that might have seeped across from the back. Two microscope slide halves were used to seal the sides of the guide by gluing them inside the valley. Other glass guides were formed by cutting cuvettes in half. They were placed into a granite V-block for support and the rest of the procedure followed as described above.

Colloidal silica spheres with 500nm (Figure 3a) and polystyrene spheres with 5μm diameters (Figure 3b) were assembled into the FCC structure in silicon guides prepared as described elsewhere [1,2]. The spheres were assembled into the guides by placing a small amount of the colloidal dispersion on the etched wafer and centrifuging at 3000g with the centrifuge set on the slowest acceleration rate available. The progressive growth of colloidal spheres in the guides is sequentially shown in Figure 4.

Figure 3. (a) An SEM image of 500nm silica and (b) 5μm polystyrene spheres in 70.53° guides etched in a silicon wafer. Note, the defect region created at the interface of the close packed plane on the silicon wafer with the cubic plane in the guide.

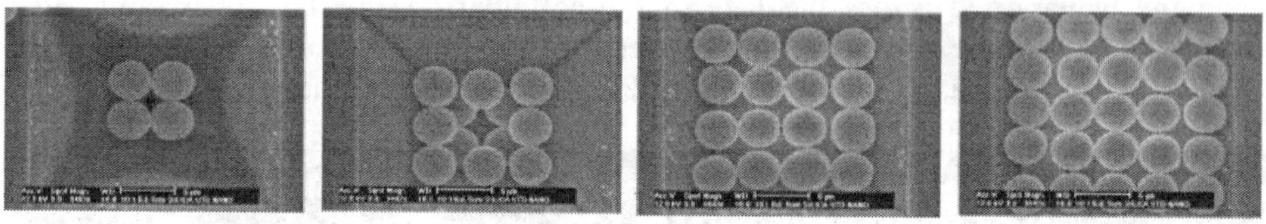

Figure 4. This series of images shows the progressive growth of the FCC lattice in the silicon guides.

3. Coating Colloidal Silica Spheres with Conjugated Polymers

Two populations of 450nm and 1000nm silica spheres were sterically stabilized by different methods. The first population of spheres had a poly(diallyldimethylammonium chloride) (PDADMAC)/ poly(sodium 4-styrenesulfonate) (PSS) bi-layer and the second had a PDADMAC/ poly(styrenesulfonate)/poly(2,3-dihydrothieno [3,4-b]-1,4-dioxin) (PSS-PEDOT) bi-layer. The bi-layers were applied as described elsewhere [7] except 250μL of a .4% w/w PDADMAC and 170μL of a .32% w/w aqueous PSS solution was added for each mL of 10% w/w colloidal silica suspension. The PSS-PEDOT was applied by the same method as PSS with 170μL of a .4% w/w aqueous PSS-PEDOT solution. The PSS and PSS-PEDOT coated spheres were dried, placed under vacuum for 5 hours and dispersed in chloroform with sphere concentrations of 3mg$(mL)^{-1}$ and 1mg$(mL)^{-1}$ respectively.

A dilute poly[2-methoxy-5-(2'-ethylhexyloxy)-1,4-phenylenevinylene] (MEH-PPV) solution in chloroform was added to dispersions of polyelectrolyte coated spheres and was allowed to adsorb for 20 minutes. The spheres were centrifuged and all solution was removed. The rinsing and redispersing process was repeated 4 times. On the last cleaning procedure all chloroform except for a small amount was removed and the spheres were redispersed in ethyl alcohol (ETOH).

The surfaces of the MEH-PPV coated spheres were recharged by adding 800μL of a 4PE mixture to each mL of colloidal dispersion. The 4PE mixture was prepared by adding 1g of PSS, poly(acrylic acid), poly(acrylic acid, sodium salt) and poly(anetholesulfonic acid, sodium salt) to 125mL of de-ionized water. After a four step rinsing and redispersing procedure, the spheres were dispersed in H_2O.

Colloidal silica spheres with 200nm, 250nm and 290nm diameters were coated with poly(anilinesulfonic acid) (PASA) and polypyrrole (PPY) by first applying poly(allamine) hydrochloride (PAH) as described above. Once coated with PAH dilute solutions of aqueous PPY and PASA were added to each colloidal dispersion, allowed to adsorb for 20min, followed by the 4 step cleaning procedure. The process was repeated 3 times yielding colloidal silica spheres coated with three layers of PAH/ PPY and PAH/ PASA.

4. Self Assembly of Colloidal Silica Spheres

Colloidal silica spheres with 200nm, 250nm, 290nm, and 450nm diameters were self assembled by sedimentation in aqueous solutions. Single crystalline grains with diameters as large a 5mm grew vertical along the (100) and (111) directions of the FCC crystalline lattice. SEM images of horizontal slices of our opals show that the (100) direction dominated as the vertical growth direction in the opals (Figure 5a). This growth direction is favored in the self assembly of charged gravitationally settling colloidal spheres when the volume fraction of spheres is less than 3% [16]. A grain boundary between a crystallite grown along the (100) direction and another orientation is presented in Figure 5b.

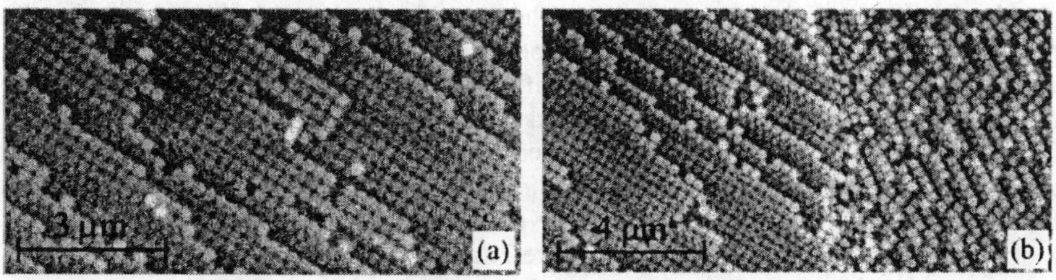

Figure 5. (a) An SEM image of cubic planes formed from 250nm colloidal silica spheres and (b) at a grain boundary. The silica spheres were oriented into the FCC lattice resulting from vertically growth along the (100) direction.

Silica spheres with 1μm, 1.2μm and 1.5μm diameters were dispersed in deionized water and 30μL of the 4PE mixture was added per mL of aqueous colloidal silica dispersion. After the polyelectrolytes absorbed onto the colloidal surfaces for 20 minutes the spheres were centrifuged, all of the solution was removed and they were redispersed in deionized water. Once the spheres settled into a polycrystalline FCC synthetic opal the excess water was removed. The opals were air dried, placed in a vacuum oven at 120°C for five hours and sintered at 750°C to give them mechanical strength (Figure 6).

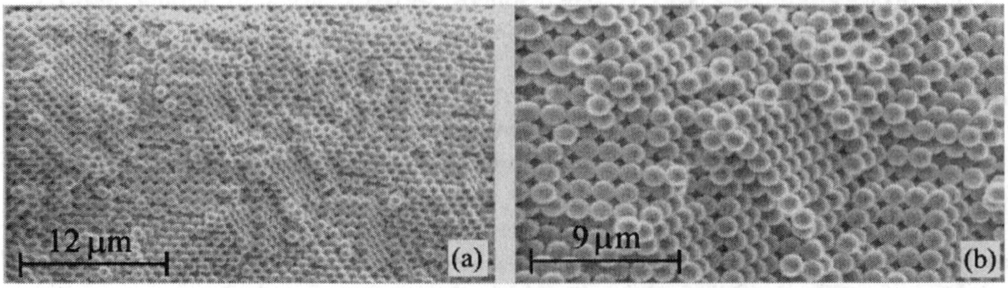

Figure 6. (a) A wide field SEM image of 1.5μm silica spheres self assembled along the (100) direction of the FCC crystalline lattice and (b) a close up image of the (100) direction in the opal.

4.1 SELF ASSEMBLY OF CONJUGATED POLYMER COATED SPHERES

Colloidal silica spheres coated with coated with PASA, PPY and re-charged MEH-PPV coated spheres were self assembled with crystalline order. The crystalline order in the coated spheres arises by controlling the polyelectrolyte coating, pH and sphere concentrations of the aqueous dispersions. Bright opalescent regions were formed with 200nm, 250nm and 290nm spheres coated with PASA and PPY as well as with 450nm and 1000nm MEH-PPV coated spheres by allowing them to settle by gravitation in aqueous solutions (Figure 7a).

Thin film opals with star shaped opalescent patterns were formed from PASA coated, polyelectrolyte coated and uncoated colloidal silica spheres with 200nm, 250nm, 290nm, 450nm, 1000nm, 1.2μm and 1.5μm diameters (Figure 7b). The opalescent star pattern arises by allowing a small amount of a colloidal dispersion to slowly dry in a relatively small vial. The star pattern is evidently created by the meniscus as the solution dries. We found that slower dried colloidal dispersions, controlled by temperature and humidity, with a relatively long Debye length and a high colloidal surface charge created more robust star shaped opalescent patterns. Colloidal dispersions of 250nm diameter spheres dried by heating in less than 5 hour showed no signs of a star shaped pattern, however, dispersions dried in 3-5 hours yielded bright blue uniform films. When the drying rate of fast drying samples was changed small radial opalescent slivers were visible (Figure 7c).

5. Inverse Opals and Photonic Crystals

Inverse opals of polypropylene (PP), gold (Au), chalcogenide and composite inverse opals of PP/ MEH-PPV, PP/ PASA, PP/ PPY and PP/ Au were fabricated from self assembled opal templates. Gold was infiltrated into silica opals from aqueous solutions of gold chloride. Gold chloride was prepared by dissolving 99.9% pure gold in Aqua

Regia, a forty one to nine mixture of concentrated hydrochloric (HCl) and nitric (HNO_3) acids, followed by distilling and re-hydrating with deionized water. Aqueous gold chloride was repeatedly infiltrated into the removable silica opal template and converted to metallic gold by heating. Due to the density difference of gold and gold chloride the maximum possible filling factor with a single infiltration and heat conversion of gold chloride is 39%. An inverse gold-polypropylene composite photonic crystal was constructed by filling 39% of the opal pores with gold followed by the infiltration of polypropylene. The silica opal template was removed with hydrofluoric acid (Figure 8).

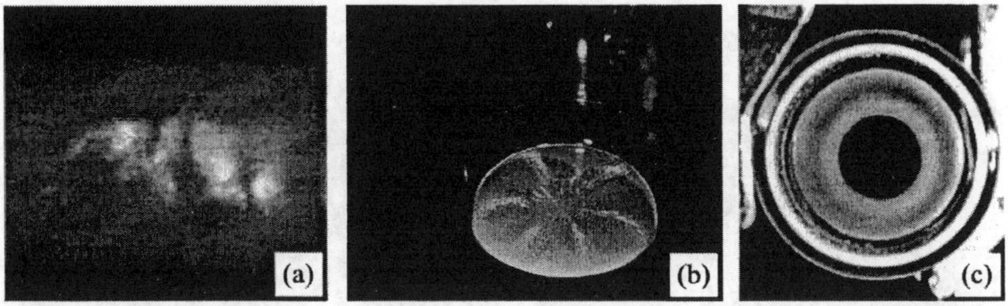

Figure 7. Opalescent crystallites in an ordered region from the self assembly of 450nm silica spheres coated with PDADMAC/PSS/MEH-PPV/4PE/PDADMAC. (b) The opalescent star shaped pattern resulting from slowly drying 200nm silica sphere coated with PAH/PASA in a 2cm wide vial. (c) A thin blue film formed from the fast drying of 250nm silica spheres and small radial opalescent slivers created when the drying rate was slowed.

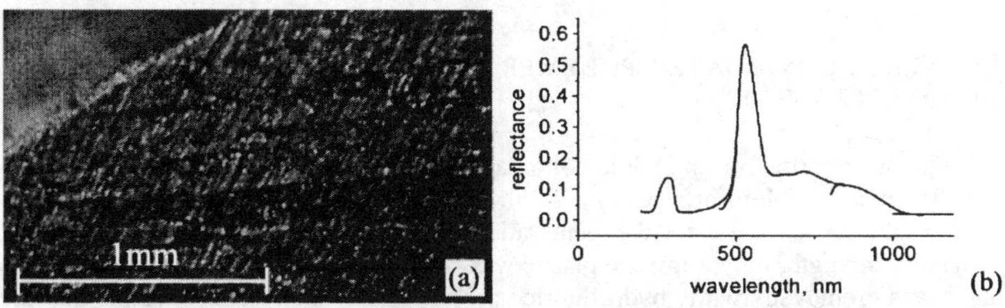

Figure 8. Optical image of an inverse PP/Au composite crystal formed from a 290nm silica template.

The filling factor of gold in the opal template can be controlled by the concentration of gold chloride in the aqueous solution and by repeated infiltrations (Figure 9). Small octahedral spheres connected in a tetrahedral structure were formed by partially infiltrating gold chloride into the opal pores (Figure 10a). The spheres at the center of the tetrahedrons are smaller than the octahedral spheres (Figure 10b). If the gold wires connecting the large octahedral spheres to the smaller spheres are considered as bonds then the cubic zinc sulfide structure is obtained. This is due to the face centered cubic orientation of the silica template.

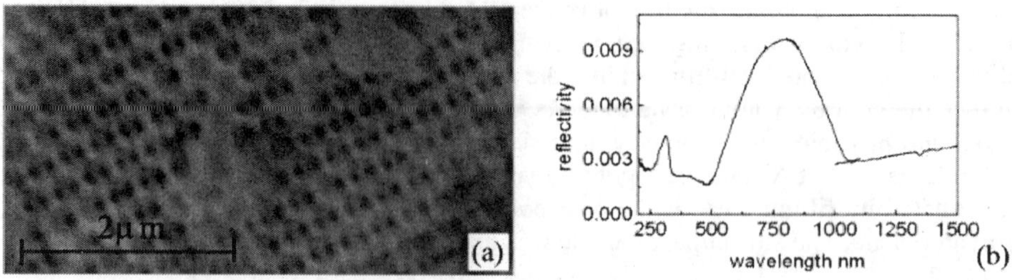

Figure 9. Inverse gold photonic crystal fabricated from a synthetic opal template constructed with 290nm silica spheres.

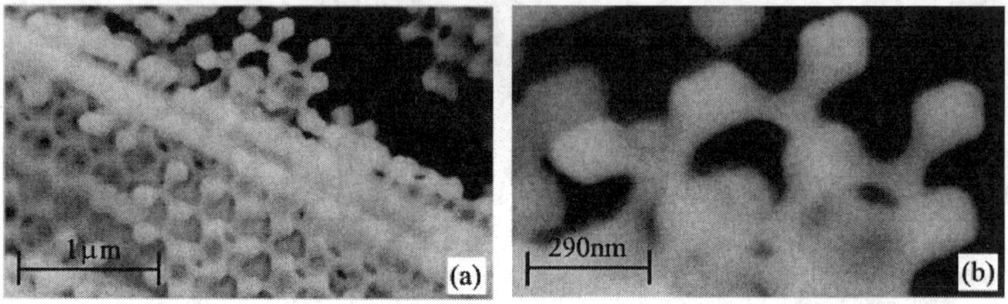

Figure 10. A zinc sulfide structure formed from large and small gold octahedral spheres held together by thin gold bonds. (b) A close up of the gold octahedral spheres held together with thin gold bonds emanating from smaller spheres.

5.1 INVERSE CONJUGATED POLYMER AND POLYPROPYLENE COMPOSITE PHOTONIC CRYSTAL

Synthetic opals formed from 290nm colloidal silica spheres were infiltrated with MEH-PPV dissolved in chloroform. Only a small amount of MEH-PPV penetrated into the pores of the opal, however, the penetration was throughout. To give the sample mechanical strength, polypropylene plastic was co-infiltrated into the opal pores and the silica was removed with hydrofluoric acid (Figure 11). The change in the photoluminescence (PL) spectrum of MEH-PPV inside the inverted opal is compared to the PL spectra of a pure MEH-PPV film in figure 11b. Since we used a large area of the crystal during our PL measurements and not a small well ordered region the partial photonic band features can not be seen in the PL spectra. A dip in the PL spectra is expected from single crystal regions in photonic crystals, as observed in [9]. However the integral PL spectra from our polycrystalline opals show a significant blue shift from that of a pure MEH-PPV film. The PL peak was shifted from 585 nm to 500 nm, which can be explained by morphological changes of polymeric chains, stretched on the inverted opal cage-walls. Similar blue shifts have been observed in conjugated polymers coated on single wall carbon nanotubes. The blue shift can be also due to interactions with polypropylene.

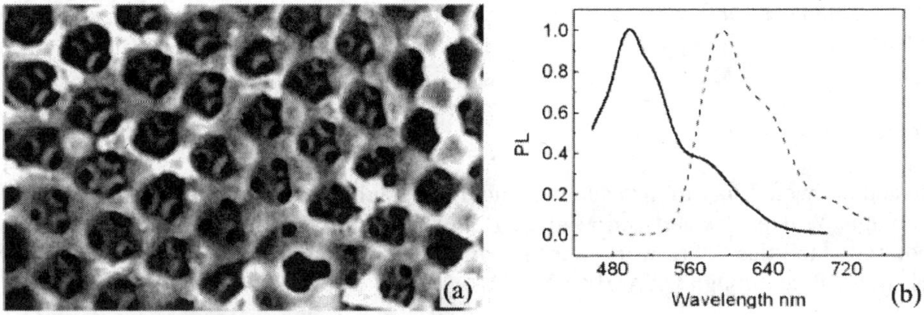

Figure 11. (a) The inverse MEH-PPV/ polypropylene opal formed from a 290nm silica template with FCC crystalline order. The image was taken along the (111) direction. (b) PL measurements of an inverse 290nm MEH-PPV/ polypropylene composite photonic crystal (dashed) and of a pure film of MEH-PPV.

Pieces of silica opals were infiltrated with chalcogenide glass $Ge_{33}As_{12}$ Se_{55} (AMTIR-1), obtained from Amorphous Materials [17], at 700°C with pressures up to 1.5 metric tons. Small pieces of silica opals (2 x 2 x 2 mm) were placed into a small stainless steel capsule with AMTIR-1 that was crushed into a powder. The capsule was then placed into a die machined from water hardening drill rod and placed into a furnace. The die was removed, placed into a hydraulic press and the capsule was squashed. The die was placed back into the furnace and the process was repeated a total of three times. Upon cooling the capsule was cut into disks with a diamond saw and the silica spheres were dissolved away in a 3% hydrofluoric acid solution for 12 hours, forming the inverse opal structure of AMTIR-1 (Figure 12).

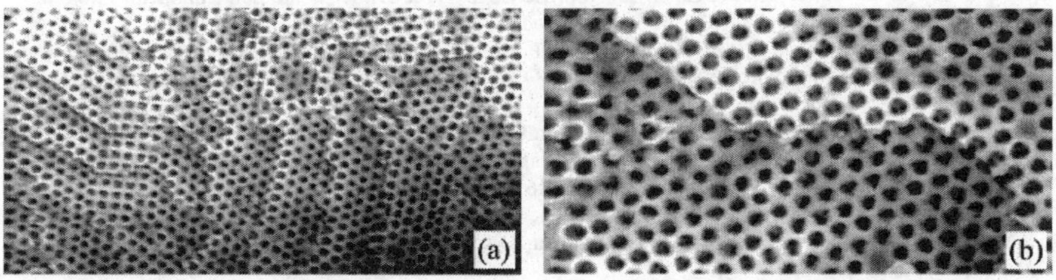

Figure 12. Inverse opals of AMTIR-1 chalcogenide glass formed with a 1.5μm silica template.

6. Conclusion

A method for the guided assembly of colloidal spheres into the face centered cubic, tetragonal and triclinic lattices using guided assembly was introduced. The method was used to make colloidal crystals with a tetragonal lattice using 5μm, 30μm and 80μm colloidal spheres. Well ordered face centered cubic colloidal crystals were grown along the (100) direction with 500nm silica spheres and 5μm polystyrene spheres. Colloidal silica sphere with various diameters were coated with conjugated polymers and polyelectrolytes prior to self assembling by sedimentation. Synthetic silica opals were used as templates for inverse opals of gold, MEH-PPV/ polypropylene composite and chalcogenide.

7. Acknowledgments

The authors would like to thank John Doug Vass, Paul Pantano, Keith Bradshaw, John Goodnight Wayne Wingert and Matt C. DeLong for their cooperation and sharing of laboratory facilities. This work was in part financially supported by the US Army Research Office through DAAD19-99-1-0316.

8. References

[1] San Ming Yang, Geoffrey A. Ozin. (2000) Opal chips: vectorial growth of colloidal crystal patterns inside silicon wafers, *Chem. Commun.*, 2507-2508.
[2] G. A. Ozin, S. M. Yang. (2001) The Race for the Photonic Chip: Colloidal Crystal Assembly in Silicon Wafers, Adv. Funct. Mater. **11**, No. 2. 2507-2508.
[3] Issue Dedicated to Photonic Crystals, (2000), Advanced Materials,**13**, 6.
[4] The Knowledge Foundation, *Advancing Materials to Control Light*, October 11 - 12, 2001, Wyndham.
[5] Robert J. Hunter, (1993) *Introduction to Modern Colloidal Science*, Oxford University Press.
[6] Robert M. Fitch, (1997) *Polymer Colloids A Comprehensive Introduction*, The University of Press, Cambridge.
[7] Frank Caruso, Rachel A. Caruso and Helmuth Möhwald, (1999) Production of Hollow Microspheres from Nanostructured Composite Particles, *Chem. Mater.*, **11**, 11, 3309-3314.
[8] Rachel A. Caruso, Andrei Susha, and Frank Caruso, (2001) Multilayered Titania, Silica, and Laponite Nanoparticle Coatings on Polystyrene Colloidal Templates and Resulting Inorganic Hollow Spheres, *Chem. Mater.* **13,** 2, 400-409.
[9] D. J. Norris, Yu. A. Vlasov, (2001) Chemical Approaches to Three-Dimensional Semiconductor Photonic Crystals, *Adv. Mater.*, **13**, No. 6, 371-376
[10] M.N. Shkunov, Z.V. Vardeny, M.C. DeLong, R.C. Polson, A.A. Zakhidov, R.H. Baughman, (2002) Tunable, Gap-State Lasing in Switchable Directions for Opal Photonic Crystals, *Adv. Funct. Mater.*, **12**, No. 1, 21-26.
[11] Benjamin J. Schwartz, Thuc-Quyen Nguyen, Junjun Wu and Sarah H. Tolbert, (2001) Interchain and intrachain exciton transport in conjugated polymers: ultrafast studies of energy migration in aligned MEH-PPV/mesoporous silica composites, *Synthetic Metals*, **116**, 1-3, 35-40.
[12] K. Landfester, R. Montenegro, U. Scherf, R. Gutner, U. Asawapirom, S. Patil, D. Neher, T. Kietzke, (2002) Semiconducting Polymer Nanospheres in Aqueous Dispersion Prepared by a Miniemulsion Process, *Adv. Mater.*, **14**, No. 9, 651-655
[13] Christina N. Aquino-Binag, Naresh Kumar, and, Robert N. Lamb, and Paul J. Pigram, (1996) Fabrication and Characterization of a Hydroquinone-Functionalized Polypyrrole Thin-Film pH Sensor, *Chem. Mater.*, **8**, 11, 2579-2585.
[14] Philip N. Bartlett , Peter R. Birkin , Mohamed A. Ghanem and Chee-Seng Toh, (2001) Electrochemical syntheses of highly ordered macroporous conducting polymers grown around self-assembled colloidal templates, *J. Mater. Chem.*, **11,** 3, 849–853
[15] Yu. A. Vlasov, D. J. Norris, Conjugated-Polymer Photonic Crystals, (2000) M. Deutsch, *Adv. Mater.*, **12**, No. 16, 1176-1180.
[16] A. K Arora and B.V.R. Tata, 1996, *Ordering and Phase Transitions in Charged Colloids*, John Wiley and Sons.
[17] Amorphous Materials 3130 Benton, Garland, Texas 75042 USA http://www.amorphousmaterials.com, AmorphousM@aol.com.

INDEX